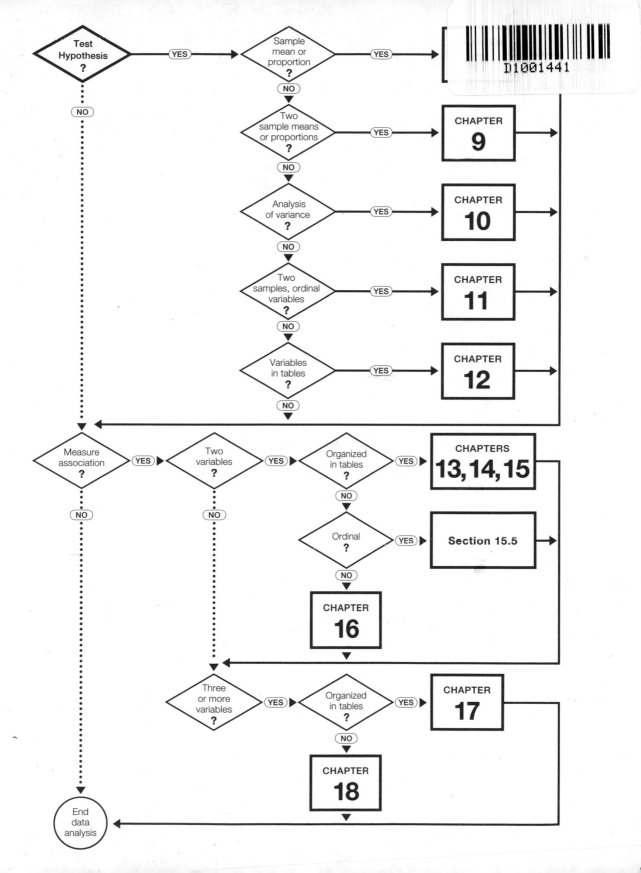

Statistics

Statistics: A Tool for Social Research

Fifth Edition

Joseph F. Healey
Christopher Newport University

 Wadsworth Publishing Company

I(T)P® An International Thomson Publishing Company

Belmont, CA • Albany, NY • Boston • Cincinnati • Johannesburg • London • Madrid • Melbourne
Mexico City • New York • Pacific Grove, CA • Scottsdale, AZ • Singapore • Tokyo • Toronto

Publisher: Eve Howard
Marketing Manager: Christine Henry
Editorial Assistant: Angela Nava
Project Editor: Jerilyn Emori
Print Buyer: Karen Hunt
Permissions Editor: Susan Walters

Production: Greg Hubit Bookworks
Cover Design: Bill Stanton/Stanton Design
Cover Image: Photo collage, Bill Stanton;
   photography, Barry Kaplan
Compositor: G & S Typesetters, Inc.
Printer: R. R. Donnelley & Sons/Crawfordsville

Printed in the United States of America
   2   3   4   5   6   7   8   9   10

For more information, contact Wadsworth Publishing Company, 10 Davis Drive, Belmont, CA 94002, or
electronically at http://www.wadsworth.com

International Thomson Publishing Europe
Berkshire House 168-173
High Holborn
London, WC1V 7AA, England

International Thomson Publishing Asia
60 Albert Street
#15-01 Albert Complex
Singapore 189969

Thomas Nelson Australia
102 Dodds Street
South Melbourne 3205
Victoria, Australia

International Thomson Publishing Japan
Hirakawacho Kyowa Building, 3F
2-2-1 Hirakawacho
Chiyoda-ku, Tokyo 102, Japan

Nelson Canada
1120 Birchmount Road
Scarborough, Ontario
Canada M1K 5G4

International Thomson Publishing
Southern Africa
Building 18, Constantia Park
240 Old Pretoria Road
Halfway House, 1685 South Africa

International Thomson Editores
Campos Eliseos 383, Piso 7
Col. Polanco
11560 México D.F. México

Library of Congress Cataloging-in-Publication Data

Healey, Joseph F.
   Statistics : a tool for social research / Joseph F. Healey.—5th ed.
   p.   cm.
   Includes index.
   ISBN 0-534-55260-9
   1. Social sciences—Statistical methods.   2. Statistics.   I. Title.
HA29.H39   1999   98-20186
519.5—dc21

Brief Contents

Detailed Contents

Preface

Sociology and the other social sciences are research-based disciplines. Social scientists test and refine their ideas against the highest quality evidence available. The research capabilities of the social sciences have grown enormously over the past several decades and, as the end of the twentieth century approaches, the essential "tools of the trade" for the social scientist now routinely include calculators and computers.

Statistics and research methodology are essential components in the education of the social scientist. However, learning (and teaching) these subjects can present challenges that are not present in other areas of the curriculum. Sociology and other social science majors typically begin a statistics course with a wide range of mathematical backgrounds and an equally diverse set of career goals. Students are often puzzled about how statistics relates to sociology in general and, not infrequently, there is some math anxiety to overcome.

This text was written to meet the challenge of introducing statistical analysis to social science majors while addressing the instructional challenges that are inherent in the course. The text makes minimal assumptions about the mathematical background of the students (the ability to read a simple formula is sufficient preparation for virtually all of the material in the text), and a variety of special features have been integrated into the text to assist students to successfully analyze data. The text has been written especially for sociology and social work programs but is sufficiently flexible to be used in statistics courses in political science or in any program with a social science base and an applied focus (for example, public administration, criminal justice, urban studies, and gerontology).

The text is written at a level intermediate between the more rigorous and sophisticated texts on one hand and the mere "cookbook" on the other. That is, while I have not sacrificed comprehensive coverage or statistical correctness, the theoretical and mathematical explanations of why statistics "do what they do" are kept at an elementary level, as is appropriate in a first exposure to social statistics. For example, I do not treat formal probability theory per se. Rather, the background necessary for an understanding of inferential statistics is introduced, informally and intuitively, in Chapters 5 and 6 while considering the concepts of the normal curve and the sampling distribution. The text makes no claim that statistics are "fun" or that the material can be mastered without considerable effort. At the same time, students are not overwhelmed with abstract proofs and mathematical theory, which at this level needlessly frustrate the learning experience.

My major goal is basic statistical literacy. The text is designed to provide a solid foundation in statistical analysis and to prepare students to be intelligent consumers of social research. More specifically, I believe that basic statistical literacy can be defined in terms of three interrelated qualities and, as a way of further describing the nature of this text, I would like to list each of these qualities and briefly summarize how the text is designed to develop them.

Computational Competence.   At a minimum, students should emerge from their first course in statistics with the ability to perform elementary forms of data analysis—to execute a series of calculations and arrive at the correct answer. Since students in social science statistics courses frequently do not have strong quantitative backgrounds, I have included a number of features to help students cope with computations:

- *Step-by-step computational algorithms* are provided for each statistic.

- *Extensive problem sets* are provided at the end of each chapter. For the most part, these problems use fictitious data and are designed for relative ease of computation.

- *Cumulative exercises* are included at the end of each part to provide practice in choosing statistics. These exercises present only data sets and research questions. Students must choose appropriate statistics as part of the exercise.

- *Solutions* to odd-numbered computational problems are provided so that students may check their answers.

- *SPSS for Windows*, the leading social science statistical software, is incorporated to give students access to the computational power of the computer. This feature is explained in more detail below.

An Appreciation of Statistics.   A statistically literate person can do much more, of course, than merely calculate correct answers. Such a person understands the relevance of statistics for social research, can select an appropriate statistic for a given purpose and a given set of data, and can analyze and interpret the meaning of that statistic. This textbook begins to develop these qualities, within the constraints imposed by the introductory nature of the course, in the following ways:

- *The relevance of statistics.* Chapter 1 includes a discussion of the role of statistics in social research and stresses the usefulness of these techniques as ways of analyzing and manipulating data and answering research questions. Each example problem is framed in the context of a research problem. A question is posed and then, with the aid of a statistic, answered. The relevance of statistics for answering questions is thus stressed throughout the text. This central theme of usefulness is further reinforced by a series of boxes labeled "Applications," each of which illustrates some specific way statistics can be used to answer questions.

  Almost all end-of-chapter problems are labeled by the social science discipline or subdiscipline from which they are drawn. The following are used as labels: $\boxed{\text{SOC}}$ for sociology, $\boxed{\text{SW}}$ for social work, $\boxed{\text{PS}}$ for political science, $\boxed{\text{CJ}}$ for criminal justice, $\boxed{\text{PA}}$ for public administration, and $\boxed{\text{GER}}$ for gerontology. By identifying problems with specific disciplines, students can more easily see the relevance of statistics to their own academic interests. (Not incidentally, they will also see that the disciplines have a large subject matter in common.)

- *Selecting appropriate statistics.* A series of flowcharts are included to help students select appropriate statistics. These flowcharts have two components. Decision points are represented by diamonds and information by rectangles. The selection process is represented in general terms on the inside front cover and at the beginning of each part. Chapters begin with detailed flowcharts that lead students to specific formulas or sections of the chapter based on a consideration of the purpose of the analysis, the format of the data, and the level of measurement criterion.

- *Interpreting statistics.* After selecting and computing a statistic, students still face difficulties in understanding what the statistic means. The ability to interpret statistics can be developed only by exposure and experience. To provide exposure, I have been careful, in the example problems, always to express the meaning of the statistic in terms of the original research question. To provide experience, the end-of-chapter problems almost always call for an interpretation of the statistic calculated. To provide examples, many of the Answers to Odd-Numbered Computational Problems in the back of the text are expressed in words as well as numbers. In particular, the answers to the cumulative exercises include discussions on the process of selecting statistics.

The Ability to Read the Professional Social Science Literature.　The statistically literate person can comprehend and critically appreciate research reports written by others. The development of this quality is a particular problem at the introductory level because of the marked disparity between the concise language of the professional researcher and the rather wordy vocabulary of the classroom. To help bridge this gap, I have included a series of boxes labeled "Reading Statistics" beginning in Chapter 1. In each box, I briefly describe the reporting style typically used for the statistic in question and try to alert students about what to expect when they approach the professional literature. These inserts include excerpts from the research literature and illustrate how statistics are actually applied and interpreted by social scientists.

Additional Features.　A number of other features make the text more meaningful for students and more useful for instructors:

- *Readability.* The writing style is informal and accessible to students without ignoring the traditional vocabulary of statistics. Problems and examples have been written to maximize student interest and to focus on issues of concern and significance. For the more difficult material (such as hypothesis testing), students are first walked through an example problem before being confronted by formal terminology and concepts. Each chapter ends with a summary of major points and formulas and a glossary of important concepts. A glossary of symbols inside the back cover can be used for quick reference.

- *Organization and coverage.* The text is divided into four parts, with most of the coverage devoted to univariate descriptive statistics, inferential statistics, and bivariate measures of association. The distinction between description and inference is introduced in the first chapter and maintained throughout the text. In selecting statistics for inclusion, I have tried to strike a balance between the essential concepts with which students must be familiar and the amount of material students can reasonably be expected to learn in their first (and perhaps only) statistics course, while bearing in mind that different instructors will naturally wish to stress different aspects of the subject. Thus, the text covers a full gamut of the usual statistics, with each chapter broken into subsections so that instructors may choose the particular statistics they wish to include.

- *Review of mathematical skills.* Appendix H provides a comprehensive review of all of the mathematical skills that will be used in this text. Students who are inexperienced or out of practice with mathematics may want to study this section early in the course and/or refer to it as needed. A self-test is included in Appendix H so students may check their level of preparation for the course.

- *Computer applications.* To help students take advantage of the power of the computer, this text integrates *SPSS*, the most widely used statistical software in the social sciences. Appendix F is an introduction to statistical packages in general and SPSS in particular. There are demonstrations at the end of each chapter that explain how to use SPSS to produce the statistics presented in the chapter. Student exercises analyzing data with SPSS are also included.

- *Realistic, up-to-date data.* The database for computer applications is the General Social Survey. This database will give students the opportunity to practice their statistical skills on "real-life" survey data. The database is described in Appendix G, and instructors may request a data diskette from the publisher.

- *Instructor's Manual.* The Instructor's Manual includes learning objectives, chapter summaries, a test item file of multiple-choice questions, answers to even-numbered computational problems, and step-by-step solutions to selected problems. In addition, the Instructor's Manual includes cumulative exercises (with answers) similar to those in the text but with different variables or scores. These may be used for testing purposes.

- *Study Guide.* The Study Guide, written by Rebecca Davis, University of Maryland–College Park, contains additional examples to illuminate basic principles, review problems with detailed answers, SPSS work, and multiple-choice questions and answers that complement but do not duplicate the test item file.

CHANGES IN THIS EDITION

Two changes are designed to further enhance computational competence:

- *Statistical techniques and end-of-chapter problems are explicitly linked.* After a technique is introduced, students are directed to specific problems for practice and review. The "how-to-do-it" aspects of calculation are reinforced immediately and clearly.

- *End-of-chapter problems are organized progressively.* Simpler problems with small data sets are presented first. Often, explicit instructions or hints accompany the first several problems in a set. The problems gradually become more challenging and require more decision making by the student (e.g., choosing the most appropriate statistic for a certain situation). Thus, each problem set develops problem-solving abilities gradually and progressively.

- To strengthen the link with the professional literature, the "Reading Statistics" boxes have been updated and increased in both number and variety.

- The database has been updated to the 1996 version of the General Social Survey (GSS), and the statistics package is the latest available version of *SPSS for Windows.* The original GSS sample includes almost 3000 respondents and hundreds of variables. Since some students may use the Student Version of *SPSS for Windows*, which limits the size of the database, I have reduced the number of variables to less than 50. Also, the number of cases has been reduced to less than 1500 by selecting a 50% random sample from the original database.

- The number of end-of-chapter problems has been increased. For the most part, the new problems feature small data sets and ease of computation.

- In response to suggestions from reviewers and instructors, the introduction to bivariate tables has been moved from Chapter 2 to Chapters 12 and 13.

- Also in response to suggestions from reviewers and instructors, the chapter on Analysis of Variance (formerly Chapter 12) has been moved so that it immediately follows Chapter 9. This tightens the connection between the t test for means in Chapter 9 and ANOVA (now Chapter 10).

The text has been thoroughly reviewed for clarity and readability. As with previous editions, my goal is to provide a comprehensive, flexible, and student-oriented text that will provide a challenging first exposure to social statistics.

Acknowledgments

This text has been in development, in one form or another, for more than fifteen years. An enormous number of people have made contributions, both great and small, to this project and, at at the risk of inadvertently omitting someone, I am bound to at least attempt to acknowledge my many debts.

This edition reflects the thoughtful guidance of both Eve Howard of Wadsworth, and Denise Simon. I thank them for their many contributions. Much of whatever integrity and quality this book has is a direct result of the contributions of colleagues who reviewed the manuscript during the various stages of writing. I have been consistently impressed by their sensitivity to the needs of the students, and I would like to thank Nelta Edwards, Arizona State University; Barry L. Johnson, Brigham Young University; David Jorgenson, Southwest Texas State University; Joseph J. Sudano, Jr., Kent State University; Keith Warren, University of Texas at Austin; and Jiping Zuo, St. Cloud State University. Whatever failings are contained in the text are, of course, my responsibility and are probably the results of my occasional decisions not to follow the advice of my colleagues.

I would like to thank the instructors who made statistics understandable to me (Professors Satoshi Ito, Noelie Herzog, and Ed Erikson) and all of my colleagues at Christopher Newport University for their support and encouragement (especially Professors F. Samuel Bauer, Robert Durel, James Forte, Ruth Kernodle, Timothy Marshall, Cheryl Mathews, Lea Pellet, Virginia Purtle, and William Winter). I would be very remiss if I did not acknowledge the constant support and excellent assistance of Jessica Ledbetter and Brigid Halpin. I also need to thank all of my students for their patience and thoughtful feedback. Also, I am grateful to the Literary Executor of the late Sir Ronald A. Fisher, F.R.S., to Dr. Frank Yates, F.R.S., and to Longman Group Ltd., London, for permission to reprint Appendixes B, C, and D, from their book *Statistical Tables for Biological, Agricultural and Medical Research* (6th edition, 1974).

Finally, I want to acknowledge the support of my family and rededicate this work to them. I have the extreme good fortune to be a member of an extended family that is remarkable in many ways and has recently increased in size. Although I cannot list everyone, I would like to especially thank the older generation (my mother, Alice T. Healey), the next generation (my sons Kevin and Christopher, my daughter-in-law Jennifer), the new members (Patricia, Christopher, Katherine, and Jennifer Schroen), and the youngest generation (Benjamin John Healey).

J. F. H.

1 Introduction

Students sometimes approach their first course in statistics with questions about the value of the subject matter. What, after all, do numbers and statistics have to do with understanding people and society? In a sense, this entire book will attempt to answer this question, and the value of statistics will become clear as we move from chapter to chapter. For now, the importance of statistics can be demonstrated, in a preliminary way, by briefly reviewing the research process as it operates in the social sciences. These disciplines are scientific in the sense that social scientists attempt to verify their ideas and theories through research. Broadly conceived, *research is any process by which information is systematically and carefully gathered for the purpose of answering questions, examining ideas, or testing theories.* Research is a disciplined inquiry that can take numerous forms. Statistical analysis is relevant only for those research projects where the information collected is represented by numbers. Numerical information of this sort is called **data**, and the sole purpose of statistics is to manipulate and analyze data. **Statistics**, then, are a set of mathematical techniques used by social scientists to organize and manipulate data for the purpose of answering questions and testing theories.

What is so important about learning how to manipulate data? On one hand, some of the most important and enlightening works in the social sciences do not utilize any statistical techniques. There is nothing magical about data and statistics. The mere presence of numbers guarantees nothing about the quality of a scientific inquiry. On the other hand, data can be the most trustworthy information available to the researcher and, consequently, deserve special attention. Data that have been carefully collected and thoughtfully analyzed are the strongest, most objective foundations for building theory and enhancing understanding. Without a firm base in data, the social sciences would lose the right to the name *science* and would be of far less value.

Thus, the social sciences rely heavily on data gathering for the advancement of knowledge. Let me be very clear about one point: it is never enough merely to gather data (or, for that matter, any kind of information). Even the most objective and carefully collected numerical information does not and cannot speak for itself. The researcher must be able to use statistics effectively. To be useful, the data must be organized, evaluated, and analyzed. Without a good understanding of the principles of statistical analysis, the researcher will be unable to make sense of the data. Without the appropriate application of statistical techniques, the data will remain mute and useless.

Statistics are an indispensable tool for the social sciences. They provide the scientist with some of the most useful techniques for evaluating ideas, testing theory, and discovering the truth. In the next section, I'll describe the relationships between theory, research, and statistics in more detail.

1.2 THE ROLE OF STATISTICS IN SCIENTIFIC INQUIRY

Figure 1.1 graphically represents the role of statistics in the research process. The diagram is based on the thinking of Walter Wallace and illustrates how the knowledge base of any scientific enterprise grows and develops. One point the diagram makes is that scientific theory and research continually shape each other. Statistics are one of the most important means by which research and theory interact. Let's take a closer look at the wheel.

Since the figure is circular, it has no beginning or end, and we could begin our discussion at any point. For the sake of convenience, let's begin at the top with Theory and follow the arrows around the circle. A **theory** is an explanation of the relationships between phenomena. People naturally (and endlessly) wonder about problems in society (like prejudice, poverty, child abuse, or serial murders) and, in their attempt to understand these phenomena, they develop explanations ("lack of education causes prejudice"). This kind of informal "theorizing" about society is no doubt very familiar to you. A major difference between our informal, everyday explanations of social phenomena and scientific theory is that the latter is subject to a rigorous testing process. Let's take the problem of racial prejudice as an example to illustrate how the research process works.

What causes racial prejudice? One possible explanation for this phenomenon is provided by a theory called the *contact hypothesis*. This theory was

FIGURE 1.1 THE WHEEL OF SCIENCE

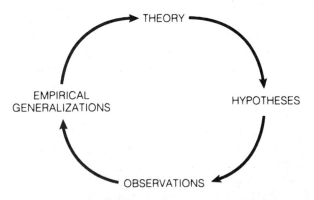

Source: Adapted from Walter Wallace, *The Logic of Science in Sociology* (Chicago: Aldine-Atherton, 1971).

stated over 40 years ago by the social psychologist Gordon Allport, and it has been tested on a number of occasions since that time.[1] The theory links prejudice to the volume of interaction between members of different racial groups. Specifically, the hypothesis asserts that contact situations in which different groups have equal status and are engaged in cooperative, non-competitive behavior will result in reductions of prejudice on all sides. The greater the extent to which contact is equal and cooperative, the more likely people will see each other as individuals and not as representatives of a particular group. For example, members of a racially mixed athletic team that cooperate with each other to achieve victory would tend to experience a decline in prejudice. On the other hand, when different groups compete for jobs, housing, or other valuable resources, prejudice would tend to increase.

The contact hypothesis is not a complete explanation of prejudice, but it will serve to illustrate the role of theory. We are trying to explain the relationship between two social phenomena: (1) prejudice and (2) equal-status, cooperative contact between different groups. People who have little contact will be more prejudiced, and those who experience more contact will be less prejudiced. Thus, we have a coherent theory that explains the relationship between two phenomena.

Before moving to the next area of Figure 1.1, let's take a moment to further examine theory. Our "theory" is simplified but, like most theories, it is stated in terms of causal relationships between variables. A **variable** is any trait that can change values from case to case. Examples of variables would be gender, age, income, or political party affiliation. In any specific theory, some variables will be identified as causes and others will be identified as effects or results. In the language of science, the causes are called **independent variables** and the "effects" or result variables are called **dependent variables.** In our theory of prejudice, contact would be the independent variable (or the cause) and prejudice would be the dependent variable (the result or effect). In other words, we are arguing that equal-status contact is a cause of prejudice or that an individual's level of prejudice depends on the extent to which he or she participates in equal-status, cooperative contacts with other groups.

So far, we have a theory of prejudice, and an independent and a dependent variable. What we don't know yet is whether the theory is true or false. To find out, we need to compare our theory with the facts: we need to do some research. The next steps in the process would be to define our terms and ideas more specifically and exactly. One problem we often face in doing research is that scientific theories are too complex and abstract to be fully

---

[1] Allport, Gordon, 1954. *The Nature of Prejudice.* Reading, Massachusetts: Addison-Wesley. For recent attempts to test this theory, see: Sigelman, Lee and Susan Welch, 1993. "The Contact Hypothesis Revisited: Black-White Interaction and Positive Racial Attitudes." *Social Forces.* 71:781–795 and Pettigrew, Thomas, 1997. "Generalized Intergroup Contact Effects on Prejudice." *Personality and Social Psychology Bulletin.* 23:173–185.

tested in a single research project. To conduct research, one or more hypotheses must be derived from the theory. A **hypothesis** is a statement about the relationship between variables that, while logically derived from the theory, is much more specific and exact.

For example, if we wished to test the contact hypothesis, we would have to say exactly what we mean by prejudice and we would need to describe "equal-status, cooperative contact" in great detail. There has been a great deal of research on the effect of contact on prejudice, and we would consult the research literature to develop and clarify our definitions of these concepts.

As our definitions develop and the hypotheses take shape, we enter the next step of the research process. Now we will design the data-gathering phase of the project. We must decide how cases will be tested, how these cases will be selected, how exactly the variables will be measured, and a host of related matters. Ultimately, these plans will lead to the observation phase (the bottom of the wheel of science), where we actually measure social reality. Before we can do this, we must have a very clear idea of what we are looking for and a well-defined strategy for conducting the search.

We might conduct a research project with people from different groups. Some of these subjects would be placed in situations that required them to cooperate with members of other groups while some were placed in competitive situations. We would need to measure levels of prejudice before and after each type of contact. We might do this by administering a survey that asked subjects to agree or disagree with statements such as "It is very important that every effort be made to racially integrate the public school system" or "Skin color is irrelevant and people are just people." Our goal would be to see if the people exposed to the cooperative contact situation actually experience a reduction in prejudice.

Now, finally, we come to statistics. As the observation phase of our research project comes to an end, we will be confronted with a large collection of numerical information or data. If our sample consisted of 100 people, we would have 200 completed surveys measuring prejudice: 100 completed before the contact situation and 100 filled out afterwards. Try to imagine dealing with 200 completed surveys. If we had asked each of the respondents just five questions to measure their prejudice, we would have a total of 1000 separate pieces of information to deal with. What do we do? We have to have some systematic way to organize and analyze this information and, at this point, statistics will become very valuable. Statistics will supply us with many ideas about "what to do" with the data; we will begin to look at some of the options in the next chapter. For now, let me stress two points about statistics.

First, statistics are crucial. Simply put, without statistics, quantitative research is impossible. Without quantitative research, the development of the social sciences would be severely impaired and perhaps even arrested. Only by the application of statistical techniques can mere data help us shape and refine our theories and understand the social world better. Second, and somewhat paradoxically, the role of statistics is rather limited. As Figure 1.1 makes clear, scientific research proceeds through several mutually interdependent stages, and statistics become directly relevant only at the end of the observation stage. Before any statistical analysis can be legitimately applied, the preceding phases of the process must have been successfully completed. If the researcher has asked poorly conceived questions or has made serious errors of design or method, then even the most sophisticated statistical analysis is valueless. As useful as they can be, statistics cannot substitute for rigorous conceptualization, detailed and careful planning, or creative use of theory. Statistics cannot salvage a poorly conceived or designed research project. They cannot make sense out of garbage.

On the other hand, inappropriate statistical applications can limit the usefulness of an otherwise carefully done project. Only by successfully completing *all* phases of the process can a quantitative research project hope to contribute to understanding. A reasonable knowledge of the uses and limitations of statistics is as essential to the education of the social scientist as is training in theory and methodology.

As the statistical analysis comes to an end, we would move on to the next stage of the process. In this phase we would primarily be concerned with assessing our theory, but we would also look for other trends in the data. Assuming that we found that equal-status, cooperative contact reduces prejudice, we might go on to ask if the pattern applies to males as well as females, to the well educated as well as the poorly educated, to older respondents as well as to the younger. As we probed the data, we might begin to develop some generalizations based on the empirical patterns we observe. For example, what if we found that contact reduced prejudice for younger respondents but not for older respondents? Could it be that younger people are less "set in their ways" and have attitudes and feelings that are more open to change? As we developed tentative explanations, we might begin to revise or elaborate our theory.

If we change the theory to take account of these findings, however, a new research project designed to test the revised theory is called for, and the wheel of science would begin to turn again. We (or perhaps some other researchers) would go through the entire process once again with this new—and, hopefully, improved—theory. This second project might result in further revisions and elaboration that would (you guessed it) require still more research projects, and the wheel of science would continue turning as long as scientists were able to suggest additional revisions or develop new insights. Every time the wheel turned, our understandings of the phenomena under consideration would (hopefully) improve.

This characterization of the research process does not include white-coated, clipboard-carrying scientists who, in a blinding flash of inspiration, discover some fundamental truth about reality and shout, "Eureka!" The truth is that, in the normal course of science, there are only rarely times when we can say with absolute certainty that a given theory or idea is definitely true or false. Rather, evidence for (or against) a theory will gradually accumulate over time, and ultimate judgments of truth will likely be the result of many years of hard work, research, and debate.

Let's briefly review our imaginary research project. We began with an idea or theory about intergroup contact and racial prejudice. We imagined some of the steps we would have to take to test the theory and took a quick look at the various stages of the research project. We wound up back at the level of theory, ready to begin a new project guided by a revised theory. We saw how theory can motivate a research project and how our observations might cause us to revise the theory and, thus, motivate a new research project. One of the great values of Wallace's wheel of science is that it clearly illustrates how theory stimulates research and how research shapes theory. This constant interaction between theory and research is the lifeblood of science and the key to enhancing our understandings of the social world.

The dialog between theory and research occurs at many levels and in multiple forms. Statistics are one of the most important links between these two realms. Statistics permit us to analyze data, to identify and probe trends and relationships, to develop generalizations, and to revise and improve our theories. As you will see throughout this text, statistics are limited in many ways. They are also an indispensable part of the research enterprise. Without statistics, the interaction between theory and research would become extremely difficult, and the progress of our disciplines would be severely retarded. *(For practice in describing the relationship between theory and research and the role of statistics in research, see problems 1.1 and 1.2.)*

1.3 THE GOALS OF THIS TEXT

In the preceding section, I argued that statistics are a crucial part of the process by which scientific investigations are carried out and that, therefore, some training in statistical analysis is a crucial component in the education of every social scientist. In this section, we will address the questions of how much training is necessary and what the purposes of that training are. First, this textbook takes the point of view that statistics are tools. They can be a very useful means of increasing our knowledge of the social world, but they are not ends in themselves. Thus, we will not take a "mathematical" approach to the subject. The techniques will be presented as a set of tools that can be used to answer important questions. To avoid misunderstandings, let me stress that we will be concerned with matters of mathematics. This text includes enough mathematical material so that you can develop a basic understanding of why statistics "do what they do." Our focus, however, will be on how these techniques are applied in the social sciences.

Second, all of you will soon become involved in advanced coursework in your major fields of study, and you will find that much of the literature used in these courses assumes at least basic statistical literacy. Furthermore, many of you, after graduation, will find yourselves in positions—either in a career or in graduate school—where some understanding of statistics will be very helpful or perhaps even required. Very few of you will become statisticians per se (and this text is not intended for the preprofessional statistician), but you must have a grasp of statistics in order to read and critically appreciate your own professional literature. As a student in the social sciences and in many careers related to the social sciences, you simply cannot realize your full potential without a background in statistics.

Within these constraints, this textbook is an introduction to statistics as they are utilized in the social sciences. The general goal of the text is to develop an appreciation—a "healthy respect"—for statistics and their place in the research process. You should emerge from this experience with the ability to use statistics intelligently and to know when other people have done so. You should be familiar with the advantages and limitations of the more commonly used statistical techniques, and you should know which techniques are appropriate for a given set of data and a given purpose. Lastly, you should develop sufficient statistical and computational skills and enough experience in the interpretation of statistics to be able to carry out some elementary forms of data analysis by yourself.

1.4 DESCRIPTIVE AND INFERENTIAL STATISTICS

Descriptive Statistics.   As noted earlier, the general function of statistics is to manipulate data so that the original research question(s) can be answered. The researcher can call upon two general classes of statistical techniques that, depending on the research situation, are available to accomplish this task. The first class of techniques is called **descriptive statistics** and is relevant (1) when the researcher needs to summarize or describe the distribution of a single variable and (2) when the researcher wishes to understand the relationship between two or more variables. If we are concerned with describing a single variable, then our goal will be to arrange the values or scores of that variable so that the relevant information can be quickly understood and appreciated. Many of the statistics that might be appropriate for this summarizing task are probably familiar to you. For example, percentages, graphs, and charts can all be used as single-variable descriptive statistics.

To illustrate briefly the usefulness of these kinds of statistics, consider the following problem: Suppose you wanted to summarize the distribution of the variable "family income" for a community of 10,000 families. How would you do it? Obviously, you couldn't simply list all incomes in the community and let it go at that. Presumably, you would want to develop some summary measures of the overall income distributions—perhaps an arithmetic average or the proportions of incomes that fall in various ranges (such as low, middle, and high). Or perhaps a graph or a chart would be more useful. Whatever specific method you choose, its function is the same: to reduce

these thousands of individual items of information into a few easily understood numbers. The process of allowing a few numbers to summarize many
numbers is called **data reduction** and is the basic goal of single-variable
descriptive statistical procedures. The first part of this text is devoted to these
statistics.

The second type of descriptive statistics is designed to help the investi-
gator understand the relationship between two or more variables. These statistics, called **measures of association**, allow the researcher to quantify the
strength and direction of a relationship. These statistics are very useful because they enable us to investigate two matters of central theoretical and practical importance to any science: causation and prediction. These techniques
help us disentangle and uncover the connections between variables. They
help us trace the ways in which some variables might have causal influences
on others; and, depending on the strength of the relationship, they enable
us to predict scores on one variable from the scores on another. Note that
measures of association cannot, by themselves, prove that two variables are
causally related. However, these techniques can provide valuable clues about
causation and are therefore extremely important for theory testing and theory
construction.

For example, suppose you were interested in the relationship between
"time spent studying statistics" and "final grade in statistics" and had gathered
the appropriate data from a group of college students. By calculating the appropriate measure of association, you could determine the strength of the
relationship and its direction. Suppose you found a strong, positive relationship between these variables. You would infer that "study time" and "grade"
were closely related (strength of the relationship) and that as one increased
in value, the other also increased (direction of the relationship). You could
make predictions from one variable to the other (for example, "the longer the
study time, the higher the grade").

Now, as a result of finding this strong, positive relationship, you might
be tempted to make causal inferences. That is, you might jump to such conclusions as "longer study time leads to (causes) higher grades." Such a conclusion might make a good deal of common sense and would certainly be
supported by your statistical analysis. However, the causal nature of the relationship is in no way proven by the statistical analysis. Measures of association can be taken as important clues about causation, but the mere existence
of a relationship should never be taken as conclusive proof of causation.

In fact, other variables might have an effect on the relationship. In the
example above, we probably would not find a perfect relationship between
"study time" and "final grade." That is, we will probably find some individuals who spend a great deal of time studying but receive low grades and some
individuals who fit the opposite pattern. We know intuitively that other variables besides study time affect grades (such as efficiency of study techniques,
amount of background in mathematics, and even random chance). Fortu-

nately, researchers can incorporate these other variables into the analysis and measure their effects. Part III of this text is devoted to bivariate (two-variable) and Part IV to multivariate (more than two variables) descriptive statistics.

Inferential Statistics.   This second class of statistical techniques becomes relevant when we wish to generalize our findings from a **sample** to a **population**. A population is the total collection of all cases in which the researcher is interested and is thus the entity that the researcher wishes to understand. Examples of possible populations would be adult male voters in the United States, all parliamentary democracies, unemployed Puerto Ricans in Atlanta, or sophomore college football players in the Midwest.

Populations can theoretically range from inconceivable in size ("all humanity") to quite small (all 35-year-old red-haired belly dancers currently residing in downtown Cleveland) but are usually fairly large. In fact, they are almost always too large to be measured. To put the problem another way, social scientists almost never have the resources or time to test every case in a population. Hence the need for **inferential statistics**, which involve using information from samples (carefully chosen subsets of the defined populations) to make inferences about populations. Samples are, of course, much cheaper to assemble, and—if the proper techniques are followed—generalizations based on these samples can be very accurate representations of the population.

Many of the concepts and procedures involved in inferential statistics may be unfamiliar. However, most of us are experienced consumers of inferential statistics—most familiarly, perhaps, in the form of public-opinion polls and election projections. When a public-opinion poll reports that 42% of the American electorate plans to vote for a certain presidential candidate, it is essentially reporting a generalization to a population ("the American electorate"—which numbers about 100 million people) from a carefully drawn sample (usually about 1500 respondents). Matters of inferential statistics will occupy our attention in Part II of this book. *(For practice in describing different statistical applications, see problems 1.3 and 1.7.)*

1.5 DISCRETE AND CONTINUOUS VARIABLES

In the next chapter, you will begin to encounter some of the broad array of statistics available to the social scientist. One of the more puzzling aspects of studying statistics is learning when to use which statistic. You will learn specific guidelines as you go along, but let me introduce some basic and general guidelines at this point. The first of these concerns discrete and continuous variables; the second, covered in the next section, concerns level of measurement.

A variable is said to be **discrete** if it has a basic unit of measurement that cannot be subdivided. The measurement process for discrete variables involves accurate counting of the number of units per case. For example, number of people per household is a discrete variable. The basic unit is people, and the fewest you can have is one. Note that the score of a given household on this variable will always be a whole number (you'll never find 2.7 people living in a specific household), and as long as we are counting accurately, the scores we report will always be exact.

A variable is **continuous** if the measurement of it can be subdivided infinitely—at least in a theoretical sense. A good example of such a variable would be time, which can be measured in nanoseconds (billionths of a second) or even smaller units. In a sense, when we measure a continuous variable, we are always approximating and rounding off the scores. We could report somebody's time in the 100-yard dash as 10.7 seconds or 10.732451 seconds, but, since time can be infinitely subdivided (if we have the technology to make the precise measurements), we will never be able to report the exact time elapsed. Since we cannot cite or work with infinitely long numbers, we must report the scores on continuous variables as if they were discrete. The distinction between the two types of variables relates more to measuring and processing the information than to the appearance of the data. This distinction between discrete and continuous variables is one of the most basic in statistics and will constitute one of the criteria by which we will choose among various statistics and graphic devices. *(For practice in distinguishing between discrete and continuous variables, see problems 1.4 through 1.8.)*

1.6 LEVEL OF MEASUREMENT

A second basic and general guideline for the selection of statistics is the **level of measurement.** Every statistical technique involves performing some mathematical operation such as adding scores or ranking cases. Before you can properly use a technique, you must measure the variable being processed in a way that justifies the required mathematical operations. For example, many statistical techniques require that scores be added together. These techniques could be legitimately used only when the variable is measured in a way that permits addition as a mathematical operation. Thus, the researcher's choice of statistical techniques is heavily dependent on the level at which the variables have been measured.

The three levels of measurement are, in order of increasing sophistication, nominal, ordinal, and interval-ratio. Since the concept of level of measurement is so central to statistics, let us consider the nature of the different levels at some length. I will make it a practice throughout this text to introduce level-of-measurement considerations for each statistical technique.

 The Nominal Level of Measurement.   The most basic and the only universal measurement procedure is to classify cases into the pre-established categories of a variable. All measurement involves classification as a minimum. In nominal measurement, classification into categories is the only measurement procedure permitted. The categories themselves are not numerical and can be compared to each other only in terms of the number of cases classified in them. The categories cannot be thought of as "higher" or "lower" than each other along some numerical scale.

Although these measurement procedures are rudimentary, we do have criteria to identify adequately measured nominal variables. In brief:

1. The categories of the variable should be mutually exclusive of each other so that no ambiguity exists concerning classification of any given case.

2. The categories should be exhaustive; a category should exist for every manifestation of the variable being measured (at least an "other" or miscellaneous category).

3. The categories should be homogeneous in terms of the specific research project being conducted. Homogeneity is, of course, a relative thing and must be evaluated in terms of the specific purpose of the research. Categories that are too broad for some purposes may be perfectly adequate for others. Thus, the categories should incorporate elements that are sufficiently similar to facilitate descriptions and generalizations that are accurate and meaningful in the context of the research goals.

Table 1.1 displays four different schemes for measuring the variable "religious preference" and demonstrates some errors of measurement. Scale A in the table violates the criterion of mutual exclusivity because of overlap between the categories Protestant and Episcopalian. Scale B does not provide a category for people with no religious preference (None), and Scale C uses a category (Non-Protestant) that would be too broad for many research purposes. Scale D represents the way religious preference is typically measured in North America.

There is a fourth criterion for judging the adequacy of a nominal variable: the categories should make theoretical sense. We could classify everything in the universe into "things that are tables" and "things that are not tables," but

TABLE 1.1 FOUR SCALES FOR MEASURING RELIGIOUS PREFERENCE

Scale A (not mutually exclusive)	Scale B (not exhaustive)	Scale C (not homogeneous)	Scale D (an adequate scale)
Protestant Episcopalian Catholic Jew None Other	Protestant Catholic Jew	Protestant Non-Protestant	Protestant Catholic Jew None Other

such a scale is of no conceivable use to a social scientist. Variables should be categorized in ways that are relevant to theory and understanding.

As a final note, numerical labels are sometimes used to identify the categories of a variable measured at the nominal level. This practice is especially common when the data are being prepared for computer analysis. For example, the various religions might be labeled with a 1 indicating Protestant, a 2 signifying Catholic, and so on. You should understand that these numbers are merely labels or names and have no numerical quality to them. They cannot be added, subtracted, multiplied, or divided. The only mathematical operation permissible with nominal variables is counting the number of occurrences that have been classified into the various categories of the variable.

The Ordinal Level of Measurement.   In addition to classifying cases into categories, variables measured at the ordinal level allow the categories to be ranked with respect to how much of the trait being measured they possess. The categories form a kind of numerical scale that can be ordered from "high" to "low." Thus, variables measured at the ordinal level are more sophisticated than nominal-level variables because, in addition to counting the number of cases in a category, we can rank the cases with respect to each other. Not only can we say that one case is different from another; we can also say that one case is higher or lower, more or less than another.

For example, the variable socioeconomic status (SES) is usually measured at the ordinal level in the social sciences. The categories of the variable are often ordered according to the following scheme:

4. Upper class
3. Middle class
2. Working class
1. Lower class

Individual cases can be compared in terms of the categories into which they are classified. Thus, an individual classified as a 4 (upper class) would be ranked higher than an individual classified as a 2 (working class). Besides SES, examples of variables measured at the ordinal level would include virtually all attitude and opinion scales such as those that measure prejudice, alienation, or political conservatism.

The major limitation of the ordinal level of measurement is that a particular score represents only position with respect to some other score. We can distinguish between high and low scores, but the distance between the scores cannot be described in precise terms. Although we know that a score of 4 is more than a score of 2, we do not know if it is twice as much as 2.

By the same token, our inability to describe distances along ordinal scales in precise terms means we cannot assume that the distances between scores are always equal. Since the operations of addition, subtraction, multiplication, and division assume equal intervals between scores, they are not per-

TABLE 1.2 BASIC CHARACTERISTICS OF THE THREE LEVELS OF MEASUREMENT

Levels	Examples	Measurement Procedures	Mathematical Operations Permitted
Nominal	Sex, race, religion, marital status	Classification into categories	Counting number of cases in each category of the variable; comparing sizes of categories
Ordinal	Social class (SES), attitude and opinion scales	Classification into categories plus ranking of categories with respect to each other	All above plus judgments of "greater than" and "less than"
Interval-Ratio	Age, number of children, income	All above plus description of distances between scores in terms of equal units	All above plus all other mathematical operations (addition, subtraction, multiplication, division, square roots, etc.)

mitted with variables measured at the ordinal level. For variables at this level, the most sophisticated mathematical operation permitted is ranking categories and cases.

The Interval-Ratio Level of Measurement.[2]   The categories of nominal-level variables have no numerical quality to them. Ordinal-level variables have categories that can be arrayed along a scale from high to low, but the exact distances between categories are unknown. Variables measured at the interval-ratio level not only permit classification and ranking but also allow the distance from category to category (or score to score) to be exactly defined.

Interval-ratio variables are measured in units that have equal intervals and a true zero point. For example, recording the ages of your respondents is a measurement procedure that would produce interval-ratio data because the unit of measurement (years) has equal intervals (the distance from year to year is 365 days) and a true zero point (it is possible to be zero years old). Other examples of interval-ratio variables would be income, number of children, weight, and years married. All mathematical operations are permitted for data measured at this level. Table 1.2 summarizes the basic characteristics of the three levels of measurement.

To summarize, different statistics require different mathematical operations and, therefore, level-of-measurement considerations are the logical first guideline to use in selecting a statistic. For example, computation of a mean

---

[2] Many statisticians distinguish between the interval level (equal intervals) and the ratio level (true zero point). I find the distinction unnecessarily cumbersome in an introductory text and will treat these two levels as one.

READING STATISTICS 1: Introduction

By this point in your education you have developed an impressive array of skills for reading words. Although you may sometimes struggle with a difficult idea or stumble over an obscure meaning, you can comprehend virtually any written work that you are likely to encounter.

As you continue your education in the social sciences, you must develop an analogous set of skills for reading numbers and statistics. To help you reach a reasonable level of literacy in statistics, I have included in this text a series of boxed inserts labeled "Reading Statistics." These will appear in most chapters and will discuss how statistical results are typically presented in the professional literature. Each installment will include an extract or quotation from the professional literature so that we can analyze a realistic example.

As you will see, professional researchers use a reporting style that is quite different from the statistical language you will find in this text. Space in research journals and other media is expensive, and the typical research project requires the analysis of many variables. Thus, a large volume of information must be summarized in very few words. Researchers may express in a word or

two a result or an interpretation that will take us a paragraph or more to state.

Because this is an introductory textbook, I have been careful to break down the computation and logic of each statistic and to identify, even to the point of redundancy, what we are doing when we use statistics. In this text we will never be concerned with more than a few variables at a time. We will have the luxury of analysis in detail and of being able to take pages or even entire chapters to develop a statistical idea or analyze a variable. Thus, a major theme of these boxed inserts will be to summarize how our comparatively long-winded (but more careful) vocabulary is translated into the concise language of the professional researcher.

When you have difficulty reading words, your tendency is (or, at least, should be) to consult reference books (especially dictionaries) to help you identify and analyze the elements (words) of the passage. When you have difficulty reading statistics, you should do exactly the same thing. I hope you will find this text a valuable reference book, but if you learn enough from this text to be able to use any source to help you read statistics, this text will have fulfilled one of its major goals.

(or arithmetic average) requires that all observations be added together and then divided by the number of observations. These operations are fully justified only when a variable is measured at the interval-ratio level.

Ideally, the researcher would utilize only those statistics that were fully justified by the level-of-measurement criteria. In this imperfect world, however, the most powerful and useful statistics (such as the mean) require interval-ratio variables, while most of the variables of interest to the social sciences are only nominal (race, sex, marital status) or at best ordinal (attitude scales). Relatively few concepts of interest to the social sciences are so precisely defined that they can be measured at the interval-ratio level. This disparity creates some very real difficulties in the research process. On one hand, the researcher should use the most sophisticated statistical procedures fully justified by the level-of-measurement criteria. To treat interval-ratio data as if they were only ordinal, for example, results in a significant loss of information and precision. Treated as interval-ratio data, the variable "age" can supply us with exact information regarding the differences between the cases (for example, "Individual A is three years and two months older than Individual B"). If this trait was treated only as an ordinal variable, however, the precision of our comparisons would suffer and we could say only that "Individual A is older (or greater than) Individual B."

On the other hand, given the nature of the disparity, researchers are more likely to treat variables as if they were higher in level of measurement than they actually are. For example, variables measured at the ordinal level might be treated as if they were interval-ratio because the statistical procedures available at the higher level are more powerful, flexible, and interesting. This practice is common but may lead to errors of logic and interpretation and, ultimately, to incorrect conclusions. In situations where level-of-measurement guidelines might have been violated, the researcher should be especially cautious and careful in assessing statistical results and developing interpretations.

At any rate, level-of-measurement considerations are important criteria, and we will always consider them when presenting statistical procedures. Level of measurement is a major organizing principle for the material that follows, and you should make sure that you are familiar with these guidelines. *(For practice in determining the level of measurement of a variable, see problems 1.4 through 1.8.)*

SUMMARY

1. Within the context of social research, the purpose of statistics is to organize, manipulate, and analyze data so that the researcher can more easily answer his or her original question. Along with theory and methodology, statistics are a basic tool by which social scientists attempt to enhance their understanding of the social world.
2. There are two general classes of statistics. Descriptive statistics are used to summarize the distribution of a single variable and the relationships between two or more variables. Inferential statistics provide us with techniques by which we can generalize to populations from random samples.
3. Two basic guidelines for selecting statistical techniques were presented. Variables may be either discrete or continuous and may be measured at any of three different levels. At the nominal level, we can classify cases into categories of the variable and compare category sizes. At the ordinal level, categories and cases can be ranked with respect to each other. At the interval-ratio level, all mathematical operations are permitted.

GLOSSARY

Continuous variable. A variable with a unit of measurement that can be subdivided infinitely.

Data. Any information collected as part of a research project and expressed as numbers.

Data reduction. Summarizing many scores with a few statistics. A major goal of descriptive statistics.

Dependent variable. A variable that is identified as an effect, result, or outcome variable. The dependent variable is thought to be caused by the independent variable.

Descriptive statistics. The branch of statistics concerned with (1) summarizing the distribution of a single variable or (2) measuring the relationship between two or more variables.

Discrete variable. A variable with a basic unit of measurement that cannot be subdivided.

Hypothesis. A statement about the relationship between variables that is derived from a theory. Hypotheses are more specific than theories, and all terms and concepts are fully defined.

Independent variable. A variable that is identified as a causal variable. The independent variable is thought to cause the dependent variable.

Inferential statistics. The branch of statistics concerned with making generalizations from samples to populations.

Level of measurement. The mathematical characteristics of a variable as determined by the measurement process. A major criterion for selecting statistical techniques. Variables can be measured at

any of three levels, each permitting certain mathematical operations and statistical techniques. The characteristics of the three levels are summarized in Table 1.2.

Measures of association. Statistics that summarize the strength and direction of the relationship between variables.

Population. The total collection of all cases in which the researcher is interested.

Sample. A carefully chosen subset of a population. In inferential statistics, information is gathered from samples and then generalized to populations.

Statistics. A set of mathematical techniques for organizing and analyzing data.

Theory. A generalized explanation of the relationship between two or more variables.

Variable. Any trait that can change values from case to case.

PROBLEMS

1.1 In your own words, describe the role of statistics in the research process. Using the "wheel of science" as a framework, explain how statistics link theory with research.

1.2 Find a research article in any social science journal. Choose an article on a subject of interest to you and don't worry about being able to understand all of the statistics that are reported.
  a. How much of the article is devoted to statistics per se (as distinct from theory, ideas, discussion, and so on)?
  b. Is the research based on a sample from some population? How large is the sample? How were subjects or cases selected? Can the findings be generalized to some population?
  c. What variables are used? Which are independent and which are dependent? For each variable, determine the level of measurement and whether the variable is discrete or continuous.
  d. What statistical techniques are used? Try to follow the statistical analysis and see how much you can understand. Save the article and read it again after you finish this course and see if you do any better.

1.3 Distinguish between descriptive and inferential statistics. Describe a research situation in which each would be useful.

1.4 Below are some items from a public-opinion survey. For each item, indicate the level of measurement and whether the variable will be discrete or continuous.
  a. What is your occupation? _____
  b. How many years of school have you completed? _____
  c. If you were asked to use one of these four names for your social class, which would you say you belonged in?
    _____ Upper _____ Middle
    _____ Working _____ Lower
  d. What is your age? _____
  e. In what country were you born? _____
  f. What is your grade-point average? _____
  g. What is your major? _____
  h. The only way to deal with the drug problem is to legalize all drugs.
    _____ Strongly agree _____ Agree
    _____ Undecided _____ Disagree
    _____ Strongly disagree

i. What is your astrological sign? _____

j. How many brothers and sisters do you have? _____

1.5 Below are brief descriptions of how researchers measured a variable. For each situation, determine the level of measurement of the variable and whether it is continuous or discrete.

a. Race. Respondents were asked to select a category from the following list:
_____ Black _____ White _____ Other

b. Honesty. Subjects were observed as they passed by a spot on campus where an apparently lost wallet was lying. The wallet contained money and complete identification. Subjects were classified into one of the following categories:
_____ Returned the wallet with money
_____ Returned the wallet but kept the money
_____ Did not return wallet.

c. Social class. Subjects were asked about their family situation when they were 16 years old. Was their family:
_____ Very well off compared to other families
_____ About average _____ Not so well off?

d. Education. Subjects were asked how many years of schooling they and each parent had completed.

e. Racial integration on campus. Students were observed during lunchtime at the cafeteria for a month. The number of students sitting with students of other races was counted for each meal period.

f. Number of children. Subjects were asked: "How many children have you ever had? Please include any that may have passed away."

g. Student seating patterns in classrooms. On the first day of class, instructors noted where each student sat. Seating patterns were remeasured every two weeks until the end of the semester. Each student was classified as
_____ same seat as last measurement;
_____ adjacent seat;
_____ different seat, not adjacent;
_____ absent.

h. Physicians per capita. The number of practicing physicians was counted in each of 50 cities, and the researchers used population data to compute the number of physicians per capita.

i. Physical Attractiveness. A panel of 10 judges rated each of 50 photos of a mixed-race sample of males and females for physical attractiveness on a scale from 0 to 20 with 20 being the highest score.

j. Number of accidents. The number of traffic accidents for each of 20 busy intersections in a city was recorded. Also, each accident was rated as
_____ minor damage, no injuries;
_____ moderate damage, personal injury requiring hospitalization;
_____ severe damage and injury.

1.6 For each of the first 20 items in the 1996 General Social Survey (see Appendix G), indicate the level of measurement and whether the variable is continuous or discrete.

1.7 For each research situation summarized below, identify the level of measurement of all variables and indicate whether they are discrete or continuous. Also, decide which statistical applications are used: descriptive statistics (single variable), descriptive statistics (two or more variables), or inferential statistics. Remember that it is quite common for a given situation to require more than one type of application.

a. The administration of your university is proposing a change in parking policy. You select a random sample of students and ask each one if he or she favors or opposes the change.

b. You ask everyone in your social research class to tell you the highest grade he or she ever received in a math course and the grade on a recent statistics test. You then compare the two sets of scores to see if there is any relationship.

c. Your aunt is running for mayor and hires you (for a huge fee, incidentally) to question a sample of voters about their concerns in local politics. Specifically, she wants a profile of the voters that will tell her what percent belong to each political party; what percent are male or female; and what percent favor or oppose the widening of the main street in town.

d. Several years ago, a state reinstituted the death penalty for first-degree homicide. Supporters of capital punishment argued that this change would reduce the homicide rate. To investigate this claim, a researcher has gathered information on number of homicides in the state for the two-year periods before and after the change.

e. A local automobile dealer is concerned about customer satisfaction. He wants to mail a survey form to all customers for the past year and ask them if they are satisfied, very satisfied, or not satisfied with their purchases.

1.8 For each research situation below, identify the independent and dependent variables. Classify each in terms of level of measurement and whether or not the variable is discrete or continuous.

a. A graduate student is studying sexual harassment on college campuses and asks 500 female students if they personally have experienced any such incidents. Each student is asked to estimate the frequency of these incidents as either "often, sometimes, rarely, or never." The researcher also gathers data on age and major to see if there is any connection between these variables and frequency of sexual harassment.

b. A supervisor in the Solid Waste Management Division of a city government is attempting to assess two different methods of trash collection. One area of the city is served by trucks with two-man crews who do "backyard" pickups, and the rest of the city is served by "hi-tech" single-person trucks with curbside pickup. The assessment measures include the number of complaints received from the two different areas over a six-month period, the amount of time per day required to service each area, and the cost per ton of trash collected.

c. The adult bookstore near campus has been raided and closed by the police. Your social research class has decided to poll the student body and get their reactions and opinions. The class decides to ask each student if he or she supports or opposes the closing of the store,

how many times each one has visited the store, and if he or she agrees or disagrees that "pornography is a direct cause of sexual assaults on women." The class also collects information on the sex, age, religious and political philosophy, and major of each student to see if opinions are related to these characteristics.

d. For a research project in a political science course, a student has collected information about the quality of life and the degree of political democracy in 50 nations. Specifically, she used infant mortality rates to measure quality of life, and the percentage of all adults who are permitted to vote in national elections as a measure of democratization. Her hypothesis is that quality of life is higher in more democratic nations.

e. A highway engineer wonders if a planned increase in speed limit on a heavily traveled local avenue will result in any change in number of accidents. He plans to collect information on traffic volume, number of accidents, and number of fatalities for the six-month periods before and after the change.

f. Students are planning a program to promote "safe sex" and awareness of a variety of other health concerns for college students. To measure the effectiveness of the program, they plan to give a survey measuring knowledge about these matters to a random sample of the student body before and after the program.

g. Several states have drastically cut their budgets for mental health care. Will this increase the number of homeless people in these states? A researcher contacts a number of agencies serving the homeless in each state and develops an estimate of the size of the population before and after the cuts.

h. Does tolerance for diversity vary by race, ethnicity, or gender? A sample of white, black, Asian, Hispanic, and Native Americans have been given a survey that measures their interest in and appreciation of cultures and groups other than their own.

Introduction to *SPSS for Windows* and the General Social Survey

The problems and exercises in this text have been written so that they can be solved with just a simple hand calculator. I've purposely kept the number of cases involved unrealistically low so that the tedium of mere calculation would not interfere unduly with the learning process. To provide a more realistic experience in the analysis of social science data, a shortened version of the 1996 General Social Survey (GSS) has been provided to your instructor. The GSS is a public-opinion poll that has been conducted on a nationally representative sample of Americans almost every year since 1972. The full survey includes hundreds of questions covering a broad range of social and political issues. The version supplied with this text has a limited number of variables and cases but is still actual, "real-life" data, so you have the opportunity to practice your statistical skills in a more realistic context.

One of the problems with reality, of course, is that it is often cumbersome and confusing. It's hard enough to do your homework with simplified problems, and you should be a little leery, in terms of your own time and effort, of promises of relevance and realism. This brings us to the second purpose of this section: computers and statistical packages. A statistical package is a set of computer programs for the analysis of data. The advantage of these packages is that, since the programs are already written, you can capitalize on the power of the computer with minimal computer literacy and virtually no programming experience.

One commonly used statistical package is called Statistical Package for the Social Sciences (SPSS). In these brief sections at the ends of chapters, I will explain how to use SPSS to manipulate and analyze the GSS data, and I will illustrate and interpret the results. If you do not have access to SPSS at all, you can use whatever package is available (there are many others) to analyze the GSS data and at least compare results. Be sure to read Appendix F before attempting any data analysis.

Part I   Descriptive Statistics

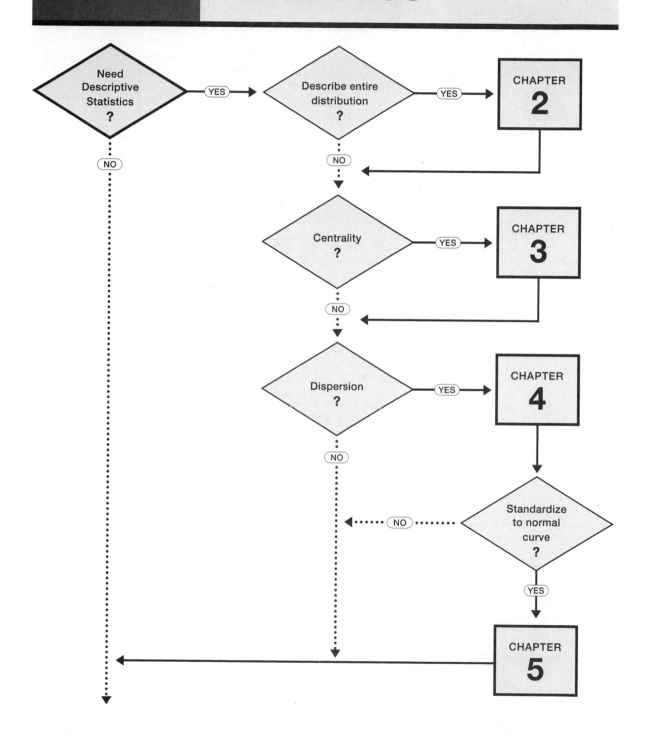

Basic Descriptive Statistics

2

Percentages, Ratios and Rates, Tables, Charts, and Graphs

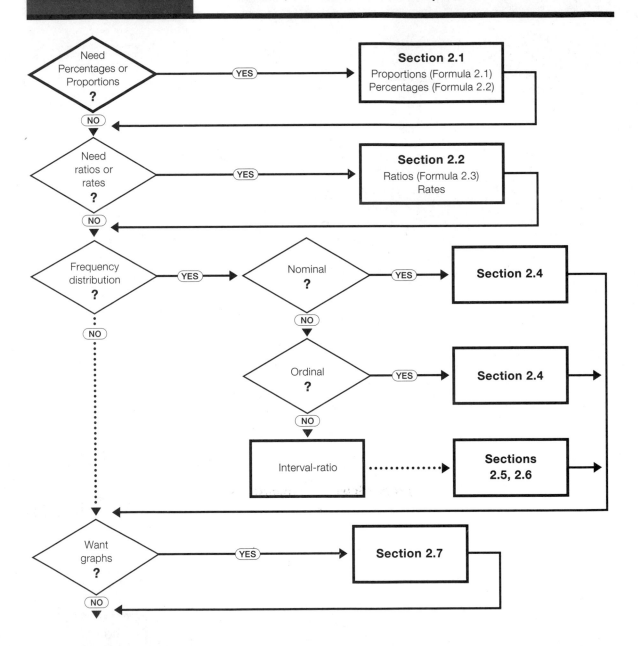

Research results do not speak for themselves. They must be arranged in ways that allow the researcher (and his or her readers) to comprehend their meaning quickly. The primary function of descriptive statistics is to present research results clearly and concisely. Researchers use a process called *data reduction* to organize data into presentable form. Data reduction involves using a few numbers, a table, or a graphic device to summarize or stand for a larger array of data.

Little about this process is mysterious or difficult, but, as a researcher or a consumer of research, you should be aware of one problem. Data reduction inevitably loses information (precision and detail); and, therefore, summarizing statistics might present a misleading picture of research results. The issue is not whether to use descriptive statistics—they are always necessary. The issue is which descriptive statistics to use. The various summarizing techniques present results in different ways, are based on different mathematical operations, and characteristically lose information in different ways. Thus, in choosing among the various summarizers, several decisions must be made: how to present the data, what kind of information to lose, and how much detail can safely be obscured.

In this chapter, we will consider several commonly used techniques for presenting research results: percentages and proportions, ratios and rates, tables, charts, and graphs.

2.1 PERCENTAGES AND PROPORTIONS

Consider the following statement: "Of the 269 cases handled by the court, 167 resulted in a prison sentence of five years or more." While there is nothing wrong with this statement, the same fact could have been more clearly conveyed if it had been reported as a percentage: "about 62% of all cases resulted in prison sentences of five or more years."

Percentages and proportions supply a frame of reference for reporting research results in the sense that they standardize the raw data: percentages to the base 100 and proportions to the base 1.00. The mathematical definitions of **proportions** and **percentages** are

FORMULA 2.1
$$\text{Proportion } (p) = \frac{f}{N}$$

FORMULA 2.2
$$\text{Percentage } (\%) = \left(\frac{f}{N}\right) \times 100$$

where f = frequency, the number of cases in any category
N = the number of cases in all categories

To illustrate the computation of percentages, consider the data presented in Table 2.1. To find the percentage of cases in the first category (sentences

TABLE 2.1   DISPOSITION OF 269 CRIMINAL CASES (fictitious data)*

Sentence	Frequency (f)	Proportion (p)	Percentage (%)
Five years or more	167	0.6208	62.08
Less than five years	72	0.2677	26.77
Suspended	20	0.0744	7.44
Acquitted	10	0.0372	3.72
	N = 269	1.0001	100.01

*The slight discrepancies in the totals of the proportion and percentage columns are due to rounding error.

of five years or more), note that there are 167 cases in the category ($f = 167$) and a total of 269 cases in all ($N = 269$). So,

$$\text{Percentage (\%)} = \left(\frac{f}{N}\right) \times 100 = \left(\frac{167}{269}\right) \times 100 = (0.6208) \times 100 = 62.08\%$$

Using the same procedures, we can also find the percentage of cases in the second category:

$$\text{Percentage (\%)} = \left(\frac{f}{N}\right) \times 100 = \left(\frac{72}{269}\right) \times 100 = (0.2677) \times 100 = 26.77\%$$

Both results could have been expressed as proportions. For example, the proportion of cases in the third category is 0.0744.

$$\text{Proportion} = \frac{f}{N} = \frac{20}{269} = 0.0744$$

Percentages and proportions are easier to read and comprehend than frequencies. This advantage is particularly obvious when attempting to compare groups of different sizes. For example, based on the information presented in Table 2.2, which college has the higher relative number of social science majors? Because the total enrollments are so different, such comparisons are difficult to conceptualize from the raw frequencies. To make comparisons easier, the difference in size can be effectively eliminated by standardizing both distributions to the common base of 100 (or, in other words, by computing percentages for both distributions). The same data are presented in percentages in Table 2.3.

The percentages in Table 2.3 make it easier to identify both differences and similarities between the two colleges. College A has a much higher percentage of social science majors (even though the absolute number of social science majors is less than at College B) and about the same percentage of humanities majors. How would you describe the differences in the remaining

Application 2.1

Not long ago, in a large social service agency, the following conversation took place between the executive director of the agency and a supervisor of one of the divisions.

Executive director: Well, I don't want to seem abrupt, but I've only got a few minutes. Tell me, as briefly as you can, about this staffing problem you claim to be having.

Supervisor: Ma'am, we just don't have enough people to handle our workload. Of the 177 full-time employees of the agency, only 50 are in my division. Yet, 6231 of the 16,722 cases handled by the agency last year were handled by my division.

Executive director (smothering a yawn): Very interesting. I'll certainly get back to you on this matter.

How could the supervisor have presented his case more effectively? Because he wants to compare two sets of numbers (his staff versus the total staff and the workload of his division versus the total workload of the agency), proportions or percentages would be a more forceful way of presenting results. What if the supervisor had said, "Only 28.25% of the staff is assigned to my division, but we handle 37.26% of the total workload of the agency"? Is this a clearer message?

The first percentage is found by

$$\% = \left(\frac{f}{N}\right) \times 100 = \left(\frac{50}{177}\right) \times 100$$

$$= (.2825) \times 100 = 28.25\%$$

and the second percentage is found by

$$\% = \left(\frac{f}{N}\right) \times 100 = \left(\frac{6231}{16,722}\right) \times 100$$

$$= (.3726) \times 100 = 37.26\%$$

TABLE 2.2   DECLARED MAJOR FIELDS OF STUDY ON TWO COLLEGE CAMPUSES (fictitious data)

Major	College A	College B
Business	103	312
Natural sciences	82	279
Social sciences	137	188
Humanities	93	217
	$N = 415$	$N = 996$

TABLE 2.3   DECLARED MAJOR FIELDS OF STUDY ON TWO COLLEGE CAMPUSES (fictitious data)

Major	College A (%)	College B (%)
Business	24.82	31.33
Natural sciences	19.76	28.01
Social sciences	33.01	18.88
Humanities	22.41	21.79
	100.00% (415)	100.01% (996)

two major fields? *(For practice in computing and interpreting percentages and proportions, see problems 2.1 and 2.2.)*

Some further rules on the use of percentages and proportions:

1. When working with a small number of cases (say, fewer than 20), it is usually preferable to report the actual frequencies rather than percentages or proportions. With a small number of cases, the percentages can change drastically with relatively minor changes in the data. For example, if you begin with a data set that includes 10 males and 10 females (that is, 50% of each gender) and then add another female, the percentage distributions will change noticeably to 52.38% female and 47.62% male. Of course, as the number of observations increases, each additional case will have a smaller impact. If we started with 500 males and females and then added one more female, the percentage of females would change by only a tenth of a percent (from 50% to 50.10%).

2. Always report the number of observations along with proportions and percentages. This permits the reader to judge the adequacy of the sample size and, conversely, helps to prevent the researcher from lying with statistics. Statements like "two out of three people questioned prefer courses in statistics to any other course" might impress you, but the claim would lose its gloss if you learned that only three people were tested. You should be extremely suspicious of reports that fail to report the number of cases that were tested.

2.2 RATIOS AND RATES

Ratios and **rates** provide two additional ways in which the distribution of a variable can be simply and dramatically summarized. Ratios are especially useful for comparing categories in terms of relative frequency. Instead of standardizing the distribution of a variable to the base 100 or 1.00, as we did in computing percentages and proportions, we determine ratios by dividing the frequency of one category by the frequency in another. Mathematically, a ratio can be defined as

FORMULA 2.3
$$\text{Ratio} = \frac{f_1}{f_2}$$

where f_1 = the number of cases in the first category
f_2 = the number of cases in the second category

To illustrate the use of ratios, suppose that you were interested in the relative sizes of the various religious denominations and found that a particular community included 1370 Protestant families and 930 Catholic families. To find the ratio of Protestants (f_1) to Catholics (f_2), divide 1370 by 930. The resultant ratio is 1.47. This number would mean that for every Catholic family, there are 1.47 Protestant families.

Note that ratios can be very economical ways of expressing the relative predominance of two categories. That Protestants outnumber Catholics in

---

Application 2.2

In Table 2.2, how many natural science majors are there compared to social science majors at College B? This question could be answered with frequencies, but a more easily understood way of expressing the answer would be with a ratio. The ratio of natural science to social science majors would be

$$\text{Ratio} = \frac{f_1}{f_2} = \frac{279}{188} = 1.48$$

For every social science major, there are 1.48 natural science majors at College B.

---

our example is obvious from the raw data. Percentages or proportions could have been used to summarize the overall distribution (that is, 59.56% of the families were Protestant, 40.44% were Catholic). Compared to these other methods, ratios are a precise measure of the relative frequency of one category per unit of the other category. They tell us in an exact way the extent to which one category outnumbers the other.

Ratios are often multiplied by some power of 10 to eliminate decimal points. For example, the ratio computed above might be reported as 147 instead of 1.47. This would mean that, for every 100 Catholic families, there are 147 Protestant families in the community. To ensure clarity, the comparison units for the ratio are often expressed as well. Based on a unit of ones, the ratio of Protestants to Catholics would be expressed as 1.47 : 1. Based on hundreds, the same statistic might be expressed as 147 : 100. *(For practice in computing and interpreting ratios, see problems 2.1 and 2.2.)*

Rates provide still another way of summarizing the distribution of a single variable. Rates are defined as the number of actual occurrences of some phenomenon divided by the number of possible occurrences per some unit of time. Rates are usually multiplied by some power of 10 to eliminate decimal points. For example, the crude death rate for a population is defined as the number of deaths in that population (actual occurrences) divided by the number of people in the population (possible occurrences) per year. This quantity is then multiplied by 1000. The formula for the crude death rate can be expressed as

$$\text{Crude death rate} = \frac{\text{Number of deaths in year}}{\text{Total population}} \times 1000$$

If there were 100 deaths during a given year in a town of 7000, the crude death rate for that year would be

$$\text{Crude death rate} = \left(\frac{100}{7000}\right) \times 1000 = (.01429) \times 1000 = 14.29$$

Or, for every 1000 people, there were 14.29 deaths during this particular year.

Application 2.3

In 1990, in a city of 167,000, there were 2500 births. In 1980, when the population of the city was only 133,000, there were 2700 births. Is the birthrate rising or falling? Although this question can be answered from the preceding information, the trend in birthrates will be much more obvious if we compute birthrates for both years. Like crude death rates, crude birthrates are usually multiplied by 1000 to eliminate decimal points. For 1980:

$$\text{Crude birthrate} = \left(\frac{2700}{133,000}\right) \times 1000 = 20.30$$

In 1980, there were 20.30 births for every 1000 people in the city.

For 1990:

$$\text{Crude birthrate} = \left(\frac{2500}{167,000}\right) \times 1000 = 14.97$$

In 1990, there were 14.97 births for every 1000 people in the city. With the help of these statistics, the decline in the birthrate is clearly expressed.

By the same token, if a city of 237,000 people experienced 120 auto thefts during a particular year, the auto theft rate would be

$$\text{Auto theft rate} = \left(\frac{120}{237,000}\right) \times 100,000 = (0.0005063) \times 100,000 = 50.63$$

Or, for every 100,000 people, there were 50.63 auto thefts during the year in question. *(For practice in computing and interpreting rates, see problems 2.3 and 2.4.)*

Up to now, we have considered three techniques (proportions and percentages, ratios, and rates) for describing and summarizing data. All three techniques express, clearly and concisely, the distribution of a single variable. They represent different ways of expressing information so that it can be quickly appreciated. All three techniques are also quite useful for purposes of comparison. For instance, the auto theft rate of 50.63 in our example might be difficult to interpret by itself (is this a high or a low rate?) but would take on much more meaning if you found that the rate for the preceding year had been 32.70. These statistics provide alternative ways of expressing relative frequency and are thus especially useful in making comparisons between different groups and/or different times.

2.3 FREQUENCY DISTRIBUTIONS: INTRODUCTION

Frequency distributions are tables that summarize the distribution of a variable by reporting the number of cases contained in each category of the variable. They are very helpful and commonly used ways of organizing and working with data. In fact, the construction of frequency distributions is almost always the first step in any statistical analysis.

TABLE 2.4   DATA FROM COUNSELING CENTER SURVEY

Student	Sex	Marital Status	Satisfaction with Services*	Age
A	Male	Single	4	18
B	Male	Married	2	19
C	Female	Single	4	18
D	Female	Single	2	19
E	Male	Married	1	20
F	Male	Single	3	20
G	Female	Married	4	18
H	Female	Single	3	21
I	Male	Single	3	19
J	Female	Divorced	3	23
K	Female	Single	3	24
L	Male	Married	3	18
M	Female	Single	1	22
N	Female	Married	3	26
O	Male	Single	3	18
P	Male	Married	4	19
Q	Female	Married	2	19
R	Male	Divorced	1	19
S	Female	Divorced	3	21
T	Male	Single	2	20

*Key: (4) Very satisfied   (2) Dissatisfied
     (3) Satisfied   (1) Very dissatisfied

To illustrate the usefulness of frequency distributions and to provide some data for examples, assume that the counseling center at a university is assessing the effectiveness of its services. Any realistic evaluation research would collect a variety of information from a large group of students but, for the sake of this example, we will confine our attention to just four variables and 20 students. The data are reported in Table 2.4.

Note that, even though the data in Table 2.4 represent an unrealistically low number of cases, it is difficult to discern any patterns or trends. For example, try to ascertain the general level of satisfaction of the students from Table 2.4. You may be able to do so with just 20 cases, but it will take some time and effort. Imagine the difficulty with 50 cases or 100 cases presented in this fashion. Clearly the data need to be organized in a format that allows the researcher (and his or her audience) to understand easily the distribution of the variables.

One general rule that applies to all frequency distributions is that *the categories of the frequency distribution must be exhaustive and mutually exclusive.* In other words, the categories must be stated in a way that permits each case to be counted in one and only one category. This basic principle applies to the construction of frequency distributions for variables measured at all three levels of measurement.

Beyond this rule, there are only guidelines to help you construct useful frequency distributions. As you will see, the researcher has a fair amount of discretion in stating the categories of the frequency distribution (especially with variables measured at the interval-ratio level). I will identify the issues to consider as you make decisions about the nature of any particular frequency distribution. Ultimately, however, the guidelines I state are aids for decision making, nothing more than helpful suggestions. As always, the researcher has the final responsibility for making sensible decisions and presenting his or her data in a meaningful way.

2.4 FREQUENCY DISTRIBUTIONS FOR VARIABLES MEASURED AT THE NOMINAL AND ORDINAL LEVELS

Nominal-Level Variables.   For nominal-level variables, construction of the frequency distribution is typically very straightforward. For each category of the variable being displayed, the occurrences are counted and the subtotals, along with the total number of cases (N), are reported. Table 2.5 displays a frequency distribution for the variable "sex" from the counseling center survey. For purposes of illustration, a column for tallies has been included in this table to illustrate how the cases would be sorted into categories. (This column would not be included in the final form of the frequency distribution.) Take a moment to notice several other features of the table. Specifically, the table has a descriptive title, clearly labeled categories (male and female), and a report of the total number of cases at the bottom of the frequency column. *These items must be included in all tables regardless of the variable or level of measurement.*

The meaning of the table is quite clear. There are 10 males and 10 females in the sample. While not exactly startling, this information is much easier to comprehend when the data have been organized into a frequency distribution.

For some nominal variables, the researcher might have to make some choices about the number of categories he or she wishes to report. For example, the distribution of the variable "marital status" could be reported using the categories listed in Table 2.4. The resultant frequency distribution is presented in Table 2.6. Although this is a perfectly fine frequency distribution, it may be too detailed for some purposes. For example, the researcher might want to focus solely on "nonmarried" as distinct from "married" students.

TABLE 2.5   SEX OF RESPONDENTS, COUNSELING CENTER SURVEY

Sex	Tallies	Frequency (f)
Male	LHT LHT	10
Female	LHT LHT	10
		$N = 20$

TABLE 2.6 MARITAL STATUS OF RESPONDENTS,
COUNSELING CENTER SURVEY

Status	Frequency
Single	10
Married	7
Divorced	3
	$N = 20$

TABLE 2.7 MARITAL STATUS OF RESPONDENTS,
COUNSELING CENTER SURVEY

Status	Frequency
Married	7
Not married	13
	$N = 20$

That is, the researcher might not be concerned with the difference between single and divorced respondents but may want to treat both as simply "not married." In that case, these categories could be grouped together and treated as a single entity, as in Table 2.7. Notice that, by this collapsing, information and detail have been lost. This latter version of the table would not allow the researcher to discriminate between the two unmarried states.

Ordinal-Level Variables. Frequency distributions for ordinal-level variables are constructed following the same routines used for nominal-level variables. Table 2.8 reports the frequency distribution of the "satisfaction" variable from the counseling center survey. Note that a column of percentages by category has been added to this table. Such columns heighten the clarity of the table (especially with larger samples) and are common adjuncts to the basic frequency distribution for variables measured at all levels.

This table reports that most students were either satisfied or very satisfied with the services of the counseling center. The most common response (nearly half the sample) was "satisfied." If the researcher wanted to emphasize this major trend, the categories could be collapsed as in Table 2.9. Again, the price paid for this increased compactness is that some information (in this case, the exact composition of satisfied and dissatisfied students) is lost. *(For practice in constructing and interpreting frequency distributions for nominal- and ordinal-level variables, see problem 2.5a.)*

TABLE 2.8 SATISFACTION WITH SERVICES, COUNSELING CENTER SURVEY

Satisfaction	Frequency (f)	Percentage (%)
(4) Very satisfied	4	20
(3) Satisfied	9	45
(2) Dissatisfied	4	20
(1) Very dissatisfied	3	15
	$N = 20$	100%

TABLE 2.9 SATISFACTION WITH SERVICES, COUNSELING CENTER SURVEY

Satisfaction	Frequency (f)	Percentage (%)
Satisfied	13	65
Dissatisfied	7	35
	$N = 20$	100%

2.5 FREQUENCY DISTRIBUTIONS FOR VARIABLES MEASURED AT THE INTERVAL-RATIO LEVEL

Basic Considerations. In general, the construction of frequency distributions for variables measured at the interval-ratio level is more complex than for nominal and ordinal variables. Interval-ratio variables usually have a large number of possible scores (that is, a wide range from the lowest to the highest score). The large number of scores requires some collapsing or grouping of categories to produce reasonably compact frequency distributions. To construct frequency distributions for interval-ratio level variables, you must decide how many categories to use and how wide these categories should be.

For example, suppose you wished to report the distribution of the variable "age" for a sample drawn from a community. Unlike the college data reported in Table 2.4, a community sample would have a very broad range of ages. If you simply reported the number of times that each year of age (or score) occurred, you could easily wind up with a frequency distribution that contained 70, 80, or even more categories. Such a large frequency distribution would not present a concise picture. The scores (years) must be grouped into larger categories to heighten clarity and ease of comprehension. How large should these categories be? How many categories should be included in the table? Although there are no hard-and-fast guidelines for making these decisions, they always involve a trade-off between more detail (a greater number of narrow categories) or more compactness (a smaller number of wide categories).

Application 2.4

The following list shows the ages of 50 prisoners enrolled in a work-release program. Is this group young or old? A frequency distribution will provide an accurate picture of the overall age structure.

18	60	57	27	19
20	32	62	26	20
25	35	75	25	21
30	45	67	41	30
37	47	65	42	25
18	51	22	52	30
22	18	27	53	38
27	23	32	35	42
32	37	32	40	45
55	42	45	50	47

We will use about 10 intervals to display these data. By inspection we see that the youngest prisoner is 18 and the oldest is 75. The range is thus 57. Interval size will be 57/10, or 5.7, which we can round off to either 5 or 6. Let's use a six-year interval beginning at 18. The stated limits of the lowest interval will be 18–23. Now we must state the limits of all other intervals, count the number of cases in

each interval, and display these counts in a frequency distribution. Columns may be added for percentages, cumulative percentages, and/or cumulative frequency. The complete distribution, with a column added for percentages, is

Ages	Frequency	Percentages
18–23	10	20
24–29	7	14
30–35	9	18
36–41	5	10
42–47	8	16
48–53	4	8
54–59	2	4
60–65	3	6
66–71	1	2
72–77	1	2
	$N = 50$	100

The prisoners seem to be fairly evenly spread across the age groups up to the 48–53 interval. There is a noticeable lack of prisoners in the oldest age groups and a concentration of prisoners in their 20s and 30s.

Constructing the Frequency Distribution. To introduce the mechanics and decision-making processes involved, we will construct a frequency distribution to display the ages of the students in the counseling center survey. Because of the narrow age range of a group of college students, we can use categories of only one year (these categories are often called **class intervals** when working with interval-ratio data). The frequency distribution is constructed by listing the ages from youngest to oldest, counting the number of times each score (year of age) occurs, and then totaling the number of scores for each category. Table 2.10 presents the information and reveals a concentration or clustering of scores in the 18 and 19 class intervals.

Even though the picture presented in this table is fairly clear, assume for the sake of illustration that you desire a more compact (less detailed) summary. To do this, you will have to group scores into wider class intervals. By increasing the interval width (say to two years), you can reduce the number

TABLE 2.10   AGE OF RESPONDENTS, COUNSELING CENTER SURVEY
(interval width = one year of age)

Class Intervals	Frequency (f)
18	5
19	6
20	3
21	2
22	1
23	1
24	1
25	0
26	1
	$N = 20$

TABLE 2.11   AGE OF RESPONDENTS, COUNSELING CENTER SURVEY
(interval width = two years)

Class Intervals	Frequency (f)	Percentage (%)
18–19	11	55
20–21	5	25
22–23	2	10
24–25	1	5
26–27	1	5
	$N = 20$	100%

of intervals and achieve a more compact expression. The grouping of scores in Table 2.11 clearly emphasizes the relative predominance of young respondents. This trend in the data can be stressed even more by the addition of a column displaying the percentage of cases in each category.

Note that the class intervals in Table 2.11 have been stated with an apparent gap between them (that is, the class intervals are separated by a distance of one unit). At first glance, these gaps may appear to violate the principle of exhaustiveness; but, since age has been measured in whole units, the gaps actually pose no problem. Given the level of precision of the measurement (in whole years as opposed to 10ths or 100ths of a year), no case could have a score falling between these class intervals. In fact, for these data, the set of class intervals contained in Table 2.11 constitutes a scale that is exhaustive and mutually exclusive. Each of the 20 respondents in the sample can be sorted into one and only one age category.

However, consider the difficulties that might have been encountered if age had been measured with greater precision. If age had been measured in

10ths of a year, into which class interval in Table 2.11 would a 19.4-year-old subject be placed? You can avoid this ambiguity by always stating the limits of the class intervals at the same level of precision as the data. Thus, if age were being measured in 10ths of a year, the limits of the class intervals in Table 2.11 would be stated in 10ths of a year. For example:

17.5–19.4
19.5–21.4
21.5–23.4
23.5–25.4
25.5–27.4

To maintain mutual exclusivity between categories, *no overlapping of class intervals is allowed*. If you state the limits of the class intervals at the same level of precision as the data and maintain a gap between intervals, you will always produce a frequency distribution where each case can be assigned to one and only one category.

Real Class Limits and Midpoints.   If the sole purpose of the frequency distribution is to display the distribution of a variable, then no additional complexities need be introduced. However, for certain purposes (such as constructing the histograms described in Section 2.7), you must conceptualize the distribution not as a series of discrete, nonoverlapping class intervals but rather as a continuous series of categories. To work with the distribution of a variable as if it were continuous, statisticians use **real class limits**. To find the real limits of any class interval, begin with the limits as stated in the frequency distribution (the **stated class limits**). Divide the distance between the stated class intervals in half and add the result to all upper stated limits and subtract it from all lower stated limits. This process is illustrated below with the class intervals stated in Table 2.11. The distance between intervals is one, so the real limits can be found by adding 0.5 to all upper limits and subtracting 0.5 from all lower limits.

Stated Limits	Real Limits
18–19	17.5–19.5
20–21	19.5–21.5
22–23	21.5–23.5
24–25	23.5–25.5
26–27	25.5–27.5

Note that, when conceptualized with real limits, the class intervals overlap with each other and the distribution can be seen as continuous.

In addition to real limits, you will need to work with the **midpoints** of the class intervals to construct some types of graphs. Midpoints are defined as the points exactly halfway between the upper and lower real limits and can be found for any interval by dividing the sum of the upper and lower real

TABLE 2.12   REAL LIMITS AND MIDPOINTS

Interval Width of Three Units		
Stated Limits	Real Limits	Midpoints
0–2	−0.5–2.5	1
3–5	2.5–5.5	4
6–8	5.5–8.5	7
9–11	8.5–11.5	10

Interval Width of Six Units		
Stated Limits	Real Limits	Midpoints
100–105	99.5–105.5	102.5
106–111	105.5–111.5	108.5
112–117	111.5–117.5	114.5
118–123	117.5–123.5 .	120.5

limits by two. Table 2.12 displays the real limits and midpoints of two different sets of class intervals.

Cumulative Frequency and Cumulative Percentage.   Two commonly used adjuncts to the basic frequency distribution for interval-ratio data are the **cumulative frequency** and **cumulative percentage** columns. Their primary purpose is to allow the researcher (and his or her audience) to tell at a glance how many cases fall below a given score or class interval in the distribution.

To construct a cumulative frequency column, begin with the lowest class interval in the distribution. The entry in the cumulative frequency columns for that interval will be the same as the number of cases in the interval. For the next higher interval, the cumulative frequency will be all cases in the interval plus all the cases in the first interval. For the third interval, the cumulative frequency will be all cases in the interval plus all cases in the first two intervals. Continue adding (or accumulating) cases until you reach the highest class interval, which will have a cumulative frequency of all the cases in the interval plus all cases in all other intervals. (For the highest interval, cumulative frequency equals the total number of cases.) Table 2.13 shows a cumulative frequency column added to Table 2.11. The arrows show the direction of addition.

The cumulative percentage column is quite similar to the cumulative frequency column. Begin by adding a column to the basic frequency distribution for percentages as in Table 2.11. This column shows the percentage of all cases in each class interval. To find cumulative percentages, follow the same addition pattern explained above for cumulative frequency. That is, the cumulative percentage for the lowest class interval will be the same as

TABLE 2.13   AGE OF RESPONDENTS, COUNSELING CENTER SURVEY

Class Interval	Frequency (f)	Cumulative Frequency
18–19	11	11
20–21	5	16
22–23	2	18
24–25	1	19
26–27	1	20
	N = 20	

same as n

TABLE 2.14   AGE OF RESPONDENTS, COUNSELING CENTER SURVEY

Class Interval	Frequency (f)	Cumulative Frequency	Percentage (%)	Cumulative Percentage (%)
18–19	11	11	55	55
20–21	5	16	25	80
22–23	2	18	10	90
24–25	1	19	5	95
26–27	1	20	5	100
	N = 20		100%	

the percentage of cases in the interval. For the next higher interval, the cumulative percentage is the percentage of cases in the interval plus the percentage of cases in the first interval, and so on. Table 2.14 shows the age data with a cumulative percentage column added. Again, the arrows show the direction of addition.

These cumulative columns are quite useful in situations where the researcher wants to make a point about how cases are spread across the range of scores. For example, Tables 2.13 and 2.14 show quite clearly that most students in the counseling center survey are less than 21 years of age. If the researcher wishes to impress this feature of the age distribution on his or her audience, then these cumulative columns are quite handy. Most realistic research situations will be concerned with many more than 20 cases and/or many more categories than our tables have. Since the cumulative percentage column is clearer and easier to interpret in such cases, it is normally preferred to the cumulative frequencies column.

Summary.   The basic guidelines for constructing frequency distributions for interval-ratio variables can now be stated:

1. Decide how many class intervals you wish to use. One reasonable convention suggests that the number of intervals should be about 10. Many research situations may require fewer than 10 intervals, and it is common to find frequency distributions with as many as 15 intervals. Only rarely will more than 15 intervals be used, since the resultant frequency distribution would not be very concise.

2. Find the size of the class interval. Once you have decided how many intervals you will use, interval size can be found by dividing the range of the scores by the number of intervals and rounding to a convenient whole number.

3. State the lowest interval so that its lower limit is equal to or below the lowest score. By the same token, your highest interval will be the one that contains the highest score. All intervals must be equal in size.

4. State the limits of the class intervals at the same level of precision as you have used to measure the data. Do not overlap intervals. You will then have defined the class intervals in such a way that each case can be sorted into one and only one category. If necessary, determine the real limits and/or the midpoints of the intervals.

5. Count the number of cases in each class interval and report these subtotals in a column labeled "frequency." Report the total number of cases (N) at the bottom of this column. The table may also include a column for percentages, cumulative frequencies, and cumulative percentages.

6. Inspect the frequency distribution carefully. Has too much detail been lost? If so, reconstruct the table with a greater number of class intervals (or smaller interval size). Is the table too detailed? If so, reconstruct the table with fewer class intervals (or use wider intervals). Remember that the frequency distribution results from a number of decisions you make in a rather arbitrary manner. If the appearance of the table seems less than optimal given the purpose of the research, redo the table until you are satisfied that you have struck the best balance between detail and conciseness.

7. Remember to give your table a clear, concise title, and number the table if your report contains more than one. All categories and columns must also be clearly labeled. *(For practice in constructing and interpreting frequency distributions for interval-ratio level variables, see problems 2.6 to 2.12 and 2.16.)*

2.6 CONSTRUCTING FREQUENCY DISTRIBUTIONS FOR INTERVAL-RATIO LEVEL VARIABLES: A REVIEW

We covered a lot of ground in the preceding section, so let's pause and review these principles by considering a specific research situation. Below are the numbers of visits received over the past year by 90 residents of a retirement community.

0	52	21	20	21	24	1	12	16	12
16	50	40	28	36	12	47	1	20	7
9	26	46	52	27	10	3	0	24	50
24	19	22	26	26	50	23	12	22	26
23	51	18	22	17	24	17	8	28	52
20	50	25	50	18	52	46	47	27	0
32	0	24	12	0	35	48	50	27	12
28	20	30	0	16	49	42	6	28	2
16	24	33	12	15	23	18	6	16	50

Listed in this format, the data are a hopeless jumble from which no one could derive much meaning. The function of the frequency distribution is to arrange and organize these data so that their meanings will be made obvious.

First, we must decide how many class intervals to use in the frequency distribution. Following the guidelines established in the previous section, let's use about 10 intervals. By inspecting the data, we can see that the lowest score is 0 and the highest is 52. The range of these scores is 52 − 0, or 52. To find the approximate interval size, divide the range (52) by the number of intervals (10). Since 52/10 = 5.2, we can set the interval size at 5.

The lowest score is 0, so the lowest class interval will be 0−4. The highest class interval will be 50−54, which will include the high score of 52. All that remains is to state the intervals in table format, count the number of scores that fall into each interval, and report the subtotals in a frequency column. These steps have been taken in Table 2.15, which also includes columns for the percentages and cumulative percentages. Note that this table is the product of several relatively arbitrary decisions. The researcher should remain aware of this fact and inspect the frequency distribution carefully. If the table is unsatisfactory for any reason, it can be reconstructed with a different number of categories and interval sizes.

Now, with the aid of the frequency distribution, some patterns in the data can be discerned. There are three distinct clusterings of scores in the table. Ten residents were visited rarely, if at all (the 0−4 visits per year interval). The single largest interval, with 18 cases, is 20−24. Combined with the intervals immediately above and below, this represents quite a sizable grouping of cases (42 out of 90, or 46.6% of all cases) and suggests that the dominant visiting rate is about twice a month, or approximately 24 visits per year. The third grouping is in the 50−54 class interval with 12 cases, reflecting a visiting rate of about once a week. The cumulative percentage column indicates that the majority of the residents (58.89%) were visited 24 or fewer times a year.

TABLE 2.15   NUMBER OF VISITS PER YEAR, 90 RETIREMENT-COMMUNITY RESIDENTS

Class Interval	Frequency (f)	Cumulative Frequency	Percentage (%)	Cumulative Percentage (%)
0–4	10	10	11.11	11.11
5–9	5	15	5.56	16.67
10–14	8	23	8.89	25.56
15–19	12	35	13.33	38.89
20–24	18	53	20.00	58.89
25–29	12	65	13.33	72.22
30–34	3	68	3.33	75.55
35–39	2	70	2.22	77.77
40–44	2	72	2.22	79.99
45–49	6	78	6.67	86.66
50–54	12	90	13.33	99.99
	N = 90		99.99%*	

*Percentage columns will occasionally fail to total to 100% because of rounding error. If the total is between 99.90% and 100.10%, ignore the discrepancy. Discrepancies of greater than plus or minus 0.10% may indicate mathematical errors, and the entire column should be computed again.

2.7 CHARTS AND GRAPHS

Researchers frequently use charts and graphs to present their data in ways that are visually more dramatic than frequency distributions. These devices are particularly useful for conveying an impression of the overall shape of a distribution and for highlighting any clustering of cases in a particular range of scores. Many graphing techniques are available, but we will examine just four. The first two, pie and bar charts, are appropriate for discrete variables at any level of measurement. The last two, histograms and line charts or frequency polygons, are used with interval-ratio variables.

The sections that follow explain how to construct graphs and charts "by hand." These days, however, computer programs are almost always used to produce graphic displays. Graphing software is sophisticated and flexible but also relatively easy to use and, if such programs are available to you, you should familiarize yourself with them. The effort required to learn these programs will be repaid in the quality of the final product. The section on *SPSS for Windows* at the end of this chapter includes a demonstration of how to use this software to produce bar chart and line charts.

Pie Charts.   To construct a **pie chart,** begin by computing the percentage of all cases that fall into each category of the variable. Then divide a circle (the pie) into segments (slices) proportional to the percentage distribution. Be sure that the chart and all segments are clearly labeled.

Figure 2.1 is a pie chart that displays the distribution of "marital status" from the counseling center survey. The frequency distribution (Table 2.6) is reproduced as Table 2.16, with a column added for the percentage distribution. Since a circle's circumference is 360°, we will apportion 180° (or 50%)

You will often need to interpret and understand tables, charts, and graphs as you become more involved in the professional literature of the social sciences. These figures will be explained in the text of the article, but you should develop the habit of reading them for yourself and coming to your own conclusions. Here are some ideas to keep in mind when reading tables and charts.

First, there are many different formats for presenting tables. Not all frequency distributions or graphs in the literature will look exactly like those presented in this chapter. Second, because of space limitations, tables and graphs will be presented with a minimum of detail. For example, the researcher may present a frequency distribution with only a percentage column.

Begin your analysis by first reading the title, all labels (that is, row and/or column headings), and any footnotes to the table. These will tell you exactly what information is contained in the figure or table. Inspect the body of the table or graph with the author's analysis in mind. See if you agree with the author's analysis. (You almost always will, but it never hurts to double-check and exercise your critical abilities.)

Finally, remember that most research projects analyze interrelationships among many variables. Because the tables, graphs, and charts covered in this chapter generally display variables one at a time, they are unlikely to be included in such research reports (or perhaps, included only as background information). Even when not reported, you can be sure that the research began with an inspection of frequency distributions or graphs for each variable. Univariate tables and graphs display a great deal of information about the variables in a compact, easily understood format and are almost universally used as descriptive devices.

Statistics in the Professional Literature

In this section, two brief excerpts from social science research are presented so that you can see how statistical information is actually interpreted and reported. The first excerpt presents a frequency distribution and the second a line chart.

Excerpt 1 Power and Gender

Are women who are unequal at home also at a disadvantage in their ability to participate in politics? Do women who have less control over family decision making and the family budget also have less impact on the political institution? Does powerlessness in one institution translate into powerlessness in another?

To find answers to these questions, researchers Nancy Burns, Kay Schlozman, and Sidney Verba interviewed 380 married couples. The interviews covered a variety of topics, including how each couple divided up the household chores. This information on domestic power relationships could then be used to analyze political power. The table on page 42, reproduced verbatim, reports some of their findings.

Note that this table presents 10 different frequency distributions (i.e., two for each of five different variables). Also note that only percentages are reported, and the tables are horizontal rather than vertical as in this text. As always, the researchers report percentage totals and the actual number of cases.

As for interpretation, here's what the authors have to say about the table:

> Table 2 shows the distribution of . . . responses for wives and husbands. Overall, a similar division of labor is reported, with wives doing more when it comes to cleaning, shopping, caring for children and, to a lesser extent, paying the bills, and with men taking primary responsibility for the car. . . . The proportion of couples in which the spouses gave contradictory answers—each claiming to do all or most of a particular chore—is relatively low. . . . The fewest contradictory responses are found in relation to traditional female chores: taking care of children and cleaning the house. Most wives—and few husbands—claim to do all or most of these tasks.

Want to find out what the researchers concluded? The citation is given below and the article is almost certainly in your library.

Source: Burns, Nancy, Kay Schlozman, and Sidney Verba. 1997. "The Public Consequences of Private Inequality: Family Life and Citizen Participation." *American Political Science Review.* 91:373–387.

(continued)

TABLE 2 DIVISION OF HOUSEHOLD CHORES

	All	Most	Some	Little	None		N	Contradictory Responses[a]
Housecleaning								
Wives say they do	41	34	17	3	5	100%	374	7%
Husbands say they do	1	10	46	31	12	100%	349	
Grocery Shopping								
Wives say they do	47	34	14	2	4	100%	374	9%
Husbands say they do	9	13	38	25	15	100%	349	
Paying Bills								
Wives say they do	48	17	13	12	11	100%	374	9%
Husbands say they do	21	20	14	20	25	100%	348	
Repairs/Car								
Wives say they do	7	13	29	27	24	100%	374	17%
Husbands say they do	69	24	5	1	1	100%	349	
Taking Care of Children								
Wives say they do	11	63	22	1	4	100%	330	5%
Husbands say they do	—	7	75	10	9	100%	224	

[a]Both members of a couple claim to do all or most

Excerpt 2 Racism in Children's Literature

How have African Americans been presented in children's books? Do these portrayals reflect patterns of prejudice, conflict, and exclusion in the larger society? To find out, researchers Bernice Pescosolido, Elizabeth Grauerholz, and Melissa Milkie analyzed thousands of children's books published in the United States between the 1930s and the 1990s. The line chart below shows some of their results.

Here's what the authors have to say about this chart:

Figure 1 shows that the visibility of Blacks in children's books varies systematically across the time period studied. The nonlinear time trend reveals roughly four phases of the dependent variable "visibility of Blacks" in these books. In the earliest phase, from 1938 to about 1957, Blacks are represented in modest proportion, with percentages decreasing

FIGURE 1 PERCENTAGE OF CHILDREN'S BOOKS THAT PORTRAY BLACK CHARACTERS, BY YEAR: ALL THREE BOOK SERIES, 1937 TO 1993

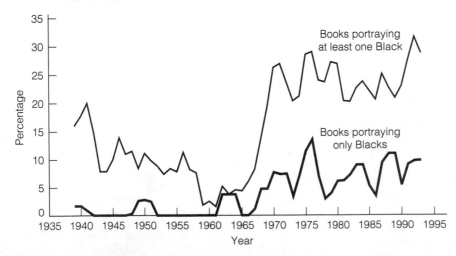

(continued)

FIGURE 2.1   SAMPLE PIE CHART: MARITAL STATUS OF RESPONDENTS (*N* = 20)

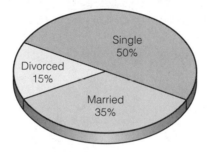

TABLE 2.16   MARITAL STATUS OF RESPONDENTS, COUNSELING CENTER SURVEY

Status	Frequency (*f*)	Percentage (%)
Single	10	50
Married	7	35
Divorced	3	15
N = 20		100%

for the first category, 126° (35%) for the second, and 54° (15%) for the last category. The pie chart visually reinforces the relative preponderance of single respondents and the relative absence of divorced students in the counseling center survey.

Bar Charts.   Like pie charts, **bar charts** are relatively straightforward. Conventionally, the categories of the variable are arrayed along the horizontal axis (or abscissa) and frequencies, or percentages if you prefer, along the vertical axis (or ordinate). For each category of the variable, construct (or draw) a rectangle of constant width with height corresponding to the number

FIGURE 2.2 SAMPLE BAR CHART: MARITAL STATUS OF RESPONDENTS, COUNSELING CENTER SURVEY (*N* = 20)

FIGURE 2.3 SAMPLE BAR CHART FOR VISITS PER YEAR, RETIREMENT COMMUNITY RESIDENTS (*N* = 90)

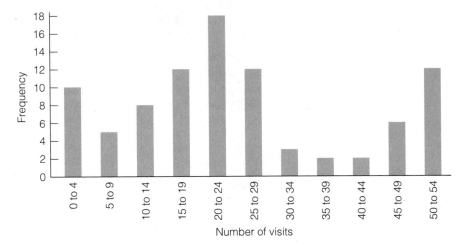

of cases in the category. The bar chart in Figure 2.2 reproduces the marital status data from Figure 2.1 and Table 2.16.

This chart would be interpreted in exactly the same way as the pie chart in Figure 2.1, and researchers are free to choose between these two methods of displaying data. However, if a variable has more than four or five categories, the bar chart would be preferred. With too many categories, the pie chart gets very crowded and loses its visual clarity. To illustrate, Figure 2.3 uses a bar chart to display the data on visiting rates for the retirement community presented in Table 2.15. A pie chart for this same data would have had 11 different "slices," a more complex or "busier" picture than that presented by the bar chart. In Figure 2.3, the clustering of scores in the "20 to 24" range (approximately two visits a year) is readily apparent, as are the groupings in the "0 to 4" and "50 to 54" ranges.

FIGURE 2.4   HOMICIDE VICTIMIZATION RATES, 1955–1993 (rates per 100,000 population, whites only)

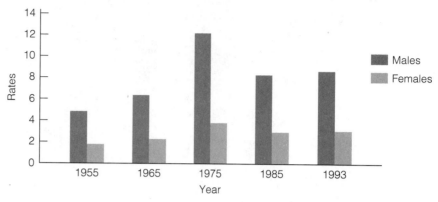

Source: U.S. Bureau of the Census, *Statistical Abstracts of the United States: 1996* (1985 & 93), p. 204; U.S. Bureau of the Census, *Statistical Abstracts of the United States: 1977* (1955, 65, & 75), p. 173.

Bar charts are particularly effective ways to display the relative frequencies for two or more categories of a variable when you want to emphasize some comparisons. Suppose, for example, that you wished to make a point about changing rates of homicide victimization for white males and females since 1955. Figure 2.4 displays the data in a dramatic and easily comprehended way. The bar chart shows that rates for males are higher than rates for females, that rates for both sexes were highest in 1975, and that rates for males rose faster than rates for females between 1955 and 1975. *(For practice in constructing and interpreting pie and bar charts, see problems 2.5 and 2.13.)*

Histograms.   **Histograms** look a lot like bar charts and, in fact, are constructed in much the same way. They are most appropriate for continuous interval-ratio level variables but are commonly used for discrete interval-ratio level variables as well. To reflect the numerical quality of the scale of measurement, the bars are contiguous to each other. To construct a histogram from a frequency distribution, follow these steps:

1. Array the class intervals or scores along the horizontal axis (abscissa) using the real class limits.
2. Array frequencies along the vertical axis (ordinate).
3. For each category in the frequency distribution, construct a bar with height corresponding to the number of cases in the category and with width corresponding to the real limits of the class intervals.
4. Label each axis of the graph.
5. Title the graph.

FIGURE 2.5   AGE OF RESPONDENTS TO GENERAL SOCIAL SURVEY, 1994

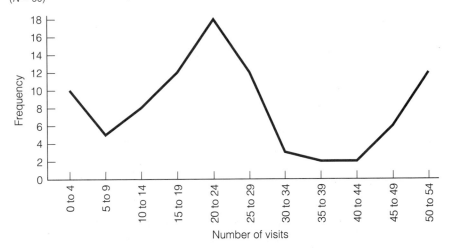

FIGURE 2.6   SAMPLE LINE CHART: NUMBER OF VISITS PER YEAR, RETIREMENT COMMUNITY RESIDENTS
(*N* = 90)

As an example, Figure 2.5 uses a histogram to display the distribution of ages for a sample of respondents to a national public-opinion poll. The graph shows that the sample is concentrated in their 30s and 40s and that the number of respondents declines with age. Note also that there are no people in the sample younger than age 18, the usual cutoff point for respondents to public-opinion polls.

Line Charts.   Construction of a **line chart** or **frequency polygon** is similar to construction of a histogram. Instead of using bars to represent the frequencies, however, use a dot at the midpoint of each interval. Straight lines then connect the dots. Figure 2.6 displays a line chart for the visiting data previously displayed in the bar chart in Figure 2.3.

FIGURE 2.7   MARRIAGE AND DIVORCE RATES, UNITED STATES

Source: U.S. Bureau of the Census, *Statistical Abstracts of the United States: 1996,* p. 104.

Line charts can also be used to display trends across time. Figure 2.7 shows both marriage and divorce rates per 100,000 population for the United States since 1980. Note that in this time period both rates were slowly falling, but the marriage rate fell faster.

Histograms and frequency polygons are alternative ways of displaying essentially the same message. Thus, the choice between the two techniques is left to the aesthetic pleasures of the researcher. *(For practice in constructing and interpreting histograms and line charts, see problems 2.7 to 2.12, 2.14, and 2.15.)*

SUMMARY

1. We considered several different ways of summarizing the distribution of a single variable and, more generally, reporting the results of our research. Our emphasis throughout was on the need to communicate our results clearly and concisely. You will often find that, as you strive to communicate statistical information to others, the meanings of the information will become clearer to you as well.
2. Percentages and proportions, ratios, and rates represent several different techniques for enhancing clarity by expressing our results in terms of relative frequency. Percentages and proportions report the relative occurrence of some category of a variable compared with the distribution as a whole. Ratios compare two categories with each other, and rates

report the actual occurrences of some phenomenon compared with the number of possible occurrences per some unit of time.
3. Frequency distributions are tables that summarize the entire distribution of some variable. It is very common to construct these tables for each variable of interest as the first step in a statistical analysis. Columns for percentages, cumulative frequency, and/or cumulative percentages often enhance the readability of frequency distributions.
4. Pie and bar charts, histograms, and line charts or frequency polygons are graphic devices used to express the basic information contained in the frequency distribution in a compact and visually dramatic way.

SUMMARY OF FORMULAS

Proportions	2.1	$p = \dfrac{f}{N}$
Percentages	2.2	$\% = \dfrac{f}{N} \times 100$
Ratios	2.3	$\text{Ratio} = \dfrac{f_1}{f_2}$

GLOSSARY

Bar chart. A graphic display device for discrete variables. Categories are represented by bars of equal width, the height of each corresponding to the number (or percentage) of cases in the category.

Class intervals. The categories used in the frequency distributions for interval-ratio variables.

Cumulative frequency. An optional column in a frequency distribution that displays the number of cases within an interval and all preceding intervals.

Cumulative percentage. An optional column in a frequency distribution that displays the percentage of cases within an interval and all preceding intervals.

Frequency distribution. A table that displays the number of cases in each category of a variable.

Frequency polygon. A graphic display device for interval-ratio variables. Class intervals are represented by dots placed over the midpoints, the height of each corresponding to the number (or percentage) of cases in the interval. All dots are connected by straight lines. Same as a line chart.

Histogram. A graphic display device for interval-ratio variables. Class intervals are represented by contiguous bars of equal width (equal to the real limits), the height of each corresponding to the number (or percentage) of cases in the interval.

Line chart. See **Frequency polygon**.

Midpoint. The point exactly halfway between the upper and lower limits of a class interval.

Percentage. The number of cases in a category of a variable divided by the number of cases in all categories of the variable, the entire quantity multiplied by 100.

Pie chart. A graphic display device especially for discrete variables with only a few categories. A circle (the pie) is divided into segments proportional in size to the percentage of cases in each category of the variable.

Proportion. The number of cases in one category of a variable divided by the number of cases in all categories of the variable.

Rate. The number of actual occurrences of some phenomenon or trait divided by the number of possible occurrences per some unit of time.

Ratio. The number of cases in one category divided by the number of cases in some other category.

Real class limits. The upper and lower limits of a class interval used when the distribution is conceptualized as a continuous distribution of categories.

Stated class limits. The upper and lower boundaries of a class interval as they appear in the frequency distribution.

PROBLEMS

2.1 [SOC] The tables below report the marital status of 20 respondents in two different apartment complexes. (*HINT: Make sure that you have the correct numbers in the numerator and denominator before solving the following problems. For example, problem a asks for "the percentage of*

*respondents who are married in each complex,"
and the denominators will be 20 for these two
fractions. Problem d, on the other hand, asks for
the "percentage of the single respondents who live
in Complex B," and the denominator for this
fraction will be 4 + 6, or 10.)*

Status	Complex A	Complex B
Married	5	10
Unmarried		
("living together")	8	2
Single	4	6
Separated	2	1
Widowed	0	1
Divorced	1	0
	20	20

a. What percentage of the respondents in each complex are married? _____
b. What is the ratio of single to married respondents at each complex? _____
c. What proportion of each sample are widowed? _____
d. What percentage of the single respondents live in Complex B? _____
e. What is the ratio of the "unmarried/living together" to the married at each complex? _____

2.2 At St. Algebra College, the numbers of males and females in the various major fields of study are as follows:

Major	Males	Females
Humanities	117	83
Social sciences	97	132
Natural sciences	72	20
Business	156	139
Nursing	3	35
Education	30	15

Read each of the following problems carefully before constructing the fraction and solving for the answer.
a. What percentage of social science majors are male?
b. What proportion of business majors are female?

c. For the humanities, what is the ratio of males to females?
d. What percentage of the total student body are males?
e. What is the ratio of males to females for the entire sample?
f. What proportion of the nursing majors are male?
g. What percentage of the sample are social science majors?
h. What is the ratio of humanities majors to business majors?
i. What is the ratio of female business majors to female nursing majors?
j. What proportion of the males are education majors?

2.3 CJ The town of Shinbone, Kansas, has a population of 211,732 and experienced 47 bank robberies, 13 murders, and 23 auto thefts during the past year. Compute a rate for each type of crime per 100,000 population. *(HINT: Make sure that you set up the fraction with size of population in the denominator.)*

2.4 CJ The numbers of major felonies reported to the police in two different towns last year are listed below. Calculate the rate for each crime per 100,000 population. Relatively speaking, which town has the greater apparent crime problem? With which crimes? Write a paragraph describing the differences.

Crime	Town A (Population = 20,109)	Town B (Population = 764,213)
Homicide	13	78
Robbery	102	617
Auto Theft	125	314
Rape	23	79
Burglary	178	537

2.5 SOC The scores of 15 respondents on three variables are reported below. These scores were taken from a public opinion survey called the General Social Survey, or the GSS. This data set, which is described in some detail in Appendix G, is used for the computer exercises in this text. Small subsamples from the GSS will be used throughout the text to provide "real" data for problems. For the actual questions and other details, see Appendix G. The numerical codes for these three variables are as follows:

Sex	Support for Gun Control	Degree
1 = Male	1 = In favor	0 = Less than HS
2 = Female	2 = Opposed	1 = HS
		2 = Jr. college
		3 = Bachelor's
		4 = Graduate

Case No.	Sex	Support for Gun Control	Level of Education
1	2	1	1.
2	1	2	1'
3	2	1	3
4	1	1	2.
5	2	1	3
6	1	1	1.
7	2	2	0.
8	1	1	1.
9	1	2	0.
10	2	1	1.
11	1	1	4
12	1	1	4
13	1	1	0.
14	2	1	1
15	1	1	1

a. Construct a frequency distribution for each of the three variables. Include a column for percentages.

b. Construct pie and bar charts to display the distributions of the three variables.

2.6 SW A local youth service agency has begun a sex education program for teenage girls who have been referred by the juvenile courts. The girls were given a 20-item test for general knowledge about sex, contraception, and anatomy and physiology upon admission to the program and again after completing the program. The scores of the first 15 girls to complete the program are listed below.

Case	Pretest	Posttest
A	8	12
B	7	13
C	10	12
D	15	19
E	10	8
F	10	17
G	3	12
H	10	11
I	5	7
J	15	12
K	13	20
L	4	5
M	10	15
N	8	11
O	12	20

Construct frequency distributions for both sets of scores. Include a column for percentages. (*HINT: There were 20 items on the test so the maximum range for these scores is 20. If you use 10 class intervals to display these scores, the interval size will be 2 and the first interval will be 0–1.*)

2.7 SOC Twenty high school students completed a class to prepare them for the College Boards. Their scores are reported below.

420 345 560 650
459 499 500 657
467 480 505 555
480 520 530 589
500 550 545 600

a. Display these data in a frequency distribution with columns for percentages and cumulative percentages.

b. Construct a histogram and frequency polygon for these data.

2.8 GER Following are reported the number of times 25 residents of a community for senior citizens left their homes for any reason during the past week.

0	2	1	7	3
7	0	2	3	17
14	15	5	0	7
5	21	4	7	6
2	0	10	5	7

a. Construct a frequency distribution to display these data.

b. What are the real limits and midpoints of the class intervals?

c. Add columns to the table to display the percentage distribution, cumulative frequency, and cumulative percentages.

d. Construct a histogram and a frequency polygon to display this distribution.

e. Write a paragraph summarizing this distribution of scores.

2.9 SOC Fifty students completed a questionnaire that measured their attitudes toward interpersonal violence. Respondents who scored high believed that in many situations a person could legitimately use physical force against another person. Respondents who scored low believed that in no situation (or very few situations) could the use of violence be justified.

52	47	17	8	92
53	23	28	9	90
17	63	17	17	23
19	66	10	20	47
20	66	5	25	17
10	82	90	40	45
8	91	82	52	20
75	32	75	60	60
80	30	70	65	52
90	29	70	66	55

a. Construct a frequency distribution to display these data.

b. What are the real limits and midpoints of the class intervals?

c. Add columns to the table to display the percentage distribution, cumulative frequency, and cumulative percentage.

d. Construct a histogram and a frequency polygon to display these data.

e. Write a paragraph summarizing this distribution of scores.

2.10 SW A psychiatric social worker rated the mental functioning of 50 patients on their release from a state mental hospital and again six months after release. He used a scale on which higher scores indicated higher levels of functioning and competence. The scores are reported below.

Case	Release Date	Six Months
1	10	12
2	40	42
3	17	16
4	35	33
5	35	45
6	17	15
7	22	47
8	22	25
9	10	15
10	15	18
11	25	20
12	29	33
13	16	35
14	27	14
15	43	35
16	45	19
17	28	35
18	41	31
19	41	30
20	30	35
21	36	31
22	21	21
23	27	32
24	12	15
25	15	25
26	21	24
27	29	32
28	34	44
29	12	18
30	34	24
31	14	27
32	16	35
33	20	32
34	19	50
35	18	28
36	28	38
37	10	20
38	11	19
39	28	38
40	10	20
41	11	19
42	12	41
43	40	42
44	25	26
45	17	25
46	35	36
47	20	22

Case	Release Date	Six Months
48	22	27
49	22	22
50	31	41

a. Construct a frequency distribution for each set of scores. Include percentages and cumulative percentages.

b. Construct histograms and frequency polygons to display these data.

c. Compare the two sets of scores and write a paragraph of interpretation.

2.11 SOC The scores below are taken from the General Social Survey. The first variable is a prestige score for the respondent's occupation, and the second is a prestige score for the respondent's father's occupation.

Case No.	Respondent's Occupational Prestige	Father's Occupational Prestige
1	26	29
2	39	45
3	47	30
4	36	45
5	60	40
6	12	23
7	32	70
8	15	51
9	58	50
10	43	42
11	78	56
12	29	50
13	17	41
14	32	41
15	40	50
16	50	39
17	47	61
18	55	41
19	50	50
20	50	27
21	26	50
22	25	61
23	62	45
24	51	47
25	34	41
26	50	32
27	61	32
28	62	51

Case No.	Respondent's Occupational Prestige	Father's Occupational Prestige
29	78	34
30	32	17
31	29	22
32	36	27
33	12	17
34	28	16
35	48	18
36	69	17
37	23	16
38	16	28
39	36	41
40	48	47
41	51	33
42	54	41
43	57	34
44	18	50
45	37	50
46	36	50
47	36	45
48	22	29
49	37	37
50	41	41

a. Construct a frequency distribution for each variable. Include a column for percentages and cumulative percentages.

b. Construct histograms and frequency polygons for each variable.

c. Compare the two variables. What can you tell about the distribution of prestige for these two generations? Write a paragraph of interpretation for these variables.

2.12 SOC Following are reported the official homicide rates for all 50 states for 1975 and 1994.*

State	Homicide Rates (per 100,000 population) 1975	1994
ME	2.8	2.3
NH	2.9	1.4
VT	2.1	1.0
MA	4.2	3.5
RI	3.0	4.1
CT	3.9	6.6
NY	11.0	11.1
NJ	6.8	5.0
PA	6.8	5.9
OH	8.1	6.0

| State | Homicide Rates (per 100,000 population) | |
	1975	1994
IN	8.5	7.9
IL	10.6	11.7
MI	11.9	9.8
WI	3.3	4.5
MN	3.3	3.2
IA	2.5	1.7
MO	10.6	10.5
ND	0.8	0.2
SD	3.7	1.4
NE	4.3	3.1
KS	5.4	5.8
DE	7.3	4.7
MD	10.7	11.6
VA	11.5	8.7
WV	7.4	5.4
NC	12.4	10.9
SC	14.7	9.6
GA	14.4	10.0
FL	13.5	8.3
KY	10.2	6.4
TN	11.4	9.3
AL	16.0	11.9
MS	13.9	15.3
AR	10.1	12.0
LA	12.6	19.8
OK	9.4	6.9
TX	13.4	11.0
MT	5.2	3.3
ID	5.2	3.5
WY	10.2	3.4
CO	7.4	3.4
NM	13.3	10.7
AZ	8.6	10.5
UT	2.7	2.9
NV	13.0	11.7
WA	5.7	5.5
OR	6.2	4.9
CA	10.4	11.8
AK	12.2	6.3
HI	7.7	4.2

*Source: United States Bureau of the Census, *Statistical Abstracts of the United States: 1977* (98th edition) and *1996* (116th edition). Washington, D.C., 1977 and 1996.

a Construct a frequency distribution for each year to display these data. Include percentage and cumulative percentage columns.
b. Construct histograms and frequency polygons for each year.
c. Compare the homicide rates for the two years. What can you tell about homicide from comparing these frequency distributions and graphs? Write a paragraph of interpretation for these data.

2.13 PA/CJ As part of an evaluation of the efficiency of your local police force, you have gathered the following data on police response time to calls for assistance during two different years. (Response times were rounded off to whole minutes.) Convert both frequency distributions into percentages and construct pie charts and bar charts to display the data. Write a paragraph comparing the changes in response time between the two years.

Response Time, 1986	f
21 minutes or more	35
16–20 minutes	75
11–15 minutes	180
6–10 minutes	375
Less than 6 minutes	210
	875

Response Time, 1996	f
21 minutes or more	45
16–20 minutes	95
11–15 minutes	155
6–10 minutes	350
Less than 6 minutes	250
	895

2.14. SOC The three line charts on pages 54–56 display trends in crime in the United States for a 10-year period. Write a paragraph describing each of these graphs. What similarities and differences can you observe among the three graphs? (For example, do crime rates always change in the same direction?) Note the differences in the vertical axes from chart to chart—for homicide the axis ranges from 0 to 12, while for burglary and auto theft the range is from 0 to 1600. The latter crimes are far more common, and a scale with smaller intervals is needed to display the rates.

2.15. PA The city's Department of Transportation has been keeping track of accidents on a particularly dangerous stretch of highway. Early in the year, the city lowered the speed limit on this highway and increased police patrols. Data on number of accidents before and after the changes are presented below. Did the changes work? Is the highway safer? Construct a line chart to display these two sets of data (use graphics software if available) and write a paragraph describing the changes.

FIGURE 2.8   HOMICIDE RATES, 1984–1994

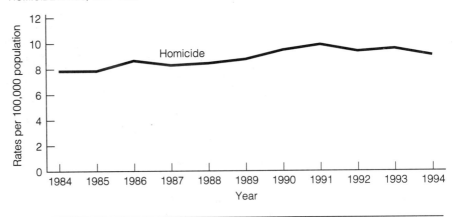

Source: U.S. Bureau of the Census, *Statistical Abstracts of the United States: 1996*, p. 201.

FIGURE 2.9   ROBBERY AND AGGRAVATED ASSAULT RATES, 1984–1994

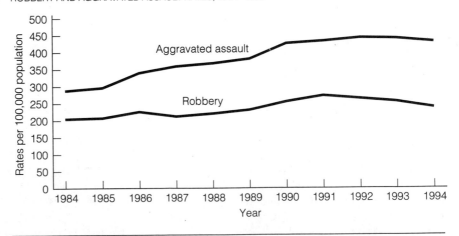

Source: U.S. Bureau of the Census, *Statistical Abstracts of the United States: 1996*, p. 201.

FIGURE 2.10 BURGLARY AND AUTO THEFT RATES, 1984–1994

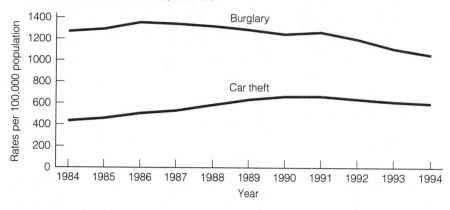

Source: U.S. Bureau of the Census, *Statistical Abstracts of the United States: 1996*, p. 201.

	12 Months Before	12 Months After
January	23	25
February	25	21
March	20	18
April	19	12
May	15	9
June	17	10
July	24	11
August	28	15
September	23	17
October	20	14
November	21	18
December	22	20

2.16 SOC You have been asked to construct a descriptive profile of the 20 volunteers who work for an organization that distributes food and clothing to the homeless. The raw data are presented below. Organize this information so that it can be easily understood by others (e.g., the board of directors for the organization), and then summarize the information in a paragraph.

Case	Sex	Age	Occupation	Hours of Service per Week	Years of Service
1	M	18	College student	4	1
2	F	43	Housewife	10	8
3	F	23	Telemarketer	5	3
4	F	30	4th grade teacher	8	5
5	M	21	Mechanic	15	3
6	F	62	Retired	4	5
7	M	18	H.S. student	12	1
8	M	35	6th grade teacher	3	2
9	M	42	Unemployed	20	6
10	F	45	Housewife	4	3
11	M	22	College student	4	4
12	M	73	Retired	15	10

Case	Sex	Age	Occupation	Hours of Service per Week	Years of Service
13	M	30	Nurse's aide	4	9
14	F	35	Accountant	10	5
15	M	22	College student	4	1
16	F	39	Social worker	4	8
17	M	37	2nd grade teacher	10	5
18	F	42	Social worker	2	5
19	F	25	Housewife	3	1
20	F	27	Graduate student	2	4

SPSS for Windows

Using *SPSS for Windows* to Produce Frequency Distributions and Graphs

Click the SPSS icon on your monitor screen to start *SPSS for Windows*. Load the 1996 GSS by clicking the file name on the first screen or by clicking **File**, **Open**, and **Data** on the **SPSS Data Editor** screen. You may have to change the drive specification to locate the 1996 GSS data supplied with this text (probably named **GSS96**). Double click the file name to open the data set. When you see the message 'SPSS Processor is Ready' on the bottom of the screen, you are ready to proceed.

DEMONSTRATION 2.1 Frequency Distributions

We produced and examined a frequency distribution for the variable *sex* in Appendix F. Use the same procedures to produce frequency distributions for the variables *age* and *marital* (marital status). From the menu bar, click **Statistics.** From the menu that drops down, click **Summarize** and **Frequencies**. The **Frequencies** window appears with the variables listed in alphabetical order by variable name (e.g. *abany, abhlth*) in the left-hand box. This window also lists variable descriptions in capital letters (e.g., ABORTION IF WOMAN WANTS FOR ANY REASON). The variable *age* (AGE OF RESPONDENT) will be visible. Click on it to highlight it and then click the arrow button in the middle of the screen to move *age* to the right-hand window.

Find *marital* in the left-hand box by using the slider button or the arrow keys on the right-hand border to scroll through the variable list. As an alternative, type **m**, and the cursor will move to the first variable description in the list that begins with that letter. In this case, the variable is *marital*, the variable in which we are interested. Once *marital* is highlighted, click the arrow button in the center of the screen to move the variable name to the **Variables** box. There should now be two variable names in the box, *age* and *marital*. SPSS will process all variables listed in the right-hand box together. Click **OK** in the upper-right-hand corner and SPSS will rush off to create the frequency distributions you requested.

The table will be in the **Output** window that will now be "closest" to you on the screen. The tables, along with other information, will be in the right-hand box of the

Output window. Click the symbol shaped like a square in the upper-right-hand corner of the **Output** window to increase the size of the window. Note that the symbol changes to two squares, one on top of the other. To return the window to its original size, click this button again.

The frequency distribution for *marital* will look like this:

Frequencies

Frequency Table

MARITAL STATUS

		Frequency	Percent	Valid Percent	Cumulative Percent
Valid	MARRIED	693	47.7	47.8	47.8
	WIDOWED	157	10.8	10.8	58.6
	DIVORCED	228	15.7	15.7	74.3
	SEPARATED	50	3.4	3.4	77.7
	NEVER MARRIED	323	22.2	22.3	100.0
	Total	1451	99.9	100.0	
Missing	NA	1	.1		
Total		1452	100.0		

Let's briefly examine the elements of this table. The variable description or label is printed at the top of the output ("MARITAL STATUS"). The various categories are printed at the far left. Moving one more column to the right, we find the actual frequencies or the number of times each score of the variable occurred. We see that 693 of the respondents were married, 157 were widowed, and so forth. Next are two columns that report percentages. The entries in the "Percent" column are based on *all* respondents who were asked this question. In this case, the denominator would include the respondent coded as "NA" (No Answer) for *marital*. The "Valid Percent" column eliminates all cases with missing values (e.g., cases coded as NA, DK for "Don't Know," or NAP for "Not Applicable"). Since we almost always ignore missing values, we will usually pay attention only to the "Valid Percent" column (even though there is little difference between the two columns in this case).

The final column is a cumulative percentage column (see Table 2.14). For nominal-level variables like *marital*, this information is not especially meaningful since the order in which the categories are stated is arbitrary. Finally, note that at the bottom of the table are totals for the number of cases, the two percentage columns, and the number of "missing" cases.

Turning to the frequency distribution for *age*, scroll the **Output** window until you get to the top of the table. Note that the table follows the same format as the table for *marital* but is much larger. This, of course, reflects the fact that *age* has many more possible scores than *marital*—so many scores, in fact, that the table is not easy to read or understand. The table needs to be made more compact by collapsing scores into fewer categories. For example, if ages were grouped into categories 10 years wide (10–19, 20–29, etc.), the number of categories would be reduced to about 8. We could collapse the categories by hand (see Section 2.5), or we could have SPSS do the work. The procedures for recoding or collapsing variables are explained in Demonstration 2.2.

DEMONSTRATION 2.2 Collapsing Categories with the Recode Command

The scores of interval-ratio level variables will of 10 have to be collapsed in order to produce readable frequency distributions. *SPSS for Windows* provides a number of ways to change the scores of a variable, and one of the most useful of these is the **Recode** command. We will use this command to create a new version of *age* that has fewer categories and is more suitable for display in a frequency distribution. When we are finished, we will have two different versions of the same variable in the data set: the original interval-ratio version with age measured in years and a new ordinal-level version with collapsed categories. If we wish, the new version of *age* can be added to the permanent data file and used in the future.

We will collapse the values of *age* into categories of 10 years each. Since the youngest respondents are 18, let's (arbitrarily) begin with an interval of 10–19. The next interval will be 20–29, followed by 30–39 and so forth until the interval 80–89. As you will see, recoding requires many small steps, so please be patient and execute the commands as they are discussed.

1. In the **SPSS Data Editor** window, click **Transform** from the menu bar and then click **Recode**. A window will open that gives us two choices: **into same variable** or **into different variable**. If we choose the former ("into same variable"), the new version of the variable will *replace* the old version—the original version of *age* (with actual years) would disappear. We definitely do *not* want this to happen, so we will choose (click on) **into different variable**. This option will allow us to keep both the old and new versions of the variable.
2. The **Recode into Different Variable** window will open. A box containing an alphabetical list of variables will appear on the left. Use the cursor to highlight *age* (AGE OF RESPONDENT) and then click on the arrow button to move the variable to the **Input Variable → Output Variable** box. The input variable is the old version of *age*, and the output variable is the new, recoded version we will soon create.
3. In the **Output Variable** box on the right, click in the **Name** box and type a name for the new (output) variable. I suggest *ager* (**age r**ecoded) for the new variable, but you can assign any name as long as it does not duplicate the name of some other variable in the data set and is no longer than eight characters. Click the **Change** button and the expression *age → ager* will appear in the **Input Variable → Output Variable** box.
4. Click on the **Old and New Values** button in the middle of the screen, and a new dialog box will open. Read down the left-hand column until you find the **Range** button. Click on the button, and the cursor will move to the small box immediately below. In these boxes we will specify the low and high points of each interval of the new variable *ager*.
5. Starting with the interval 10–19, type 10 into the left-hand **Range** dialog box and then click on the right-hand box and type 19. In the **New Value** box in the upper-right-hand corner of the screen, click the **Value** button. Type 1 in the **Value** dialog box and then click the **Add** button directly below. The expression 10–19 → 1 will appear in the **Old → New** dialog box. This completes the first recode instruction to SPSS.
6. Continue recoding by returning to the **Range** dialog boxes on the left. Type 20

in the left-hand box and 29 in the right-hand box and then click the **Value** button in the **New Values** box. Type 2 in the **Value** dialog box and then click the **Add** button. The expression 20 − 29 → 2 appears in the **Old → New** dialog box.

7. Continue this sequence of operations until all old values of *age* have been recoded into new values. The last expression in the **Old → New** dialog box should be 80 − 89 → 8. Now click the **Continue** button at the bottom of the screen, and you will return to the **Recode into Different Variable** dialog box. Click **OK**, and SPSS will execute the transformation.

You now have a data set with one more variable named *ager* (or whatever name you gave the recoded variable). SPSS adds the new variable to the data set, and you can find it in the last column to the right in the data window. If you saved the data set (click any part of the data window to move it "closest" to you and then click **File** and **Save**), the new variable would become a permanent addition to the data file. If you do not want to save the new, expanded data file, when you close SPSS at the end of this session (not now) click **No** when the system asks if you want to save the data file. If you are using the student version of *SPSS for Windows*, remember that the program is limited to a maximum of 50 variables.

To produce a frequency distribution for the new variable, click **Statistics**, **Summary**, and **Frequency**. Find *ager* in the left-hand box and click the arrow key to move it to the right-hand box. Click **OK**, and you will soon be looking at the following table:

AGER

		Frequency	Percent	Valid Percent	Cumulative Percent
Valid	1.00	17	1.2	1.2	1.2
	2.00	264	18.2	18.2	19.4
	3.00	329	22.7	22.7	42.1
	4.00	314	21.6	21.7	63.8
	5.00	209	14.4	14.4	78.2
	6.00	155	10.7	10.7	88.9
	7.00	112	7.7	7.7	96.6
	8.00	49	3.4	3.4	100.0
	Total	1449	99.8	100.0	
Missing	System	3	.2		
Total		1452	100.0		

There are few respondents in their teens and their 80s. The sample is clustered in their 30s and 40s (about 45% of the sample is in the categories 3 and 4 or in their 30s and 40s). We are missing data on only 3 respondents for this variable.

DEMONSTRATION 2.3 Graphs and Charts

SPSS for Windows can produce a variety of different kinds of graphs and gives the user a great deal of control over their final appearance, especially if you have access to a color printer. In this demonstration, we will use the program to produce a bar chart and a line chart (frequency polygon), and you will see some of the choices that can be made. To conserve space, I will keep the choices as simple as possible, but you should explore the options for yourself. For any questions you might have that are not answered in this demonstration, click on '**Help**' on the main menu bar.

To produce a bar chart, first click on **Graphs** on the main menu bar and then click **Bar**. The **Bar Chart** dialog box appears with three choices for the type of graph we want. The **Simple** option is already highlighted, and this is the one we want, so just click **Define**. The **Define Simple Bar** dialog box appears with variable names listed on the left. Choose *degree* from the variable list. (As usual, you can move through the list by any of three methods: you can use the slider buttons, the arrow keys, or type the first letter of the label of the variable you want to locate. Move the cursor to highlight *degree*.) Click the arrow button in the middle of the screen to move *degree* to the **Category Axis** box.

Note the **Bars Represent** box above the **Category Axis** box. This dialog box gives you control over the vertical axis of the graph, which can be calibrated in frequencies, percentages, or cumulative frequencies or percentages. Let's choose **N of cases or frequencies**, the option that is already selected.

Click the **Options** button in the lower-right-hand corner. In the **Options** dialog box, make sure that the button next to the **Display groups defined by missing values** option is *not* selected (or checked). Click **Continue** and then click **OK** in the **Define Simple Bar** dialog box, and the following bar chart will be produced:

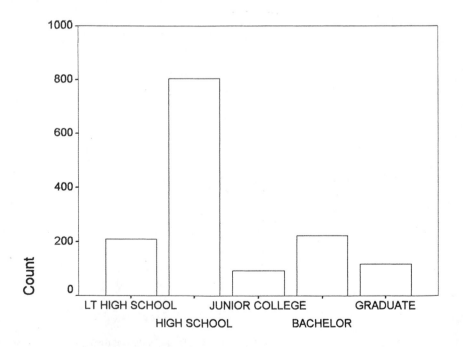

Click on any part of the graph and the **SPSS Chart Editor** window will appear. The menu bar across the top of the window gives you a wide array of options for the final appearance of the chart including, under the **Format** command, the color of the bars and their borders. Explore these options at your leisure using the **Help** function as necessary.

Turning to the chart itself, by far the most common level of education for this sample was high school, with college and less than high school running close to each other for second place. When you are ready, the bar chart can be printed or saved by selecting the appropriate command from the **File** menu.

The procedures for producing a line chart are very similar to those for a bar chart. Click **Graphs** and then **Line**. The **Line Charts** dialog box will appear with the **Simple** option already chosen. This is the one we want, so click **Define**, and the **Define Simple Line** dialog box will appear. Your choices here are the same as in the **Define Simple Bar** dialog box. Line charts are appropriate for interval-ratio data, so let's use *educ* (HIGHEST YEAR OF SCHOOL COMPLETED) for this demonstration. You might think of this variable as a noncollapsed form of *degree*, and we should be able to compare the two variables quickly and easily.

Select *educ* from the variable list on the left and click the arrow button in the middle of the screen to move *educ* to the **Category Axis** box. Note that the **Lines Represent** box is the same as in the **Simple Bar** dialog box. We want the bars to represent *frequencies*, and this option is already selected. Click the **Options** button and make sure that the **Display Groups Defined by Missing Values** option is *not* selected (or checked). Click **Continue** and then click **OK** in the **Define Simple Line** dialog box, and the following line chart will be produced:

HIGHEST YEAR OF SCHOOL COMPLETED

The line chart can be edited by clicking on any part of the chart and activating the Chart Editor window. For example, to change the color of the line, click anywhere on the line in the Chart Editor window, then click **Format** and **Color**. Click on the new color and then click **Apply**. Don't forget to **Save** or **Print** the chart if you wish.

The chart has a very high peak at 12 years of education (high school) and another peak at 16 (college). This essentially duplicates the information in the bar chart for *degree*.

Exercises

2.1 Get frequency distributions for *degree* and *educ*. Use the **Recode** command to collapse *educ* into categories comparable to *degree*. That is, use the categories 0–11 for less than high school, 12 for high school graduates, 13–15 to approximate the number of junior college graduates, 16 for college graduates, and 17–20 for graduate degrees. Get a frequency distribution for recoded *educ* and compare it to the frequency distribution for *degree*. Write a paragraph describing the three frequency distributions.

2.2 Get frequency distributions for 5 or 10 nominal or ordinal variables, including *race* and *relig*. Get bar charts for each variable and write a sentence or two summarizing each frequency distribution.

2.3 Get a frequency distribution for *prestg80* (respondent's occupational prestige). Find the range of this variable and use this information to determine reasonable class intervals by following the procedures described in Section 2.5. Use the **Recode** command to produce a final frequency distribution for the recoded variable and write a sentence or two of description and interpretation.

2.4 Choose five or six more variables of interest to you and produce bar charts or line charts for each. Remember that bar charts are more appropriate for nominal and ordinal variables and line charts for interval-ratio variables. Write a sentence or two of interpretation for each chart.

3 Measures of Central Tendency

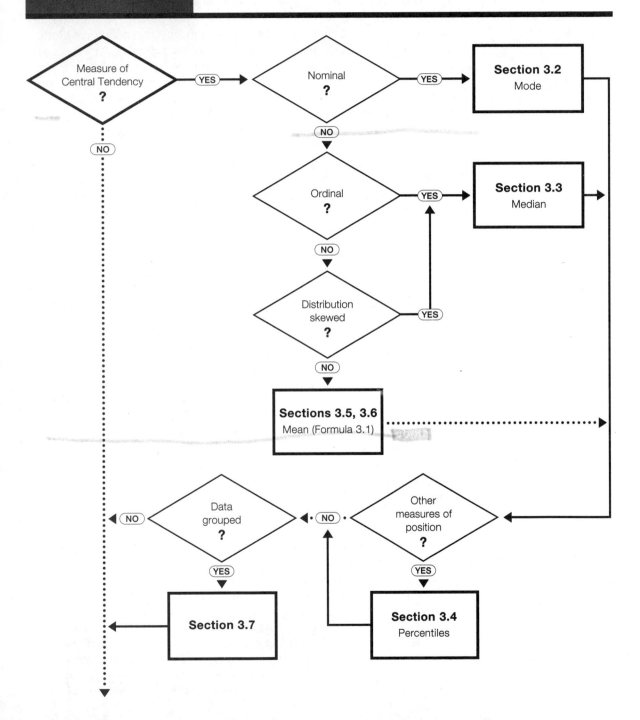

3.1 INTRODUCTION

One clear benefit of frequency distributions, graphs, and charts is that they summarize the overall shape of a distribution of scores in a way that can be quickly comprehended. Often, however, you will need to report more detailed information about the distribution. Specifically, two additional kinds of statistics are almost always useful: some idea of the typical or average case in the distribution (for example, "the average starting salary for social workers is $22,000 per year") and some idea of how much variety or heterogeneity there is in the distribution ("starting salaries for social workers range from $17,000 per year to $25,000 per year"). The first kind of statistic, called **measures of central tendency**, will be the subject of this chapter. The second kind of statistic, measures of dispersion, will be presented in Chapter 4.

The three commonly used measures of central tendency—the mode, median, and mean—are all probably familiar to you. All three summarize an entire distribution of scores by describing the most typical, or central, or representative value, respectively, of that distribution. These statistics are powerful because they can reduce huge arrays of data to a single, easily understood number. Remember that the central purpose of descriptive statistics is to summarize or "reduce" data.

Even though they share a common purpose, the three measures of central tendency are quite different from each other. In fact, they will have the same value only under specific and limited conditions. As we shall see, they vary in terms of level-of-measurement considerations and, perhaps more importantly, in terms of how they define *typical* (that is, they will not necessarily identify the same score or category as "the average"). Thus, your choice of an appropriate measure of central tendency will depend in part on the way you measure your variable and in part on the purpose of the research.

3.2 THE MODE

The **mode** of any distribution is the value that occurs most frequently. For example, in the set of scores 58, 82, 82, 90, 98, the mode is 82 since it occurs twice and the other scores occur only once.

The mode is, relatively speaking, a rather simple statistic, most useful when you want a "quick and easy" indicator of central tendency and when you are working with nominal-level variables. In fact, the mode is the only measure of central tendency that can be used with nominal-level variables. Such variables do not, of course, have numerical "scores" per se, and the mode of a nominally measured variable would be its largest category. For example, Table 3.1 reports the religious affiliations of a fictitious sample of 242 respondents. The mode of this distribution, the single largest category, is Protestant.

If a researcher desires to report only the most popular or common value of a distribution, or if the variable under consideration is nominal, then the mode is the appropriate measure of central tendency. However, keep in mind that the mode does have several limitations. First, some distributions have no mode at all (see Table 2.5) or so many modes that the statistic loses all mean-

TABLE 3.1   RELIGIOUS PREFERENCE (fictitious data)

Denomination	Frequency
Protestant	128
Catholic	57
Jew	10
None	32
Other	15
	$N = 242$

ing. Second, with ordinal and interval-ratio data, the modal score may not be central to the distribution as a whole. That is, *most common* does not necessarily mean "typical" in the sense of identifying the center of the distribution. For example, consider the following rather unusual (but not impossible) distribution of scores on a statistics test:

Scores (% correct)	Frequency
58	2
60	2
62	3
64	2
66	3
67	4
68	1
69	1
70	1
93	5
	$N = 24$

The mode of the distribution is 93. Is this score very close to the majority of the scores? If the instructor summarized this distribution by reporting only the modal score, would he or she be conveying an accurate picture of the distribution as a whole? *(For practice in finding and interpreting the mode, see problems 3.1 to 3.8.)*

3.3 THE MEDIAN

Unlike the mode, the **median (Md)** always represents the exact center of a distribution of scores. The median is the score of the case that is in the exact middle of a distribution: half the cases have scores higher and half the cases have scores lower than the case with the median score. Thus, if the median family income for a community is $30,000, half the families earn more than $30,000 and half earn less.

Before finding the median, the cases *must* be placed in order from the highest to the lowest score—or from the lowest to the highest. Once this is done, find the central or middle case. The median is the score associated with that case. If five students received grades of 93, 87, 80, 75, and 61 on a test, the median would be 80, the score that splits the distribution into two equal halves.

When the number of cases (N) is odd, the value of the median is unambiguous because there will always be a middle case. With an even number of cases, however, there will be two middle cases; and, in this situation, the median is defined as the score exactly halfway between the scores of the two middle cases.

To illustrate, assume that seven students were asked to indicate their level of support for the intercollegiate athletic program at their universities on a scale ranging from 10 (indicating great support) to 0 (no support). After arranging their responses from high to low, you can find the median by locating the case that divides the distribution into two equal halves. With a total of seven cases, the middle case would be the fourth case, since there will be three cases above and three cases below this case. If the seven scores were 10, 10, 8, 7, 5, 4, and 2, then the median is 7, the score of the fourth case. To find the middle case when N is odd, increase the value of N by 1 and divide by 2. With an N of 7, the median is the score associated with the $(7 + 1)/2$, or fourth, case. If N had been 25, the median would be the score associated with the $(25 + 1)/2$, or 13th, case.

Now, if we added a subject whose support for athletics was measured as a 1 to the group above and thereby made N an even number (8), we would no longer have a single middle case. The ordered distribution of scores would now be 10, 10, 8, 7, 5, 4, 2, 1; any value between 7 and 5 would technically satisfy the definition of a median (that is, would split the distribution into two equal halves of four cases each). This ambiguity is resolved by defining the median as the average of the scores of the two middle cases. In the example above, the median would be defined as $(7 + 5)/2$, or 6.

To identify the two middle cases when N is an even number, divide N by 2 to find the first middle case and then increase that number by 1 to find the second middle case. In the example above with eight cases, the first middle case would be the fourth case ($N/2 = 4$) and the second middle case would be the $(N/2) + 1$, or fifth, case. If N had been 142, the first middle case would have been the 71st case and the second the 72nd case. Remember that the median is defined as the average of the scores associated with the two middle cases.*

As I have implied throughout this discussion, the median cannot be calculated for variables measured at the nominal level; such variables do

---

*If the middle cases have the same score, that score is defined as the median. In the distribution 10, 10, 8, 6, 6, 4, 2, 1 the middle cases both have scores of 6 and, thus, the median would be defined as 6.

not have scores that can be ordered or ranked. This measure can be calculated for either ordinal or interval-ratio data but is more commonly used with the former. *(The median may be found for any problem at the end of this chapter. Problems with smaller data sets, such as 3.1 or 3.2, are recommended for practice in finding and interpreting this statistic. Attempt problems with larger data sets, such as 3.18, only when you are comfortable with these procedures.)*

3.4 OTHER MEASURES OF POSITION: PERCENTILES, DECILES, AND QUARTILES*

In addition to serving as a measure of central tendency, the median is also a member of a class of statistics that measure position or location. The median identifies the exact middle of a distribution, but it is sometimes useful to locate other points as well. We may want to know, for example, the scores that split the distribution into thirds or fourths or the point below which a given percentage of the cases fall. A familiar application of these measures would be scores on standardized tests, which are often reported in terms of location (for example, "a score of 476 is higher than 46% of the scores").

One commonly used statistic for reporting position is the **percentile**. A percentile identifies the point below which a specific percentage of cases fall. If a score of 476 is reported as the 46th percentile, this means that 46% of the cases had scores lower than 476. To find a percentile for ungrouped data, multiply the number of cases (*N*) by the proportional value of the percentile. For example, the proportional value for the 46th percentile would be 0.46. The resultant value identifies the number of the case that marks the percentile. If we had a sample of 78 cases and wanted to find the 37th percentile, we would multiply 78 by .37. The result is 28.86, and the 37th percentile would be 86/100 of the distance between the scores of the 28th and 29th cases. In most cases, we would probably avoid this extra calculation, round off 28.86 to 29, and call the score of the 29th case the 37th percentile. The slight inaccuracy would be worth the considerable savings in time and calculational effort.

Note that, if we think in terms of percentiles, then the median is simply the 50th percentile and, in our example above, we would find the median by multiplying 78 by .50, finding the 39th case, and declaring the score of that case to be the 50th percentile. Notice again that we are cutting some corners here. Technically, the median would be the score halfway between the two middle cases (the 39th and 40th cases), but it is unlikely that this inaccuracy would be very significant.

Some other commonly used measures of position are **deciles** and **quartiles**. Deciles divide the distribution of scores into tenths. So, the first decile

---

*This section is optional.

is the point below which 10% of the cases fall and is equivalent to the 10th percentile. The fifth decile is also the same as the 50th percentile, which is the same as (you guessed it) the median. Quartiles divide the distribution into quarters, and the first quartile is the same as the 25th percentile, the second quartile is the 50th percentile, and the third quartile is the 75th percentile. Any of these measures can be found by the method described above for percentiles. Remember that multiplying N by the proportional value of the percentile, decile, or quartile gives the number of the appropriate *case*, and it's the *score* of the case that actually marks the location. Also remember that this technique cuts some (probably minor) computational corners, and use it with caution. *(For practice in finding and interpreting quartiles, deciles, and percentiles, see problem 3.17c.)*

3.5 THE MEAN

The **mean** ($\bar{X}$, read this as "ex-bar")*, or arithmetic average, is by far the most commonly used measure of central tendency. It reports the average score of a distribution, and its calculation is straightforward and probably familiar to most of you. To compute the mean, add the scores and then divide by the number of scores (N). To illustrate: a birth control clinic administered a 20-item test of general knowledge about contraception to 10 clients. The number of correct responses was 2, 10, 15, 11, 9, 16, 18, 10, 11, 7. To find the mean of this distribution, add the scores (total = 109) and divide by the number of scores (10). The result (10.9) is the average score on the test.

The mathematical formula for the mean is

FORMULA 3.1

$$\bar{X} = \frac{\Sigma(X_i)}{N}$$

where $\bar{X}$ = the mean
$\Sigma(X_i)$ = the summation of the scores
N = the number of scores

Since this formula introduces some new symbols, let us take a moment to consider it. First, the symbol Σ (uppercase Greek letter **sigma**) is a mathematical operator just like the plus sign ($+$) or divide sign ($\div$). It stands for "the summation of" and directs us to add whatever quantities are stated immediately following it. The second new symbol is X_i ("X sub i"), which refers to any single score—the "ith" score. If we wished to refer to a particular score in the distribution, the specific number of the score could replace the subscript. Thus, X_1 would refer to the first score, X_2 to the second, X_{26} to the

---

*This is the symbol for the mean of a sample. The mean of a population is symbolized with the Greek letter mu (μ—read this symbol as "mew").

26th, and so forth. The operation of adding all the scores is symbolized as $\Sigma(X_i)$.

   This combination of symbols directs us to sum the scores, beginning with the first score and ending with the last score in the distribution. Thus, Formula 3.1 states in symbols what has already been stated in words (to calculate the mean, add the scores and divide by the number of scores), but in a very succinct and precise way.* *(For practice in computing the mean, see any of the problems at the end of this chapter. Problems with smaller data sets, such as 3.1 or 3.2, are recommended for practicing computations until you are comfortable with these procedures.)*

   Since the computation of a mean requires the mathematical procedures of addition and division, its use is fully justified only when working with interval-ratio data. However, researchers sometimes calculate the mean for variables measured at the ordinal level, because the mean is much more flexible than the median and is a central feature of many interesting and powerful advanced statistical techniques. Thus, if the researcher plans to do any more than merely describe his or her data, the mean will probably be the preferable measure of central tendency. When we violate level-of-measurement considerations, of course, we are obligated to demonstrate that our conclusions still make statistical and logical sense.

3.6 SOME CHARACTERISTICS OF THE MEAN

Because the mean is the most commonly used measure of central tendency, let us consider its mathematical and statistical characteristics in some detail. First, the mean is always the center of any distribution of scores. The mean is the point around which all of the scores (X_i) cancel out. Symbolically:

$$\Sigma(X_i - \bar{X}) = 0$$

Or, if we take each score in a distribution, subtract the mean from it, and add all of the differences, the resultant sum will always be zero. To illustrate, consider the following set of test scores: 65, 73, 77, 85, and 90. The mean of these five scores is 390/5, or 78. So

X	$(X_i - \bar{X})$
65	$65 - 78 = -13$
73	$73 - 78 = -5$
77	$77 - 78 = -1$
85	$85 - 78 = 7$
90	$90 - 78 = 12$
$\Sigma X = 390$	$\Sigma(X_i - \bar{X}) = 0$

---

*See Appendix H for a further review of the summation sign.

The total of the negative differences (-19) is exactly equal to the total of the positive differences ($+19$) and will always be equal regardless of the distribution. This algebraic relationship between the scores and the mean indicates that the mean is a good descriptive measure of the centrality of scores. You may think of the mean as a fulcrum that exactly balances all of the scores.

A second characteristic of the mean will only be mentioned at this point but is of great importance in later chapters. This characteristic is expressed in the statement

$$\Sigma(X_i - \bar{X})^2 = \text{minimum}^*$$

which says that, if the differences between the scores and the mean are squared and then added, the resultant sum will be less than the sum of the squared differences between the scores and any other point in the distribution. That is, the mean is the point in a distribution around which the variation of the scores (as indicated by the squared differences) is minimized. In a sense, this algebraic property merely underlines the fact that the mean is closer to all of the scores than the other measures of central tendency. However, this characteristic of the mean is of central importance when we take up the topics of correlation and regression.

The final important characteristic of the mean is that every score in the distribution affects it. The mode (which is only the most common score) and the median (which deals only with the middle score or scores) are not so affected. This quality is both an advantage and a disadvantage. On one hand, the mean utilizes all the available information—every score in the distribution affects the mean. On the other hand, when a distribution has a few extreme cases (very high or low scores), the mean may become very misleading as a measure of centrality.

To illustrate, consider the following set of five scores: 15, 20, 25, 30, 35. Both the mean and median of this distribution are 25. ($\bar{X} = 125/5 = 25$. Md = score of third case = 25.) What will happen if we change the last score from 35 to 3500? Note that nothing happens to the median. It remains at exactly 25. The median is based only on the score of the middle case and is not affected by changes in the scores of other cases in the distribution. The mean, since it takes all scores into account, is very much affected. It becomes 3590/5, or 718. Clearly, the one extreme score in the data set disproportion-

---

*To illustrate this principle, let us use the distribution of five test scores mentioned above. The differences between the scores and the mean have already been found. If we square and sum these differences, we would get $(-13)^2 + (-5)^2 + (-1)^2 + (7)^2 + (12)^2$, or $(169 + 25 + 1 + 49 + 144)$, or 388. So, $\Sigma(X_i - \bar{X})^2 = 388$. If we performed the same operation with any other number, the resultant sum would be greater than 388. For example, if we used 77 instead of the mean, the sum of the squared differences would be $(65 - 77)^2 + (73 - 77)^2 + (77 - 77)^2 + (85 - 77)^2 + (90 - 77)^2$, or $(-12)^2 + (-4)^2 + (0)^2 + (8)^2 + (13)^2$, or $(144 + 16 + 0 + 64 + 169)$, or 393.

Application 3.1

Ten students have been asked how many hours they spent in the college library during the past week. What is the average "library time" for these students? The hours are reported in the following list, and we will find the mode, the median, and the mean for these data.

$$X_i$$

X_i
0
2
5
5
7
10
14
14
20
30
107

By scanning the scores, we can see that two scores, 5 and 14, occurred twice, and no other score occurred more than once. This distribution has two modes, 5 and 14.

Because the number of cases is even, the median will be the average of the two middle cases after all cases have been ranked in order. With 10 cases, the first middle case will be the ($N/2$), or (10/2), or fifth case. The second middle case is the ($N/2$) + 1, or (10/2) + 1, or sixth case. The median will be the score halfway between the scores of the fifth and sixth cases. Counting down from the top, we find that the score of the fifth case is 7 and the score of the sixth case is 10. The median for these data is (7 + 10)/2, or (17/2), or 8.5. The median, the score that divides this distribution in half, is 8.5.

The mean is found by first adding all the scores and then dividing by the number of scores. The sum of the scores is 107, so the mean is

$$\bar{X} = \sum \frac{X_i}{N} = \frac{107}{10} = 10.7$$

These 10 students spent an average of 10.7 hours in the library during the week in question.

Note that the mean is a higher value than the median. This indicates a positive skew in the distribution (a few extremely high scores). By inspection we can see that the positive skew is caused by the two students who spent many more hours (20 hours and 30 hours) in the library than the other eight students.

ately affects the mean. In this case, which of the two measures presents the more accurate or truer measure of central tendency for this distribution? *(For practice in dealing with the effects of extreme cases on means and medians, see problems 3.14b, 3.15b and c, 3.16b, 3.17a, 3.18a, 3.20, and 3.21.)*

What general point can we derive from these considerations? Relative to the median, the mean is always pulled in the direction of extreme scores. The two measures of central tendency will have the same value when and only when a distribution is symmetrical. When a distribution has some extremely high scores (a positive **skew**), the mean will always have a greater numerical value than the median. If the distribution has some very low scores (a negative skew), the mean will be lower in value than the median. Figures 3.1 to 3.3 depict three different frequency polygons that demonstrate these relationships.

FIGURE 3.1 A POSITIVELY SKEWED DISTRIBUTION (The mean is greater in value than the median.)

FIGURE 3.2 A NEGATIVELY SKEWED DISTRIBUTION (The mean is less than the median.)

FIGURE 3.3 AN UNSKEWED, SYMMETRICAL DISTRIBUTION (The mean and median are equal.)

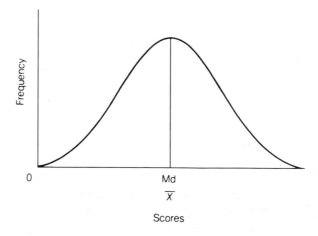

These relationships between medians and means also have a practical value. For one thing, a quick comparison of the median and mean will always tell you if a distribution is skewed and the direction of the skew. Second, they provide a simple and effective way to "lie" with statistics. For example, if you want to minimize the average score of a positively skewed distribution, report the median. Income data usually have a positive skew (there are only a few very wealthy people). If you want to impress someone with the general affluence of a mixed-income community, report the mean. If you want a lower figure, report the median. For the good and honest researcher, the selection of a measure of central tendency for a badly skewed distribution will hinge on what he or she wishes to show and, in most cases, either both statistics or the median alone should be reported.

3.7 COMPUTING MEASURES OF CENTRAL TENDENCY FOR GROUPED DATA*

In this section, we'll consider the basic techniques for computing means and medians for data presented in frequency distribution form. These techniques are useful because they can result in a significant savings in time and energy when you are working with a large data set and do not have access to a computer or calculator. Also, there are occasions when the data of interest have been organized into a frequency distribution and you do not have access to the data in ungrouped form. Under this latter condition, which occurs most often when the data are presented in the reports of other researchers, there simply is no other way to calculate means or medians. Nevertheless, a small price is attached to using these techniques. Specifically, certain assumptions must be made about the way the scores are distributed within the intervals of the frequency distribution. These assumptions will not exactly reflect the way in which the scores are actually distributed and, technically, the statistics we calculate based on them should be regarded as approximations to the true mean or median.

To illustrate, consider the following situation: as a new employee of a Youth Service Bureau, you are faced with the task of preparing a report on the truancy levels in local high schools. The report is due in two weeks, leaving you no time to gather data. Fortunately, you stumble across some truancy data left behind by the person you replaced, but the data are in the form of a frequency distribution (Table 3.2), and the ungrouped data are not available. What will you do?

Computing the Mean for Grouped Data.   To compute a mean for any distribution, we must find two values: the summation of the scores and the number of scores. For the distribution in Table 3.2, we know only the latter ($N = 124$). Because the data have been grouped, we do not know the exact distribution of the original scores. We know from the table that 25 students were truant five days or fewer, but we do not know exactly how many days. For all we know, these 25 students could have been truant for only one day, or three or five days, or any combination of days between one and five, inclusive.

---

*This section is optional.

TABLE 3.2   TOTAL DAYS TRUANT FOR ALL HIGH-SCHOOL STUDENTS
WHO WERE TRUANT AT LEAST ONCE DURING THE PAST
SCHOOL YEAR

Days	Frequency
1–5	25
6–10	30
11–15	30
16–20	20
21–25	10
26–30	5
31–35	4
	$N = 124$

TABLE 3.3   COMPUTING A MEAN FOR GROUPED DATA

Days	Frequency (f)	Midpoint (m)	Frequency × Midpoint (fm)
1–5	25	3	75
6–10	30	8	240
11–15	30	13	390
16–20	20	18	360
21–25	10	23	230
26–30	5	28	140
31–35	4	33	132
	$N = 124$		$\Sigma(fm) = 1567$

We can bypass this dilemma by assuming that all the scores in each interval are located at the midpoint of the interval. Then, by multiplying each midpoint by the number of cases in the interval, we can obtain an approximation of the total original scores in the interval. To do this, add two columns to the original frequency distribution, one for the midpoints and one for the midpoints times the frequencies. This is done for the truancy data in Table 3.3. The summation of the last column can be labeled $\Sigma(fm)$ and will be approximately the same value as the summation of the original scores or $\Sigma(X_i)$. So, we have*

$$\bar{X} = \frac{\Sigma(X_i)}{N} \approx \frac{\Sigma(fm)}{N}$$

*The symbol $\cong$ means "approximately equal to."

TABLE 3.4 COMPUTING A MEDIAN FOR GROUPED DATA

Days	Frequency (f)	Cumulative Frequency (cf)
1–5	25	25
6–10	30	55
11–15	30	85
16–20	20	105
21–25	10	115
26–30	5	120
31–35	4	124
	N = 124	

The final step in computing the mean for grouped data is to divide the total of the last column, $\Sigma(fm)$, by the number of cases:

$$\bar{X} = \frac{\Sigma(fm)}{N} = \frac{1567}{124} = 12.64$$

Thus, these 124 students were truant an average of 12.64 days over the year. To put it another way, the students who were truant at least once missed an average of about two and a half weeks of school.

Computing the Median for Grouped Data. To locate the median, we must first find the middle case of the distribution. With 124 cases in the truancy sample, we will identify the middle case as the $(N/2)$, or 62nd case.* Our problem is first to locate this case and then find the score associated with it. To do so, we make the assumption that the cases are evenly spaced throughout each interval in the frequency distribution. We then locate the interval that contains the middle case and, based on the assumption of equal spacing, find the score associated with that case.

To locate the interval that contains the median, it is helpful to add a cumulative frequency column to the original frequency distribution, as in Table 3.4. Looking at the cumulative frequency column, we see that 55 cases are accumulated below the upper limit of the 6–10 interval and that 85 cases have been accumulated below the upper limit of the 11–15 interval. We know now that the median is some value between 10.5 (the lower real limit of this interval) and 15.5 (the upper real limit), but we do not know the exact value.

To resolve this dilemma, assume that all 30 of the cases in this interval are evenly spaced throughout the interval, with case 56 located at the lower

---

*Technically, with an even N, the median would be the average score of the two middle cases—the 62nd and 63rd. Designating only the $(N/2)$ case as the middle case will, however, result in a significant simplification of the computations without much loss in accuracy.

real limit (10.5) and case 85 at the upper real limit (15.5). Case 62 (our middle case) is the seventh case of the 30 in the interval. If the scores are evenly spaced, then the 62nd case will be 7/30 of the distance from 10.5 to 15.5. Now, 7/30 of 5 (the interval size) is 1.17, so we can approximate the median at 10.5 + 1.17, or 11.67. The formula below summarizes these steps.

FORMULA 3.2

$$\text{Md} = \text{Real lower limit} + \left(\frac{N(.50) - cf \text{ below}}{f}\right)i$$

where cf below = cumulative frequency below the interval containing the median
f = number of cases in the interval containing the median
i = interval width

Applying this formula to our example, we would have

$$\text{Md} = 10.5 + \left(\frac{124(.50) - 55}{30}\right)5$$

$$\text{Md} = 10.5 + \left(\frac{62 - 55}{30}\right)5$$

$$\text{Md} = 10.5 + \left(\frac{7}{30}\right)5$$

$$\text{Md} = 10.5 + 1.17$$

$$\text{Md} = 11.67$$

Thus, half of this sample was truant fewer than 11.67 days and half more than 11.67 days.

Since the value of the median (11.67) is almost a full day lower than the value of the mean (12.6), we are justified in concluding that this distribution has a positive skew or a few very high scores. This skew is reflected in the frequency distribution itself, with most cases in the lower intervals. For purposes of description, the median may be the preferred measure of central tendency for these data, since the mean is affected by the relatively few very high scores.

It will be helpful to restate the procedures above as a series of steps. Assuming that all cases are evenly spaced through the intervals, the median for grouped data is found by these steps:

1. Find the middle case, given by $N(.50)$.
2. Find the interval that contains the middle case. A cumulative frequency column will be quite helpful for this.
3. Find the number of cases you will have to go into the interval to locate the middle case. This is given by $N(.50)$ minus the cumulative frequency below (cf below) the interval that contains the median.
4. Divide the number you found in step 3 by the number of cases in the interval (f).

5. Multiply the number you found in step 4 by the interval size (i).
6. Add the number you found in step 5 to the lower real limit of the interval that contains the median. The result is the median.

These same techniques can be used to locate any percentile, decile, or quartile in a distribution of scores. Simply change the proportional value in the "$N(.50)$" expression in Formula 3.2 to correspond to the location you are seeking and solve the formula. To illustrate, suppose we wanted to find the first quartile (or 25th percentile) for the data displayed in Table 3.4. Making the appropriate change, Formula 3.2 would become

$$\text{First quartile} = \text{Real lower limit} + \left(\frac{(N(.25) - cf\,\text{below})}{f}\right)i$$

The case marking the first quartile is given by $N(.25)$, or $(124)(.25) = 31$. By inspection we see that the 31st case is in the interval 6–10. Substituting the proper values, we would have

$$\text{First quartile} = 5.5 + \left(\frac{31 - 25}{30}\right)5$$

$$= 5.5 + \left(\frac{6}{30}\right)5$$

$$= 5.5 + (.20)5$$

$$= 5.5 + 1$$

$$\text{First quartile} = 6.5$$

Twenty-five percent of the sample had scores lower than 6.5, and 75% of the sample had scores higher than this value.

Admittedly, this procedure is rather formidable the first time it is attempted. With a little practice, however, it will become routine. Of course, if you do not have access to the ungrouped data, there simply is no other way to find the median. *(For practice in finding means and medians with grouped data, see problems 3.14c and 3.17d.)*

3.8 CHOOSING A MEASURE OF CENTRAL TENDENCY

The selection of a measure of central tendency should, in general, be based on level-of-measurement considerations and on an evaluation of what each of the three statistics shows. Remember that the mode, median, and mean are three different statistics that will be the same value only under certain, specific conditions (that is, for symmetrical distributions with one mode). Each of the three has its own message to report and, in many circumstances, you might want to report all three. When choosing a single measure of central tendency, the following guidelines may be helpful:

Use the mode when

1. Variables are measured at the nominal level.
2. You want a quick and easy measure for ordinal and interval-ratio variables.
3. You want to report the most common score.

Use the median when

1. Variables are measured at the ordinal level.
2. Variables measured at the interval-ratio level have highly skewed distributions.
3. You want to report the central score. The median always lies at the exact center of a distribution.

Use the mean when

1. Variables are measured at the interval-ratio level (except for highly skewed distributions).
2. You want to report the typical score. The mean is "the fulcrum that exactly balances all of the scores."
3. You anticipate additional statistical analysis.

SUMMARY

1. The three measures of central tendency presented in this chapter share a common purpose. Each reports some information about the most typical or representative value in a distribution. Appropriate use of these statistics permits the researcher to report important information about an entire distribution of scores in a single, easily understood number.
2. The mode reports the most common score and is used most appropriately with nominally measured variables.
3. The median (Md) reports the score that is the exact center of the distribution. It is most appropriately used with variables measured at the ordinal level and with variables measured at the interval-ratio level when the distribution is skewed.

4. The mean ($\bar{X}$), the most frequently used of the three measures, reports the most typical score. It is used most appropriately with variables measured at the interval-ratio level (except when the distribution is highly skewed).
5. The mean has a number of mathematical characteristics that are significant for statisticians. First, it is the point in a distribution of scores around which all other scores cancel out. Second, the mean is the point of minimized variation. Last, as distinct from the mode or median, the mean is affected by every score in the distribution and is therefore pulled in the direction of extreme scores.

SUMMARY OF FORMULAS

Mean	3.1	$\bar{X} = \dfrac{\Sigma(X_i)}{N}$
Median (grouped data)	3.2	$Md = \text{Real lower limit} + \left(\dfrac{N(.50) - cf\,\text{below}}{f}\right)i$

GLOSSARY

Deciles. The points that divide a distribution of scores into 10ths.

Mean. The arithmetic average of the scores. $\bar{X}$ represents the mean of a sample, and μ, the mean of a population.

Measures of central tendency. Statistics that summarize a distribution of scores by reporting the most typical or representative value of the distribution.

Median (Md). The point in a distribution of scores above and below which exactly half of the cases fall.

Mode. The most common value in a distribution or the largest category of a variable.

Percentile. A point below which a specific percentage of the cases fall.

Quartiles. The points that divide a distribution into quarters.

Σ (uppercase Greek letter **sigma**). "The summation of."

Skew. The extent to which a distribution of scores has a few scores that are extremely high (positive skew) or extremely low (negative skew).

X_i ("X sub i"). Any score in a distribution.

PROBLEMS

An asterisk (*) indicates an optional problem.

3.1 [SOC] A variety of information has been gathered from a sample of college freshmen and seniors including their region of birth, the extent to which they support legalization of marijuana (measured on a scale on which 7 = strong support, 4 = neutral, and 1 = strong opposition), and the amount of money they spend each week out-of-pocket for food, drinks, and entertainment. Some results are presented below. Find the *most appropriate* measure of central tendency for each variable for freshmen and then for seniors.

(HINT: Determine the level of measurement for each variable first. In general, this will tell you which measure of central tendency is appropriate. See Section 3.8 to review the relationship between measure of central tendency and level of measurement. Also, remember that the mode is the most common *score and, especially, remember to array scores from high to low before finding the median.)*

FRESHMEN

Student	Region of Birth	Legal-ization	Out-of-Pocket Expenses
A	North	7 ~	23 ~
B	North	4 ~	29 ~
C	South	3 ~	35 ~

Student	Region of Birth	Legal-ization	Out-of-Pocket Expenses
D	Midwest	2 ~	37 ~
E	North	3 ~	52 ~
F	North	5 ~	38 ~
G	South	1 ~	42 ~
H	South	4 ~	55 ~
I	Midwest	1 ~	41 ~
J	West	2 ~	33 ~

Measure of Central Tendency
(Report the measure you selected as well as its value.)

Birth Region: _____
Legalization: _____
Expenses: _____

SENIORS

Student	Region of Birth	Legal-ization	Out-of-Pocket Expenses
K	North	7	55
L	Midwest	6	52
M	North	7	50
N	North	5	80
O	South	1	52
P	South	5	47

Student	Region of Birth	Legal- ization	Out-of- Pocket Expenses
Q	West	6	30
R	West	7	39
S	North	3	35
T	West	5	75
U	North	4	68

Measure of Central Tendency

(Report the measure you selected as well as its value.

Birth Region: _____

Legalization: _____

Expenses: _____

Respon- dent	Sex	Social Class	Number of Years in the Party
A	M	High	32
B	M	Medium	17
C	M	Low	32
D	M	Low	50
E	M	Low	25
F	M	Medium	25
G	F	High	12
H	F	High	10
I	F	Medium	21
J	F	Medium	33
K	M	Low	37
L	F	Low	15
M	F	Low	31

Measure of Central Tendency

(Report the measure you selected as well as its value.)

Sex: _____

Social Class: _____

Years: _____

3.2 A variety of information has been collected for each of the nine high schools in a district. Find the most appropriate measure of central tendency for each variable and summarize this information in a paragraph. (*HINT: The level of measurement of the variable will generally tell you which measure of central tendency is appropriate. Remember to organize the scores from high to low before finding the median.*)

High School	Total Enrollment	Largest Racial/ Ethnic Group	Percent College Bound	Most Popular Sport	Condition of Physical Plant (scale of 1–10, 10 = high)
1	1400	White	25	Football	10
2	1223	White	77	Baseball	7
3	876	Black	52	Football	5
4	1567	Hispanic	29	Football	8
5	778	White	43	Basketball	4
6	1690	Black	35	Basketball	5
7	1250	White	66	Soccer	6
8	970	White	54	Football	9
9	1109	Hispanic	64	Soccer	3

3.3 PS You have been observing the local Democratic Party in a large city and have compiled some information about a small sample of party regulars. Find the appropriate measure of central tendency for each variable.

3.4 SOC You have compiled the information below on each of the graduates voted "most likely to succeed" by a local high school for a 10-year period. For each variable, find the appropriate measure of central tendency.

Case	Present Income	Marital Status	Owns a BMW?	Years of Schooling Completed after HS
A	24,000	Single	No	8
B	48,000	Divorced	No	4
C	54,000	Married	Yes	4
D	45,000	Married	No	4
E	30,000	Single	No	4
F	35,000	Separated	Yes	8
G	30,000	Married	No	3
H	17,000	Married	No	1
I	33,000	Married	Yes	6
J	48,000	Single	Yes	4

**Measure of
Central Tendency**
(Report the measure you
selected as well as its value.)

Income: _____

Marital Status: _____

BMW?: _____

Years of school: _____

3.5 SOC For 15 respondents, data have been gathered on four variables (table below). Find and report the appropriate measure of central tendency for each of the four variables.

Respondent	Marital Status	Racial or Ethnic Group	Age	Attitude on Abortion Scale*
A	Single	White	18	10
B	Single	Hispanic	20	9
C	Widowed	White	21	8
D	Married	White	30	10
E	Married	Hispanic	25	7
F	Married	White	26	7
G	Divorced	Black	19	9
H	Widowed	White	29	6
I	Divorced	White	31	10
J	Married	Black	55	5
K	Widowed	Asian American	32	4
L	Married	American Indian	28	3
M	Divorced	White	23	2
N	Married	White	24	1
O	Divorced	Black	32	9

*This scale is constructed so that a high score indicates strong opposition to abortion under any circumstances.

Measure of Central Tendency

(Report the measure you selected as well as its value.)

Marital Status: _____

Group: _____

Age: _____

Attitude: _____

3.6 [SOC] Below are four variables for 30 cases from the General Social Survey. Age is reported in years. The variable "happy" consists of answers to the question "Taken all together, would you say that you are (1) very happy (2) pretty happy, or (3) not too happy?" Respondents were asked how many sex partners they had over the past five years. Responses were measured on the following scale: 0–4 = actual numbers; 5 = 5–10 partners; 6 = 11–20 partners; 7 = 21–100 partners; 8 = more than 100. For each variable, find the appropriate measure of central tendency and write a sentence reporting this statistical information as you would in a research report.

Measure of Central Tendency

(Report the measure you selected as well as its value.)

Age: _____

Happiness: _____

No. of Partners: _____

Religion: _____

Respondent	Age	Happiness	Number of Partners	Religion
1	20	1	2	Protestant
2	32	1	1	Protestant
3	31	1	1	Catholic
4	34	2	5	Protestant
5	34	2	3	Protestant
6	31	3	0	Jew
7	35	1	4	None
8	42	1	3	Protestant
9	48	1	1	Catholic
10	27	2	1	None
11	41	1	1	Protestant
12	42	2	0	Other
13	29	1	8	None
14	28	1	1	Jew
15	47	2	1	Protestant
16	69	2	2	Catholic
17	44	1	4	Other
18	21	3	1	Protestant
19	33	2	1	None
20	56	1	2	Protestant
21	73	2	0	Catholic
22	31	1	1	Catholic
23	53	2	3	None
24	78	1	0	Protestant
25	47	2	3	Protestant
26	88	3	0	Catholic
27	43	1	2	Protestant
28	24	1	1	None
29	24	2	3	None
30	60	1	1	Protestant

3.7 Find the appropriate measure of central tendency for each of the four variables displayed in Table 2.4. Report each statistic as you would in a formal research report.

**Measure of
Central Tendency**
(Report the measure you
selected as well as its value.)

Sex: _____

Marital Status: _____

Satisfaction: _____

Age: _____

3.8 [SOC] For problem 2.5, find the most appropriate measure of central tendency for each of the three variables.

**Measure of
Central Tendency**
(Report the measure you
selected as well as its value.)

Sex: _____

Gun Control: _____

Education: _____

3.9 [SOC] The administration is considering a total ban on student automobiles. You have conducted a poll on this issue of fellow students and the neighbors who live around the campus and have calculated scores for your respondents. On the scale you used, a high score indicates strong opposition to the proposed ban. The scores are presented below for both groups. Calculate an appropriate measure of central tendency and compare the two groups in a sentence or two.

Students		Neighbors	
10	11	0	7
10	9	1	6
10	8	0	0
10	11	1	3
9	8	7	4
10	11	11	0
9	7	0	0

Students		Neighbors	
5	1	1	10
5	2	10	9
0	10	10	0

**Measure of
Central Tendency**
(Report the measure you
selected as well as its value.)

Students: _____

Neighbors: _____

3.10 [SW] As the head of a social services agency, you believe that your staff of 20 social workers is very much overworked compared to 10 years ago. The case loads for each worker are reported below for each of the two years in question. Has the average case load increased? What measure of central tendency is most appropriate to answer this question? Why?

1987		1997	
52	55	42	82
50	49	75	50
57	50	69	52
49	52	65	50
45	59	58	55
65	60	64	65
60	65	69	60
55	68	60	60
42	60	50	60
50	42	60	60

**Measure of
Central Tendency**
(Report the measure you
selected as well as its value.)

1987: _____

1997: _____

3.11 [SOC] Compute the median and mean for the data presented in problem 2.7. Is there a positive or negative skew in the distribution of these data?

3.12 [SW] For the test scores presented in problem 2.6, compute a median and mean for both the pretest and posttest. Interpret these statistics.

3.13 [SOC] A sample of 25 freshmen at a major university completed a survey that measured their degree of racial prejudice (the higher the score, the greater the prejudice).

a. Compute the median and mean scores for these data.

10	43	30	30	45
40	12	40	42	35
45	25	10	33	50
42	32	38	11	47
22	26	37	38	10

b. These same students completed the same survey during their senior year. Compute the median and mean for this second set of scores and compare them to the earlier set. What happened?

10	40	35	27	50
35	10	50	40	30
40	10	10	37	10
40	15	30	20	43
23	25	30	40	10

3.14 [PS] The data below represent the percentage of eligible voters who turned out in each of 50 precincts for a recent election.

20	42	57	47	40
40	17	49	72	80
21	65	56	18	25
45	30	36	60	24
62	55	60	55	50
25	60	46	43	56
59	57	45	67	25
55	56	70	56	46
89	87	58	80	45
53	54	42	47	30

a. Calculate the mean and median of this distribution.

b. Compare the mean and median. Which is the higher value? Why?

*c. Group these data into a frequency distribution and calculate the mean and median using the grouped method. Compare these measures with those you calculated in part a.

d. Report the mean and median as you would in a formal research report.

3.15 [PA] The data below represent the percentage of all workers in each city who use public transportation to commute to work.

City	Workers Using Public Transportation (%)
New York	61.8
Chicago	36.2
Los Angeles	9.3
Philadelphia	37.1
Detroit	18.4
Houston	7.8
Washington, D.C.	37.8
Dallas	10.6
San Francisco	35.7
Boston	39.3
Cleveland	22.1
Baltimore	27.0
St. Louis	21.3
Atlanta	21.3
San Diego	5.5
Milwaukee	19.2
Indianapolis	7.9
Seattle	14.9
Pittsburgh	29.5
Denver	8.2
Kansas City	10.5
Minneapolis	19.0

Source: United States Bureau of the Census, *Statistical Abstracts of the United States: 1977* (98th edition). Washington, D.C., 1977.

a. Calculate the mean and median of this distribution.

b. Compare the mean and median. Which is the higher value? Why?

c. If you removed New York from this distribution and recalculated, what would happen to the mean? to the median? Why?

d. Report the mean and median as you would in a formal research report.

3.16 [SOC] The data below represent the percentage of the population living in large cities for 48 nations in 1990.

12.7	3.3	27.5	21.3	5.8
54.3	43.1	14.4	51.5	10.3
57.9	0	0	10.9	25.7
37.6	9.6	34.2	27.5	28.9
28.5	0	12.1	12.4	40.2
11.6	31.5	17.6	3.6	21.7
25.6	10.6	9.7	6.0	5.5
14.0	24.3	2.7	8.2	35.6
5.4	6.4	20.1	22.0	16.2
21.9	34.0	9.0		

a. Calculate the mean and median of this distribution.

b. Compare the mean and median. Which is the higher value? Why?

c. Report the mean and median as you would in a formal research report.

3.17 GER/SOC Compute the mode, median, and mean from the ungrouped data presented in problems 2.8 and 2.9. (Remember to order the data from high to low to find the median.)

a. Compare the mean and median for each data set. Can you tell if the distribution is skewed? In what direction?

b. For each distribution, write a sentence or two reporting the central tendency as you would in a formal research report.

*c. For each distribution, find the third quartile, the fourth decile, and the 23rd percentile.

*d. For each distribution, compute the mean and the median from the frequency distribution you constructed as part of the original exercise. Compare the grouped measures to the ungrouped (true) measures. How accurate are the grouped techniques for approximating measures of central tendency?

3.18 SOC Compute the median and mean homicide rates for each of the years presented in problem 2.12.

a. For each year, compare the mean and the median. Are these distributions skewed? How?

b. Compare the statistics for 1975 with 1994. Describe what happened to the homicide rates between these two years.

3.19 SW For problem 2.10, compute the mean and median for both sets of scores. Describe what happened to the sample between release and six months later.

3.20 SOC For problem 2.11, compute the mean and median for both variables. Which is higher: respondent's occupational prestige or the occupational prestige of the respondent's father? Are the distributions of either variable skewed? How can you tell?

3.21 Professional athletes are threatening to strike because they claim that they are underpaid. The team owners have released a statement that says, in part, "the average salary for players was $1.2 million last year." The players counter by issuing their own statement that says, in part, "the average player earned only $753,000 last year." Is either side necessarily lying? If you were a sports reporter and had just read Chapter 3 of this text, what questions would you ask about these statistics?

SPSS for Windows

Using *SPSS for Windows* for Measures of Central Tendency and Percentiles

Start *SPSS for Windows* by clicking the SPSS icon on your monitor screen. Load the 1996 GSS and when you see the message "SPSS Processor is Ready" on the bottom of the "closest" screen, you are ready to proceed.

DEMONSTRATION 3.1 Producing Measures of Central Tendency

The only procedure in SPSS that will produce all three commonly used measures of central tendency (mode, median, and mean) is **Frequencies**. We used this procedure to produce frequency distributions in Demonstrations 2.1 and 2.2 and in Appendix F. Here we will use **Frequencies** to calculate measures of central tendency for three variables: *relig* (religious denomination), *attend* (frequency of attendance at religious services), and *age*.

The three variables vary in level of measurement, and we could request only the appropriate measure of central tendency for each variable. That is, we could request the mode for *relig* (nominal), the median for *attend* (ordinal), and the mean for *age* (interval-ratio). While this would be reasonable, it's actually more convenient to get all three measures for each variable and ignore the irrelevant output. Statistical packages typically generate more information than necessary, and it is common to disregard some of the output.

To produce modes, median, and means, begin by clicking **Statistics** from the menu bar and then click **Summarize** and **Frequencies**. In the **Frequencies** dialog box, find the variable names in the list on the left and click the arrow button in the middle of the screen to move the names (*relig*, *attend*, and *age*) to the **Variables** box on the right.

To request specific statistics, click the **Statistics** button at the bottom of the **Frequencies** dialog box, and the **Frequencies: Statistics** dialog box will open. Find the **Central Tendency** box on the right and click **Mean**, **Median**, and **Mode**. Click **Continue**, and you will be returned to the **Frequencies** dialog box, where you might want to click the **Display Frequency Tables** box. When this box is *not* checked, SPSS will not produce frequency distribution tables, and only the statistics we request (mode, median, and mean) will appear in the **Output** window. Click **OK**, and SPSS will produce the following output:

Statistics

		RS RELIGIOUS PREFERENCE	HOW OFTEN R ATTENDS RELIGIOUS SERVICES	AGE OF RESPONDENT
N	Valid	1450	1419	1499
	Missing	2	33	3
Mean		1.84	3.75	45.51
Median		1.00	3.00	43.00
Mode		1	7	38

Looking only at the most appropriate measures for each variable, the mode for *relig* is "1.000" (see the bottom line of output). What does the value mean? You can find out by consulting either the code book in Appendix G of the text or the online code book. To use the latter, click **Utilities** and then click **Variables** and find *relig* in the variable list on the left. Either way, you will find that a score of 1 indicates

Protestant. This was the most common religious affiliation in the sample and, thus, is the mode.

The median for *attend* is a value of "3.00." Again, use either Appendix G or the on-line code book, and you will see that the category associated with this score is "several times a year." This means that the middle case in this distribution of 1419 cases has a score of 3 (or that the middle case is in the interval "several times a year").

The output for *age* indicates that the respondents were, on the average, 45.51 years of age. Since age is an interval-ratio variable that has been measured in a defined unit (years), the value of the mean is numerically meaningful, and we do not need to consult the code book to interpret its meaning.

Note that SPSS did not hesitate to compute means and medians for the two variables that were not interval-ratio. The program cannot distinguish between numerical codes (such as the scores for *relig*) and actual numbers—to SPSS, all numbers are the same. Also, SPSS cannot screen your commands to see if they are statistically appropriate. If you request nonsensical statistics (average sex or median religious denomination), SPSS will carry out your instructions without hesitation. The blind willingness of SPSS to simply obey commands makes it easy for you, the user, to request statistics that are completely meaningless. Computers don't care about meaning; they just crunch the numbers.

In this case, it was more convenient for us to produce statistics indiscriminately and then ignore the ones that are nonsensical. This will not always be the case, and the point of all this, of course, is to caution you to use SPSS wisely. As the manager of your local computer center will be quick to remind you, computer resources (including paper) are not unlimited.

Before closing this demonstration, note that the mean for *age* is greater than the median. Recalling Section 3.6, this indicates that the variable has a positive skew or a few extremely high (old) cases. Verify this by doing a line chart for *age*: Click **Graphs** and then click **Line** and then **Define**. The **Define Simple Line** dialog box appears with variable names listed on the left. Choose *age* from the variable list and click the arrow button in the middle of the screen to move the variable name to the **Category Axis** box. Click the **Options** button in the lower-left-hand corner and make sure that the button next to the **Display groups defined by missing values** option is *not* selected (or checked). Click **Continue** and then click **OK**, and the line chart for *age* will be produced.

Note how the chart peaks in the 30s and 40s age groups and then gradually declines into the 70s and 80s. Although this figure is not as smooth as Figure 3.1, it does indicate that *age* is skewed in a positive direction (i.e., has a few very high scores).

DEMONSTRATION 3.2 The Descriptives Command

The **Descriptives** command in *SPSS for Windows* is designed to provide summary statistics for continuous interval-ratio level variables. By default (i.e., unless you tell it otherwise), **Descriptives** produces the mean, the minimum and maximum scores (i.e., the lowest and highest scores), and the standard deviation (which will be discussed in Chapter 4). This procedure will not produce frequency distributions (unlike **Frequencies**), and it is most appropriate for continuous, interval-ratio variables with many different scores.

To illustrate the use of **Descriptives**, let's run the procedure for *age*, *educ* (HIGHEST YEAR OF SCHOOL COMPLETED), and *tvhours* (HOURS PER DAY WATCHING TV). Click **Statistics**, **Summarize**, and **Descriptives**, and the **Descriptives** dialog box will open. This dialog box looks just like the **Frequencies** dialog box and works in the same way. Find the variable names in the list on the left and, once they are highlighted, click the arrow button in the middle of the screen to transfer them to the **Variables** box on the right. Click **OK**, and the following output will be produced:

Descriptives

Descriptive Statistics

	N	Minimum	Maximum	Mean	Std. Deviation
HOURS PER DAY WATCHING TV	989	0	24	2.97	2.40
HIGHEST YEAR OF SCHOOL COMPLETED	1446	0	20	13.41	2.97
AGE OF RESPONDENT	1449	18	89	45.51	16.82
Valid N (listwise)	982				

On the average, the sample is 45.51 years of age (this duplicates the **Frequencies** output in Demonstration 3.1), has over 13 years of education, and watches almost three hours of TV daily.

DEMONSTRATION 3.3 Finding Percentiles, Deciles, and Quartiles

The **Frequencies** procedure can be used to find percentiles, deciles, and quartiles for any variable (see Section 3.4). These statistics are especially useful for continuous variables with broad ranges of scores. One of the few variables that fits this description in the GSS data file is *age*, and we will once again use this variable for our illustrations.

Click **Statistics**, **Summarize**, and **Frequencies** to get the **Frequencies** dialog box. Click the **Statistics** button to get the **Frequencies: Statistics** dialog box. For purposes of illustration, let's find quartiles, deciles, and the 23rd and 47th percentiles. In the **Percentile Values** box, check the boxes next to Quartiles and next to Cut points for 10 equal groups. The latter instruction will find deciles (ten equal groups) and could be changed to split the distribution in other ways (e.g., into thirds).

To get specific percentiles, check the box next to Percentiles, type the desired percentile values in the box to the right, and click the **add** button after each value. Click **Continue**, and you will return to the **Frequencies** dialog box, where you might want to click the **Display Frequency Tables** box. Recall that when this box is *not* checked, SPSS will not produce frequency distribution tables. Click **OK**, and SPSS will produce the following output:

```
                  Statistics
```

AGE OF RESPONDENT

N	Valid	1449
	Missing	3
Mean		45.51
Median		43.00
Mode		38
Percentiles	10	25.00
	20	30.00
	23	31.00
	25	32.00
	30	35.00
	40	39.00
	47	42.00
	50	43.00
	60	48.00
	70	53.00
	75	56.00
	80	61.00
	90	70.00

The output prints the deciles, quartiles, and the percentiles we requested in order. Find the 10th percentile (noted as 10.00 in the column labeled "Percentiles"). The score associated with this percentile is 25 (noted as a value of 25.00). This indicates that 10% of the sample were younger than 25. Similarly, 20% were younger than age 30, 23% (the percentile we requested) were less than 31, 25% (the first quartile) were younger than 32, and so forth. Note that the fifth decile, 2nd quartile, and 50th percentile are all associated with the age of 43, which is also the value of the median.

Exercises

3.1 Use the **Frequencies** procedure to get all three measures of central tendency for *marital, income91, race, region, polviews, cappun, prestg80,* and five more variables of your own choosing. Select the most appropriate measure for each of these variables and write a sentence or two reporting and summarizing the measure.

3.2 Use the **Descriptives** procedure for *attend, income91, partnrs5, and sexfreq*. Note that these variables are ordinal in level of measurement and that, therefore, calculation of the mean is not completely justified. Such violations of strict level-of-measurement considerations are common in social science research. Write a sentence or two reporting and summarizing the mean for each variable.

3.3 Revise the command you wrote in exercise 3.1 to also find the quartiles and the deciles for each of the variables listed.

Measures of Dispersion

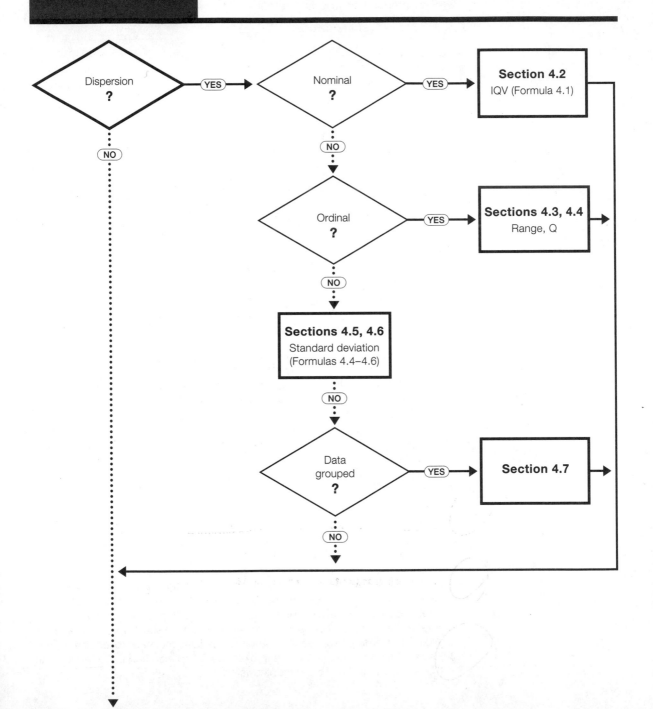

4.1 INTRODUCTION

By themselves, measures of central tendency cannot summarize data completely. For a full description of a distribution of scores, measures of central tendency must be paired with **measures of dispersion**. Whereas the mean, median, and mode are designed to locate the typical and/or central scores, measures of dispersion provide an indication of the amount of heterogeneity or variety within a distribution of scores.

The importance of the concept of **dispersion** might be easier to grasp if we consider a brief example. Suppose that the director of public safety for a city has been given the task of evaluating two ambulance services that have contracted with the city to provide emergency medical aid. As a part of the investigation, she has collected data on the response time of both services to calls for assistance. Data collected for the past year show that the mean response time is 7.4 minutes for Service A and 7.6 minutes for Service B. Since the average response times are essentially the same, there would seem to be no grounds for judging one service more or less efficient than the other. However, measures of dispersion might reveal substantial differences in the underlying distributions even when the measures of central tendency are equivalent. For example, consider Figure 4.1, which displays the distribution of response times for the two services.

Compare the shapes of these two figures. Note that the frequency polygon for Service B is much flatter than that for Service A. The scores for Service B are more spread out over the range of scores, while the scores for Service A are clustered or grouped around the mean. Comparatively speaking, Service B was much more variable in response time. Thus, even though both distributions have essentially the same mean, the distribution of scores for Service B is more heterogeneous. Had the dispersion of these distributions not been investigated, a possibly important difference in the performance of the two ambulance services might have gone unnoticed.

Although you can derive a general notion of what is meant by dispersion from Figure 4.1, the concept is not easily described in words alone. In this chapter we will introduce some of the more common measures of dispersion, each providing a precise, quantitative indication of heterogeneity. We will begin with the index of qualitative variation, mention two measures—the range and the interquartile range—quite briefly, and devote most of our attention to the standard deviation and the variance.

4.2 THE INDEX OF QUALITATIVE VARIATION (IQV)

The **index of qualitative variation (IQV)** is essentially the ratio of the amount of variation actually observed in a distribution of scores to the maximum variation that could exist in that distribution. The index varies from 0.00 (no variation) to 1.00 (maximum variation) and is used most commonly with variables measured at the nominal level. However, the IQV can be used with any variable when scores have been grouped into a frequency distribution.

To illustrate the logic of this statistic, assume that a researcher is interested in comparing the racial heterogeneity of the three small neighborhoods

FIGURE 4.1   RESPONSE TIME FOR TWO AMBULANCE SERVICES

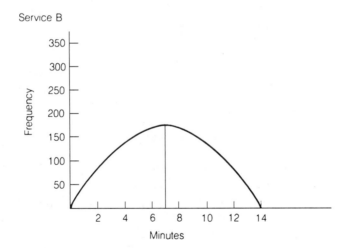

displayed in Table 4.1. By inspection, you see that neighborhood A is the least heterogeneous of the three. In fact, since all residents of A are white, there is no variation at all in this distribution. Neighborhood B is more heterogeneous than A, and neighborhood C is the most heterogeneous of the three. Let us see how the IQV substantiates these observations. The computational formula for the IQV is

FORMULA 4.1

$$IQV = \frac{k(N^2 - \Sigma f^2)}{N^2(k - 1)}$$

where k = the number of categories
N = the number of cases
Σf^2 = the sum of the squared frequencies

TABLE 4.1 RACIAL COMPOSITION OF THREE NEIGHBORHOODS

Neighborhood A		Neighborhood B		Neighborhood C	
Race	Frequency	Race	Frequency	Race	Frequency
White	90	White	60	White	30
Black	0	Black	20	Black	30
Other	0	Other	10	Other	30
	$N = 90$		$N = 90$		$N = 90$

TABLE 4.2 FINDING THE SUM OF THE SQUARED FREQUENCIES

Neighborhood A		
Race	Frequency	Squared Frequency
White	90	8100
Black	0	0
Other	0	0
	$N = 90$	8100

Neighborhood B		
Race	Frequency	Squared Frequency
White	60	3600
Black	20	400
Other	10	100
	$N = 90$	4100

Neighborhood C		
Race	Frequency	Squared Frequency
White	30	900
Black	30	900
Other	30	900
	$N = 90$	2700

To use this formula, the sum of the squared frequencies must first be computed. Do so by adding a column to the frequency distribution for the squared frequencies and then sum this column. This procedure is illustrated in Table 4.2.

The sum of the frequency column is, of course, N, and the sum of the squared frequency (Σf^2) is the total of the second column. Substituting these values into Formula 4.1 for neighborhood A, we would have an IQV of 0.00:

$$IQV = \frac{3(8100 - 8100)}{8100(2)}$$

$$IQV = \frac{3(0)}{16,200}$$

$$IQV = \frac{0}{16,200}$$

$$IQV = 0.00$$

Since the values of k and N are the same for all three tables, the IQV for the other two neighborhoods can be found by changing the values for Σf^2. For neighborhood B,

$$IQV = \frac{3(8100 - 4100)}{8100(2)}$$

$$IQV = \frac{12,000}{16,200}$$

$$IQV = 0.74$$

and, similarly, for neighborhood C,

$$IQV = \frac{3(8100 - 2700)}{16,200}$$

$$IQV = \frac{16,200}{16,200}$$

$$IQV = 1.00$$

Thus, the IQV, in a quantitative and precise way, substantiates our impressions. Neighborhood A exhibits no variation on the variable "race" (IQV = 0.00), neighborhood B has substantial variation (IQV = 0.74), and neighborhood C exhibits the maximum amount of variation possible (IQV = 1.00) for these data. *(For practice in calculating and interpreting the IQV, see problems 4.1 and 4.2.)*

4.3 THE RANGE (R) AND INTERQUARTILE RANGE (Q)

H-L

The range (R) is defined as the distance between the highest and lowest scores in a distribution. It is quite easy to calculate (high score minus low score) and is perhaps most useful to gain a quick and general notion of variability while scanning many distributions.

Unfortunately, the range is often deceptive as a measure of dispersion. It is based on only two scores in the distribution (the highest and lowest scores). Since almost any sizable distribution will contain some atypically high and low scores, the range might be quite misleading. Also, R yields no information about the nature of the scores between these two extremes.

The **interquartile range (Q)** is a kind of range. It avoids some of the problems associated with R by considering only the middle 50% of the cases in a distribution. To find Q, arrange the scores from highest to lowest and then divide the distribution into quarters (as distinct from halves, as in locating the median). The first quartile (Q_1) is the point below which 25% of the cases fall and above which 75% of the cases fall. The second quartile (Q_2) divides the distribution into halves (thus, Q_2 is equal in value to the median). The third quartile (Q_3) is the point below which 75% of the cases fall and above which 25% of the cases fall. Thus, if line LH represents a distribution of scores, the quartiles are located as shown:

The interquartile range is defined as the distance from the third to the first quartile ($Q = Q_3 - Q_1$). Thus, Q essentially extracts the middle 50% of the cases and, like R, is based on only two scores. While Q avoids the problem of being based on the most extreme scores, it has all the other disadvantages associated with R. Most importantly, Q also fails to yield any information about the nature of the scores other than the two upon which it is based.

4.4 COMPUTING THE RANGE AND INTERQUARTILE RANGE*

Table 4.3 presents per capita school expenditures for 20 states. What are the range and interquartile range of these data?

Note that the scores have already been ordered from high to low. This makes the range easy to calculate and is necessary for finding the interquartile range. Of the 20 states listed in Table 4.3, New Jersey spent the most, per capita, on public education ($1426) and North Carolina spent the least ($815). The range is therefore $1426 - 815$, or $611 (R = $611).

To find Q, we must locate the first and third quartiles (Q_1 and Q_3). In a manner analogous to the technique for finding the median, we can define both of these points in terms of the scores associated with certain cases. Q_1 is determined by multiplying N by (.25). Since (20) × (.25) is 5, Q_1 is the score associated with the fifth case, counting up from the lowest score. The fifth case is California, with a score of 897. So, $Q_1 = 897$.

The case that lies at the third quartile (Q_3) is given by multiplying N by (.75): (20) × (.75) = 15th case. The 15th case is Maine, with a score of 1110 ($Q_3 = 1110$). Therefore:

*This section is optional.

$$Q = Q_3 - Q_1$$
$$Q = 1110 - 897$$
$$Q = 213$$

In most situations, the locations of Q_1 and Q_3 will not be as obvious as they are when $N = 20$. For example, if N had been 157, then Q_1 would be (157)(.25), or the score associated with the 39.25th case, and Q_3 would be (157)(.75), or the score associated with the 117.75th case. Since fractions of cases are impossible, these numbers present some problems. The easy solution to this difficulty is to take the score of the closest case to the numbers that mark the quartiles. Thus, Q_1 would be defined as the score of the 39th case and Q_3 as the score of the 118th case.

The more accurate solution would be to take the fractions of cases into account. For example, Q_1 could be defined as the score that is one-quarter of the distance between the scores of the 39th and 40th cases, and Q_3 could be defined as the score that is three-quarters of the distance between the scores of the 117th and 118th cases. (This procedure could be analogous to defining the median—which is also Q_2—as halfway between the two middle scores when N is even.) In most cases, the differences in the values of Q for these methods would be quite small. *(For practice in finding and interpreting Q, see problems 4.5 and 4.6. The range may be found for any of the problems at the end of this chapter except 4.1 and 4.2.)*

TABLE 4.3   PER CAPITA EXPENDITURES ON PUBLIC EDUCATION, 1995

State	Expenditures per Capita
New Jersey	1426
Wyoming	1320
Michigan	1264
Texas	1147
Oregon	1118
Maine	1110
Pennsylvania	1025
New Hampshire	1012
Ohio	1008
Florida	985
Virginia	982
Nebraska	981
Idaho	980
Arizona	956
Illinois	912
California	897
Louisiana	865
Alabama	826
Mississippi	819
North Carolina	815

Source: United States Bureau of the Census, *Statistical Abstracts of the United States: 1996* (116th edition). Washington, D.C., 1996. P. 170.

4.5 THE STANDARD DEVIATION

The basic limitation of both Q and R is their failure to use all the scores in the distribution—they do not capitalize on all the available information. Also, neither Q nor R yields any information about the average or typical deviation of the scores. In that context, we might stipulate some characteristics of a good measure of dispersion. It should

1. Use all the scores in the distribution.
2. Describe the average or typical deviation of the scores.
3. Increase in value as the distribution of scores becomes more heterogeneous.

One way to develop a statistic to meet these criteria would be to find the distances between each score and some central point (that is, the mean) and sum these individual distances. The distances between the scores and the mean are called **deviations**, and their sum can be symbolized as $\Sigma(X_i - \bar{X})$. A measure of dispersion based on these deviations would meet all the criteria above. Most importantly, it would increase in value as the distribution of scores became more heterogeneous (that is, the greater the distances, the greater the sum of the deviations). This characteristic would enable us to judge the relative dispersion of different distributions by merely glancing at the summed deviations.

Unfortunately, there are two disadvantages to $\Sigma(X_i - \bar{X})$, by itself, as a measure of dispersion. First, regardless of how much variability there was in the distribution, the sum of the deviations increases with sample size (that is, the larger the number of scores, the greater the value of the measure). Comparing the relative variability of distributions of different sizes would be very difficult. Dividing the measure by N (sample size) and thus standardizing for samples of different sizes can solve this problem. The measure now is $\Sigma(X_i - \bar{X})/N$.

The second disadvantage is more serious. Algebraically, the deviations of scores around the mean will always total zero—thus making this eminently logical measure completely useless (see Section 3.6). To illustrate, consider a distribution of five scores: 10, 20, 30, 40, and 50. If the deviations of the scores from the mean were summed, we would have:

Scores (X_i)	Deviations $(X_i - \bar{X})$
10	$(10 - 30) = -20$
20	$(20 - 30) = -10$
30	$(30 - 30) = 0$
40	$(40 - 30) = 10$
50	$(50 - 30) = 20$
$\Sigma(X_i) = 150$	$\Sigma(X_i - \bar{X}) = 0$

$$\bar{X} = 150/5 = 30$$

Since the basic logic introduced above still has some appeal, we should search for a way to bypass this algebraic law by eliminating the minus signs of the deviations. There are two ways to do this. First, we can ignore the signs of the deviations. Second, we can square the deviations. Since any negative number will become positive when multiplied by itself, this will eliminate all negative values.

Our first solution yields the formula $\Sigma|X_i - \bar{X}|/N$, which directs us to sum the absolute values of the deviations (that is, disregard signs during the addition). This formula is the arithmetic definition of a statistic known as the **average deviation (AD)**:

FORMULA 4.2
$$AD = \frac{\Sigma|X_i - \bar{X}|}{N}$$

For the sample problem above, AD would be

$$AD = \frac{|20| + |10| + |0| + |10| + |20|}{5}$$

$$AD = \frac{60}{5} = 12$$

The average deviation indicates that, on the average, the scores lie 12 units from the mean.

The average deviation fulfills all three of our criteria for a good measure of dispersion. It uses all available information and is, by definition, a measure of the typical or average distance from the mean to the scores. Also, AD will increase in value as the distribution becomes more heterogeneous and the deviations from the mean increase. Unfortunately, however, absolute values are difficult to manipulate algebraically and, for this reason, the average deviation is not often used in the social sciences.

The second method for eliminating negative signs (squaring all deviations) yields a statistic known as the **variance**, which is symbolized as s^2 when referring to a sample and σ^2 (lowercase Greek letter sigma squared) when referring to a population. The variance is used primarily in inferential statistics, although it is a central concept in the design of some measures of association. For purposes of describing the dispersion of a distribution, a closely related statistic called the **standard deviation** (symbolized s for samples and σ for populations) is typically used, and this statistic will be our focus for the remainder of the chapter. The formulas for the sample variance and standard deviation are

FORMULA 4.3
$$s^2 = \frac{\Sigma(X_i - \bar{X})^2}{N-1}$$

FORMULA 4.4
$$s = \sqrt{\frac{\Sigma(X_i - \bar{X})^2}{N-1}}$$

Strictly speaking, Formulas 4.3 and 4.4 are for the variance and standard deviation of a population. Slightly different formulas, with $N - 1$ instead of

N in the denominator, should be used when we are working with random samples rather than entire populations. This is important because many of the electronic calculators and statistical software packages that you might be using use the "$N - 1$" formulas and produce results that are at least slightly different from ours. Some calculators offer the choice of "$N - 1$" or "N" in the denominator. If you use the latter, the values calculated for the standard deviation should match the values in this text.

At any rate, for the problem above, the standard deviation is

$$s = \sqrt{\frac{\Sigma(X_i - \bar{X})^2}{N}}$$

$$s = \sqrt{\frac{(-20)^2 + (-10)^2 + (0)^2 + (10)^2 + (20)^2}{5}}$$

$$s = \sqrt{\frac{400 + 100 + 100 + 400}{5}}$$

$$s = \sqrt{\frac{1,000}{5}}$$

$$s = \sqrt{200}$$

$$s = 14.14$$

To find the variance, square the standard deviation. For this problem, the variance is $s^2 = (14.14)^2 = 200$.

4.6 COMPUTING THE STANDARD DEVIATION

Formula 4.4 is the definitional formula for the standard deviation, but it is tedious to use in actual computations. For greater speed and ease of calculation, a number of computational formulas can be derived from Formula 4.4. Two of the most commonly used are Formulas 4.5 and 4.6:

FORMULA 4.5

$$s = \sqrt{\frac{\Sigma X_i^2}{N} - \bar{X}^2}$$

same

FORMULA 4.6

$$s = \frac{1}{N}\sqrt{N\Sigma X_i^2 - (\Sigma X_i)^2}$$

These formulas are algebraically equivalent to Formula 4.4 but are easier to use, since they eliminate computational steps. The formulas also introduce two new symbols, which we should consider before proceeding. The first new symbol is ΣX_i^2, which is contained in both formulas, and the second is $(\Sigma X_i)^2$, which is in Formula 4.6. Although these symbols might seem quite similar, they refer to two very different operations. The first—ΣX_i^2—is read as "the sum of the squared scores" and is found by first squaring and then summing the scores. On the other hand, $(\Sigma X_i)^2$ is read as "the sum of the scores, squared" and is found by first summing the scores and then squaring the sum. See Appendix H, Section 3 for a more comprehensive treatment of the differences between these symbols.

TABLE 4.4 COMPUTING THE STANDARD DEVIATION BY THE DEFINITIONAL FORMULA (Formula 4.4)

Test Score (X_i)	$(X_i - \bar{X})$	$(X_i - \bar{X})^2$
57	−19	361
60	−16	256
65	−11	121
70	− 6	36
78	2	4
80	4	16
82	6	36
85	9	81
90	14	196
93	17	289
$\Sigma(X_i) = 760$	$\Sigma(X_i - \bar{X}) = 0$	$\Sigma(X_i - \bar{X})^2 = 1396$

$$\bar{X} = 760/10 = 76.0$$

To illustrate, we will work through the computation of the standard deviation using the definitional formula and computational Formula 4.6. The data in Table 4.4 represent the scores of 10 students on a statistics test. Using the figures calculated in Table 4.4 and Formula 4.4, we get

$$s = \sqrt{\frac{\Sigma(X_i - \bar{X})^2}{N}}$$

$$s = \sqrt{\frac{1396}{10}}$$

$$s = \sqrt{139.6}$$

$$s = 11.82$$

Formula 4.6 eliminates the need to subtract the mean from each score and, thus, is a significantly faster computational routine. Using data from Table 4.5 and Formula 4.6, we get

$$s = \frac{1}{N}\sqrt{N\Sigma X_i^2 - (\Sigma X_i)^2}$$

$$s = \frac{1}{10}\sqrt{(10)(59,156) - (760)^2}$$

$$s = \frac{1}{10}\sqrt{591,560 - 577,600}$$

$$s = \frac{1}{10}\sqrt{13,960}$$

$$s = \frac{1}{10}(118.15)$$

$$s = 11.82$$

TABLE 4.5 COMPUTING THE STANDARD DEVIATION BY THE
COMPUTATIONAL FORMULA (Formula 4.6)

Test Score (X_i)	X_i^2
57	3249
60	3600
65	4225
70	4900
78	6084
80	6400
82	6724
85	7225
90	8100
93	8649
$\Sigma X_i = 760$	$\Sigma X_i^2 = 59{,}156$

Note that the value of the standard deviation is exactly the same as that computed using the definitional formula. Formula 4.6 (like 4.5) is equivalent to, but significantly easier to use than, 4.4. Also note that the variance can be found by squaring the value of the standard deviation. In the example above, the variance would be $(11.82)^2$, or 139.71. *(For practice in computing and interpreting s and s^2, see any of the problems at the end of this chapter except 4.1 and 4.2. Problems with smaller data sets, such as 4.3 to 4.5, are recommended for practicing computations until you are comfortable with these procedures.)*

4.7 COMPUTING THE STANDARD DEVIATION FROM GROUPED DATA*

When data have been grouped, we face the same problem in computing the standard deviation that we faced with the mean and median: the exact value of the scores is no longer known. You can resolve this problem by using the midpoints of the intervals as approximations of the original scores.

To solve for the standard deviation by Formula 4.6, three quantities are required: N, ΣX_i, and ΣX_i^2. The number of cases is the total of the frequency (f) column. The sum of the scores (ΣX_i) is approximated by multiplying each midpoint (m) by the number of cases in the interval and then adding these products (Σfm). The sum of the squared scores (ΣX_i^2) is found by squaring the midpoints, multiplying by the number of cases in the interval, and then summing (Σfm^2). We thus have

$$\Sigma X_i \cong \Sigma fm$$

$$\Sigma X_i^2 \cong \Sigma fm^2$$

and Formula 4.6 can now be solved by making the appropriate substitutions.

*This section is optional.

TABLE 4.6   COMPUTING THE STANDARD DEVIATION FROM GROUPED DATA

Correct Responses	Frequency (f)	Midpoints (m)	(fm)	m^2	fm^2
0–2	1	1	1	1	1
3–5	2	4	8	16	32
6–8	3	7	21	49	147
9–11	4	10	40	100	400
12–14	3	13	39	169	507
15–17	2	16	32	256	512
18–20	2	19	38	361	722
21–23	2	22	44	484	968
24–26	1	25	25	625	625
Totals	N = 20		248		3914

To illustrate, Table 4.6 is a frequency distribution with several columns added for the purpose of organizing the required computations. The data are the number of correct responses on a 25-item arithmetic test by a sample of 20 students. The sum of the third column (Σfm) will be used in place of the sum of the scores (ΣX_i), and the sum of the last column (Σfm^2) will replace the sum of the squared scores (ΣX_i^2). Formula 4.6 can be restated using our approximations in place of the original symbols.

FORMULA 4.7

$$s = \frac{1}{N} \sqrt{N(\Sigma fm^2) - (\Sigma fm)^2}$$

Substituting the proper numbers, we would have

$$s = \frac{1}{20} \sqrt{(20)(3914) - (248)^2}$$

$$s = \frac{1}{20} \sqrt{78,280 - 61,504}$$

$$s = \frac{1}{20} \sqrt{16,776}$$

$$s = \frac{1}{20}(129.52)$$

$$s = 6.48$$

The mean of this distribution can also be calculated from Table 4.6 by dividing the sum of column 3 (Σfm) by N (see Section 3.7). The mean is 248/20, or 12.4. These 20 students answered an average of 12.4 items correctly, and the distribution of correct responses has a standard deviation of 6.48. (*For practice in computing s with grouped data, see problems 4.5b and 4.6b.*)

Application 4.1

At a local preschool, 10 children were observed for 1 hour, and the number of aggressive acts committed by each was recorded in the following list. What is the standard deviation of this distribution? We will use Formula 4.6 to compute the standard deviation.

Number of Agressive Acts (X_i)	X_i^2
0	0
3	9
5	25
2	4
7	49
10	100
0	0
8	64
2	4
0	0
$\Sigma X_i = 37$	$\Sigma X_i^2 = 255$

Substituting these values into Formula 4.6, we have

$$s = \frac{1}{N} \sqrt{N\Sigma X_i^2 - (\Sigma X_i)^2}$$

$$s = \frac{1}{10} \sqrt{(10)(255) - (37)^2}$$

$$s = \frac{1}{10} \sqrt{2550 - 1369}$$

$$s = \frac{1}{10} \sqrt{1181}$$

$$s = \frac{1}{10} \sqrt{(34.37)}$$

$$s = 3.44$$

The standard deviation for these data is 3.44.

Application 4.2

You have just won the state lottery and are, overnight, fabulously wealthy. Even after a brief visit with the IRS you still have enough money to live wherever you want and completely ignore such mundane considerations as work. One of your first decisions is to find the "nicest place to live" in all the world. Because you are somewhat eccentric, your only criterion for "nicest place" is climate. Specifically, you want to locate a city where the temperature is exactly 78°F. After much research you find three cities where the average daily temperature is exactly 78°F. Which of the three cities will you choose as your permanent residence?

The preceding information reports central tendency (the average temperature) and does not provide you with a basis for deciding among the three cities. What if you also discovered that the standard deviation and the range of the daily temperature were, respectively, .7 and 3 degrees for City A; 10.3 and 30 degrees in City B; and 25.8 and 103 degrees in City C? Can you choose a permanent residence now? Which city would you choose? Why?

As was the case with frequency distributions, measures of central tendency and dispersion may not be presented in research reports in the professional literature. Given the large number of variables included in a typical research project and the space limitations in the journals and other media, there may not be room for the researcher to describe each variable fully. Furthermore, the great majority of research reports focus on relationships between variables rather than the distribution of single variables. In this sense, univariate descriptive statistics will be irrelevant to the main focus of the report.

This does not mean, of course, that univariate descriptive statistics are irrelevant to the research project; nor do I mean to imply that researchers do not calculate and interpret these statistics. Measures of central tendency and dispersion will be calculated and interpreted for virtually every variable, in virtually every research project. However, in the interest of efficiency, these statistics are less likely to be included in final research reports than the more analytical statistical techniques to be presented in the remainder of this text. Furthermore, some of these statistics (for example, the mean and standard deviation) serve a dual function. They not only are valuable descriptive statistics but also form the basis for many analytical techniques. Thus, they may be reported in the latter role if not in the former.

When included in research reports, measures of central tendency and dispersion will most often be presented in some summary form for all relevant variables—often in the form of a table. Means and standard deviations for many variables might, for example, be presented in the following table format.

Variable	$\bar{X}$	s	N
Age	33.2	1.3	1078
Number of children	2.3	.7	1078
Years married	7.8	1.5	1052
Income	30,078	982	978
.	.	.	.
.	.	.	.
.	.	.	.

These tables describe the overall characteristics of the sample succinctly and clearly. If you inspect the table carefully, you will have a good sense of the nature of the sample on the traits relevant to

the project. Note that the number of cases varies from variable to variable in the preceding table. This is normal and is caused by missing data or incomplete information on some of the cases.

Statistics in the Professional Literature

This section presents two brief excerpts from social science research. In both cases, the descriptive statistics are not the focus of the article but, rather, the basis for a statistical test that we will consider in Chapter 9. The first article presents means and standard deviations and the second presents means only.

Excerpt 1 Gender Differences in the Productivity of Scientists

Professor J. Scott Long notes that "The average female scientist publishes fewer articles and receives fewer citations than the average male scientist. This finding persists across disciplines and through time." (p. 1297) What factors account for the lower productivity of female scientists? In an attempt to find out, Professor Long assembled a sample of 556 males and 603 females who earned Ph.D.s in biochemistry between 1950 and 1967. He gathered information about their scholarly careers, their research productivity, and their personal lives. Some of his results are presented in the table below. Inspect the table carefully and see if you can distill the essential meanings of the information presented.

Sex Differences in the Levels of Major Variables

	Sex	Mean	Std. Dev.
Age at BS degree	♀	22.089	2.339
	♂	22.728	2.220
Age at Ph.D.	♀	31.299	5.709
	♂	29.651	3.821
Articles by student— three year count	♀	1.529	1.627
	♂	1.932	2.232
Baccalaureate selectivity	♀	2.958	1.819
	♂	3.066	1.694
Number of children at Ph.D	♀	0.286	0.710
	♂	0.844	1.084
Interruption between BS and Ph.D. (in years)	♀	0.527	0.500
	♂	0.344	0.476

Professor Long uses means and standard deviations to summarize and describe the sample and also uses these two statistics as the basis for a test that helps to identify important or consequential differences by gender. Here's what the author has to say about the differences:

Males and females come from undergraduate institutions of nearly identical selectivity. While females are about three-fourths of a year younger than males at completion of the baccalaureate, they are more than a year and one-half older than males upon completion of the Ph.D. The older age of females upon completion of the Ph.D. is the result of females interrupting their education between the baccalaureate and the Ph.D. more frequently and longer than males. While males are more likely to be married and to have children before receiving the Ph.D., females are more likely to have had their educations interrupted, possibly as a result of family obligations.

The author concludes that the key factor accounting for the productivity difference is that females have fewer and weaker relationships with faculty mentors because they more often interrupt their education for child bearing and child rearing. Professor Long concludes: "For females, having young children strongly decreases the odds of collaborating with the mentor; this in turn decreases predoctoral productivity." Curious and want to learn more? Complete bibliographic information is given below.

J. Scott Long: 1990. "The Origins of Sex Differences in Science." *Social Forces*, Vol. 68 (4): 1297–1315. Copyright © The University of North Carolina Press. Reprinted by permission.

Excerpt 2 The Adjustment to the United States of Immigrants from Africa: Implications for Research and for Social Work Practice

Professor Hugo Kamya notes that researchers have largely ignored immigrant groups from Africa. As a result, little information is available to social workers and other service providers to help them understand the specific problems, stresses, and coping mechanisms of these groups.

Professor Kamya tested a sample of 52 African immigrants on their levels of stress, hardiness (or ability to cope with stress), spiritual well-being, and level of self-esteem. He was particularly interested in the role of religion in the adjustment process and divided the sample into two groups in terms of whether they saw religion as a minor or major influence. Some of his results are reported in the following table:

Mean Scores for Influence of Religion in Relation to Hardiness, Stress, Self-Esteem, Spiritual Well-Being, and Coping Resources

	Religion Is	
Scale	Minor Influence ($N = 15$)	Major Influence ($N = 37$)
Hardiness overall	45.2	45.4
Stress	37.2	41.7
Self-Esteem	66.8	76.1
Spiritual well-being overall	88.8	101.8
Coping resources Total	50.8	54.9

Here's what Professor Kamya has to say about these results:

Immigrants for whom religion was a major influence reported higher scores on the spiritual well-being scale and the coping resources scale. This finding supports the literature that suggests that religion is a major coping resource for immigrants. And threat, therefore, to a person's cultural or religious beliefs poses a threat to the person's ability to cope.

If you would like to learn more, bibliographic information is given below.

Hugo A. Kamya. 1997. "African Immigrants in the United States: The Challenge for Research and Practice." *Social Work*. 42: 154–163.

4.8 INTERPRETING THE STANDARD DEVIATION

It is very possible that the meaning of the standard deviation (i.e., why we calculate it) is not obvious to you at this point. You might be asking: "Once I've gone to the trouble of calculating the standard deviation, what do I have?" The meaning of this measure of dispersion can be expressed in three ways. The first and most important involves the normal curve, and we will defer this interpretation until the next chapter.

A second way of thinking about the standard deviation is as an index of variability. As we saw earlier, the standard deviation increases in value as the distribution becomes more variable. The reverse relationship also holds: the less the variability in the distribution, the lower the value of the standard deviation. The limiting case in this direction would be a distribution with no dispersion—a distribution where every case had exactly the same score. In such a case, the standard deviation would be 0. Thus, 0 is the lowest value possible for the standard deviation (although there is no upper limit).

A third way to get a feel for the meaning of the standard deviation is by comparing one distribution with another. Assume we had the following information on the distribution of household income in two cities:

City A	City B
$\bar{X} = 31,233$	$\bar{X} = 31,233$
$s = 1729$	$s = 5268$

The two cities are equal in central tendency, but income is much more variable from household to household in City B. This would be the case, for example, if City B has more families in both the higher and lower income brackets and families in City A were more clustered around the mean income. The standard deviation is extremely useful for making comparisons of this sort between distributions of scores.

SUMMARY

1. Measures of dispersion summarize information about the heterogeneity or variety in a distribution of scores. When combined with an appropriate measure of central tendency, these statistics convey a large volume of information in just a few numbers. While measures of central tendency locate the central points of the distribution, measures of dispersion indicate the amount of diversity in the distribution.

2. The index of qualitative variation (IQV) can be computed for any variable that has been organized into a frequency distribution. It is the ratio of the amount of variation observed in the distribution to the maximum variation possible in the distribution. The IQV is most appropriate for variables measured at the nominal level.

3. The range (R) is the distance from the highest to the lowest score in the distribution. The interquartile range (Q) is the distance from the third to the first quartile (the "range" of the middle 50% of the scores). These two ranges can be used with variables measured at either the ordinal or interval-ratio level.

4. The standard deviation (s) is the most important measure of dispersion because of its central role in many more advanced statistical applications. The standard deviation has a minimum value of zero

(indicating no variation in the distribution) and increases in value as the variability of the distribution increases. It is used most appropriately with variables measured at the interval-ratio level.

5. The variance (s^2) is used primarily in inferential statistics and in the design of some measures of association.

SUMMARY OF FORMULAS

Index of Qualitative Variation	4.1	$IQV = \dfrac{k(N^2 - \Sigma f^2)}{N^2(k - 1)}$
Average Deviation	4.2	$AD = \dfrac{\Sigma\|X_i - \bar{X}\|}{N}$
Variance (population)	4.3	$s^2 = \dfrac{\Sigma(X_i - \bar{X})^2}{N}$
Standard Deviation (population)	4.4	$s = \sqrt{\dfrac{\Sigma(X_i - \bar{X})^2}{N}}$
Standard Deviation (computational)	4.5	$s = \sqrt{\Sigma\dfrac{X_i^2}{N} - \bar{X}^2}$
Standard Deviation (computational)	4.6	$s = \dfrac{1}{N}\sqrt{N\Sigma X_i^2 - (\Sigma X_i)^2}$

GLOSSARY

Average deviation (AD). The average of the absolute deviations of the scores around the mean.

Deviations. The distances between the scores and the mean.

Dispersion. The amount of variety or heterogeneity in a distribution of scores.

Index of qualitative variation (IQV). A measure of dispersion for variables that have been organized into frequency distributions.

Interquartile range (Q). The distance from the third quartile to the first quartile.

Measures of dispersion. Statistics that indicate the amount of variety or heterogeneity in a distribution of scores.

Range (R). The highest score minus the lowest score.

Standard deviation. The square root of the squared deviations of the scores around the mean, divided by N. The most important and useful descriptive measure of dispersion; s represents standard deviation of a sample; σ, the standard deviation of a population.

Variance. The squared deviations of the scores around the mean divided by N. A measure of dispersion used primarily in inferential statistics and also in correlation and regression techniques; s^2 represents the variance of a sample; σ^2, the variance of a population.

PROBLEMS

Asterisks (*) indicate optional problems.

4.1 [SOC] The marital status of residents of four apartment complexes is reported below. Compute the index of qualitative variation (IQV) for each neighborhood. Which is the most heterogeneous of the four? Which is the least? (HINT: It may be helpful to organize your computations as in Table 4.2.)

	Complex A
Marital Status	Frequency
Single	26
Married	31
Divorced	12
Widowed	5
	$N = 74$

	Complex B
Marital Status	Frequency
Single	10
Married	12
Divorced	8
Widowed	7
	$N = 37$

	Complex C
Marital Status	Frequency
Single	20
Married	30
Divorced	2
Widowed	1
	$N = 53$

	Complex D
Marital Status	Frequency
Single	52
Married	3
Divorced	20
Widowed	10
	$N = 85$

4.2 [SOC/CJ] Compute the Index of Qualitative Variation for the distributions presented in problems 2.1 and 2.4.

4.3 Compute the range and standard deviation of 10 scores reported below. Use both the definitional and one of the computational formulas for computing *s*. *(HINT: It may be helpful to organize your computations as in Tables 4.4 and 4.5.)*

10, 12, 15, 20, 25, 30, 32, 35, 40, 50

4.4 For the 10 test scores below, compute the standard deviation using both the definitional and computational formulas.

77, 83, 69, 72, 85, 90, 95, 75, 55, 45

4.5 [GER]
a. Compute the range, interquartile range, standard deviation, and variance for the *ungrouped* data presented in problem 2.8.
*b. Compute the standard deviation from the frequency distribution you constructed for problem 2.8. Use the grouped method presented in Section 4.7.

4.6 [SOC]
a. Compute the range, interquartile range, standard deviation, and variance for the *ungrouped* data presented in problem 2.9.
*b. Compute the standard deviation from the frequency distribution you constructed for problem 2.9. Use the grouped method presented in Section 4.7.

4.7 For problem 3.2, find the standard deviation for enrollment, percent college bound, and condition of physical plant. What information does the standard deviation add to what you already knew about these variables?

4.8 [SOC] Labor force participation rates (percent employed), percent high school graduates, and mean income for males and females in 10 states are reported below. Calculate a mean and a standard deviation for both groups for each variable and describe the differences. Are males and females unequal on any of these variables? How great is the gender inequality?

State	Labor Force Participation		% High School Graduates		Mean Income	
	Male	Female	Male	Female	Male	Female
A	73.9	54.2	65	67	35,623	27,345
B	81.1	63.2	57	60	32,345	28,134
C	80.9	59.9	72	76	35,789	30,546
D	76.6	59.9	77	75	38,907	31,788
E	80.0	60.8	75	74	42,023	35,560
F	74.1	51.5	70	72	34,000	35,980
G	73.5	50.6	68	66	25,800	19,001
H	77.6	54.9	70	71	29,000	26,603
I	77.4	54.1	66	66	31,145	30,550
J	80.1	75.2	72	75	34,334	29,117

4.9 [SW] Compute the standard deviation for the pretest and posttest scores presented in problem 2.6. Considering the means you calculated in problem 3.12, briefly describe how the sample changed from test to test. What does the standard deviation add to the information you already had?

4.10 [SOC] Compute a standard deviation for the two sets of scores presented in problem 3.9. What does the standard deviation add to the information you already had?

4.11 [CJ] Per capita expenditures for police protection for 20 cities are reported below for 1990 and 1995. Compute a mean and standard deviation for each year and describe the differences in expenditures for the five-year period.

1990		1995	
180	167	210	225
95	101	110	209
87	120	124	201
101	78	131	141
52	107	197	94
117	55	200	248
115	78	119	140
88	92	87	131
85	99	125	152
100	103	150	178

4.12 [PA] Compute the range and standard deviation for the data presented in problem 3.15. What would happen to the value of the standard deviation if you removed New York City from this distribution and recalculated? Why?

4.13 [SOC] Below are listed the rates of abortion per 100,000 women for 20 states in 1973 and 1975. Describe what happened to these distributions over the two-year period. Did the average rate increase or decrease? What happened to the dispersion of this distribution? What happened between 1973 and 1975 that might explain these changes in central tendency and dispersion? *(HINT: It was a Supreme Court decision.)*

State	1973	1975
Maine	3.5	9.5
Massachusetts	10.0	25.7
New York	53.5	40.7
Pennsylvania	12.1	18.5

State	1973	1975
Ohio	7.3	17.9
Michigan	18.7	20.3
Iowa	8.8	14.7
Nebraska	7.3	14.3
Virginia	7.8	18.0
South Carolina	3.8	10.3
Florida	15.8	30.5
Tennessee	4.2	19.2
Mississippi	0.2	0.6
Arkansas	2.9	6.3
Texas	6.8	19.1
Montana	3.1	9.9
Colorado	14.4	24.6
Arizona	6.9	15.8
California	30.8	33.6
Hawaii	26.3	31.6

Source: United States Bureau of the Census, *Statistical Abstracts of the United States: 1977* (98th edition). Washington, D.C., 1977.

4.14 [SW] One of your goals as the new chief administrator of a large social service bureau is to equalize work loads within the various divisions of the agency. You have gathered data on case loads per worker within each division. Which division comes closest to the ideal of an equalized work load? Which is farthest away?

A	B	C	D
50	60	60	75
51	59	61	80
55	58	58	74
60	55	59	70
68	56	59	69
59	61	60	82
60	62	61	85
57	63	60	83
50	60	59	65
55	59	58	60

4.15 [SOC] Below are the scores of 25 individuals on four variables taken from the General Social Survey. For each variable, compute the mean and the standard deviation. Write a sentence or two of summary and interpretation for each variable. See Appendix G for exact wording of questions and meaning of codes.

Case	Number of Children	Educational Level	Degree of Political Liberalism	Frequency of Church Attendance
1	1	10	4	2
2	0	12	6	5
3	0	12	6	7
4	0	11	4	7
5	2	12	1	4
6	2	13	3	5
7	3	8	4	4
8	1	4	5	7
9	2	12	7	0
10	2	10	4	7
11	0	16	5	7
12	2	16	4	7
13	0	13	1	1
14	0	14	4	0
15	1	12	4	6
16	2	5	3	2
17	1	8	7	3
18	1	12	3	2
19	0	9	4	1
20	0	8	5	0
21	2	12	3	7
22	2	16	5	5
23	0	16	6	0
24	2	16	5	2
25	5	12	4	7

4.16 [SOC] Compute the standard deviation for the age of the 30 respondents in problem 3.6. Taking into account the mean you computed in problem 3.6, write a sentence summarizing this variable.

4.17 [SW] Compute the standard deviation for both data sets presented in problem 3.10. Compare the standard deviations for the two time periods along with the changes in the mean. What happened? Why?

4.18 [SOC] Compute the standard deviation for both sets of data presented in problem 3.13. Compare the standard deviation computed for freshmen with the standard deviation computed for seniors. What happened? Why? Does this change relate at all to what happened to the mean over the four-year period? How? What happened to the shapes of the underlying distributions?

4.19 [SOC] Compute the standard deviation for the data presented in problem 3.16. Taking into account the mean you computed in problem 3.16, write a sentence summarizing this variable.

4.20 [PS] Compute the range and standard deviation for the data presented in problem 3.14. Taking into account the mean you computed in problem 3.14, write a sentence summarizing this variable.

4.21 [SOC] The occupational prestige scores of 50 respondents and their fathers were presented in problem 2.11. Find the range and standard deviation for each of these variables. Combining these measures of dispersions with the mean (problem 3.20) and the frequency distributions (problem 2.11), describe these two distributions as you would in a research report.

4.22 SW Compute the standard deviation for each data set presented in problem 2.10. Considering also the mean and median you calculated in problem 3.19, write a few sentences of interpretation and summary for these data.

4.23 SOC Compute the standard deviation for each year's homicide rates as presented in problem 2.12. Considering also the mean and median you calculated in problem 3.18, write a few sentences of interpretation and summary for these data. What happened to homicide rates over the time period?

4.24 At St. Algebra College, the math department ran some special sections of the freshman math course using a variety of innovative teaching techniques. Students were randomly assigned to either the traditional sections or the experimental sections, and all students were given the same final exam. The results of the final are summarized below. What was the effect of the experimental course?

Traditional	Experimental
$\bar{X} = 77.8$	$\bar{X} = 76.8$
$s = 12.3$	$s = 6.2$
$N = 478$	$N = 465$

4.25 CJ You're the governor of the state and must decide which of four metropolitan police departments will win the annual award for efficiency. The performance of each department is summarized in monthly arrest statistics as reported below. Which department will win the award? Why?

	Departments			
	A	B	C	D
$\bar{X} =$	601.30	633.17	592.70	599.99
$s =$	2.30	27.32	40.17	60.23

SPSS for Windows

Using *SPSS for Windows* to Produce Measures of Dispersion

Start *SPSS for Windows* by clicking the SPSS icon on your monitor screen and load the 1996 GSS data set.

DEMONSTRATION 4.1 Producing the Range and the Standard Deviation

Most of the statistics discussed in this chapter are available from either the **Frequencies** or **Descriptives** procedures that you are already familiar with. In this demonstration, we will use **Descriptives** to find the range and standard deviation for *age* (years of age), *educ* (HIGHEST YEAR OF SCHOOL COMPLETED), and *tvhours* (HOURS PER DAY WATCHING TV). These three variables were also used in Demonstration 3.2.

From the main menu, click **Statistics**, **Summarize**, and **Descriptives**. The **Descriptives** dialog box will open. Use the cursor to find the names of the three variables in the list on the left and click the right arrow button to transfer them to the **Variables** box. Click **OK**, and SPSS will produce the same output we analyzed in Demonstration 3.2. Now, however, we will consider dispersion rather than central tendency. The output looks like this:

Descriptive Statistics

	N	Minimum	Maximum	Mean	Std. Deviation
HOURS PER DAY WATCHING TV	989	0	24	2.97	2.40
HIGHEST YEAR OF SCHOOL COMPLETED	1446	0	20	13.41	2.97
AGE OF RESPONDENT	1449	18	89	45.51	16.82
Valid N (listwise)	982				

The standard deviation for each variable is reported in the column labeled "Std.Deviation" and the range can be computed from the values given in the Minimum and Maximum columns. For *tvhours*, the standard deviation is 2.40, and scores ranged from 0 hours of television watching to 24 (the maximum possible in a day). For *educ*, the standard deviation was almost 3, and scores ranged from zero to 20. Respondents with scores of 20 have completed four years of formal education beyond the bachelor's level. The standard deviation for *age* was almost 17 years, and the youngest and oldest respondents were 18 and 89, respectively.

At this point, the range is probably easier to understand and interpret than the standard deviation. As we saw in Section 4.8, the latter is more meaningful when we have a point of comparison. For example, suppose we were interested in the variable *tvhours* and how television-viewing habits have changed over the years. The output we produced earlier shows that in 1996, people watched an average of almost three hours a day with a standard deviation of 2.4. Suppose that a sample from 1976 showed an average of 2.7 hours of television viewing a day with a standard deviation of 1.1. You could conclude that television watching had, on the average, increased slightly over the 20-year period but that Americans had also become much more diverse in their frequency of viewing.

DEMONSTRATION 4.2 Using the COMPUTE Command to Create an "Attitude Toward Abortion" Scale

SPSS provides a variety of ways to transform and manipulate variables. In the SPSS exercises at the end of Chapter 2, the **Recode** command was introduced as a way of changing the values associated with a variable. In this demonstration, I will present the **Compute** command as a way to create new variables and summary scales.

Let's begin by considering the two questions from the 1996 GSS that measure attitudes toward abortion, *abany* and *abhlth*. The items present two different situations under which an abortion might be desired and ask the respondent to react to each situation independently. *Abany* asks if an abortion should be possible for "any reason," and *abhlth* stipulates that the women's health is in danger. Since the situations are distinct, even though both relate to abortion, each item should be analyzed in its own right. Suppose, however, that you wanted to create a summary scale that indicated a person's *overall* feelings about abortion.

One way to do this would be to add the scores on the two variables together. This would create a new variable with three possible scores: a score of 2 could be acquired only by answering "yes" (coded as "1") to both items. Such a score would indicate a consistent "pro-abortion" position. A score of 3 would occur when a respondent answered "yes" to one item and "no" to the other. This might be labeled a "moderate" position. The final possibility would be a score of 4, if the respondent answered "no" to both items. This would be a consistent "anti-abortion" position.

The variable created by adding the scores from these two items would summarize each respondent's overall position on abortion. Once created, the new variable could be analyzed, transformed, and manipulated exactly like a variable actually recorded in the data file.

To use the **Compute** command, click **Transform** and then **Compute** from the main menu. The **Compute Variable** window will appear. Find the **Target Variable** box in the upper-left-hand corner of this window. The first thing we need to do is assign a name to the new variable we are about to compute—let's call it *abscale*—and type that name in this box. Next, we need to tell SPSS how to compute the new variable. In this case, *abscale* will be computed by adding the scores of *abany* and *abhlth*. Find *abany* in the variable list on the left and click the arrow button in the middle of the screen to transfer the variable name to the **Numeric Expression** box. Next, click the plus sign (+) on the calculator pad under the **Numeric Expression** box, and the sign will appear next to *abany*. Finally, highlight *abhlth* in the variable list and click the arrow button to transfer the variable name to the **Numeric Expression** box.

The expression in the **Numeric Expression** box should now read

```
abany + abhlth
```

Click **OK**, and *abscale* will be created and added to the data set. If you want to keep this new variable permanently, click **Save** from the **File** menu, and the updated data set with *abscale* added will be saved to disk. If you are using the student version of SPSS, remember that your data set is limited to 50 variables.

We now have three variables that measure attitudes towards abortion, two items referring to specific situations, and a more general, summary item. It is always a good idea to check the frequency distribution for computed and recoded variables and make sure that the computations were carried out correctly. Use the **Frequencies** procedure (click **Statistics**, **Summarize**, and **Frequencies**) to get tables for *abany*, *abhlth*, and *abscale*. Your output will look like this:

Statistics

		ABORTION IF WOMAN WANTS FOR ANY REASON	WOMANS HEALTH SERIOUSLY ENDANGERED	ABSCALE
N	Valid	904	911	877
	Missing	548	541	575

Frequency Table

ABORTION IF WOMAN WANTS FOR ANY REASON

		Frequency	Percent	Valid Percent	Cumulative Percent
Valid	YES	409	28.2	45.2	45.2
	NO	495	34.1	54.8	100.0
	Total	904	62.3	100.0	
Missing	NAP	503	34.6		
	DK	41	2.8		
	NA	4	.3		
	Total	548	37.7		
Total		1452	100.0		

WOMANS HEALTH SERIOUSLY ENDANGERED

		Frequency	Percent	Valid Percent	Cumulative Percent
Valid	YES	829	57.1	91.0	91.0
	NO	82	5.6	9.0	100.0
	Total	911	62.7	100.0	
Missing	NAP	503	34.6		
	DK	36	2.5		
	NA	2	.1		
	Total	541	37.3		
Total		1452	100.0		

ABSCALE

		Frequency	Percent	Valid Percent	Cumulative Percent
Valid	2.00	401	27.6	45.7	45.7
	3.00	401	27.6	45.7	91.4
	4.00	75	5.2	8.6	100.0
	Total	877	60.4	100.0	
Missing	System	575	39.6		
Total		1452	100.0		

Note that the level of approval for the two specific situations varies sharply. While an overwhelming majority (91%) of the sample approved of the legal right to an abortion when the woman's health was in danger, a slight majority disapproved when the woman wants an abortion "for any reason." Looking at the combined scores on *abscale*, we see that about 46% of the sample approved of the legal right to an abortion in both cases (scored 2), and less than 9% disapproved in both situations (scored 4). A very large percentage (46%) had a score of 3, which means that they approved in one situation (when health is at stake) but not in the other.

Note that only about 900 respondents answered the original two abortion items. Remember that no respondent is given the entire GSS, and the vast majority of the "missing cases" received a form of the GSS that did not include these two items. Now look at *abscale*, the summary scale, and note that fewer cases (877) are included in the summary scale than in either of the two original items. When

SPSS executes a **Compute** statement, it automatically eliminates any cases that are missing scores on any of the constituent items. If these cases were not eliminated, a variety of errors and misclassifications could result. For example, if cases with missing scores were included, a person who scored a 2 ("anti-abortion") on *abany* and then failed to respond to *abhlth* would have a total score of 2 on *abscale*. Thus, this case would be treated as "pro-abortion" when, actually, the only information we have indicates that this respondent is "anti-abortion." To eliminate this kind of error, cases with missing scores on any of the constituent variables are deleted from calculations.

Exercises

4.1 Use the **Descriptives** procedure to produce univariate descriptive statistics for *paeduc* (respondent's father's years of education), *prestg80* (respondent's occupational prestige), and *papres80* (respondent's father's occupational prestige).

    a. Compare the statistics for *prestg80* and *papres80*. Describe the difference in occupation prestige between the two generations. Are the respondents higher or lower in prestige than their fathers? Is the sample more or less homogeneous on this variable than their fathers?

    b. Compare the statistics for *educ* (produced in Demonstration 4.1) and *paeduc*. Describe the differences between the two generations. Are the respondents more or less educated than their fathers? Is the sample more or less homogeneous on this variable than their fathers?

4.2 Use the **Compute** command to create a summary scale for *fehelp* and *fefam*. Use **Descriptives** to get a univariate descriptive statistics on the summary scale. (*HINT*: The expression in the **Numerical Expression** box will look like this: *fehelp + fefam*.)

The Normal Curve

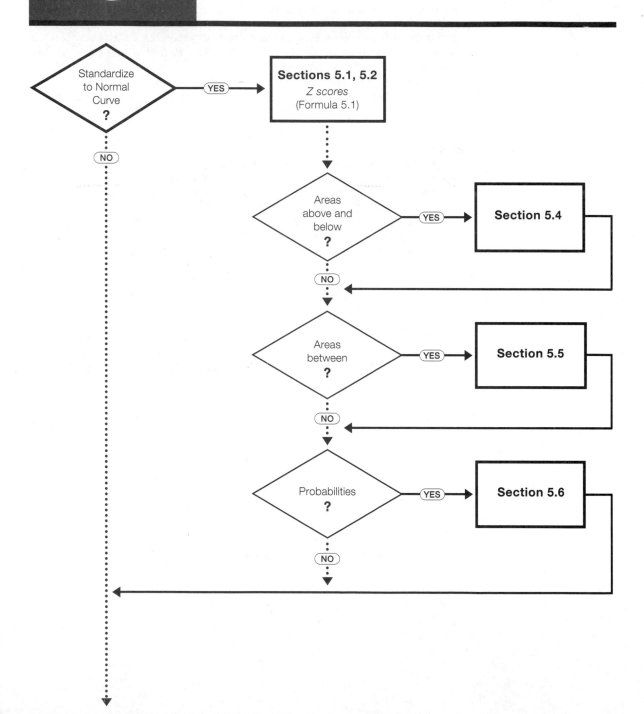

5

5.1 INTRODUCTION

The **normal curve** is a concept of great significance in statistics. As we shall see in Part II, it is central to the theory that underlies inferential statistics. Also, in combination with the standard deviation, the normal curve can be used to construct precise descriptive statements about empirical distributions. This chapter will conclude our treatment of statistics as descriptive devices in Part I and lay important groundwork for Part II.

The normal curve is a theoretical model, a kind of frequency polygon that is perfectly symmetrical and smooth. It is bell shaped and unimodal, with its tails extending infinitely in both directions. Of course, no empirical distribution has a shape that perfectly matches this ideal model, but many are close enough to permit the assumption of normality. In turn, this assumption makes possible one of the most important uses of the normal curve—description of empirical distributions based on our knowledge of the theoretical normal curve.

The crucial point about the normal curve is that distances along the abscissa (horizontal axis) of the distribution, when measured in standard deviations from the mean, always encompass the same proportion of the total area under the curve. In other words, regardless of the precise shape of the normal curve and the nature of the underlying distribution, the distance from any given point to the mean (when measured in standard deviations) will cut off exactly the same proportion of the total area.

To illustrate, Figures 5.1 and 5.2 present two hypothetical distributions of IQ scores, one for a group of males and one for a group of females, both normally distributed (or nearly so) such that

Males	Females
$\bar{X} = 100$	$\bar{X} = 100$
$s = 20$	$s = 10$
$N = 1000$	$N = 1000$

Figures 5.1 and 5.2 are drawn with two scales on the horizontal axis (or abscissa). The first scale is stated in "IQ units" and the second in standard deviations from the mean. These scales should help you to visualize the relationships explained below between distances from the mean as measured in units of the standard deviation and areas under the curve.

On any normal curve, the distance between ±1 standard deviation encompasses exactly 68.26% of the total area under the curve. The standard deviation for males is 20 so in Figure 5.1, 68.26% of the total area lies between the score of 80 (−1 standard deviation) and 120 (+1 standard deviation). The standard deviation for females is 10, so the same percentage of the area lies between the scores of 90 and 110. As long as an empirical distribution is normal (or nearly so), 68.26% of the total area will always be encompassed between ±1 standard deviation—regardless of the trait being measured and the number values of the mean and standard deviation.

FIGURE 5.1   IQ SCORES FOR A SAMPLE OF MALES

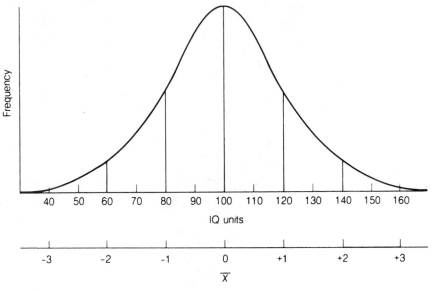

Standard deviations from the mean

FIGURE 5.2   IQ SCORES FOR A SAMPLE OF FEMALES

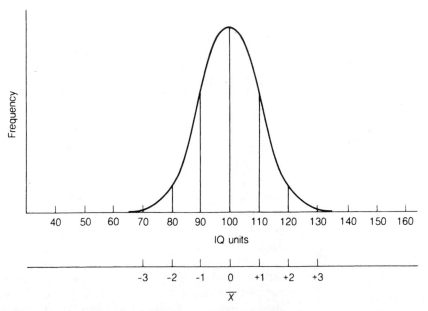

Standard deviations from the mean

FIGURE 5.3   AREAS UNDER THE THEORETICAL NORMAL CURVE

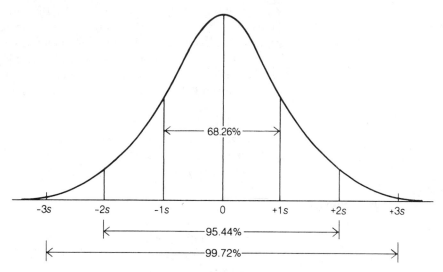

It will be useful to familiarize yourself with the following relationships between distances from the mean and areas under the curve:

between ±1 standard deviation lies 68.26% of the area

between ±2 standard deviations lies 95.44% of the area

between ±3 standard deviations lies 99.72% of the area

These relationships are displayed graphically in Figure 5.3.

The relationship between distance from the mean and area allows us to describe empirical distributions that are at least approximately normal. The position of individual scores can be described with respect to the mean, the distribution as a whole, or any other score in the distribution.

The areas between scores can also be expressed, if desired, in numbers of cases rather than percentage of total area. For example, a normal distribution of 1000 cases will contain about 683 cases (68.26% of 1000 cases) between ±1 standard deviation of the mean, about 954 between ±2 standard deviations, and about 997 between ±3 standard deviations. Thus, for any normal distribution, only a few cases will be farther away from the mean than ±3 standard deviations.

5.2 COMPUTING *Z* SCORES   To find the percentage of the total area (or number of cases) above, below, or between scores in an empirical distribution, the original scores must first be expressed in units of the standard deviation or converted into **Z scores**. The original scores could be in any unit of measurement (feet, IQ, dollars), but Z scores always have the same values for their mean and standard deviation. Think of converting the original scores into Z scores as a process of changing value scales—similar to changing from meters to yards, kilometers to miles, or gallons to liters. These units are different but equally valid ways of expressing distance or volume. The original (or "raw") scores and Z scores are two equally valid but different ways of measuring distances under the normal curve.

The standardized theoretical normal curve has a mean of 0 and a standard deviation of 1. When an empirical normal distribution is standardized, its mean will be converted to 0, its standard deviation to 1, and all values will be expressed in Z-score form. The formula for computing Z scores is

FORMULA 5.1

$$Z = \frac{X_i - \bar{X}}{s}$$

This formula will convert any score (X_i) from an empirical normal distribution into the equivalent Z score. To illustrate with the men's IQ data (Figure 5.1), the Z-score equivalent of a raw score of 120 would be

$$Z = \frac{120 - 100}{20} = +1.00$$

The equivalent Z score of positive 1.00 indicates that this score lies one standard deviation unit above (to the right of) the mean. A negative score would fall below (to the left of) the mean. *(For practice in computing Z scores, see any of the problems at the end of this chapter.)*

5.3 THE NORMAL CURVE TABLE

The theoretical normal curve has been very thoroughly analyzed and described by statisticians. The areas related to any Z score have been precisely determined and organized into a table format. This **normal curve table** or Z-score table is presented as Appendix A in this text. The table consists of three columns, with Z scores in the left-hand column, areas between the Z score and the mean of the normal curve in the middle column, and areas beyond the Z score in the right-hand column. To find the area between any Z score and the mean, go down the column labeled "Z" until you find the Z score. For our example, go down the column until you find the score 1.00. The entry in the column labeled "Area Between Mean and Z" is 0.3413. The table presents all areas in the form of proportions, but we can easily translate these into percentages by multiplying them by 100 (see Chapter 2). We could say either that "a proportion of 0.3413 of the total area under the curve lies between a Z score of 1.00 and the mean," or "34.13% of the total area lies between a score of 1.00 and the mean."

The third column in the table presents "Areas Beyond Z." These are areas above positive scores or below negative scores. This column will be used when we want to find an area above or below certain Z scores, an application that will be explained in Section 5.4.

To conserve space, the table includes only positive Z scores. Since the normal curve is perfectly symmetrical, however, the areas related to any negative score will be exactly the same as those of the corresponding positive score. For example, the area between a Z score of -1.00 and the mean will also be 34.13%, exactly the same as the area we found previously for a score of $+1.00$. As will be repeatedly demonstrated below, however, the sign of the Z score is extremely important and should be carefully noted.

For practice in using Appendix A to describe areas under an empirical normal curve, verify that the Z scores and areas given below are correct for the men's IQ distribution. For each IQ score, the equivalent Z score is computed using Formula 5.1, and then Appendix A is used to find areas between the score and the mean. ($\bar{X} = 100$, $s = 20$ throughout.)

IQ Score	Z Score	Area Between Z and the Mean
110	+0.50	19.15%
125	+1.25	39.44%
133	+1.65	45.05%
138	+1.90	47.13%

The same procedures apply when the Z-score equivalent of an actual score happens to be a minus value (that is, when the raw score lies below the mean).

IQ Score	Z Score	Area Between Z and the Mean
93	−0.35	13.68%
85	−0.75	27.34%
67	−1.65	45.05%
62	−1.90	47.13%

Remember that the areas in Appendix A will be the same for Z scores of the same numerical value regardless of sign. The area between the score of 138 ($+1.90$) and the mean is the same as the area between 62 (-1.90) and the mean. *(For practice in using the normal curve table, see any of the problems at the end of this chapter.)*

5.4 FINDING TOTAL AREA ABOVE AND BELOW A SCORE

To this point, we have seen how the normal curve table can be used to find areas between a Z score and the mean. The information presented in the table can also be used to find other kinds of areas in empirical distributions that are at least approximately normal in shape. For example, suppose you

FIGURE 5.4   FINDING THE AREA BELOW A POSITIVE *Z* SCORE

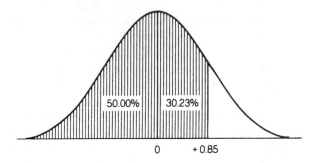

need to determine the total area below the scores of two male subjects in the distribution described in Figure 5.1. The first subject has a score of 117 ($X_1 = 117$), which is equivalent to a *Z* score of +0.85:

$$Z_1 = \frac{X_1 - \bar{X}}{s} = \frac{117 - 100}{20} = \frac{17}{20} = +0.85$$

The plus sign of the *Z* score indicates that the score should be placed above (to the right of) the mean. To find the area below a positive *Z* score, the area between the score and the mean must be added to the area below the mean. Since the normal curve is symmetrical (unskewed), the mean will be equal to the median, and the area below the mean will therefore be 50%. Study Figure 5.4 carefully. We are interested in the lined area.

By consulting the normal curve table, we find that the area between the score and the mean is 30.23% of the total area. The area below a *Z* score of +0.85 is therefore 80.23% (50% + 30.23%). This subject scored higher than 80.23% of the persons tested.

The second subject has an IQ score of 73 ($X_2 = 73$), which is equivalent to a *Z* score of −1.35:

$$Z_2 = \frac{X_2 - \bar{X}}{s} = \frac{73 - 100}{20} = -\frac{27}{20} = -1.35$$

To find the area below a negative score, we use the column labeled "Area Beyond *Z*." The area of interest is depicted in Figure 5.5, and we must determine the size of the lined area. The area beyond a score of −1.35 is given as 0.0885, which we can express as 8.85%. The second subject ($X_2 = 73$) scored higher than 8.85% of the tested group.

In the examples above, we use the techniques for finding the area below a score. Essentially the same techniques are used to find the area above a score. If we need to determine the area above an IQ score of 108, for example, we would first convert to a *Z* score,

$$Z = \frac{108 - 100}{20} = \frac{8}{20} = +0.40$$

FIGURE 5.5   FINDING THE AREA BELOW A NEGATIVE Z SCORE

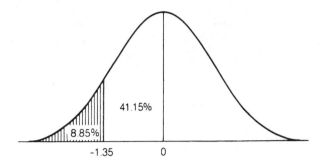

FIGURE 5.6   FINDING THE AREA ABOVE A POSITIVE Z SCORE

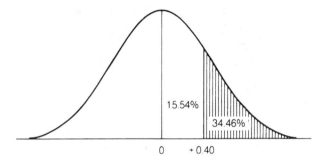

and then proceed to Appendix A. The lined area in Figure 5.6 represents the area in which we are interested. The area above a positive score is found in the "Area Beyond Z" column, and, in this case, the area is 0.3446, or 34.46%.

These techniques may be summarized as follows. To find the total area above a positive Z score or below a negative Z score, go down the "Z" column until you find the score. The area you are seeking will be in the "Area Beyond Z" column. To find the total area below a positive Z score or above a negative score, locate the score and then add the area in the "Area Between Mean and Z" to either .5000 (for proportions) or 50.00 (for percentages). These techniques might be confusing at first, and you will find it helpful to draw the curve and shade in the areas in which you are interested. *(For practice in finding areas above or below Z scores, see problems 5.1 to 5.5 and 5.7 to 5.9a.)*

5.5 FINDING AREAS BETWEEN TWO SCORES

On occasion, you will need to determine the area between two scores rather than the total area above or below one score. In the case where the scores are on opposite sides of the mean, the area between the scores can be found by adding the areas between each score and the mean. Using the men's IQ data as an example, if we wished to know the area between the IQ scores of 93 and 112, we would convert both scores to Z scores, find the area between each score and the mean from Appendix A, and add these two areas together. The first IQ score of 93 converts to a Z score of -0.35:

$$Z_1 = \frac{X_1 - \bar{X}}{s} = \frac{93 - 100}{20} = -\frac{7}{20} = -0.35$$

The second IQ score (112) converts to $+0.60$:

$$Z_2 = \frac{112 - 100}{20} = \frac{12}{20} = +0.60$$

Both scores are placed on Figure 5.7. We are interested in the total lined area. The total area between these two scores is 13.68% + 22.57%, or 36.25%. Therefore, 36.25% of the total area (or about 363 of the 1000 cases) lies between the IQ scores of 93 and 112.

When the scores of interest are on the same side of the mean, a different procedure must be followed to determine the area between them. For example, if we were interested in the area between the scores of 113 and 121, we would begin by converting these scores into Z scores:

$$Z_1 = \frac{X_1 - \bar{X}}{s} = \frac{113 - 100}{20} = \frac{13}{20} = +0.65$$

$$Z_2 = \frac{X_2 - \bar{X}}{s} = \frac{121 - 100}{20} = \frac{21}{20} = +1.05$$

The scores are noted in Figure 5.8; we are interested in the lined area. To find the area between two scores on the same side of the mean, find the area between each score and the mean (given in Appendix A) and then subtract the smaller area from the larger. Between the Z score of $+0.65$ and the mean lies 24.22% of the total area. Between $+1.05$ and the mean lies 35.31% of the total area. Therefore, the area between these two scores is 35.31% − 24.22%, or

FIGURE 5.7   FINDING THE AREA BETWEEN TWO SCORES

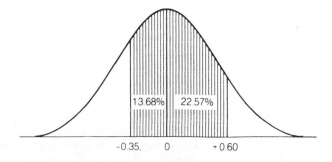

13.68%   22.57%

-0.35.   0   +0.60

APPLICATION 5.1

You have just received your score on a test of intelligence. If your score was 78 and you know that the mean score on the test was 67 with a standard deviation of 5, how does your score compare with the distribution of all test scores?

If you can assume that the test scores are normally distributed, you can compute a Z score and find the area below or above your score. The Z-score equivalent of your raw score would be

$$Z = \frac{X_i - \bar{X}}{s}$$

$$Z = \frac{78 - 67}{5}$$

$$Z = \frac{11}{5}$$

$$Z = 2.20$$

Turning to Appendix A, we find that the "Area Between Mean and Z" is 0.4861, which could also be expressed as 48.61%. Since this is a positive Z score, we need to add this area to 50.00% to find the total area below. Your score is higher than (48.61 + 50.00), or 98.61%, of all the test scores. You did pretty well!

FIGURE 5.8 FINDING THE AREA BETWEEN TWO SCORES

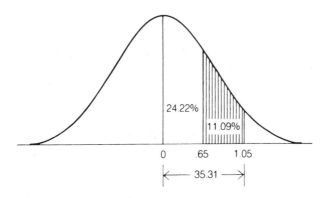

11.09% of the total area. The same technique would be followed if both scores had been below the mean. *(For practice in finding areas between two scores, see problems 5.4, 5.5, 5.7, 5.8, and 5.9b.)*

5.6 USING THE NORMAL CURVE TO ESTIMATE PROBABILITIES

To this point, we have thought of the theoretical normal curve as a way of describing the percentage of total area above, below, and between scores in an empirical distribution. We have also seen that these areas can be converted into the number of cases above, below, and between scores. The theoretical normal curve may also be thought of as a distribution of probabilities.

Specifically, we may use the properties of the theoretical normal curve (Appendix A) to estimate the probability that a case randomly selected from an empirical normal distribution will have a score that falls in a certain range. In terms of techniques, these probabilities will be found in exactly the same way as areas were found. Before we consider these mechanics, however, let us examine what is meant by the concept of probability.

Although we are rarely systematic or rigorous about it, we all attempt to deal with probabilities every day and, indeed, we base our behavior on our estimates of the likelihood that certain events will occur. We constantly ask (and answer) questions such as, What is the probability of precipitation? Of drawing to an inside straight in poker? Of the cheap retreads on my car going flat? Of passing the final exam in Introduction to Nuclear Physics if I don't study?

To estimate the probability of events such as these, we must first be able to define what would constitute a "success." The examples above contain several different definitions of a success (that is, precipitation, drawing a certain card, flat tires, and passing grades). To determine a probability per se, a fraction must be established, with the numerator equaling the number of events that would constitute a success and the denominator equaling the total number of possible events where a success could theoretically occur.

To illustrate, assume that we wish to know the probability of selecting a specific card—say, the king of hearts—in one draw from a well-shuffled deck of cards. Our definition of a success is quite specific (drawing the king of hearts); and with the information given, we can establish a fraction. Only one card satisfies our definition of success, so the number of events that would constitute a success is 1; this value will be the numerator of the fraction. There are 52 possible events (that is, 52 cards in the deck), so the denominator will be 52. The fraction is thus 1/52, which represents the probability of selecting the king of hearts on one draw from a well-shuffled deck of cards. Our probability of success is 1 out of 52.

We can leave the fraction established above as it is, or we can express it in several other ways. For example, we can express it as an odds ratio by inverting the fraction, showing that the odds of selecting the king of hearts on a single draw are 52:1 (or fifty-two to one). We can express the fraction as a proportion by dividing the numerator by the denominator. For our example above, the corresponding proportion is .0192, which is the proportion of all possible events that would satisfy our definition of a success. In the social sciences, probabilities are usually expressed as proportions, and we will follow this convention throughout the remainder of this section.

As conceptualized here, probabilities have an exact meaning: over the long run, the events that we define as successes will bear a certain proportional relationship to the total number of events. The probability of .0192 for selecting the king of hearts in a single draw really means that, over thousands of selections of one card at a time from a full deck of 52 cards, the proportion of successful draws would be .0192. Or, for every 10,000 draws, 192 would be the king of hearts, and the remaining 9808 selections would be other cards. Thus, when we say that the probability of drawing the king of hearts in one draw is .0192, we are essentially applying our knowledge of what would happen over thousands of draws to a single draw.

Like proportions, probabilities range from 0.00 (meaning that the event has absolutely no chance of occurrence) to 1.00 (a certainty). As the value of the probability increases, the likelihood that the defined event will occur also increases. A probability of .0192 is close to zero, and this means that the event (drawing the king of hearts) is unlikely or improbable.

Combining this way of thinking about probability with our knowledge of the theoretical normal curve allows us to estimate the likelihood of selecting a case that has a score within a certain range. For example, suppose we wished to estimate the probability that a randomly chosen subject from the distribution of men's IQ scores would have an IQ score between 95 and the mean score of 100. Our definition of a success here would be the selection of any subject with a score in the specified range. Normally, we would next establish a fraction with the numerator equal to the number of subjects with scores in the defined range and the denominator equal to the total number of subjects. However, if the empirical distribution is normal in form, we can skip this step since the probabilities, in proportion form, are already stated in Appendix A. That is, the areas in Appendix A can be interpreted as probabilities.

To determine the probability that a randomly selected case will have a score between 95 and the mean, we would convert the original score to a Z score:

$$Z = \frac{95 - 100}{20} = -\frac{5}{20} = -0.25$$

The area between this score and the mean (0.0987) is the probability we are seeking. The probability that a randomly selected case will have a score between 95 and 100 is 0.0987 (or, rounded off, 0.1, or one out of 10). In the same fashion, the probability of selecting a subject from any range of scores can be estimated. Note that the techniques by which probabilities are estimated are exactly the same as those by which areas are found. The only new information introduced in this section is the idea that the areas in the normal curve table can also be thought of as probabilities.

Let us close by stressing a very important point about probabilities and the normal curve. The probability is very high that any case randomly selected from a normal distribution will have a score close in value to that of the mean. The shape of the normal curve is such that most cases are clustered around the mean and decline in frequency as we move farther away from the mean value. In fact, given what we know about the normal curve, the probability that a randomly selected case will have a score within ±1 standard deviation of the mean is 0.6826, whereas the probability of the case having a score beyond ±3 standard deviations from the mean is 0.0028 (1.000 − 0.9972). Thus, if we randomly select a number of cases from a normal distribution, we will often select cases that have scores close to the mean but rarely select cases that have scores far above or below the mean. *(For practice in using the normal curve table to find probabilities, see problems 5.9c to 5.14 and problem 5.17.)*

SUMMARY

1. The normal curve, in combination with the mean and standard deviation, can be used to construct precise descriptive statements about empirical distributions that are normally distributed. This chapter also lays some important groundwork for Part II.
2. To work with the theoretical normal curve, raw scores must be transformed into their equivalent Z scores. Z scores allow us to find areas under the theoretical normal curve (Appendix A).

3. We considered three uses of the theoretical normal curve: finding total areas above and below a score, finding areas between two scores, and expressing these areas as probabilities. This last use of the normal curve is especially germane because inferential statistics are centrally concerned with estimating the probabilities of defined events in a fashion very similar to the process introduced in Section 5.6.

SUMMARY OF FORMULAS

Z scores   5.1 $$Z = \frac{X_i - \bar{X}}{s}$$

GLOSSARY

Normal curve. A theoretical distribution of scores that is symmetrical, unimodal, and bell shaped. The standard normal curve always has a mean of 0 and a standard deviation of 1.

Normal curve table. Appendix A; a detailed descrip-

tion of the area between a Z score and the mean of any standardized normal distribution.

Z scores. Standard scores; the way scores are expressed after they have been standardized to the theoretical normal curve.

PROBLEMS

5.1 Scores on a quiz were normally distributed and had a mean of 10 and a standard deviation of 3. For each score below, find the Z score and the percentage of area above and below the score.

X_i	Z Score	% Area Above	% Area Below
5			
6			

X_i	Z Score	% Area Above	% Area Below
7			
8			
9			
11			
12			
14			
15			
16			
18			

5.2 Assume that the distribution of a college entrance exam is normal with a mean of 500 and a standard deviation of 100. For each score below, find the equivalent Z score, the percentage of the area above the score, and the percentage of the area below the score.

X_i	Z Score	% Area Above	% Area Below
650			
400			
375			
586			
437			
526			
621			
498			
517			
398			

5.3 SOC The senior class has just finished a battery of standardized tests designed to assess their learning experience in college. The summary statistics and the scores of 15 members of the class are reported below. Assuming that the three distributions are approximately normal, what percentage of their classmates did each student score higher than on each test?

	Math and Science	Language Skills and Writing Ability	History and Social Science
	$\bar{X} = 58.90$	$\bar{X} = 39.06$	$\bar{X} = 74.77$
	$s = 3.90$	$s = 2.76$	$s = 10.11$

X_i	Z	% Area Below	X_i	Z	% Area Below	X_i	Z	% Area Below
55			44			64		
52			45			60		

X_i	Z	% Area Below	X_i	Z	% Area Below	X_i	Z	% Area Below
49			41			58		
48			43			55		
45			42			50		
44			38			46		
43			35			40		
47			50			75		
58			40			77		
60			37			80		
61			39			81		
65			34			89		
67			33			90		
70			30			94		
71			29			88		

5.4 SOC For the history and social science test scores reported in problem 5.3, what percentage of the students had scores

$\bar{X} = 74.77$

a. between 75 and 85? _____
b. between 80 and 85? _____
c. above 80? _____
d. above 83? _____
e. between 80 and 70? _____
f. between 75 and 70? _____
g. below 75? _____
h. below 77? _____
i. below 80? _____
j. below 85? _____

5.5 For a normal distribution where the mean is 50 and the standard deviation is 10, what percentage of the area is
a. between the scores of 40 and 47? _____
b. above a score of 47? _____
c. below a score of 53? _____
d. between the scores of 35 and 65? _____
e. above a score of 72? _____
f. below a score of 31 and above a score of 69? _____
g. between the scores of 55 and 62? _____
h. between the scores of 32 and 47? _____

5.6 At St. Algebra College, the 200 freshmen enrolled in Introductory Biology took a final exam on which their mean score was 72 and their standard deviation was 6. Below are the grades of 10 students. Convert each into a Z score and determine the *number of people* who scored higher or lower than each of the 10 students. (*HINT*: Multiply the appropriate proportion by N and round the result.)

X_i	Z Score	Number of Students Above	Number of Students Below
60			
57			
55			
67			
70			
72			
78			
82			
90			
95			

5.7 If a distribution of test scores is normal with a mean of 78 and a standard deviation of 11, what percentage of the area lies
a. below 60? _____
b. below 70? _____
c. below 80? _____
d. below 90? _____
e. between 60 and 65? _____
f. between 65 and 79? _____
g. between 70 and 95? _____
h. between 80 and 90? _____
i. above 99? _____
j. above 89? _____
k. above 75? _____
l. above 65? _____

5.8 SOC A scale measuring prejudice has been administered to a large sample of respondents. The distribution of scores is approximately normal with a mean of 31 and a standard deviation of 5. What percentage of the sample had scores
a. below 20? _____
b. below 40? _____
c. between 30 and 40? _____
d. between 35 and 45? _____
e. above 25? _____
f. above 35? _____

5.9 PS A 25-item scale measuring political conservatism has been administered to a sample of 500 respondents. The scores of eight respondents are listed below.

a. For each score, determine the equivalent Z score and the percentage of cases above and below.

$$\bar{X} = 17$$
$$s = 3$$
$$N = 500$$

X_i	Z Score	Percentage of Cases Below	Percentage of Cases Above
19			
10			
14			
15			
18			
20			
22			
23			

b. Find areas between the following scores.

Scores	Z Scores	% Area Between
8 and 12		
9 and 13		
11 and 17		
15 and 19		
16 and 20		
17 and 23		
18 and 19		
19 and 22		

c. If you randomly selected subjects from the sample one at a time, what is the probability that their scores would be in each of the following categories?

Scores	Z Scores	Probability
Less than 17		
Less than 24		
Less than 10		
Less than 8		
Between 8 and 12		
Between 11 and 17		
Between 16 and 18		
Between 20 and 24		

Scores	Z Scores	Probability
More than 24		
More than 20		
More than 15		
More than 9		

5.10 For the past 10 years, your community has averaged 47 house fires per month, with a standard deviation of 5. Assuming the factors that cause house fires are constant, what is the probability that next month the number of house fires will be
  a. 60 or more? _____
  b. 47 or more? _____
  c. between 42 and 52? _____
  d. fewer than 30? _____
  e. fewer than 55? _____
  f. between 45 and 50? _____
  g. 50 or more? _____

5.11 One thousand students have taken a language placement test, and the average score was 35 with a standard deviation of 4. What is the probability that a randomly selected student from this sample will have a score
  a. between 31 and 37? _____
  b. between 41 and 46? _____
  c. less than 40? _____
  d. less than 33? _____
  e. more than 45? _____
  f. more than 29? _____

5.12 [CJ] The average burglary rate for a jurisdiction has been 311 per year with a standard deviation of 50. What is the likelihood that next year the number of burglaries will be
  a. less than 250? _____
  b. less than 300? _____
  c. more than 350? _____
  d. more than 400? _____
  e. between 250 and 350? _____
  f. between 300 and 350? _____
  g. between 350 and 375? _____

5.13 [SOC] For the math and science test scores reported in problem 5.3, the mean score was 58.9 and the standard deviation was 3.9. What is the probability that a student randomly selected from this class will have a score
  a. between 55 and 65? _____
  b. between 60 and 65? _____
  c. above 65? _____
  d. between 60 and 50? _____
  e. between 55 and 50? _____
  f. below 55? _____

5.14 [SOC] On the scale mentioned in problem 5.8, if a score of 40 or more is considered "highly prejudiced," what is the probability that a person selected at random will have a score in that range?

5.15 The local police force in Shinbone, Kansas, gives all applicants an entrance exam and accepts only those applicants who score in the top 15% on this test. If the mean score this year is 87 and the standard deviation is 8, would an individual with a score of 110 be accepted?

5.16 After taking the state merit examinations for the positions of social worker and employment counselor, you receive the following information on the tests and on your performance. On which of the tests did you do better?

Social Worker	Employment Counselor
$\bar{X} = 118$	$\bar{X} = 27$
$s = 17$	$s = 3$
Your score = 127	Your score = 29

5.17 In a distribution of scores with a mean of 35 and a standard deviation of 4, which event is more likely: that a randomly selected score will be between 29 and 31 or that a randomly selected score will be between 40 and 42?

5.18 To be accepted into an honor society, students must have GPAs in the top 10% of the school. If the mean GPA is 2.78 and the standard deviation is .33, which of the following GPAs would qualify?

    3.20   3.21   3.25   3.30   3.35

Using *SPSS for Windows* to Transform Raw Scores into *Z* Scores

DEMONSTRATION 5.1 Computing *Z* Scores

The **Descriptives** program introduced at the end of Chapter 4 can also be used to compute *Z* scores for any variable. These *Z* scores are then available for further operations and may be used in other tasks. SPSS will create a new variable consisting of the transformed scores of the original variable. The program uses the letter *Z* and the first seven letters of the variable name to designate the normalized scores of a variable.

In this demonstration, we will have SPSS compute *Z* scores for *age*. First, load the 1996 GSS data set and then click **Statistics**, **Summarize**, and **Descriptives**. Find *age* in the variable list and click the arrow to move the variable to the **Variable(s):** window. Find the "Save standardized values as variables" option below the variable list and click the box next to it. With this option selected for **Descriptives**, SPSS will compute *Z* scores for all variables listed in the **Variable(s):** window. Click **OK**, and SPSS will produce the usual set of descriptive statistics for *age*. It will also add the new variable (called *zage*) that contains the standardized scores for *age* to the data set. To verify this, run the **Descriptives** command again, and you will find *zage* in the variable list. Transfer *zage* to the **Variable(s):** window with *age*, click **OK**, and the following output will be produced:

Descriptives

Descriptive Statistics

	N	Minimum	Maximum	Mean	Std. Deviation
AGE OF RESPONDENT	1449	18	89	45.51	16.82
Zscore: AGE OF RESPONDENT	1449	-1.63515	2.58560	1.13E-15	1.0000000
Valid N (listwise)	1449				

Note that the mean for *zage* is reported as 1.13E$-$15. This is a form of scientific notation and can be read by moving the decimal point to the left the number of places that follows "E$-$" (.00000000000000113). In other words, like any set of *Z* scores, *zage* effectively has a mean of zero and a standard deviation of one. The new variable *zage* can be treated just like any other variable and used in any SPSS procedure.

If you would like to inspect the scores of *zage*, use the **Case Summaries** procedure. Click **Statistics**, **Summarize**, and then **Case Summaries.** Move both *age* and *zage* to the **Variable(s):** window. Find the "Limit cases to first" option at the bottom of the window. This option can be used to set the number of cases included in the output. For this exercise, let's set a limit of 10 cases. Make sure the box to the left of the option is checked and type 10 in the box to the right. Click **OK**, and the following output will be produced:

Summarize

Case Processing Summary[a]

| | Cases | | | | | |
| | Included | | Excluded | | Total | |
	N	Percent	N	Percent	N	Percent
AGE OF RESPONDENT	10	100.0%	0	0.0%	10	100.0%
Zscore(AGE) AGE OF RESPONDENT	10	100.0%	0	0.0%	10	100.0%

a. Limited to first 10 cases.

Case Summaries[a]

	AGE OF RESPONDENT	Zscore(AGE) AGE OF RESPONDENT
1	79	1.99113
2	32	-.80289
3	50	.26716
4	56	.62385
5	48	.14827
6	29	-.98123
7	40	-.32731
8	46	.02937
9	37	-.50565
10	43	-.14897
Total N	10	10

a. Limited to first 10 cases.

Scan the list of scores and note that the scores that are close in value to the mean of *age* (45.51) are very close to the mean of *zage* (0.00), and the further away the score is from 45.51, the greater the numerical value of the Z score. Also note that, of course, scores below the mean (less than 45.51) have negative signs and scores above the mean (greater than 45.51) have positive signs.

Exercises

5.1 Use SPSS to compute Z scores for *prestg80, papres80, and tvhours*. Use the **Case Summaries** procedure to display the normalized and "raw" scores for each variable for 15 cases.

5.2 Use the **Graphs** command to get simple line charts of *age* and then *zage*. How close are these charts to smooth, bell-shaped, normal curves?

1. To what extent do people apply religion to the problems of everyday living? Fifteen people have responded to a series of questions including the following:

   1. What is your religion?
      1. Protestant
      2. Catholic
      3. Jewish
      4. None
      5. Other (Muslim, Hindu, etc.)
   2. On a scale of 1 to 10 (with 10 being the highest), how strong would you say your faith is?
   3. How many times a day do you pray?
   4. When things aren't going well, my religion is my major source of comfort.
      1. Strongly agree
      2. Slightly agree
      3. Neither agree nor disagree
      4. Slightly disagree
      5. Strongly disagree
   5. How old are you?

Case	Religion	Strength	Pray	Comfort	Age
1	1	2	0	2	30
2	2	8	1	1	67
3	1	8	3	1	45
4	1	6	0	1	43
5	3	9	3	1	32
6	3	3	0	3	18
7	4	0	0	5	52
8	5	9	6	1	37
9	1	5	0	2	54
10	2	8	2	1	55
11	2	3	0	5	33
12	1	6	1	3	45
13	1	8	2	2	37
14	1	7	2	1	50
15	2	9	1	1	25

   For each variable, construct a frequency distribution and calculate appropriate measures of central tendency and dispersion. Write a sentence summarizing each variable.

2. A survey measuring attitudes toward interracial dating was administered to 1000 people. The survey asked the following questions:

   1. What is your age? _____
   2. What is your sex?
      1. Male
      2. Female

3. Marriages between people of different racial groups just don't work out and should be banned by law.
    1. Strongly agree
    2. Agree
    3. Undecided
    4. Disagree
    5. Strongly disagree
4. How many years of schooling have you completed? _____
5. Which category below best describes the place where you grew up?
    1. Large city
    2. Medium-size city
    3. Suburbs of a city
    4. Small town
    5. Rural area
6. What is your marital status?
    1. Married
    2. Separated or divorced
    3. Widowed
    4. Never married

The scores of 30 respondents are reproduced below.

Case	Age	Sex	Attitude on Interracial Dating	Years of School	Area	Marital Status
1	17	1	5	12	1	4
2	25	2	3	12	2	1
3	55	2	3	14	2	1
4	45	1	1	12	3	1
5	38	2	1	10	3	1
6	21	1	1	16	5	1
7	29	2	2	16	2	2
8	30	2	1	12	4	1
9	37	1	1	12	2	1
10	42	2	3	18	5	4
11	57	2	4	12	2	3
12	24	2	2	12	4	1
13	27	1	2	18	3	2
14	44	1	1	15	1	1
15	37	1	1	10	5	4
16	35	1	1	12	4	1
17	41	2	2	15	3	1
18	42	2	1	10	2	4
19	20	2	1	16	1	4
20	21	2	1	16	1	4
21	25	2	1	16	1	4
22	65	1	1	16	5	1
23	70	2	2	12	5	1
24	68	1	3	12	3	1

Case	Age	Sex	Attitude on Interracial Dating	Years of School	Area	Marital Status
25	42	1	4	8	5	3
26	39	2	4	16	2	1
27	26	1	2	12	3	1
28	21	2	3	12	3	2
29	33	1	1	16	4	4
30	45	2	5	16	5	4

a. For each variable, construct a frequency distribution and select and calculate an appropriate measure of central tendency and a measure of dispersion. Summarize each variable in a sentence.

b. For all 1000 respondents, the mean age was 34.70 with a standard deviation of 3.4 years. Assuming the distribution of age is approximately normal, compute Z scores for each of the first 10 respondents above and determine the percentage of the area below (younger than) each respondent.

3. The data set below is taken from the General Social Survey. Abbreviated versions of the questions along with the meanings of the codes are also presented. See Appendix G for the codes and the complete question wordings. For each variable, construct a frequency distribution and select and calculate an appropriate measure of central tendency and a measure of dispersion. Summarize each variable in a sentence.

1. How many children have you ever had? (Values are actual numbers.)
2. Respondent's educational level:
    0. Less than HS
    1. HS
    2. Jr. college
    3. Bachelor's degree
    4. Graduate school
3. Race:
    1. White
    2. Black
    3. Other
4. ". . . methods of birth control should be available to teenagers . . ."
    1. Strongly agree
    2. Agree
    3. Disagree
    4. Strongly disagree
5. Number of hours of TV watched per day. (Values are actual numbers of hours.)
6. What is your religious preference?
    1. Protestant
    2. Catholic
    3. Jewish
    4. None
    5. Other

Case	Number of Children	Years of School	Race	Birth Control	TV	Religion
1	3	1	1	3	3	1
2	2	0	1	4	1	1
3	4	2	1	2	3	1
4	0	3	1	1	2	1
5	5	1	1	3	2	1
6	1	1	1	3	3	1
7	9	0	1	1	6	1
8	6	1	2	3	4	1
9	4	3	1	1	2	4
10	2	1	3	1	1	1
11	2	0	1	2	4	1
12	4	1	2	1	5	2
13	0	1	1	3	2	2
14	2	1	1	4	2	1
15	3	1	2	3	4	1
16	2	0	1	2	2	1
17	2	1	1	2	2	1
18	0	3	1	3	2	1
19	3	0	1	3	5	2
20	2	1	2	1	10	1
21	2	1	1	3	4	1
22	1	0	1	3	5	1
23	0	2	1	1	2	2
24	0	1	1	2	0	4
25	2	4	1	1	1	2
26	1	0	1	1	10	2
27	4	4	1	4	3	2
28	0	2	1	2	3	2
29	2	3	1	4	1	1
30	3	0	1	4	9	1
31	4	1	1	2	3	1
32	0	4	2	3	4	1
33	2	1	2	1	1	1
34	2	1	1	3	2	1
35	1	1	1	4	4	1
36	4	0	1	2	1	1
37	3	1	2	4	0	1
38	0	1	2	2	2	5
39	3	4	1	2	3	2
40	0	1	1	3	0	1
41	4	1	1	1	2	1
42	0	4	1	2	0	4
43	2	1	1	1	6	2
44	3	0	1	2	3	1
45	6	1	1	4	7	2

Introduction to Inferential Statistics
Sampling and the Sampling Distribution

6.1 INTRODUCTION

Researchers in the social sciences constantly confront the problem of populations that are too large to test. The researcher almost never has enough time or money to collect data on every case of concern. Indeed, even for relatively small populations (such as a local community) the logistics of testing every case are staggering to contemplate. To deal with this problem, social scientists select samples, or subsets of cases, from the populations of interest. Our goal in inferential statistics is to generalize to the characteristics of a population (these are often called **parameters**), based on what we can learn from our samples.

Two applications of inferential statistics are covered in this text. In estimation procedures, a "guess" of the population parameter is made, based on what is known about the sample. In hypothesis testing, the validity of a hypothesis about the population is tested against sample outcomes. We will consider estimation techniques in Chapter 7 and the various forms of hypothesis testing in Chapters 8 through 12. In the present chapter, we will look briefly at sampling and then introduce the concept of the sampling distribution.

6.2 TECHNIQUES FOR PROBABILITY SAMPLING

Social scientists have developed a variety of techniques for selecting samples from populations. In this chapter, we will review the basic procedures for selecting probability samples, the only type of sample that fully supports the use of inferential statistical techniques to generalize to populations. These types of samples are often described as "random," and you may be more familiar with this terminology. Because of its greater familiarity, I will often use the phrase "random sample" in the following chapters. The term "probability sample" is preferred, however, because, in everyday language, "random" is often used to mean "by coincidence" or to give a connotation of unpredictability. As you will see, probability samples are selected by techniques that are careful and methodical and leave no room for haphazardness. Interviewing the people you happen to meet in a mall one afternoon may be "random" in some sense, but this technique will not result in a sample that could support inferential statistics.

Before considering probability sampling techniques, let me point out that social scientists often use nonprobability samples and that such samples are very useful for a number of purposes. Nonprobability samples are typically less costly and easier to assemble and are appropriate in many different research situations. The major limitation of nonprobability samples is that they do not permit the use of inferential statistics to generalize to populations.

With probability sampling techniques, our goal is to select cases so that the final sample is **representative** of the population from which it was drawn. A sample is representative if it reproduces the important characteristics of the population. For example, if the population consists of 60% females and 40% males, the sample should contain essentially the same proportions. Representativeness is crucial for inferential statistics. If the sample is not representative, generalizing to the population becomes, at best, extremely hazardous.

The fundamental principle of probability sampling is that a sample is very likely to be representative if it is selected by a principle called **EPSEM**, which stands for the "**E**qual **P**robability of **SE**lection **M**ethod." The EPSEM principle is that every element or case in the population must have an equal probability of being selected for the sample. So, our goal is to select a representative sample, and the technique we use to achieve that goal is to follow EPSEM, or the equal probability of selection method.

I should stress that the EPSEM selection technique and the representativeness of the sample are two different things. In other words, the fact that a sample is properly selected does not guarantee that it will be an exact representation or microcosm of the population. The probability is very high that an EPSEM sample will be representative but, just as a perfectly honest coin will sometimes show 10 heads in a row when flipped, an EPSEM sample will occasionally present an inaccurate picture of the population. One great strength of inferential statistics is that they allow the researcher to estimate the probability of this type of error and interpret results accordingly.

6.3 EPSEM SAMPLING TECHNIQUES

The most basic EPSEM sampling technique produces a **simple random sample**. To implement the EPSEM principle, we need a list of all elements or cases in the population and a system for selecting cases from the list that will guarantee that every case has an equal chance of being selected for the sample. The selection process could be based on a number of different kinds of operations (for example, drawing cards from a well-shuffled deck, flipping coins, throwing dice, drawing numbers from a hat, and so on). Cases are often selected by using tables of random numbers. These tables are lists of numbers that have no pattern to them (that is, they are random), and an example of such a table is Appendix E.

To use the table of random numbers, first assign each case on the population list a unique identification number. Then, select cases for the sample when their identification number corresponds to the number chosen from the table. Since the numbers in the table are in random order and any number is just as likely as any other number, we will meet the EPSEM criteria by following this procedure. Stop selecting cases when you have reached your desired sample size and, if an identification number is selected more than once, ignore the repeats.*

Rigidly following the procedures above will generate random samples as long as you select from a complete list of the population. However, these procedures can be very cumbersome when there is a long list of cases. Consider the situation when your population numbers 10,000 cases. It is perfectly possible that the first case you select will come from the front of the list, the second from the back, the third from the front again, and so on—leading to a great deal of paper shuffling and a fair amount of confusion. To save time and money in such a situation, researchers often use a technique called **systematic sampling**, where only the first case is randomly selected. Thereafter, every kth case is selected. For example, if you are drawing from a list of 10,000 and desire a sample of 200, select the first case randomly and every 10,000/200th, or 50th, case thereafter. If you randomly start with case 013, then your second case will be 063, your third 113, and so on, until you reach the end of the list.

Note that systematic sampling does not strictly conform to the criterion of EPSEM. That is, once the first case has been selected, the other cases no longer have an equal probability of being chosen. In our example above, cases other than the 13th, the 63rd, the 113th, and so on will not be selected for the sample. In general, this increased probability of error is very slight as long as the list from which cases are chosen is itself random, or at least non-cyclical with respect to the traits you wish to measure. For example, if you are concerned with ethnicity and are drawing your sample from an alphabetical list, you might encounter difficulties because there is a tendency for certain ethnic names to begin with the same letter (for example, the Irish prefix *O*). Therefore, when using systematic sampling, pay careful attention to the nature of the population list as well as your sampling technique.

A third type of EPSEM sampling produces the **stratified sample**. This technique is very desirable because it guarantees that the sample will be representative on the selected traits. To apply this technique, you first stratify (or divide) the population list into sublists according to some relevant trait and then sample from the sublists. If you select a number of cases from each sublist proportional to the numbers for that characteristic in the population, the sample will be representative of the population.

For example, suppose that you are drawing a sample of 300 of your classmates and you wish to have proportional representation from every major field on campus. If only 10% of the student body is majoring in zoology, the sampling techniques discussed above could conceivably result in a sample

---

*Ignoring identification numbers when they are repeated is called "sampling without replacement." Technically, this practice compromises the randomness of the selection process. However, if the sample is a small fraction of the total population, we will be unlikely to select the same case twice, and ignoring repeats will not bias our conclusions.

with very few (or even no) zoologists. If, however, you first divide the population into sublists by major, you can use EPSEM to select exactly 30 zoologists from the appropriate sublist. Following the same procedure with other majors will create a sample that is, by definition, representative of the population on this characteristic. Thus, stratified samples are guaranteed to meet the all-important criterion of representativeness (at least for the traits that are used to stratify the samples).

A major limitation of stratified random sampling is that the exact composition of the population is often unknown. If we have no information about the population, we will be unable to establish a scheme for stratification and determine how many cases should be taken from each sublist.

To this point, sampling techniques have been presented as straightforward processes of randomly selecting cases from a list or sublists of the population. However, sampling is rarely as uncomplicated as I have implied, and the major difficulty almost always centers on what might appear, at first glance, to be the easiest part: establishing the list of the population. For many of the populations of interest to the social sciences, there are no complete, up-to-date lists. There is no list of United States citizens, no list of the residents of any given state, and no complete, up-to-date list of residents of your local community. Devices such as telephone books or city directories might appear to contain complete lists of local residents. However, the former will omit unlisted numbers and a disproportionate number of low-income households, and the latter is very likely to be outdated.

Social scientists have devised several ways of dealing with the limitations imposed by the scarcity of lists. Probably the most significant of these is **cluster sampling**, which involves selecting groups of cases (clusters) rather than single cases. The clusters are often based on geography, and the selection of clusters often proceeds in stages. For example, you might draw a cluster sample of your city or town by first numbering all of the voting precincts within the political boundaries. Next, you would use EPSEM to select a sample of precincts. The second stage of selection would involve numbering the blocks within each of the selected precincts and, following EPSEM, selecting a sample of blocks. A third stage might involve the selection of households within each selected block. When these stages are completed, you would have a sample that had a very high probability of being representative of the entire city without ever using a list of residents of the city.

Cluster sampling is less trustworthy than the other techniques summarized above. A cluster sample is a less accurate representation of the population than a simple random sample of comparable size. In part, this decreased accuracy is a result of the multiple selection stages described above. With a simple random sample, the sample is drawn in one selection from the list of the population. In a multistage cluster sample, each stage in the selection process (e.g., first the precincts, then the blocks, and then the households) has a probability of error. That is, each time we sample, we run the risk of selecting an unrepresentative sample. In simple random sampling, we run this risk once; with cluster sampling we will run the risk anew at each stage.

Although we have to treat inferences to populations based on cluster samples with some additional caution, we often have no alternative method of sampling. While it may be extremely difficult (or even impossible) to construct an accurate list of an entire city population, all you need for compiling a cluster sample is a map (or a list of voting precincts, census tracts, and so forth).

By way of summary, let me return to a major point. The purpose of inferential statistics is to acquire knowledge about populations, based on the information derived from samples drawn from that population. Each of the statistics to be presented in the following chapters requires that samples be selected according to EPSEM. While even the most painstaking and sophisticated sampling techniques will not guarantee representativeness, the probability is high that EPSEM samples will be representative of the populations from which they are selected.

6.4 THE SAMPLING DISTRIBUTION

In inferential statistics, researchers face a puzzling dilemma. On one hand, they have a great deal of information about the sample distribution. On the other hand, they know virtually nothing about the population—and let me stress, it is the population that is of interest. The sample distribution is interesting primarily insofar as it allows the researcher to generalize to the population.

Generally, the information necessary to characterize the distribution of a variable adequately would include (1) the shape of the distribution, (2) some measure of central tendency, and (3) some measure of dispersion. Clearly, all three kinds of information can be gathered (or computed) on any variable for the sample. Just as clearly, none of the information can be gathered for the population. Except in rare cases (for example, IQ and height are thought to be approximately normal in distribution), nothing can be known about the exact shape of the distribution of a variable in the population. Its mean and standard deviation are also unknown. Indeed, if any of this information were available for the population, inferential statistics would probably be unnecessary.

In statistics, a device known as the **sampling distribution** bridges this vast ocean of ignorance. Although we might be totally ignorant about the population distribution, the characteristics of the sampling distribution are based on the laws of probability, not on empirical information, and are very well known indeed. In fact, the sampling distribution is the central concept in inferential statistics, and a prolonged examination of its characteristics is certainly in order.

The general strategy of all applications of inferential statistics is to move from the sample to the population via the sampling distribution. Thus, three separate and distinct distributions are involved in every application of inferential statistics:

1. The sample distribution, which is empirical and known. Indeed, it is collected by researchers themselves but is important primarily insofar as it allows the researcher to learn about the population.

2. The population distribution, which, while empirical (that is, it exists in reality), is unknown. Amassing information about or making inferences to the population is the sole purpose of inferential statistics.

3. The sampling distribution, which is nonempirical (theoretical). Because of the laws of probability, a great deal is known about this distribution. Specifically, the shape, central tendency, and dispersion of the distribution can be deduced and, therefore, the distribution can be adequately characterized.

The sampling distribution can be formally defined as a theoretical, probabilistic distribution of all possible sample outcomes (with constant sample size, N) for the statistic that is to be generalized to the population. The utility of the distribution is implied in the definition. Because it encompasses all possible sample outcomes, the sampling distribution enables us to estimate the probability of any particular sample outcome for that statistic. Exactly how this is done and why it is important forms, in fact, the subject matter of the next six chapters.

The sampling distribution is theoretical, which means that it is never obtained in reality by the researcher. However, to understand better the structure and function of the distribution, let's consider how one might be constructed. Suppose that a researcher is concerned with a specific trait—for example, the average age of a particular population. Following EPSEM, the researcher draws a sample ($N = 100$) from the population, asks all respondents their age, computes the mean, and notes the resultant score on a graph. For the sake of illustration, assume that the average age of this sample is 27 ($\bar{X} = 27$). This score is noted on the graph in Figure 6.1.

Now, replace the 100 respondents in the first sample and draw another sample ($N = 100$) from the same population and again compute the average age. Assume that the mean for the second sample is 32 ($\bar{X} = 32$) and note this sample outcome on Figure 6.1. Replace the respondents from the second

FIGURE 6.1 CONSTRUCTING A SAMPLING DISTRIBUTION

sample and draw still another sample ($N = 100$), calculate and note the $\bar{X}$, replace this third sample, and draw a fourth sample, continuing these operations an infinite number of times, calculating and noting the mean of each sample. Now, try to imagine what Figure 6.1 would look like after thousands of individual samples had been collected and the mean had been computed for each sample. What would be the shape, mean, and standard deviation of this distribution?

For one thing, we know that each sample will be at least slightly different from every other sample, since we will probably never sample exactly the same 100 people twice. Hence, each sample mean will be slightly different. We also know that even though the samples are random, they are not always exact representations of the population. Some of the sample means will be much lower than the population mean; others will be much higher. Common sense suggests, however, that across these thousands of individual samples, most of the sample means will cluster around the true population value.

To illustrate further, assume that we come to know that the true mean age of the population is 30. As we have seen above, most of the sample means will also be approximately 30. Thus, the sampling distribution of these sample outcomes ($\bar{X}$'s) should peak at 30. Some of the sample means will "miss the mark," but the frequency of such misses should decline as we get farther away from 30. That is, the distribution should slope to the base as we get farther away from the population value (sample means of 29 or 31 should be common; means of 20 or 40 should be rare). Since the samples are random, the means should miss an equal number of times on either side of the population value, and the distribution itself should therefore be roughly symmetrical. In other words, the sampling distribution of all possible sample means should be approximately normal and will resemble the distribution presented in Figure 6.2.

These common-sense notions about the shape of the sampling distribution and other very important information about central tendency and dispersion are stated in two theorems. The first of these theorems states that

> if repeated random samples of size N are drawn from a normal population with mean μ and standard deviation σ, then the sampling distribution of sample means will be normal with a mean μ and a standard deviation of $\sigma/\sqrt{N}$.

To translate: if we begin with a trait that is normally distributed across a population (IQ, height, or weight, for example) and take an infinite number of equally sized random samples from that population, then the sampling distribution of sample means will be normal. If it is known that the trait is distributed normally in the population, it can be assumed that the sampling distribution will be normal.

The theorem tells us more than the shape of the sampling distribution, however. It also defines its mean and standard deviation. In fact, it says that the mean of the sampling distribution will be exactly the same value as the mean of the population. That is, if we know that the mean IQ of the entire

FIGURE 6.2 A SAMPLING DISTRIBUTION OF SAMPLE MEANS

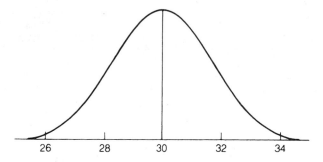

population is 100, then we know that the mean of any sampling distribution of sample mean IQ's will also be 100. Exactly why this should be so is not a matter that can be fully explicated at this level. Recall, however, that most sample means will cluster around the population value over the long run. Thus, the fact that these two values are equal should have intuitive appeal. As for dispersion, the theorem says that the standard deviation of the sampling distribution, also called the **standard error of the mean**, will be equal to the standard deviation of the population divided by the square root of N (symbolically, $\sigma/\sqrt{N}$).

If the mean and standard deviation of a normally distributed population are known, the theorem allows us to compute the mean and standard deviation of the sampling distribution.* Thus, we will know exactly as much about the sampling distribution (shape, central tendency, and dispersion) as we ever knew about any normal curve. But what if the population distribution is not normal? This eventuality (very common, in fact) is covered by a second theorem, called the **Central Limit Theorem**:

> If repeated random samples of size N are drawn from any population, with mean μ and standard deviation σ, then, as N becomes large, the sampling distribution of sample means will approach normality, with mean μ and standard deviation $\sigma/\sqrt{N}$.

The importance of the Central Limit Theorem is that it removes the constraint of normality in the population. Whenever sample size is large, it can be assumed that the sampling distribution is normal, with a mean equal to the population mean and a standard deviation equal to $\sigma/\sqrt{N}$ regardless of the shape of the population. Thus, even if we are working with a characteristic with a skewed distribution (such as income), we can still assume a normal sampling distribution.

---

*In the typical research situation, the values of the population mean and standard deviation are, of course, unknown. However, these values can be estimated from sample statistics, as we shall see in the chapters that follow.

TABLE 6.1 SYMBOLS FOR MEANS AND STANDARD DEVIATIONS OF THREE DISTRIBUTIONS

	Mean	Standard Deviation	Proportion
1. Samples	$\bar{X}$	s	P_s
2. Populations	μ	σ	P_u
3. Sampling distributions			
of means	$\mu_{\bar{x}}$	$\sigma_{\bar{x}}$	
of proportions	μ_p	σ_p	

The issue remaining, of course, is to define what is meant by a large sample. A good rule of thumb is that if sample size (N) is 100 or more, the Central Limit Theorem applies, and you can assume that the sampling distribution is normal in shape. When N is less than 100, you must have good evidence of a normal population distribution before you can assume that the sampling distribution is normal. Thus, a normal sampling distribution can be ensured by the expedient of using fairly large samples.

6.5 SYMBOLS AND TERMINOLOGY

In the following chapters keep in mind that we will be working with three entirely different distributions. The purpose of inferential statistics is to acquire knowledge of the population distribution from the sample distribution by means of the sampling distribution. Furthermore, we will be concerned with several different kinds of sampling distributions—including the sampling distribution of sample means and the sampling distribution of sample proportions.*

To distinguish clearly among these various distributions, we will often use symbols. The symbols used for the means and standard deviations of samples and populations have already been introduced in Chapters 3 and 4. Here, I would like to introduce some of the symbols that will be used for the sampling distribution and to present the relevant symbols in a summary table that you can use for reference. Basically, as you can see from Table 6.1, the sampling distribution is denoted with Greek letter symbols that are subscripted according to the sample statistic of interest.

SUMMARY

1. Since populations are almost always too large to test, a fundamental strategy of social science research is to select a sample from the defined population and then use information from the sample to generalize to the population. This is done either by estimation or by hypothesis testing.

---

*Symbols for the sampling distributions of statistics other than means and proportions will be introduced at the appropriate time.

2. Several techniques are commonly used for selecting random samples. Each of these techniques involves selecting cases for the sample according to EPSEM. Even the most rigorous technique, however, cannot guarantee representativeness. One of the great strengths of inferential statistics is that the probability of this kind of error (nonrepresentativeness) can be estimated.

3. The sampling distribution, the central concept in inferential statistics, is a theoretical distribution of all possible sample outcomes. Since its overall shape, mean, and standard deviation are known (under the conditions specified in the two theorems), the sampling distribution can be adequately characterized and utilized by researchers.

4. The two theorems that were introduced in this chapter state that when the variable of interest is normally distributed in the population or when sample size is large, the sampling distribution will be normal in shape, its mean will be equal to the population mean, and its standard deviation will be equal to the population standard deviation divided by the square root of N.

5. All applications of inferential statistics involve generalizing from the sample to the population by means of the sampling distribution. Both estimation procedures and hypothesis testing incorporate the three distributions, and it is crucial that you develop a clear understanding of each distribution and its role in inferential statistics.

GLOSSARY

Central Limit Theorem. A theorem that specifies the mean, standard deviation, and shape of the sampling distribution, given that the sample is large.

Cluster sample. A method of sampling by which geographical units are randomly selected and all cases within each selected unit are tested.

EPSEM. The Equal Probability of SElection Method for selecting samples. Every element or case in the population must have an equal probability of selection for the sample.

μ. The mean of a population.

$\mu_{\bar{x}}$. The mean of a sampling distribution of sample means.

μ_p. The mean of a sampling distribution of sample proportions.

Parameter. A characteristic of a population.

P_s. (P-sub-s) Any sample proportion.

P_u. (P-sub-u) Any population proportion.

Representative. The quality a sample is said to have if it reproduces the major characteristics of the population from which it was drawn.

Sampling distribution. The distribution of all possible sample outcomes of a given statistic. Under specified conditions, the sampling distribution will be normal in shape with a mean equal to the population value and a standard deviation equal to the population standard deviation divided by the square root of N.

Simple random sample. A method for choosing cases from a population by which every case and every combination of cases has an equal chance of being included.

Standard error of the mean. The standard deviation of a sampling distribution of sample means.

Stratified sample. A method of sampling by which cases are selected from sublists of the population.

Systematic sampling. A method of sampling by which the first case from a list of the population is randomly selected. Thereafter, every kth case is selected.

PROBLEMS

6.1 Imagine that you had to gather a random sample ($N = 300$) of the student body at your school. How would you acquire a list of the population? Would the list be complete and accurate? What procedure would you follow in selecting cases (that is, simple or systematic random sampling)? Would cluster sampling be an appropriate technique (assuming that no list was available)? Describe in detail how you would construct a cluster sample.

6.2 This exercise is extremely tedious and hardly ever works out the way it ought to (mostly because not many people have the patience to draw an "infinite" number of even very small samples). However, if you want a more concrete and tangible understanding of sampling distributions and the two theorems presented in this chapter, then this exercise may have a significant payoff. Below are listed the ages of a population of college students ($N = 50$). By a random method (such as a table of random numbers), draw at least 50 samples of size 2 (that is, 50 pairs of cases), compute a mean for each sample, and plot the means on a frequency polygon. (Incidentally, this exercise will work better if you draw 100 or 200 samples and/or use larger samples than $N = 2$.)

a. The curve you've just produced is a sampling distribution. Observe its shape; after 50 samples, it should be approaching normality. What is your estimate of the population mean (μ) based on the shape of the curve?

b. Calculate the mean of the sampling distribution ($\mu_{\bar{x}}$). Be careful to do this by summing the sample means (not the scores) and dividing by the number of samples you've drawn. Now compute the population mean (μ). These two

means should be very close in value because $\mu = \mu_{\bar{x}}$ by the Central Limit Theorem.

c. Calculate the standard deviation of the sampling distribution (use the means as scores) and the standard deviation of the population. Compare these two values. You should find that $\sigma_{\bar{x}} = \sigma/\sqrt{N}$.

d. If none of the above exercises turned out as they should have, it is for one or more of the following reasons:

1. You didn't take enough samples. You may need as many as 100 or 200 (or more) samples to see the curve begin to look "normal."
2. Sample size (2) is too small. An N of 5 or 10 would work much better.
3. Your sampling method is not truly random and/or the population is not arranged in random fashion.

17	20	20	19	20
18	21	19	20	19
19	22	19	23	20
20	23	18	20	20
22	19	19	20	20
23	17	18	21	20
20	18	20	19	20
22	17	21	21	21
21	20	20	20	22
18	21	20	22	21

Using *SPSS for Windows* to Draw Random Samples

DEMONSTRATION 6.1 Estimating Average Age

SPSS for Windows includes a procedure for drawing random samples from a database. We can use this procedure to illustrate some points about sampling and to convince the skeptics in the crowd that properly selected samples will produce statistics that are close approximations of the corresponding population values or parameters. The instructions below will calculate a sample mean for *age* for three random samples of different sizes. The actual average age of the entire 1996 GSS sample (μ) is 45.51 (see Demonstration 5.1). The samples are roughly 10%, 25%, and 50% of the population size, and the program selects them by a process that is quite similar to a table of random numbers. Therefore, these samples may be considered "simple random samples."

As a part of this procedure we also request the "standard error of the mean" or S.E. MEAN. This is the standard deviation of the sampling distribution ($\sigma/\sqrt{N}$) for a sample of this size. This statistic will be of interest because we can expect our sample means to be within this distance of the population value.

I've included the sample means I found to illustrate some results. When you run this task, you will, of course, draw samples that are at least slightly different from mine, so your results will be at least slightly different.

With the 1996 GSS loaded, click **Data** from the menu bar of the Data Editor and then click **Select Cases**. The **Select Cases** window appears and presents a number of different options. To select random samples, click the radio button next to "Random sample of cases" and then click on the **Sample** button. The **Select Cases: Random Sample** window will open. We can specify the size of the sample in two different ways. If we use the first option, we can specify that the sample will include a certain percentage of cases in the database. The second option allows us to specify the exact number of cases in the sample. Let's use the first option and request a 10% sample by typing 10 into the box on the first line. Click **Continue**, and then click **OK** on the **Select Cases** window. The sample will be selected and can now be processed.

To find the mean age for the 10% sample, click **Statistics**, **Summarize**, and then **Descriptives**. Find *age* in the variable list and transfer it to the **Variable(s)**: window. On the **Descriptives** menu, click the **Options** button and select S.E. MEAn in addition to the usual statistics. Click **Continue** and then **OK**, and the requested statistics will appear in the output window.

Now, to produce a 25% sample, return to the **Select Cases** window by clicking **Data** and **Select Cases**. Click the **Reset** button at the bottom of the window, and the full data set ($N = 1449$) will be restored. Repeat the procedure we followed for selecting the 10% sample. Click the radio button next to "Random sample of cases" and then click on the **Sample** button. The **Select Cases: Random Sample** window will open. Request a 25% sample by typing 25 in the box, click **Continue** and **OK**, and the new sample will be selected.

Run the **Descriptives** procedure for the 25% sample (don't forget S.E.MEAn) and note the results. Finally, repeat these steps for a 50% sample. My results (yours will be at least slightly different) are summarized below:

Sample %	Sample Size	Sample Mean	Standard Error	Sample Mean Plus and Minus Standard Error
10%	151	46.75	1.42	45.33–48.17
25%	328	45.29	.96	44.33–46.25
50%	729	45.40	.62	44.78–46.02

Note that standard error (or the standard deviation of the sampling distribution) decreases as sample size increases. This should reinforce the common-sense notion that larger samples will provide more accurate estimates of population values. All three samples produced estimates (sample means) that are quite close in value to the population value of 45.51. All three sample means are within a standard error of the population mean.

Furthermore, the smallest sample is the least accurate (46.75 is 1.21 years above the true population value of 45.51), the 25% sample is closer (.22 of a year too low), and the largest sample is very close (only .11 year too low). This relationship between accuracy and size of sample should make sense intuitively, but remember that random samples will not always be so cooperative and it is perfectly possible that your samples will behave quite differently.

This demonstration should reinforce one of the main points of this chapter: statistics calculated on samples that have been selected according to the principle of EPSEM will (almost always) be reasonable approximations of their population counterparts.

Estimation Procedures

7

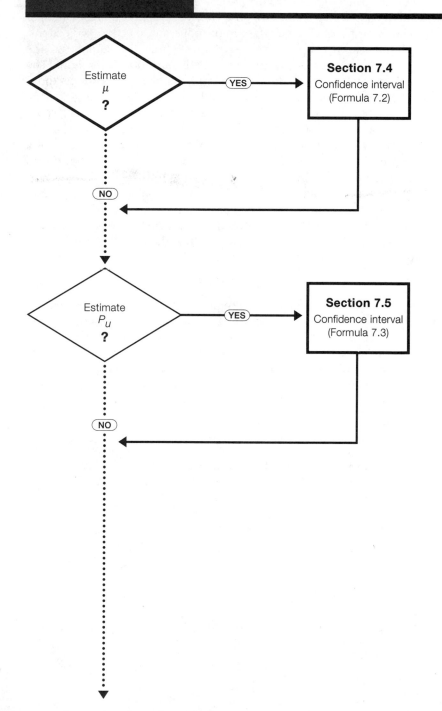

7.1 INTRODUCTION

The object of this branch of inferential statistics is to estimate population values from statistics computed from samples. Although the techniques presented in this chapter may be new to you, you are certainly familiar with their most common applications: public-opinion polls and election projections. Polls and surveys on every conceivable issue—from the sublime to the petty—have become a staple of the mass media and popular culture. The techniques you will learn in this chapter are essentially the same as those used by the most reputable and scientific pollsters.

There are two kinds of estimation procedures. First, point estimates consist of single-sample statistics that are used to guess a population value. A polling organization that projects that 42% of the electorate will vote for a particular presidential candidate is reporting a **point estimate**. The second kind of estimation procedure involves **confidence intervals**, which consist of a range of values (an interval) instead of a single point. Rather than estimating a specific figure as in a point estimate, an interval estimate might be phrased as "from 39% to 45% of the electorate will vote for the candidate." In this latter estimate, we are essentially guessing that the population value falls between 39% and 45%, but we do not specify where.

7.2 BIAS AND EFFICIENCY

In both point and interval estimation, sample statistics serve as the estimators. Which of the many available sample statistics should be used? Estimators can be selected according to two criteria: **bias** and **efficiency**. An estimator is unbiased if and only if the mean of its sampling distribution is equal to the population value of interest. We know from the theorems presented in Chapter 6 that sample means conform to this criterion. The mean of the sampling distribution of sample means (which we will note symbolically as $\mu_{\bar{X}}$) is the same as the population mean (μ).

In addition, sample proportions (P_s) are also unbiased. That is, if we take repeated random samples of size N and calculate and array the resultant proportions, the sampling distribution of sample proportions will have a mean (μ_p) equal to the population proportion (P_u). Thus, if we are concerned with coin flips and sample honest coins 10 at a time ($N = 10$), the sampling distribution will have a mean equal to 0.5, which is the probability that an honest coin will be heads (or tails) when flipped. All other sample statistics are biased (that is, have sampling distributions with means not equal to the population value).*

---

*In particular, the sample standard deviation (s) is a biased estimator of the population standard deviation (σ). As you might expect, there is less dispersion in a sample than in a population and, as a consequence, s will underestimate σ. As we shall see, however, sample standard deviation can be corrected for this bias and still serve as an estimate of the population standard deviation for large samples.

Bias is important for a number of reasons. For one thing, knowing that an estimator is unbiased means that we can determine the probability that our sample statistic lies within a given distance of the population value we are trying to estimate. To illustrate, consider a specific problem. Assume that we wish to estimate the average income of a population. A random sample ($N = 500$) is taken and a sample mean of $20,000 is computed. Note that we have no idea what the population value (μ) is (if we did, we wouldn't need the sample), but it is μ that we are interested in. The sample mean of $20,000 is important and interesting primarily insofar as it can give us information about the population. However, because of the two theorems presented in Chapter 6, we do know that the sampling distribution of all possible sample means is normal and that its mean is equal to the population mean. We also know that all normal curves contain about 68% of the cases (the cases here are sample means) within ± 1 Z, 95% of the cases within ± 2 Z's, and 99% of the cases within ± 3 Z's of the mean. Remember that we are discussing the sampling distribution here—the distribution of all possible sample outcomes or, in this instance, sample means. Thus, the probabilities are very good (approximately 68 out of 100 chances) that our sample mean of $20,000 is within ± 1 Z, excellent (95 out of 100) that it is within ± 2 Z's, and overwhelming (99 out of 100) that it is within ± 3 Z's of the mean of the sampling distribution (which is the same value as the population mean). These relationships are graphically depicted in Figure 7.1.

If an estimator is unbiased, it is probably an accurate estimate of the population parameter (μ in this case). Note that in less than 1% of the cases, a sample mean will be more than ± 3 Z's away from the mean of the sampling distribution (very inaccurate) by random chance alone. We literally have no

FIGURE 7.1 AREAS UNDER THE SAMPLING DISTRIBUTION OF SAMPLE MEANS

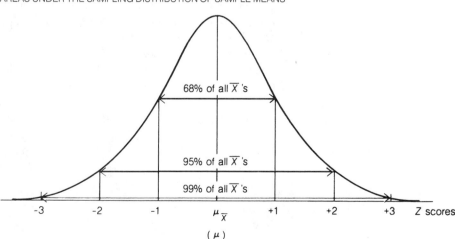

idea if our sample mean of $20,000 is in this small minority. We do know, however, that the odds are high that our sample mean is considerably closer than ±3 Z's to the mean of the sampling distribution and, thus, to the population mean.

The second desirable characteristic of an estimator is efficiency, which is the extent to which the sampling distribution is clustered about its mean. Efficiency or clustering is essentially a matter of dispersion. The smaller the standard deviation of a sampling distribution, the greater the clustering and the higher the efficiency. Remember that the standard deviation of the sampling distribution of sample means, or the standard error of the mean, is equal to the population standard deviation divided by the square root of N ($\sigma_{\bar{X}} = \sigma/\sqrt{N}$). Therefore, the standard deviation of this sampling distribution is an inverse function of N. As sample size increases, $\sigma_{\bar{X}}$ will decrease. We can improve the efficiency (or decrease the standard deviation of the sampling distribution) for any estimator by increasing sample size.

An example should make this clearer. Consider two samples of different sizes:

Sample 1	Sample 2
$\bar{X}_1 = \$20,000$	$\bar{X}_2 = \$20,000$
$N_1 = 100$	$N_2 = 1000$

Both sample means are unbiased, but which is the more efficient estimator? Consider sample 1 and assume, for the sake of illustration, that the population standard deviation (σ) is $500.* In this case, the standard deviation of the sampling distribution of all possible sample means with an N of 100 would be $\sigma/\sqrt{N}$, or $500/\sqrt{100}$, or $50.00. For sample 2, the standard deviation of all possible sample means with an N of 1000 would be much smaller. Specifically, it would be equal to $500/\sqrt{1000}$ or $15.81.

Sampling distribution 2 is much more clustered than sampling distribution 1. In fact, distribution 2 contains 68% of all possible sample means within ±15.81 of the μ while distribution 1 requires a much broader interval of ±50 to do the same. The estimate based on a sample with 1000 cases is much more likely to approximate the population value than is an estimate based on a sample of 100 cases. Figures 7.2 and 7.3 illustrate these relationships graphically.

To summarize, since the standard deviation of all sampling distributions is an inverse function of N, the larger the sample, the greater the clustering and the higher the efficiency. In part, these relationships between sample size

---

*In reality, of course, the value of σ would be unknown.

FIGURE 7.2   A SAMPLING DISTRIBUTION WITH $N = 100$ AND $\bar{X} = \$50.00$

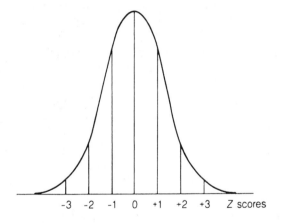

	-150	-100	-50		50	100	150	
	-3	-2	-1	0	+1	+2	+3	Z scores

FIGURE 7.3   A SAMPLING DISTRIBUTION WITH $N = 1000$ AND $\bar{X} = \$15.81$

-3 -2 -1 0 +1 +2 +3   Z scores

and the standard deviation of the sampling distribution do nothing more than underscore our common-sense notion that much more confidence can be placed in large samples than in small (as long as both have been randomly selected).

7.3 ESTIMATION PROCEDURES: INTRODUCTION

The procedure for constructing a point estimate is straightforward. Draw an EPSEM sample, calculate either a proportion or a mean, and estimate that the population value is the same as the sample statistic. Remember that the larger the sample, the greater the efficiency and the more likely that the estimator is approximately the same as the population value. Also remember that, no matter how rigid the sampling procedure or how large the sample, there is always some chance that the estimator is very inaccurate.

Compared to point estimates, interval estimates are more complicated but safer in the sense that, by guessing a range of values, we are more likely to include the population parameter. The first step in constructing an interval estimate is to decide on the risk you are willing to take of being wrong. An interval estimate is wrong if it does not include the population parameter. This probability of error is called **alpha** (symbolized α). The exact value of alpha will depend on the nature of the research situation, but a 0.05 probability is commonly used. Setting alpha equal to 0.05, also called using the 95% **confidence level**, means that over the long run the researcher is willing to be wrong only 5% of the time. Or, to put it another way, if an infinite number of intervals were constructed at this alpha level (and with all other things being equal), 95% of them would contain the population value and 5% would not. Since, in reality, only one interval is constructed, by setting the probability of error very low, we are setting the odds in our favor that the interval will include the population value.

The second step is to picture the sampling distribution and divide the probability of error equally into the upper and lower tails of the distribution and then find the corresponding Z score. For example, if we decided to set alpha equal to 0.05, we would place half (0.025) of this probability in the lower tail and half in the upper tail of the distribution. The sampling distribution would thus be divided as illustrated in Figure 7.4.

We need to find the Z score that marks the beginnings of the lined areas. In Chapter 5, we learned how to calculate Z scores and find areas under the normal curve. Here, we will simply reverse that process. We need to find the Z score beyond which lies a proportion of .0250 of the total area. To do this, go down column (c) of Appendix A until you find this proportion (.0250). The associated Z score is 1.96. Since the curve is symmetrical and we are

FIGURE 7.4   THE SAMPLING DISTRIBUTION WITH ALPHA (α) EQUAL TO 0.05

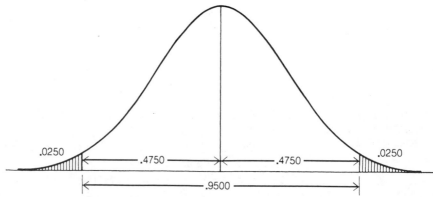

.0250   .4750   .4750   .0250

.9500

interested in both the upper and lower tails, we designate the Z score that corresponds to an alpha of .05 as ±1.96 (see Figure 7.5).

We now know that 95% of all possible sample outcomes fall within ±1.96 Z-score units of the population value. In reality, of course, there is only one sample outcome but, if we construct an interval estimate based on ±1.96 Z's, the probabilities are that 95% of all such intervals will trap the population value. Thus, we can be 95% confident that our interval contains the population value.

Besides the 95% level, there are three other commonly used confidence levels: the 90% level (α = .10), the 99% level (α = 0.01), and the 99.9% level (α = .001). To find the corresponding Z scores for these levels, follow the procedures outlined above for an alpha of 0.05. Table 7.1 summarizes all the information you will need.

You should turn to the Z-score table and confirm for yourself that these scores do indeed correspond to the alpha levels. As you do, note that, in the cases where alpha is set at 0.10 and 0.01, the precise areas we seek do not appear in the table. For example, with an alpha of 0.10, we are looking for the area .0500. Instead we find an area of .0505 (Z = ±1.64) or an area of .0495 (Z = ±1.65). The Z score we are seeking is somewhere between these

FIGURE 7.5 FINDING THE Z SCORE THAT CORRESPONDS TO AN ALPHA (α) OF 0.05

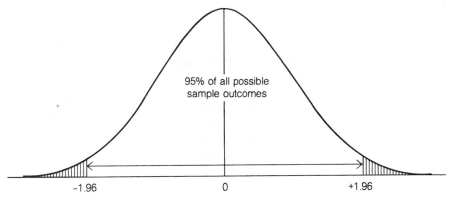

95% of all possible sample outcomes

-1.96   0   +1.96

TABLE 7.1 Z SCORES FOR VARIOUS LEVELS OF ALPHA (α)

Confidence Level	Alpha	$\alpha/2$	Z Score
90%	.10	.0500	±1.65
95%	.05	.0250	±1.96
99%	.01	.0050	±2.58
99.9%	.001	.0005	±3.29

two other scores. When this condition occurs, take the larger of the two scores as *Z*. This will make the interval as wide as possible under the circumstances and is thus the most conservative course of action. In the case of an alpha of 0.01 we encounter the same problem (the exact area .0050 is not in the table), resolve it the same way, and take the larger score as *Z*. Finally, note that in the case where alpha is set at .001, we can choose from several *Z* scores. Although our table is not detailed enough to show it, the closest *Z* score to the exact area we want is ±3.291, which we can round off to ±3.29. *(For practice in finding Z scores, see problem 7.3.)*

The third step is to actually construct the confidence interval. In the sections that follow, we illustrate how to construct an interval estimate first with sample means and then with sample proportions.

7.4 INTERVAL ESTIMATION PROCEDURES FOR SAMPLE MEANS

The formula for constructing a confidence interval based on sample means is given in Formula 7.1:

FORMULA 7.1

$$\text{c.i.} = \bar{X} \pm Z\frac{\sigma}{\sqrt{N}}$$

where c.i. = confidence interval
$\bar{X}$ = the sample mean
Z = the Z score as determined by the alpha level
$\dfrac{\sigma}{\sqrt{N}}$ = the standard deviation of the sampling distribution or the standard error of the mean

As an example, suppose you wanted to estimate the average IQ of a community and had randomly selected a sample of 200 residents, with a sample mean IQ of 105. Assume that the population standard deviation for IQ scores is about 15, so we can set σ equal to 15. If we are willing to run a 5% chance of being wrong and set alpha at 0.05, the corresponding Z score will be ±1.96. These values can be directly substituted into Formula 7.1 and an interval can be constructed:

$$\text{c.i.} = \bar{X} \pm Z\frac{\sigma}{\sqrt{N}}$$

$$\text{c.i.} = 105 \pm 1.96\frac{15}{\sqrt{200}}$$

$$\text{c.i.} = 105 \pm 1.96\frac{15}{14.14}$$

$$\text{c.i.} = 105 \pm (1.96)(1.06)$$

$$\text{c.i.} = 105 \pm 2.08$$

That is, our estimate is that the average IQ for the population in question is somewhere between 102.92 (105 − 2.08) and 107.08 (105 + 2.08). Since 95% of all possible sample means are within ±1.96 Z's (or 2.08 IQ units in this case) of the mean of the sampling distribution, the odds are very high that our interval will contain the population mean. In fact, even if the sample mean is as far off as ±1.96 Z's (which is unlikely), our interval will still contain $\mu_{\bar{X}}$ and, thus, μ. Only if our sample mean is one of the few that is more than ±1.96 Z's from the mean of the sampling distribution will we have failed to include the population mean.

Note that in the example above, the value of the population standard deviation was supplied. Needless to say, it is unusual to have such information about a population. In the great majority of cases, we will have no knowledge of σ. In such cases, however, we can estimate σ with s, the sample standard deviation. Unfortunately, s is a biased estimator of σ and the formula must be changed slightly to correct for the bias. For larger samples, the bias of s will not affect the interval very much. The revised formula for cases in which σ is unknown is

 FORMULA 7.2

$$\text{c.i.} = \bar{X} \pm Z\frac{s}{\sqrt{N-1}}$$

In comparing this formula with 7.1, note that there are two changes. First, σ is replaced by s and, second, the denominator of the last term is the square root of $N − 1$ rather than the square root of N. The latter change is the correction for the fact that s is biased.

Let me stress here that the substitution of s for σ is permitted only for large samples (that is, samples with 100 or more cases). For smaller samples, when the value of the population standard deviation is unknown, the standardized normal distribution summarized in Appendix A cannot be used in the estimation process. It is perfectly possible to construct meaningful interval estimates for samples smaller than 100 but, to do so, we must use a different theoretical distribution, called the Student's t distribution, to find areas under the sampling distribution. We will defer the presentation of the t distribution until Chapter 8 and confine our attention here to estimation procedures for large samples only.

Let us close this section by working through a sample problem with Formula 7.2. Average income for a random sample of a particular community is $20,000, with a standard deviation of $200. What is the 95% interval estimate of the population mean, μ?

Given that

$$\bar{X} = \$20,000$$
$$s = \$200$$
$$N = 500$$

and using an alpha of 0.05, the interval can be constructed.

$$\text{c.i.} = \bar{X} \pm Z\frac{s}{\sqrt{N-1}}$$

$$\text{c.i.} = 20,000 \pm 1.96\frac{200}{\sqrt{499}}$$

$$\text{c.i.} = 20,000 \pm 17.55$$

The average income for the community as a whole is between \$19,982.45 (20,000 − 17.55) and \$20,017.55 (20,000 + 17.55). Remember that this interval has only a 5% chance of being wrong (that is, of not containing the population mean). *(For practice with confidence intervals for sample means, see problems 7.1, 7.4– 7.7, 7.20, and 7.21.)*

7.5 INTERVAL ESTIMATION PROCEDURES FOR SAMPLE PROPORTIONS (LARGE SAMPLES)

Estimation procedures for sample proportions are essentially the same as those for sample means. The major difference is that, since proportions are different statistics, we must use a different sampling distribution. In fact, again based on the Central Limit Theorem, we know that sample proportions have sampling distributions that are normal in shape with means (μ_p) equal to the population value (P_u) and standard deviations (σ_p) equal to $\sqrt{P_u(1-P_u)/N}$. The formula for constructing confidence intervals based on sample proportions is

FORMULA 7.3

$$\text{c.i.} = P_s \pm Z\sqrt{\frac{P_u(1-P_u)}{N}}$$

The values for P_s and N come directly from the sample. As was the case when discussing estimation with sample means, the confidence level determines the value of Z. This leaves one unknown in the formula, P_u—the value we are trying to estimate. Practically, this dilemma can be resolved by setting the value of P_u at 0.5. Since the second term in the numerator under the radical $(1 - P_u)$ is the reciprocal of P_u, the entire expression will always have a value of 0.5 × 0.5, or 0.25, which is the maximum value this expression can attain. That is, if we set P_u at any value other than 0.5, the expression $P_u(1 - P_u)$ will decrease in value. If we set P_u at 0.4, for example, the second term $(1 - P_u)$ would be 0.6, and the value of the entire expression would decrease to 0.24. Setting P_u at 0.5 ensures that the expression $P_u(1 - P_u)$ will be at its maximum possible value and, consequently, the interval will be at maximum width. This is the most conservative solution possible to the dilemma posed by having to assign a value to P_u in the estimation equation.

To illustrate these procedures, assume that you wish to estimate the proportion of students at your university who missed at least one day of classes because of illness last semester. Out of a random sample of 200 students, 60 reported that they had been sick enough to miss classes at least once during the previous semester. The sample proportion upon which we

Application 7.1

A study of the leisure activities of Americans was conducted on a sample of 1000 households. The respondents identified television viewing as a major form of leisure activity. If the sample reported an average of 6.2 hours of television viewing a day, what is the estimate of the population mean? The information from the sample is

$$\bar{X} = 6.2$$
$$s = 0.7$$
$$N = 1000$$

If we set alpha at 0.05, the corresponding Z score will be ±1.96, and the 95% confidence interval will be

$$\text{c.i.} = \bar{X} \pm Z \frac{s}{\sqrt{N-1}}$$

$$\text{c.i.} = 6.2 \pm 1.96 \frac{0.7}{\sqrt{1000-1}}$$

$$\text{c.i.} = 6.2 \pm 1.96 \frac{0.7}{\sqrt{31.61}}$$

$$\text{c.i.} = 6.2 \pm (1.96)(.022)$$

$$\text{c.i.} = 6.2 \pm .04$$

Based on this result, we would estimate that the population spends an average of 6.2 ± .04 hours per day viewing television. The lower limit of our interval estimate (6.2 − .04) is 6.16, and the upper limit (6.2 + .04) is 6.24. Thus, another way to state the interval would be

$$6.16 \leq \mu \leq 6.24$$

The population mean is greater than or equal to 6.16 and less than or equal to 6.24. This estimate has a 5% chance of being wrong (that is, of not containing the population mean).

will base our estimate is thus 60/200, or 0.30. At the 95% level, the interval estimate will be

$$\text{c.i.} = P_s \pm Z \sqrt{\frac{P_u(1-P_u)}{N}}$$

$$\text{c.i.} = 0.30 \pm 1.96 \sqrt{\frac{(0.5)(0.5)}{200}}$$

$$\text{c.i.} = 0.30 \pm 1.96 \sqrt{\frac{.25}{200}}$$

$$\text{c.i.} = 0.30 \pm (1.96)(0.035)$$

$$\text{c.i.} = 0.30 \pm 0.07$$

Based on this sample proportion of 0.30, you would estimate that the proportion of students who missed at least one day of classes because of illness was between 0.23 and 0.37. The estimate could, of course, also be phrased in percentages by reporting that between 23% and 37% of the student body was affected by illness at least once during the past semester. *(For practice with confidence intervals for sample proportions, see problems 7.2, 7.8–7.12, 7.20, and 7.21.)*

You are most likely to encounter estimates to the population in the mass media in the form of public-opinion polls, election projections, and the like. Professional polling firms typically use interval estimates. Responsible reporting by the media will usually emphasize the estimate itself (for example, "In a recently conducted survey of the American public, 45% of the respondents said that they approve of the president's performance") but also will report the width of the interval ("This estimate is accurate to within ±3%," or "Figures from this poll are subject to a sampling error of ±3%"), the alpha level (usually as the confidence level of 95%), and the size of the sample ("1458 households were surveyed").

Election projections and voter analyses, at least at the presidential level, have been common for several decades. More recently, public opinion polling has been increasingly used to gauge reactions to everything from the newest movies to the hottest gossip to the president's conduct during the latest crisis. Newsmagazines routinely report poll results as an adjunct to news stories, and similar stories are regular features of TV news and newspapers. I would include an example or two of these applications here, but polls have become so pervasive that you can do your own example. Just pick up a newsmagazine or newspaper, leaf through it casually, and I bet that you'll find at least one poll. Read the story and see if you can identify the population, the confidence interval width, the sample size, and the confidence level. Bring the news item to class and dazzle your instructor.

As a citizen, as well as a social scientist, you should be extremely suspicious of estimates that do not include such vital information as sample size or interval width. You should also check to see how the sample was assembled. If the sample was not selected in some random fashion from the population as a whole (for example, a magazine polling only its subscribers), it cannot be regarded as representative of the American public.

Advertisements, commercials, and reports published by partisan groups sometimes report statistics that seem to be estimates to the population. Often, however, such estimates are based on woefully inadequate sampling sizes and biased sampling procedures, and the data are collected under circumstances that evoke a desired response. "Person in the street" (or shopper in the grocery store) interviews have a certain folksy appeal but must not be accorded the same credibility as surveys conducted by reputable polling firms.

Public Opinion Surveys in the Professional Literature

For the social sciences, probably the single most important consequence of the growth in opinion polling is that many nationally representative databases are now available for research purposes. These high-quality databases are often available for a nominal fee, and they make it possible to conduct "state-of-the-art" research without the expense and difficulty of collecting data yourself. This is an important development because we can now easily test our theories against very high-quality data, and our conclusions will thus have a stronger empirical basis. Our research efforts will have greater credibility with our colleagues, with policy makers, and with the public at large.

One of the more important and widely used databases of this sort is called the General Social Survey, or the GSS. In virtually every year since 1972, the National Opinion Research Council has questioned a nationally representative sample of Americans about a wide variety of issues and concerns. Since many of the questions are asked every year, the GSS offers a longitudinal record of American sentiment and opinion about a large variety of topics. Each year, new topics of current concern are added and explored, and the variety of information available thus continues to expand. Like other nationally representative samples, the GSS sample is chosen by a complex probability design that resembles cluster sampling (see Chapter 6). Sample size varies from 1400 to over 3000, and estimates based on samples this large will be accurate to within ±3% (see Table 7.4 and Section 7.7). The computer exercises in this text are based on the 1993 GSS, and this database is described more fully in Appendix F.

7.6 CONTROLLING THE WIDTH OF INTERVAL ESTIMATES

The width of a confidence interval for either sample means or sample proportions can be partly controlled by manipulating two terms in the equation. First, the confidence level can be raised or lowered and, second, the interval can be widened or narrowed by gathering samples of different size. The researcher alone determines the risk he or she is willing to take of being wrong (that is, of not including the population value in the interval estimate). The exact confidence level (or alpha level) will depend, in part, on the purpose of the research. For example, if potentially harmful drugs were being tested, the researcher would naturally demand very high levels of confidence (99.9% or 99.99%). On the other hand, if intervals are being constructed only for loose "guesstimates," then much lower confidence levels can be tolerated (such as 90%). The relationship between interval size and confidence level is that intervals widen as confidence levels increase. This should make intuitive sense. Wider intervals are more likely to trap the population value; hence, more confidence can be placed in them.

To illustrate this relationship, let us return to the example where we estimated the average income for a community. In this problem, we were working with a sample of 500 residents, and the average income for this sample was $20,000, with a standard deviation of $200. We constructed the 95% confidence interval and found that it extended 17.55 around the sample mean (that is, the interval was $20,000 ± 17.55).

If we had constructed the 90% confidence interval for these sample data (a lower confidence level), the Z score in the formula would have decreased to ±1.65, and the interval would have been narrower:

$$\text{c.i.} = 20,000 \pm 1.65 \frac{200}{\sqrt{499}}$$

$$\text{c.i.} = 20,000 \pm (1.65)(8.95)$$

$$\text{c.i.} = 20,000 \pm 14.77$$

On the other hand, if we had constructed the 99% confidence interval, the Z score would have increased to ±2.58, and the interval would have been wider:

$$\text{c.i.} = 20,000 \pm 2.58 \frac{200}{\sqrt{499}}$$

$$\text{c.i.} = 20,000 \pm (2.58)(8.95)$$

$$\text{c.i.} = 20,000 \pm 23.09$$

At the 99.9% confidence level, the Z score would be ±3.29, and the interval would be even wider:

$$\text{c.i.} = 20,000 \pm 3.29 \frac{200}{\sqrt{499}}$$

$$\text{c.i.} = 20,000 \pm (3.29)(8.95)$$

$$\text{c.i.} = 20,000 \pm 29.45$$

These four intervals are grouped together in Table 7.2, and the increase in interval size can be readily observed. Although sample means have been used to illustrate the relationship between interval width and confidence level, exactly the same relationships apply to sample proportions. *(To further explore the relationship between alpha and interval width, see problem 7.13.)*

Sample size bears the opposite relationship to interval width. As sample size increases, interval width decreases. Larger samples give more precise (narrower) estimates. Again, an example should make this clearer. In Table 7.3, confidence intervals for four samples of various sizes are constructed and then grouped together for purposes of comparison. The sample data are the same as in Table 7.2, and the confidence level is 95% throughout. The relationships illustrated in Table 7.3 also hold true, of course, for sample proportions. *(To further explore the relationship between sample size and interval width, see problem 7.14.)*

Notice that the decrease in interval width (or, increase in precision) does not bear a constant or linear relationship with sample size. With sample 2 as

TABLE 7.2   INTERVAL ESTIMATES FOR FOUR CONFIDENCE LEVELS
($\bar{X}$ = $20,000, s = $200, N = 500 throughout)

Alpha	Confidence Level	Interval	Interval Width
.10	90%	$20,000 ± 14.77	$29.54
.05	95%	$20,000 ± 17.55	$35.10
.01	99%	$20,000 ± 23.09	$46.18
.001	99.9%	$20,000 ± 29.45	$58.90

TABLE 7.3   INTERVAL ESTIMATES FOR FOUR DIFFERENT SAMPLES
($\bar{X}$ = $20,000, s = $200, alpha = 0.05 throughout)

Sample 1 (N = 100)	Sample 2 (N = 500)
c.i. = $20,000 ± 1.96 $(200/\sqrt{99})$ c.i. = $20,000 ± 39.40	c.i. = $20,000 ± 1.96 $(200/\sqrt{499})$ c.i. = $20,000 ± 17.55

Sample 3 (N = 1000)	Sample 4 (N = 10,000)
c.i. = $20,000 ± 1.96 $(200/\sqrt{999})$ c.i. = $20,000 ± 12.40	c.i. = $20,000 ± 1.96 $(200/\sqrt{9999})$ c.i. = $20,000 ± 3.92

Sample	N	Interval Width
1	100	$78.80
2	500	$35.10
3	1,000	$24.80
4	10,000	$ 7.84

Application 7.2

If 45% of a random sample of 1000 Americans reports that walking is their major physical activity, what is the estimate of the population value? The sample information is

$$P_s = 0.45$$
$$N = 1000$$

Note that the percentage of "walkers" has been stated as a proportion. If we set alpha at 0.05, the corresponding Z score will be ± 1.96, and the interval estimate of the population proportion will be

$$\text{c.i.} = P_s \pm Z \sqrt{\frac{P_u(1 - P_u)}{N}}$$

$$\text{c.i.} = 0.45 \pm 1.96 \sqrt{\frac{(0.5)(0.5)}{1000}}$$

$$\text{c.i.} = 0.45 \pm 1.96 \sqrt{0.00025}$$

$$\text{c.i.} = 0.45 \pm (1.96)(0.016)$$

$$\text{c.i.} = 0.45 \pm 0.03$$

We can now estimate that the proportion of the population for which walking is the major form of physical exercise is between 0.42 and 0.48. That is, the lower limit of the interval estimate is (0.45 − 0.03) or 0.42, and the upper limit is (0.45 + 0.03) or 0.48. We may also express this result in percentages and say that between 42% and 48% of the population walk as their major form of physical exercise. This interval has a 5% chance of not containing the population value.

---

compared to sample 1, the sample size was increased by a factor of 5, but the interval is not five times as narrow. This is an important relationship because it means that N might have to be increased many times to improve the accuracy of an estimate. Since the cost of a research project is a direct function of sample size, this relationship implies a point of diminishing returns in estimation procedures. A sample of 10,000 will cost about twice as much as a sample of 5000, but estimates based on the larger sample will not be twice as precise.

7.7 ESTIMATING SAMPLE SIZE*

The relationships displayed in Table 7.3 suggest a useful way to estimate sample size for any desired level of precision. These techniques would be most useful in the planning stage of a research project because they could yield information about approximate costs.

As an example, suppose a national polling organization wants to be able to predict the outcome of presidential races to within $\pm 3\%$. How large a sample will be needed to ensure this level of precision? In setting the precision, the researcher is essentially predetermining the width of the interval. In

---

*This section is optional.

READING STATISTICS 5: Using Representative Samples to Track National Trends

In Reading Statistics 4, I mentioned nationally representative databases such as the General Social Survey (GSS) and the Gallup poll. Among many other uses, this information can be used to track national trends and shifts in public opinion over time (at least since 1972, the first year the GSS was administered). In a recent article, Professor Mark Warr used these polls to assess the idea that Americans are "fed up with crime" and "paralyzed by fear" about their own vulnerability to victimization. Although such characterizations are frequently found in the mass media, do they accurately describe the level of fear in society?

The Gallup poll and the GSS have, since the mid-1960s, repeatedly included an item that measures fear of victimization: "Is there an area right around here—that is, within a mile—where you would be afraid to walk at night?" Professor Warr constructed a line chart (see Chapter 2) that allowed him to observe changes in the level of fear along with changes in the actual rate of violent crime:

Remember that, although the data on trends in "fear" were collected from random samples of about 1500 respondents, they can be used to infer the characteristics of the population to within about three percentage points. Here's what Professor Warr has to say about the chart:

The most striking feature of [the responses to the survey items] is the relative constancy of fear over the past 2 decades. After rising moderately during the 1960s, fear has remained relatively constant since the 1970s, varying no more than about 10 percent. This stability stands in sharp contrast to recurring media accounts of "skyrocketing" or "epidemic" fear in the United States.*

Professor Warr goes on to review other pieces of evidence and concludes that American public opinion about crime—although harsh and punitive in some regards—is "often more informed and accurate than one might surmise from journalistic accounts."

*Mark Warr. 1995. "Public Opinion on Crime and Punishment." *Public Opinion Quarterly*. 59:296–310. Copyright 1995 by the University of Chicago Press. Reprinted by permisssion of the author and publisher.

FIGURE 1 PERCENTAGE OF RESPONDENTS AFRAID TO WALK ALONE AT NIGHT, 1965–93, AND NATIONAL CRIME SURVEY VIOLENT CRIME RATE, 1973–92

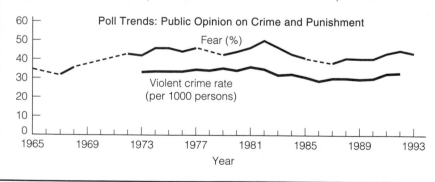

this case, the interval must be no wider than ±3%. Starting with the formula for estimating confidence intervals for sample proportions,

$$\text{c.i.} = P_s \pm Z\sqrt{\frac{P_u(1 - P_u)}{N}}$$

we are predetermining the value of the second and third terms in the equation. Therefore, a new equation can be written:

$$0.03 = Z\sqrt{\frac{P_u(1 - P_u)}{N}}$$

This equation can be algebraically manipulated to isolate the unknown of interest (N). The formula for estimating N for sample proportions when the desired accuracy is ±3% would be

FORMULA 7.4
$$N = \frac{(Z^2)(P_u)(1 - P_u)}{(0.03)^2}$$

Setting alpha at 0.05 ($Z = \pm1.96$) and P_u equal to 0.5, we can solve for N:

$$N = \frac{(1.96)^2(0.5)(0.5)}{0.0009}$$

$$N = 1067.11$$

Thus, to get precision (or interval width) of ±3%, a sample size of about 1000 is needed. Table 7.4 summarizes the relationship between precision and sample size for sample proportions. *(For practice in estimating sample size, see problem 7.19.)*

Note that sample size and precision do not increase at the same rate. In fact, sample size increases much faster. To double the precision of the estimate, the sample size must be increased by a factor of 4. To double the precision from 10% to 5%, for example, the sample size must be quadrupled. This relationship is significant for planning purposes because sample size is a rough index of cost, and what is suggested here is a point of diminishing

TABLE 7.4 PRECISION AND SAMPLE SIZE
(alpha = 0.05, P_u = 0.5 throughout)

Precision (Interval Width)	Approximate Sample Size
±10%	100
±7%	200
±5%	400
±3%	1000
±2%	2400
±1%	9600

returns. Samples that are impressively large might be a waste of resources, with no real gain in precision over a much smaller (less costly) sample. Of course, the same relationships apply to sample means.

SUMMARY

1. Population values can be estimated with sample values. With point estimates, a single sample statistic is used to estimate the corresponding population value. With confidence intervals, we estimate that the population value falls within a certain range of values.
2. Estimates based on sample statistics must be unbiased and relatively efficient. Of all the sample statistics, only means and proportions are unbiased. The means of the sampling distributions of these statistics are equal to the respective population values. Efficiency is largely a matter of sample size. The greater the sample size, the lower the value of the standard deviation of the sampling distribution, the more tightly clustered the sample outcomes will be around the mean of the sampling distribution, and the more efficient the estimate.

3. With point estimates, we estimate that the population value is the same as the sample statistic (either a mean or proportion). With interval estimates, we construct a confidence interval, a range of values into which we estimate that the population value falls. The width of the interval is a function of the risk we are willing to take of being wrong (the alpha level) and the sample size. The interval widens as our probability of being wrong decreases and as sample size decreases.
4. These techniques also provide us with a way of estimating sample size for any desired level of precision. This application of estimation techniques is especially useful in the planning stages of a research project.

SUMMARY OF FORMULAS

Confidence interval for a sample mean, population standard deviation known:

$$7.1 \quad \text{c.i.} = \bar{X} \pm Z \frac{\sigma}{\sqrt{N}}$$

Confidence interval for a sample mean, population standard deviation unknown:

$$7.2 \quad \text{c.i.} = \bar{X} \pm Z \frac{s}{\sqrt{N-1}}$$

Confidence interval for a sample proportion, large samples:

$$7.3 \quad \text{c.i.} = P_s \pm Z \sqrt{\frac{P_u(1 - P_u)}{N}}$$

Estimating N for an accuracy of $\pm 3\%$:

$$7.4 \quad N = \frac{(Z^2)(P_u)(1 - P_u)}{(0.03)^2}$$

GLOSSARY

Alpha (α). The probability of error or the probability that a confidence interval does not contain the population value. Alpha levels are usually set at 0.10, 0.05, 0.01, or 0.001.

Bias. A criterion used to select sample statistics as estimators. A statistic is unbiased if the mean of

its sampling distribution is equal to the population value of interest.

Confidence interval. An estimate of a population value in which a range of values is specified.

Confidence level. A frequently used alternate way of expressing alpha, the probability that an interval estimate will not contain the population value. Confidence levels of 90%, 95%, 99%, and 99.9% correspond to alphas of 0.10, 0.05, 0.01, and 0.001, respectively.

Efficiency. The extent to which the sample outcomes are clustered around the mean of the sampling distribution.

Point estimate. An estimate of a population value where a single value is specified.

PROBLEMS

Problem 7.19 is optional (*).

7.1 For each set of sample outcomes below, construct the 95% confidence interval for estimating μ, the population mean.

a. $\bar{X} = 5.2$ **b.** $\bar{X} = 100$ **c.** $\bar{X} = 20$
$s = .7$ $s = 9$ $s = 3$
$N = 157$ $N = 620$ $N = 220$
d. $\bar{X} = 1020$ **e.** $\bar{X} = 7.3$ **f.** $\bar{X} = 33$
$s = 50$ $s = 1.2$ $s = 6$
$N = 329$ $N = 105$ $N = 220$

7.2 For each set of sample outcomes below, construct the 99% confidence interval for estimating P_u.

a. $P_s = .14$ **b.** $P_s = .37$ **c.** $P_s = .79$
$N = 100$ $N = 522$ $N = 121$
d. $P_s = .43$ **e.** $P_s = .40$ **f.** $P_s = .63$
$N = 1049$ $N = 548$ $N = 300$

7.3 For each confidence level below, determine the corresponding Z score.

Confidence Level	Alpha	Area Beyond Z	Z score
95%	.05	.0250	±1.96
94%			
92%			
97%			
98%			
99.9%			

7.4 SW You have developed a series of questions that measures "burnout" in social workers. A random sample of 100 social workers working in greater metropolitan Shinbone, Kansas, has an average score of 10.3, with a standard deviation of 2.7. What is your estimate of the average burnout score for the population as a whole? Use the 95% confidence level.

7.5 SOC A researcher has gathered information from a random sample of 178 households. For each variable below, construct confidence intervals to estimate the population mean. Use the 90% level.

a. An average of 2.3 people reside in each household. Standard deviation is .35.

b. There was an average of 2.1 television sets ($s = .10$) and .78 telephones ($s = .55$) per household.

c. The households averaged 6.0 hours of television viewing per day ($s = 3.0$).

7.6 SOC A random sample of 100 television programs contained an average of 2.37 acts of physical violence per program. At the 99% level, what is your estimate of the population value?

$$\bar{X} = 2.37$$
$$s = .30$$
$$N = 100$$

7.7 SOC A random sample of 429 college students was interviewed about a number of matters.

a. They reported that they had spent an average of $178.23 on textbooks during the previous semester. If the sample standard deviation for these data is $15.78, construct an estimate of the population mean at the 99% level.

b. They also reported that they had visited the health clinic an average of 1.5 times a semester. If the sample standard deviation is 0.3, construct an estimate of the population mean at the 99% level.

c. On the average, the sample had missed 2.8 days of classes per semester because of illness. If the sample standard deviation is 1.0, construct an estimate of the population mean at the 99% level.

d. On the average, the sample had missed 3.5 days of classes per semester for reasons other then illness. If the sample standard deviation is 1.5, construct an estimate of the population mean at the 99% level.

7.8 [CJ] A random sample of 500 residents of Shinbone, Kansas, shows that exactly 50 of the respondents had been the victims of violent crime over the past year. Estimate the proportion of victims for the population as a whole, using the 90% confidence level. (*HINT: Calculate the sample proportion P_s before using Formula 7.3. Remember that proportions are equal to frequency divided by N.*)

7.9 [SOC] The survey mentioned in problem 7.5 found that 25 of the 178 households consisted of unmarried couples who were living together. What is your estimate of the population proportion? Use the 95% level.

7.10 [PA] A random sample of 324 residents of a community revealed that 30% were very satisfied with the quality of trash collection. At the 99% level, what is your estimate of the population value?

7.11 [SOC] A random sample of 1496 respondents was questioned about their sexual attitudes and behaviors. Construct estimates to the population at the 90% level for each of the results reported below. Express the final confidence interval in percentages (e.g. "between 40 and 45% agreed that premarital sex was always wrong").
  a. When asked to agree or disagree with the statement "Explicit sexual books and magazines lead to rape and other sex crimes," 823 agreed.
  b. When asked to agree or disagree with the statement "There is far too much explicit sex on television these days," 650 agreed.
  c. 375 of the sample said that they had sex with someone other than their spouse at least once while married.
  d. 1023 of the sample said that they had sex before they were first married.
  e. 800 agreed that premarital sex was "always wrong."

7.12 [SW] A random sample of 100 patients treated in a program for alcoholism and drug dependency

over the past 10 years was selected. It was determined that 53 of the patients had been readmitted to the program at least once. At the 95% level, construct an estimate to the population proportion.

7.13 For the sample data below, construct four different interval estimates of the population mean, one each for the 90%, 95%, 99%, and 99.9% level. What happens to the interval width as confidence level increases? Why?

$$\bar{X} = 100$$
$$s = 10$$
$$N = 500$$

7.14 For each of the three sample sizes below, construct the 95% confidence interval. Use a sample proportion of 0.40 throughout. What happens to interval width as sample size increases? Why?

$$P_s = 0.40$$
Sample A: $N = 100$
Sample B: $N = 1000$
Sample C: $N = 10,000$

7.15 [PS] Two individuals are running for mayor of Shinbone. You conduct an election survey a week before the election and find that 51% of the respondents prefer candidate A. Can you predict a winner? Use the 99% level. (*HINT: In a two-candidate race, what percentage of the vote would the winner need? Does the confidence interval indicate that candidate A has a sure margin of victory? Remember that while the population parameter is probably ($\alpha = .01$) in the confidence interval, it may be anywhere in the interval.*)

$$P_s = 0.51$$
$$N = 578$$

7.16 [SOC] A random sample of 1000 registered voters in a state were asked if they favored a proposal to restrict the sale of handguns. If 510 approved of the legislation, is it safe, at the 95% level, to conclude that "a majority" of the population approves?

7.17 [SOC] The fraternities and sororities at St. Algebra College have been plagued by declining membership over the past several years and want to know if the incoming freshman class will be a fertile recruiting ground. Not having enough money to survey all 1600 freshmen, they commission you to survey the interests of a random sample. You find that 35 of your 150 respondents are "extremely" interested in social clubs. At the 95% level, what is your estimate of the *number* of freshmen who would be extremely interested? (*HINT: The high and low values of your final confidence interval are proportions. How can proportions also be expressed as numbers?*)

7.18 SOC You are the consumer-affairs reporter for a daily newspaper. Part of your job is to investigate the claims of manufacturers, and you are particularly suspicious of a new economy car that the manufacturer claims will get 78 miles per gallon. After checking the mileage figures for a random sample of 120 owners of this car, you find an average miles per gallon of 75.5 with a standard deviation of 3.7. At the 99% level, do your results tend to confirm or refute the manufacturer's claims?

*7.19 GER You are the head of the research division for a local agency that serves senior citizens. The agency wants to conduct a survey to determine the number of potential clients it has for its various services. The agency head wants the estimates to be accurate to within ±3%. How large a sample will you need to draw?

7.20 GER A sample of 103 residents of a retirement community was interviewed to assess the level of satisfaction with present services and the need for additional services. For each finding below, construct the *appropriate* confidence interval at the 95% level.

a. A total of 54 residents reported that they were "very satisfied" with the quality of health care in the facility.

b. The sample reported using the shuttle bus to town an average of 5.7 times per week with a standard deviation of .70.

c. Only 17 of the respondents reported that they had ever attended any of the free crafts classes.

d. The sample reported watching TV an average of 6.9 hours per day with a standard deviation

of 1.3. Also, 37 reported that they were "very satisfied" with the quality of programming available through the cable system.

e. When asked what services or facilities they would like to see added, 50 wanted a heated swimming pool, 40 wanted expanded health education programs, and 13 wanted a dating service established within the community.

7.21 SOC The General Social Survey (GSS) summarized in Appendix G and used for computer exercises in this text was given to an EPSEM sample that should be representative of American society. The entire sample consists of about 3000 respondents, but the version used here has fewer than 1500 respondents. To judge the accuracy of our estimates, information on population parameters is given where possible in the Answers to Odd-Numbered Problems. See Appendix G for the codes and the actual wording of the questions. For each sample statistic reported below, construct the 95% confidence interval estimate of the population value. (Note that we will be using a "sample of a sample" as the basis for the estimates and we should correct for the additional source of error. As you will see, however, the sample still produces reasonable estimates in those cases where the population parameter is known.)

a. The average occupational prestige score for 1400 respondents was 43.35, with a standard deviation of 13.83.

b. The respondents ($N = 990$) reported watching an average of 3.07 hours of TV per day with a standard deviation of 3.86.

c. The average number of children for 1445 respondents was 1.86, with a standard deviation of 1.68.

d. Of the 1450 respondents who answered the question, 362 identified themselves as Catholic.

e. Three hundred twenty-three of the 1451 respondents said that they had never married.

f. Of the 949 respondents for whom we have information, 479 said that they voted for Clinton in the 1992 presidential election.

g. When asked about capital punishment, 1020 of the 1446 respondents said that they favored the death penalty for murder.

Using SPSS for Windows Procedures to Produce Confidence Intervals

DEMONSTRATION 7.1 Generating Statistical Information for Use in Constructing Confidence Intervals

SPSS does not provide any programs specifically for constructing confidence intervals, although some of the procedures we'll cover in future chapters do include confidence intervals as part of the output. Rather than make use of these programs, I want to use this section to confirm something you may already suspect: fancy computer programs are not always helpful or even particularly useful. The arithmetic required by estimation procedures (let's face it) is not particularly difficult, and you could probably complete the formulas faster by hand than by SPSS.

On the other hand, who wants to do all the calculations it would take to get the summary statistics on which the estimates will be based? Calculating the mean age for the GSS sample would require the addition of almost 1500 numbers. The dimensions of that task should suggest the proper role of the computer. Once you know the mean age, the rest of the calculations for confidence intervals are not very formidable. So, how do you get the sample statistics?

Take a look at problem 7.21. I used a number of variables from the GSS data set for this problem and got the information I needed by using **Descriptives** for the interval-ratio variables and **Frequencies** for the nominal variables. **Descriptives** gives means and standard deviations, and **Frequencies** produces frequency distributions with a column for percentages.

Exercises

7.1 Use the **Descriptives** command to calculate means and standard deviations for *papres80, age,* and *income.* Use the output to formulate confidence interval estimates for the population values. *(NOTE: your estimate based on age will not reflect the total American population. The GSS sample is restricted to people age 18 and older, and the mean of the sample is much higher than the mean of the population as a whole.)*

7.2 Use the **Frequencies** command to get sample proportions (convert the percentages in the frequency distributions) to estimate the population parameter for each of the following: proportion with college degree (*degree*), proportion black (*race*), and proportion favoring gun control (*gunlaw*).

Hypothesis Testing I
The One-Sample Case

8

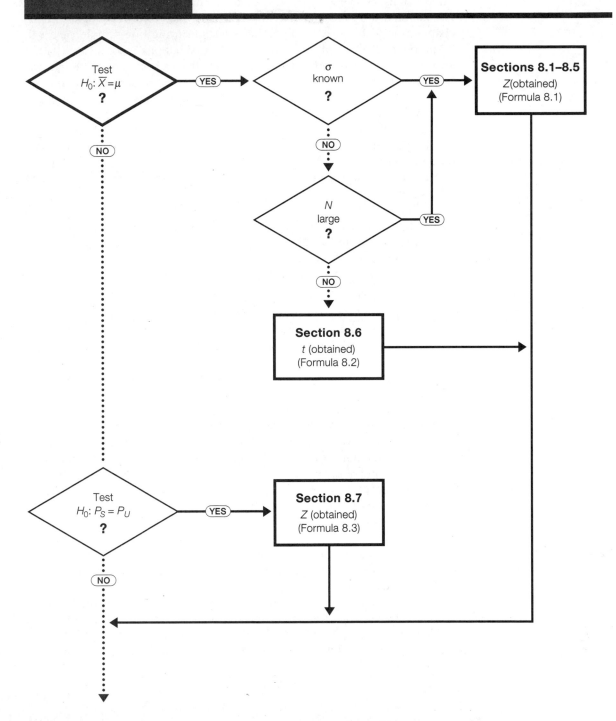

8.1 INTRODUCTION

Chapter 7 introduced the techniques for estimating population values from sample statistics. We will now investigate a rather different application of inferential statistics called **hypothesis testing** or **significance testing.** In this chapter, the techniques for hypothesis testing in the one-sample case will be introduced. The central problem in these situations is to decide (with a known probability of error) if a sample came from a population that has certain characteristics. We will have some information about a population and some statistics calculated from a sample. If the differences between the population values and the sample values are large enough, we will conclude that the sample did not come from that specific population. If the differences are small, or if there are no differences, we'll conclude that the sample did come from the defined population.

These techniques might be used in such research situations as the following:

(a) A researcher has selected a sample of 789 older citizens who live in a particular city and also has some information on the rate of criminal victimization for the city as a whole. Do older citizens, as represented by the sample, have the same rate of victimization as the city as a whole?

(b) Are the GPAs of college athletes different from the GPAs of the general student body? To investigate, the academic records of a random sample of 129 student athletes from a large state university are compared with the characteristics of all students at the school.

In each of these situations, we would compare sample and population information and decide if the sample had been drawn from a population with the specified characteristic.

No matter what our final decision, there is a chance that we will make an error. We cannot avoid risk because, as we saw in Chapter 6, even the most painstaking sampling methods will occasionally produce a nonrepresentative sample. One of the great advantages of inferential statistics, however, is that we will be able to estimate the probability of this type of error and evaluate our decisions accordingly.

8.2 AN OVERVIEW OF HYPOTHESIS TESTING

Let's begin by considering a situation in which a single sample test of hypotheses would be appropriate. Suppose that a researcher has been hired to evaluate a treatment program for alcoholics. The program serves a very large metropolitan area, and the researcher is unable to test every client. A sample of 127 is drawn from a list of all those who have completed the program. While examining the work records of the subjects, the researcher notes that they appear to have lower absentee rates from work than the community as a whole. Specifically,

Community	Sample
μ = 7.2 days per year	$\bar{X}$ = 6.8 days per year
σ = 1.43	N = 127

Consider the information we have at this point. We know the average absentee rate for the population (μ = 7.2) and for the sample ($\bar{X}$ = 6.8). Our question is this: Do treated alcoholics, as represented by this sample of 127, have lower absentee rates than the community as a whole? Clearly, we would be in a better position to answer this question if we had information about *everyone* who had completed the program (the population of all treated alcoholics). Then, we could compare one mean (the community) with another mean (all treated alcoholics), and we would know, without doubt, if the program was effective on this criterion. By using a sample, we create the possibility of error since samples are sometimes not representative of the populations from which they are selected.

Given the fact that we are working with a sample rather than a population, there are two ways to account for the difference in absentee rates. The first, which we shall label explanation A, is that the absentee rates of the treated alcoholics really are different from those of the community as a whole! Another way to say this would be "the difference in absentee rates is statistically significant." The phrase "statistically significant" means that the difference is very unlikely to have occurred by random chance alone. If explanation A is true, the sample of treated alcoholics did *not* come from a population with a mean absentee rate of 7.2 days.

In direct contradiction to explanation A is explanation B: There is really no difference between alcoholics treated in this program and the community as a whole in terms of absenteeism. The difference noted above is trivial and due to the effects of random chance. If explanation B is true, the sample *did* come from a population with a mean absentee rate of 7.2 days.

Which explanation for the observed difference is correct? As long as we are working with a sample rather than a population, we cannot know which explanation is correct on an absolute basis. However, we can set up a decision-making procedure so conservative that one of the two explanations can be chosen, with the knowledge that the probability of making an incorrect choice is very low.

This decision-making process, in broad outline, begins with the assumption that explanation B is correct. Symbolically, this assumption can be stated as

$$\mu = 7.2 \text{ days per year}$$

Remember that μ refers to the mean of the population of all alcoholics who have completed the program. This assumption, μ = 7.2, can be tested statistically.

If explanation B is true, then the probability of getting the observed sample outcome ($\bar{X} = 6.8$) can be found. Let us add an objective decision rule in advance. If the odds of getting the observed difference are less than .05 (5 out of 100, or 1 in 20), we will declare explanation B unlikely. If this explanation were true, a difference of this size (7.2 days vs. 6.8 days) would be a very rare event, and in hypothesis testing, we always bet against rare events.

The remaining problem is to determine the probability of the observed sample outcome if explanation B is correct and there really is no difference between the sample and the population. We can determine this probability with precision because of what we know about the sampling distribution of all possible sample outcomes. Looking back at the information we have, we can see that we know a great deal about the sampling distribution of sample means. Based on the Central Limit Theorem and on our assumption that explanation B is true, we may assume that the sampling distribution is normal in shape, has a mean of 7.2 ($\mu_{\bar{X}} = \mu$), and a standard deviation of 1.43/127 because $\sigma_{\bar{X}} = \sigma/\sqrt{N}$. We also know that the standard normal distribution can be interpreted as a distribution of probabilities and that the particular sample outcome noted above ($\bar{X} = 6.8$) is one of thousands of possible sample outcomes. The sampling distribution, with the sample outcome noted, is depicted in Figure 8.1.

Using our knowledge of the standardized normal distribution, we can add further useful information to this sampling distribution of sample means. Specifically, with Z scores, we can depict the decision rule stated previously: any sample outcome with probability less than 0.05 if explanation B is true, will be cause to declare explanation B unlikely. The probability of 0.05 can be translated into an area and divided equally into the upper and lower tails

FIGURE 8.1   THE SAMPLING DISTRIBUTION OF ALL POSSIBLE SAMPLE MEANS

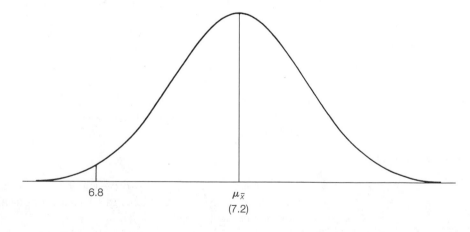

FIGURE 8.2   THE SAMPLING DISTRIBUTION OF ALL POSSIBLE SAMPLE MEANS

of the sampling distribution. Using Appendix A, we find that the Z-score equivalent of this area is ± 1.96. The areas and Z scores are depicted in Figure 8.2.

The decision rule can now be rephrased. Any sample outcome falling in the lined areas depicted in Figure 8.2 by definition has a probability of occurrence of less than 0.05. Such an outcome would be a rare event and would cause us to reject explanation B. All that remains is to translate our sample outcome into a Z score so we can see where it falls on the curve. To do this, we use the standard formula for locating any particular raw score under a normal distribution. When we use known or empirical distributions, this formula is expressed as

$$Z = \frac{X_i - \bar{X}}{s}$$

Or, to find the equivalent Z score for any raw score, subtract the mean of the distribution from the raw score and divide by the standard deviation of the distribution. Since we are now concerned with the sampling distribution of all sample means rather than an empirical distribution, the symbols in the formula will change, but the form remains exactly the same:

FORMULA 8.1

$$Z = \frac{\bar{X} - \mu_{\bar{X}}}{\sigma/\sqrt{N}}$$

Or, to find the equivalent Z score for any sample mean, subtract the mean of the sampling distribution from the sample mean and divide by the standard deviation of the sampling distribution.

Recalling the data given on this problem, we can now find the Z-score equivalent of the sample mean:

Community	Sample
$\mu = 7.2 \; (= \mu_{\bar{X}})$	$\bar{X} = 6.8$
$\sigma = 1.43$	$N = 127$

$$Z = \frac{\bar{X} - \mu}{\sigma/\sqrt{N}}$$

$$Z = \frac{6.8 - 7.2}{1.43/\sqrt{127}}$$

$$Z = -.400/.127$$

$$Z = -3.15$$

In Figure 8.3, this Z score of -3.15 is noted on the distribution of all possible sample means, and we see that the sample outcome does fall in the shaded area. If explanation B is true, this particular sample outcome has a probability of occurrence of less than 0.05 (how much less than 0.05 is irrelevant, since the decision rule was established in advance). The sample outcome ($\bar{X} = 6.8$ or $Z = -3.15$) would therefore be rare if explanation B were true, and the researcher is justified in rejecting explanation B. If B were true, this sample outcome would be extremely unlikely. Therefore, the researcher would bet that B is not true and that our 127 treated alcoholics come from a population that is significantly different from the community on the trait of absenteeism.

FIGURE 8.3 THE SAMPLING DISTRIBUTION OF SAMPLE MEANS WITH THE SAMPLE OUTCOME ($\bar{X} = 6.8$) NOTED IN Z SCORES

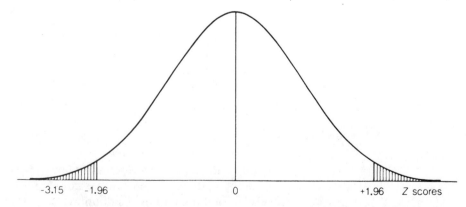

Keep in mind that in inferential statistics, our conclusions must be based on information gathered from random samples. On rare occasions, a given sample may not be representative of the population from which it was selected. Let me repeat that the decision-making process outlined above has a very high probability of resulting in correct decisions, but, as long as we must work with samples rather than populations, we face an element of risk in the process. That is, the decision with respect to explanation B might be incorrect if this sample happens to be one of the few that is unrepresentative of the population of alcoholics treated in this program. One important strength of hypothesis testing is that the probability of making an incorrect decision can be estimated. In the example at hand, explanation B was rejected and the probability of this decision being incorrect is 0.05—the decision rule established at the beginning of the process. To say that the probability of having rejected B incorrectly is 0.05 means that, if we repeated this same test an infinite number of times, we would incorrectly reject B only 5 times out of every 100.

8.3 THE FIVE-STEP MODEL FOR HYPOTHESIS TESTING

All the formal elements and concepts used in hypothesis testing were surreptitiously sneaked into the preceding discussion. This section presents their proper names and introduces a **five-step model** for organizing all hypothesis testing:

Step 1. Making assumptions

Step 2. Stating the null hypothesis

Step 3. Selecting the sampling distribution and establishing the critical region

Step 4. Computing the test statistic

Step 5. Making a decision

We will look at each step individually, using the problem from Section 8.2 as an example throughout.

Step 1. Making Assumptions. Any application of a statistical technique requires that certain assumptions be made about the data. When conducting a test of hypothesis with a single sample mean, we must assume that the sample has been randomly selected from the defined population. Also, to justify computation of a mean, we must assume that the data are interval-ratio in level of measurement. Finally, we must assume that the sampling distribution of all possible sample means is normal in shape so that we may use the standardized normal distribution to find areas under the sampling distribution. We will refer to these assumptions about the sampling process, the level of measurement, and the shape of the sampling distribution as the "mathematical model." You should be very certain of the correctness of these model assumptions.

Usually, we will state these assumptions in abbreviated form. For example:

> Model: Random sampling
> Level of measurement is interval-ratio
> Sampling distribution is normal

Step 2. Stating the Null Hypothesis (H_0). The **null hypothesis** is the formal name for explanation B and is always a statement of "no difference." The exact form of the null hypothesis will vary depending on the test being conducted. In the single-sample case, the null hypothesis states that the sample comes from a population with a certain characteristic. In our example, the null is that treated alcoholics are "no different" from the community as a whole, that the average days of absenteeism for this population will also be 7.2, and that the difference between 7.2 and the sample mean of 6.8 is caused by random chance. Symbolically, the null would be stated as

$$H_0: \mu = 7.2$$

where μ refers to the mean of the population of treated alcoholics. The null hypothesis is the central element in any test of hypothesis because the entire process is aimed at rejecting or failing to reject the H_0.

Usually, the researcher believes that there is a significant difference and desires to reject the null hypothesis. At this point in the five-step model, the researcher's belief is stated in a **research hypothesis** (H_1), a statement that directly contradicts the null hypothesis. Thus, the researcher's goal in hypothesis testing is often to gather evidence for the research hypothesis by rejecting the null hypothesis. The research hypothesis can be stated in several ways. One form would simply assert that the sample was selected from a population that did not have a certain characteristic or, in terms of our example, had a mean that was not equal to a specific value:

$$(H_1: \mu \neq 7.2)$$

where $\neq$ means "not equal to"

Symbolically, this statement asserts that the sample does not come from a population with a mean of 7.2, or that treated alcoholics are different from the community as a whole. The research hypothesis is enclosed in parentheses to emphasize that it has no formal standing or role in the hypothesis-testing process (except, as we shall see in the next section, in choosing between one-tailed and two-tailed tests). It serves as a reminder of what the researcher believes to be the truth.

Step 3. Selecting the Sampling Distribution and Establishing the Critical Region. The sampling distribution is, as always, the probabilistic yardstick against which a particular sample outcome is measured. By assuming that the null hypothesis is true (and *only* by this assumption), we can attach values to the mean and standard deviation of the sampling distribution and thus measure the probability of any specific sample outcome. There are several different sampling distributions, but for now we will confine our attention to the sampling distribution of sample means. We can measure areas under this distribution by using the standard normal curve as summarized in Appendix A.

The **critical region** consists of the areas under the sampling distribution that include all unlikely sample outcomes. Prior to the test of hypothesis, we must define what is meant by *unlikely*. That is, we must specify in advance those sample outcomes so unlikely that they will lead us to reject the H_0. This decision rule will establish the critical region or region of rejection. The word *region* is used because, essentially, we are describing those areas under the sampling distribution that contain unlikely sample outcomes. In the example above, this area corresponded to a Z score of ± 1.96, called **Z (critical)**, that was graphically displayed in Figure 8.2. The lined area was the critical region. Any sample outcome for which the Z-score equivalent fell in this area (that is, below -1.96 or above $+1.96$) would have caused us to reject the null hypothesis.

By convention, the size of the critical region is reported as alpha (α), the proportion of all of the area included in the critical region. In the example above, our **alpha level** was 0.05. Other commonly used alphas are 0.10, 0.01, and 0.001.

In abbreviated form, all the decisions made in this step are noted below. The critical region is noted by the Z scores that mark its beginnings.

$$\text{Sampling distribution} = Z \text{ distribution}$$

$$\alpha = 0.05$$

$$Z \text{ (critical)} = \pm 1.96$$

(For practice in finding Z (critical) scores, see problem 8.1a.)

Step 4. Computing the Test Statistic. To evaluate the probability of any given sample outcome, the sample value must be converted into a Z score. Solving the equation for Z-score equivalents is called computing the **test statistic**, and the resultant value will be referred to as **Z (obtained)** in order to differentiate the test statistic from the critical region. In our example above, we found a Z (obtained) of -3.15. *(For practice in computing obtained Z scores for means, see problems 8.1c, 8.2 to 8.7, 8.15e and f, and 8.16a to c.)*

Step 5. Making a Decision. As the last step in the hypothesis-testing process, the test statistic is compared with the critical region. If the test statistic falls in the critical region, our decision will be to reject the null hypothesis. If the test statistic does not fall in the critical region, we fail to reject the null hypothesis. In our example, the two values were

$$Z \text{ (critical)} = \pm 1.96$$

$$Z \text{ (obtained)} = -3.15$$

and we saw that the Z (obtained) fell in the critical region whose beginnings are marked by Z (critical). Our decision was to reject the null hypothesis and conclude that the difference observed with respect to absenteeism was unlikely to have occurred by chance alone. On the trait of absenteeism, treated alcoholics are different from the community as a whole.

This five-step model will serve us as a framework for decision making throughout the hypothesis-testing chapters. The exact nature and method of expression for our decisions will, of course, change as we encounter the various tests. However, familiarity with this model will assist you in mastering this material by providing a common frame of reference for all significance testing.

8.4 ONE-TAILED AND TWO-TAILED TESTS OF HYPOTHESIS

The five-step model for hypothesis testing is fairly rigid, and the researcher has little room for making choices. Follow the steps one by one as specified above and you cannot make a mistake. Nonetheless, the researcher must still make two crucial choices. First, he or she must decide between a one-tailed and a two-tailed test. Second, an alpha level must be selected. In this section, we will discuss the former decision and, in Section 8.5, the latter.

The choice between a one- and two-tailed test is based on the researcher's expectations about the population from which the sample was selected. These expectations are reflected in the research hypothesis (H_1), which usually states what the researcher believes to be "the truth" and is contradictory to the null hypothesis. In most situations, the researcher will wish to support the research hypothesis by rejecting the null hypothesis.

The format for the research hypothesis may take either of two forms, depending on the relationship between what the null hypothesis states and what the researcher believes to be the truth. The null hypothesis states that the population has a specific characteristic. In the example that has served us throughout this chapter, the null stated, in symbols, that "treated alcoholics have the same absentee rate (7.2 days) as the community." The researcher might believe that the population value is actually greater than or less than the value stated in the null hypothesis, or he or she might be unsure about the direction of the difference.

If the researcher is unsure about the direction, the research hypothesis would state only that the population is "not equal" to the value or does not have the specific characteristic stated in the null. The research hypothesis stated in Section 8.3 ($\mu \neq 7.2$) was in this format. This is called a **two-tailed test** of significance. The researcher will be equally concerned with the possibility that the true population value is greater than the value specified in the null hypothesis and the possibility that the true population value is less than the value specified in the null hypothesis.

In other situations, the researcher might be concerned only with differences in a specific direction. If the direction of the difference can be predicted, or if the researcher is concerned only with differences in one direction, a **one-tailed test** can be used. A one-tailed test may take one of two forms, depending on the researcher's expectations about the specific direction of the difference. If the researcher believes that the true population value is higher than the value specified in the null, the research hypothesis would reflect that belief. In our example, if we had predicted that treated alcoholics had higher absentee rates than the community, our research hypothesis would have been

$$(H_1: \mu > 7.2)$$

where $>$ signifies "greater than"

For a one-tailed test in the opposite direction, the research hypothesis would be

$$(H_1: \mu < 7.2)$$

where $<$ signifies "less than"

This research hypothesis reflects the belief that treated alcoholics have lower absentee rates.

An example of a situation in which a one-tailed test is called for might be in program evaluation. Consider the evaluation of a random sample of programs designed to reduce unemployment. The evaluators would be concerned only with outcomes that show a decrease in the unemployment rate. If the rate shows no change, or if unemployment increases, the program is a failure and both of these outcomes might be considered equally negative by the researchers.

In terms of the five-step model, the choice of a one-tailed or two-tailed test determines what we do with the critical region under the sampling distribution in step 3. As you recall, in a two-tailed test, we split the critical region equally into the upper and lower tails. In a one-tailed test, we place the entire critical area in one tail of the sampling distribution. If we believe that the population characteristic is "greater than" the value stated in the null (if the H_1 includes the $>$ symbol), we place the entire critical region in the upper tail. If we believe that the difference is "less than" the null value (if the H_1 includes the $<$ symbol), the entire critical region goes in the lower tail.

For example, in a two-tailed test with alpha equal to 0.05, the critical region begins at Z (critical) $= \pm 1.96$. In a one-tailed test at the same alpha level, the Z (critical) is $+1.65$ if the upper tail is specified and -1.65 if the lower tail is specified. The contrast in the placement of the critical region is graphically summarized in Figure 8.4.

FIGURE 8.4 ESTABLISHING THE CRITICAL REGION, ONE-TAILED TESTS VERSUS
TWO-TAILED TESTS (alpha = 0.05)

A. The two-tailed test, Z (critical) = ± 1.96

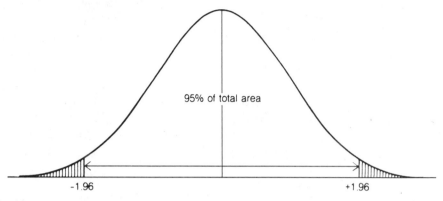

B. The one-tailed test for upper tail, Z (critical) = $+1.65$

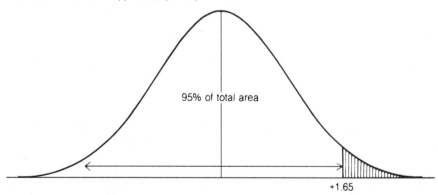

You should immediately see that a one-tailed test is advantageous because it moves the critical region closer to the mean of the sampling distribution and thus improves the probability of rejecting the H_0 without affecting the alpha level. Note also that, if the wrong tail has been specified, the probability of rejection is zero. Thus, there is a risk involved in using the one-tailed test. But, assuming that the proper tail has been specified, one-tailed tests are a way of statistically both having and eating your cake and should be used whenever (1) the direction of the difference can be confidently predicted or (2) the researcher is concerned only with differences in one tail of the sampling distribution.

An example will clarify how the one-tailed test is used. After many years of work, a sociologist has noted that sociology majors seem more sophisticated, charming, and cosmopolitan than the rest of the student body. A "Sophistication Scale" test has been administered to the entire student body and to a random sample of 100 sociology majors, and these results have been obtained:

Student Body	Sociology Majors
$\mu = 17.3$	$\bar{X} = 19.2$
$\sigma = 7.4$	$N = 100$

To test the H_0 of no difference between sociology majors and the general student body, we will use the five-step model.

Step 1. Making Assumptions. Since we are using a mean to summarize the sample outcome, we must assume that the Sophistication Scale generates interval-ratio level data. With a sample size of 100, the Central Limit Theorem applies, and we can assume that the sampling distribution is normal in shape.

> Model: Random sampling
> Level of measurement is interval-ratio
> Sampling distribution is normal

Step 2. Stating the Null Hypothesis (H_0). The null hypothesis states that there is no difference between sociology majors and the general student body. The research hypothesis (H_1) will also be stated at this point. The researcher has predicted a direction for the difference ("Sociology majors are more sophisticated"), so a one-tailed test is justified. The two hypotheses may be stated as

$$H_0: \mu = 17.3$$

$$(H_1: \mu > 17.3)$$

Step 3. Selecting the Sampling Distribution and Establishing the Critical Region. We will use the standardized normal distribution (Appendix A) to find areas under the sampling distribution. If alpha is set at 0.05, the critical region will begin at the Z score $+1.65$. These decisions may be summarized as

$$\text{Sampling distribution} = Z \text{ distribution}$$

$$\alpha = 0.05 \text{ (one-tailed)}$$

$$Z \text{ (critical)} = +1.65$$

Step 4. Computing the Test Statistic.

$$Z \text{ (obtained)} = \frac{\bar{X} - \mu}{\sigma/\sqrt{N}}$$

$$Z \text{ (obtained)} = \frac{19.2 - 17.3}{7.4/\sqrt{100}}$$

$$Z \text{ (obtained)} = +2.57$$

FIGURE 8.5   *Z* (OBTAINED) VERSUS *Z* (CRITICAL) (alpha = 0.05, one-tailed test)

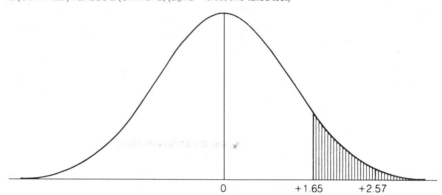

Step 5. ~~Making a Decision.~~ Comparing the *Z* (obtained) with the *Z* (critical):

$$Z \text{ (critical)} = +1.65$$

$$Z \text{ (obtained)} = +2.57$$

We see that the test statistic falls in the critical region. This outcome is depicted graphically in Figure 8.5. We will reject the null hypothesis because, if the H_0 were true, a difference of this size would be very unlikely. There is a significant difference between sociology majors and the general student body in terms of sophistication. Since the null hypothesis has been rejected, the research hypothesis (sociology majors are more sophisticated) is supported. *(For practice in dealing with tests of significance for means that may call for one-tailed tests, see problems 8.2, 8.3, 8.6, and 8.8.)*

8.5 SELECTING AN ALPHA LEVEL

In addition to deciding between one-tailed and two-tailed tests, the researcher must also select an alpha level. We have seen that the alpha level plays a crucial role in hypothesis testing. By assigning a value to alpha, we in effect define what we will mean by an "unlikely" sample outcome. If the probability of the observed sample outcome is lower than the alpha level (if the test statistic falls in the critical region), we reject the null hypothesis as untrue. Thus, the alpha level will have important consequences for our decision in step 5.

How can reasonable decisions be made with respect to the value of alpha? Recall that, in addition to defining what will be meant by *unlikely*, the alpha level is also the probability that, if the test statistic falls in the critical region, the decision to reject the null will be incorrect. In hypothesis testing, this kind of error is called **Type I error**, or **alpha error**. Type I error can be defined as the rejection of a null hypothesis that is, in fact, true or, more simply if somewhat redundantly, as falsely rejecting a true null. To minimize this type of error, very small alphas should be used.

To elaborate, when an alpha level is specified, the sampling distribution is divided into two sets of possible sample outcomes. The critical region includes all sample outcomes that we will define as unlikely or rare and that will cause us to reject the null hypothesis. The remainder of the area consists of all sample outcomes that we have defined as "nonrare." The lower the alpha level, the smaller the critical region and the greater the distance between the mean of the sampling distribution and the beginnings of the critical region. Compare, for the sake of illustration, the following alpha levels and Z (criticals) for two-tailed tests.

If Alpha Equals	The Two-Tailed Critical Region Will Begin at Z (critical) Equal to
0.10	±1.65
0.05	±1.96
0.01	±2.58
0.001	±3.29

As alpha goes down, the critical region moves farther away from the mean of the sampling distribution. The lower the alpha level, the harder it will be to reject the null and, since a Type I error can be made only if our decision in step 5 is to reject, the lower the probability of Type I error. To minimize the probability of rejecting a null hypothesis that is in fact true, use very low alpha levels.

However, there is a complication. As the critical region decreases in size (as alpha levels decrease), the noncritical region must become larger. All other things being equal, the lower the alpha level, the less likely that the sample outcome will fall in the critical region. This raises the possibility of a second type of incorrect decision, called **Type II error**, or **beta error:** failing to reject a null that is, in fact, false. Although the probability of Type I error decreases as alpha level decreases, the probability of Type II error increases. Thus, the two types of error are inversely related, and it is not possible to minimize both in the same test. As the probability of one type of error decreases, the other increases, and vice versa.

What all of this means, finally, is that the selection of an alpha level must be conceived of as an attempt to balance the two types of error. Higher alpha levels will minimize the probability of Type II error, and lower alpha levels will minimize the probability of Type I error. Normally, in social science research, we will want to minimize Type I error, and lower alpha levels will be used.

As a final note, social scientists conventionally set alpha at 0.05 or, somewhat less frequently, 0.10, 0.01, or 0.001. The 0.05 level in particular seems to have emerged as a generally recognized indicator of a significant result. However, the widespread use of the 0.05 level is simply a convention, and there is no reason that alpha cannot be set at virtually any sensible level (such as 0.04, 0.027, 0.083). The researcher has the responsibility of selecting the alpha level that seems most reasonable in terms of the goals of the research project.

8.6 THE STUDENT'S t DISTRIBUTION

To this point, we have considered only one type of hypothesis test. Specifically, we have focused on situations involving single sample means where the value of the population standard deviation (σ) was known. Needless to say, in most research situations, the value of σ will not be known. However, a value for σ is required in order to compute the standard error of the mean ($\sigma/\sqrt{N}$), convert our sample outcome into a Z score, and place the Z (obtained) on the sampling distribution (step 4). How can a value for the population standard deviation reasonably be obtained?

It might seem sensible to estimate σ with s, the sample standard deviation. As we noted in Chapter 7, s is a biased estimator of σ, but the degree of bias decreases as sample size increases. For large samples (that is, samples with 100 or more cases), the sample standard deviation yields an adequate estimate of σ. Thus, for large samples, we simply substitute s for σ in the formula for Z (obtained) in step 4 and continue to use the standard normal curve to find areas under the sampling distribution.*

For smaller samples, however, when σ is unknown, an alternative distribution called the **Student's t distribution** must be used to find areas under the sampling distribution and establish the critical region. The shape of the t distribution varies as a function of sample size. The relative shapes of the t and Z distributions are depicted in Figure 8.6. For small samples, the t distribution is much flatter than the Z distribution, but, as sample size increases, the t distribution comes to resemble the Z distribution more and more until the two are essentially identical when sample size is greater than 120. As N increases, the sample standard deviation (s) becomes a more and more adequate estimator of the population standard deviation (σ), and the t distribution becomes more and more like the Z distribution.

The t distribution is summarized in Appendix B. The t table differs from the Z table in several ways. First, there is a column at the left of the table labeled df for "degrees of freedom."† As mentioned above, the exact shape

---

*Even though its effect will be minor and will decrease with sample size, we will always correct for the bias in s by using the term $N - 1$ rather than N in the computation for the standard deviation of the sampling distribution when σ is unknown.

†Degrees of freedom refer to the number of values in a distribution that are free to vary. For a sample mean, a distribution has $N - 1$ degrees of freedom. This means that, for a specific value of $\bar{X}$ and N, $N - 1$ scores are free to vary. For example, if $\bar{X} = 3$ and $N = 5$, the distribution of five scores would have $N - 1$, or four, degrees of freedom. When the value of four of the scores is known, the value of the fifth is fixed. If four scores are 1, 2, 3, and 4, the fifth must be 5 and no other value.

FIGURE 8.6   THE *t* DISTRIBUTION AND THE *Z* DISTRIBUTION

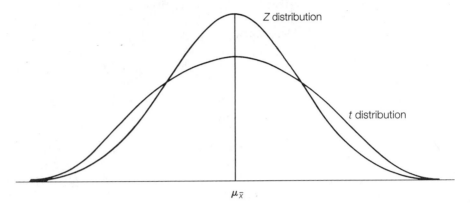

of the *t* distribution and thus the exact location of the critical region for any alpha level varies as a function of sample size. Degrees of freedom, which are equal to $N - 1$ in the case of a single-sample mean, must first be computed before the critical region for any alpha can be located. Second, alpha levels are arrayed across the top of Appendix B in two rows, one row for the one-tailed tests and one for two-tailed tests. To use the table, begin by locating the selected alpha level in the appropriate row.

The third difference is that the entries in the table are the actual scores, called *t* **(critical)**, that mark the beginnings of the critical regions and not areas under the sampling distribution. To illustrate the use of this table with single-sample means, find the critical region for alpha equal to 0.05, two-tailed test, for $N = 30$. The degrees of freedom will be $N - 1$, or 29; reading down the proper column, you should find a value of 2.045. Thus, the critical region for this test will begin at *t* (critical) = ±2.045.

Take a moment to notice some additional features of the *t* distribution. First, note that the *t* (critical) we found above is larger in value than the comparable *Z* (critical), which—for a two-tailed test at an alpha of 0.05—would be ±1.96. This relationship reflects the fact that the *t* distribution is flatter than the *Z* distribution (see Figure 8.6). When you use the *t* distribution, the critical regions will begin farther away from the mean of the sampling distribution and, therefore, the null hypothesis will be harder to reject. Furthermore, the smaller the sample size (the lower the degrees of freedom), the larger the value of *t* (obtained) necessary for a rejection of the H_0.

Second, scan the column for an alpha of 0.05, two-tailed test. Note that, for one degree of freedom, the *t* (critical) is ±12.706 and that the value of *t* (critical) decreases as degrees of freedom increase. For degrees of freedom greater than 120, the value of *t* (critical) is the same as the comparable value of *Z* (critical), or ±1.96. As sample size increases, the *t* distribution comes to resemble the Z distribution more and more until, with sample sizes greater than 120, the two distributions are essentially identical.*

To demonstrate the uses of the t distribution in more detail, we will work through an example problem. Note that, in terms of the five-step model, the changes required by using t scores occur mostly in steps 3 and 4. In step 3, the sampling distribution will be the t distribution, and degrees of freedom (df) must be computed before locating the critical region as marked by t (critical). In step 4, a slightly different formula for computing the test statistic, **t (obtained)**, will be used. As compared with the formula for Z (obtained), s will replace σ and $N - 1$ will replace N.

Specifically,

FORMULA 8.2
$$t \text{ (obtained)} = \frac{\bar{X} - \mu}{s/\sqrt{N - 1}}$$

A researcher wonders if commuter students are different from the general student body in terms of academic achievement. She has gathered a random sample of 30 commuter students and has learned from the registrar that the mean grade-point average for all students is 2.00 ($\mu = 2.00$), but the standard deviation of the population (σ) has never been computed. Sample data are reported below. Is the sample from a population that has a mean of 2.00?

Student Body	Commuter Students
$\mu = 2.00 \ (= \mu_{\bar{x}})$	$\bar{X} = 2.58$
$\sigma = ?$	$s = 1.23$
	$N = 30$

Step 1. Making Assumptions.

> Model: Random sampling
> Level of measurement is interval-ratio
> Sampling distribution is normal

Step 2. Stating the Null Hypothesis.

$$H_0\text{: } \mu = 2.00$$

$$(H_1\text{: } \mu \neq 2.00)$$

You can see from the research hypothesis that the researcher has not predicted a direction for the difference. This will be a two-tailed test.

---

*Appendix B abbreviates the t distribution by presenting a limited number of t (critical)s for degrees of freedom between 31 and 120. If the degrees of freedom for a specific problem equal 77 and alpha equals 0.05, two-tailed, we have a choice between a t (critical) of ± 2.000 (df = 60) and a t (critical) of ± 1.980 (df = 120). In situations such as these, take the larger table value as t (critical). This will make rejection of the H_0 less likely and is therefore the more conservative course of action.

Step 3. Selecting the Sampling Distribution and Establishing the Critical Region. Since σ is unknown and sample size is small, the t distribution will be used to find the critical region. Alpha will be set at 0.01.

$$\text{Sampling distribution} = t \text{ distribution}$$

$$\alpha = 0.01, \text{ two-tailed test}$$

$$\text{df} = (N - 1) = 29$$

$$t \text{ (critical)} = \pm 2.756$$

Step 4. Computing the Test Statistic.

$$t \text{ (obtained)} = \frac{\bar{X} - \mu}{s/\sqrt{N - 1}}$$

$$t \text{ (obtained)} = \frac{2.58 - 2.00}{1.23/\sqrt{29}}$$

$$t \text{ (obtained)} = \frac{.58}{.23}$$

$$t \text{ (obtained)} = +2.52$$

Step 5. Making a Decision. The test statistic does not fall in the critical region. Therefore, the researcher fails to reject the H_0. The difference between the sample mean (2.58) and the population mean (2.00) is no greater than what would be expected if only random chance were operating. The test statistic and critical regions are displayed in Figure 8.7.

To summarize, when testing single-sample means we must make a choice regarding the theoretical distribution we will use to establish the critical

FIGURE 8.7   SAMPLING DISTRIBUTION SHOWING t (obtained) VERSUS t (critical) ($\alpha = 0.05$, two-tailed test, df = 29)

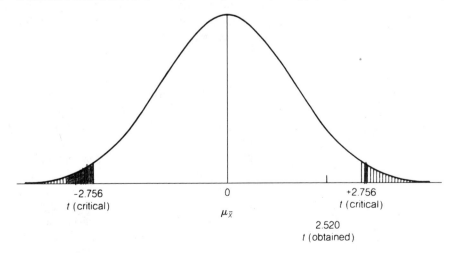

-2.756
t (critical)

0
$\mu_{\bar{X}}$

+2.756
t (critical)

2.520
t (obtained)

Application 8.1

For a random sample of 152 felony cases tried in a local court, the average prison sentence was 27.3 months. Is this significantly different from the average prison term for felons nationally? The necessary information for conducting a test of the null hypothesis is

$$\bar{X} = 27.3 \qquad \mu = 28.7$$
$$s = 3.7$$
$$N = 152$$

The null hypothesis would be

$$H_0: \mu = 28.7$$

The null states that the sample comes from a population that has a mean of 28.7.

The test statistic, Z (obtained), would be

$$Z \text{ (obtained)} = \frac{\bar{X} - \mu}{s/\sqrt{N-1}}$$

$$Z \text{ (obtained)} = \frac{27.3 - 28.7}{3.7/\sqrt{152 - 1}}$$

$$Z \text{ (obtained)} = \frac{-1.40}{3.7/\sqrt{151}}$$

$$Z \text{ (obtained)} = -\frac{1.40}{0.30}$$

$$Z \text{ (obtained)} = -4.67$$

If alpha were set at the 0.05 level, the critical region would begin at Z (critical) $= \pm 1.96$. With an obtained Z score of -4.67, the null would be rejected. The difference between the prison sentences of felons convicted in the local court and felons convicted nationally is statistically significant. The difference is so large that we may conclude that it did not occur by random chance. The decision to reject the null hypothesis has a 0.05 probability of being wrong.

region. The choice is straightforward. If the population standard deviation (σ) is known or sample size is large, the Z distribution (summarized in Appendix A) will be used. If σ is unknown and the sample is small, the t distribution (summarized in Appendix B) will be used. *(For practice in using the t distribution in a test of hypothesis, see problems 8.8 to 8.10 and 8.18. A one-tailed test may be appropriate for problem 8.18.)*

8.7 TESTS OF HYPOTHESES FOR SINGLE-SAMPLE PROPORTIONS (LARGE SAMPLES)

In many cases, the characteristic of interest in the sample will not be measured in a way that justifies the assumption of interval-ratio level of measurement. One alternative in this situation would be to use a sample proportion (P_s) rather than a sample mean as the test statistic. As we shall see below, the overall procedures for testing single-sample proportions are the same as those for testing means. The central question is still "Does the population from which the sample was drawn have a certain characteristic?" We still conduct the test based on the assumption that the null hypothesis is true, and we still evaluate the probability of the obtained sample outcome against a sampling distribution of all possible sample outcomes. Our decision at the end of the test is also the same. If the obtained test statistic falls in the critical region (is unlikely, given the assumption that the H_0 is true), we reject the H_0.

Having stressed the continuity in procedures and logic, I must hastily point out the important differences as well. These differences are best related in terms of the five-step model for hypothesis testing. In step 1, when working with sample proportions, we assume that the variable is measured at the nominal level of measurement. In step 2, the symbols used to state the null hypothesis are different even though the null is still a statement of "no difference."

In step 3, we will use only the standardized normal curve (the Z distribution) to find areas under the sampling distribution and locate the critical region. This will be appropriate as long as sample size is large. We will not consider small-sample tests of hypothesis for proportions in this text.

In step 4, computing the test statistic, the form of the formula remains the same. That is, the test statistic, Z (obtained), equals the sample statistic minus the mean of the sampling distribution, divided by the standard deviation of the sampling distribution. However, the symbols will change because we are basing the tests on sample proportions. The formula can be stated as

FORMULA 8.3
$$Z \text{ (obtained)} = \frac{P_s - P_u}{\sqrt{P_u(1 - P_u)/N}}$$

Step 5, making a decision, is exactly the same as before. If the test statistic, Z (obtained), falls in the critical region, as marked by Z (critical), reject the H_0.

An example should clarify these procedures. A random sample of 122 households in a low-income neighborhood revealed that 53, or a proportion of 0.43, of the households were headed by females. In the city as a whole, the proportion of female-headed households is .39. Are households in the lower-income neighborhood significantly different from the city as a whole in terms of this characteristic?

Step 1. Making Assumptions.

> Model: Random sampling
> Level of measurement is nominal
> Sampling distribution is normal in shape

Step 2. Stating the Null Hypothesis. The research question, as stated above, asks only if the sample proportion is different from the population proportion. Since no direction is predicted for the difference, a two-tailed test will be used.

$$H_0: P_u = .39$$

$$(H_1: P_u \neq .39)$$

Application 8.2

In a random sample drawn from the most affluent neighborhood in a community, 76% of the respondents reported that they had voted Republican in the most recent presidential election. For the community as a whole, 66% of the electorate voted Republican. Was the affluent neighborhood significantly more likely to have voted Republican? The information necessary for a test of the null hypothesis, expressed in the form of proportions, is

Neighborhood	Community
$P_s = 0.76$	$P_u = 0.66$
$N = 103$	

Our null hypothesis would be

$$H_0: P_u = 0.66$$

or, the sample was taken from a population that voted 66% Republican.

The test statistic, Z (obtained), would be

$$Z \text{ (obtained)} = \frac{P_s - P_u}{\sqrt{\dfrac{P_u(1 - P_u)}{N}}}$$

$$Z \text{ (obtained)} = \frac{0.76 - 0.66}{\sqrt{\dfrac{(0.66)(1 - 0.66)}{103}}}$$

$$Z \text{ (obtained)} = \frac{0.100}{0.047}$$

$$Z \text{ (obtained)} = 2.13$$

Because we want to see if the affluent neighborhood was "significantly more likely to have voted Republican," a one-tailed test of hypothesis is called for. With alpha set at the 0.05 level, our critical region will begin at Z (critical) $= +1.65$. The critical region is placed in the upper tail because we are predicting that the sample comes from a population that voted Republican at a higher rate than the community.

With an obtained Z score of 2.13, the null hypothesis would be rejected. The difference between the affluent neighborhood and the community as a whole is statistically significant and in the predicted direction. Residents of the affluent neighborhood were significantly more likely to have voted Republican in the last presidential election.

Step 3. Selecting the Sampling Distribution and Establishing the Critical Region.

$$\text{Sampling distribution} = Z \text{ distribution}$$

$$\alpha = 0.10, \text{ two-tailed}$$

$$Z \text{ (critical)} = \pm 1.65$$

Step 4. Computing the Test Statistic.

$$Z \text{ (obtained)} = \frac{P_s - P_u}{\sqrt{P_u(1 - P_u)/N}}$$

$$Z \text{ (obtained)} = \frac{0.43 - 0.39}{\sqrt{(0.39)(0.61)/122}}$$

$$Z \text{ (obtained)} = +0.91$$

Step 5. Making a Decision. The test statistic, Z (obtained), does not fall in the critical region. Therefore, we fail to reject the H_0. There is no statistically

FIGURE 8.8   SAMPLING DISTRIBUTION SHOWING Z (obtained) VERSUS Z (critical) ($\alpha = 0.10$, two-tailed test)

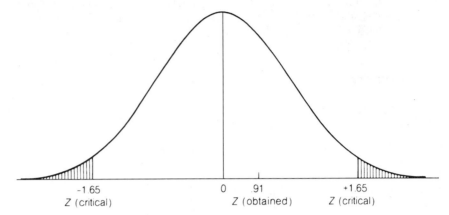

significant difference between the low-income community and the city as a whole in terms of the proportion of households headed by females. Figure 8.8 displays the sampling distribution, the critical region, and the Z (obtained). *(For practice in tests of significance using sample proportions, see problems 8.11 to 8.14, 8.15 a to d, 8.16d and e, and 8.17. A one-tailed test may be appropriate for problems 8.11, 8.14, and 8.17.)*

SUMMARY

1. All the basic concepts and techniques for testing hypotheses were presented. We saw how to test the null hypothesis of "no difference" for both sample means and proportions.

2. All tests of a hypothesis involve finding the probability of the observed sample outcome, given that the null hypothesis is true. If the outcomes have low probabilities, we reject the null hypothesis. In the usual research situation, we will wish to reject the null and thereby support the research hypothesis.

3. The five-step model will be our framework for decision making throughout the hypothesis-testing chapters. We will always (1) make assumptions, (2) state the null hypothesis, (3) select a sampling distribution, specify alpha, and find the critical region, (4) compute a test statistic, and (5) make a decision. What we do during each step, however, will vary, depending on the specific test being conducted.

4. If we can predict a direction for the difference in stating the research hypothesis, a one-tailed test is

called for. If no direction can be predicted, a two-tailed test is appropriate. There are two kinds of errors in hypothesis testing. Type I, or alpha, error is rejecting a true null; Type II, or beta, error is failing to reject a false null. The probabilities of committing these two types of error are inversely related and cannot be simultaneously minimized in the same test. By selecting an alpha level, we try to balance the probability of these two kinds of error.

5. When testing sample means, the t distribution must be used to find the critical region when the population standard deviation is unknown and sample size is small.

6. Sample proportions can also be tested for significance. Tests are conducted using the five-step model. Compared to the test for the sample mean, the major differences lie in the level-of-measurement assumption (step 1), the statement of the null (step 2), and the computation of the test statistic (step 4).

7. If you are still confused about the uses of inferential statistics described in this chapter, don't be

alarmed or discouraged. A sizable volume of rather complex material has been presented and only rarely will a beginning student fully comprehend the unique logic of hypothesis testing on the first exposure. After

all, it is not every day that you learn how to test a statement you don't believe (the null hypothesis) against a distribution that doesn't exist (the sampling distribution)!

SUMMARY OF FORMULAS

Single-sample means, large samples:

8.1 $Z \text{ (obtained)} = \dfrac{\bar{X} - \mu_{\bar{X}}}{\sigma/\sqrt{N}}$

Single-sample means when samples are small and population standard deviation is unknown:

8.2 $t \text{ (obtained)} = \dfrac{\bar{X} - \mu}{s/\sqrt{N - 1}}$

Single-sample proportions, large samples:

8.3 $Z \text{ (obtained)} = \dfrac{P_s - P_u}{\sqrt{P_u(1 - P_u)/N}}$

GLOSSARY

Alpha level (α). The proportion of area under the sampling distribution that contains unlikely sample outcomes, given that the null hypothesis is true. Also, the probability of Type I error.

Critical region (region of rejection). The area under the sampling distribution that, in advance of the test itself, is defined as including unlikely sample outcomes, given that the null hypothesis is true.

Five-step model. A step-by-step guideline for conducting tests of hypotheses. A framework that organizes decisions and computations for all tests of significance.

Hypothesis testing. Statistical tests that estimate the probability of sample outcomes if assumptions about the population (the null hypothesis) are true.

Null hypothesis (H_0). A statement of "no difference." In the context of single-sample tests of significance, the population from which the sample was drawn is assumed to have a certain characteristic or value.

One-tailed test. A type of hypothesis test used when (1) the direction of the difference can be predicted or (2) concern focuses on outcomes in only one tail of the sampling distribution.

Research hypothesis (H_1). A statement that contradicts the null hypothesis. In the context of single-sample tests of significance, the population from

which the sample was drawn does not have a certain characteristic or value.

Significance testing. See Hypothesis testing

t (critical). The t score that marks the beginning of the critical region of a t distribution.

t distribution. A distribution used to find the critical region for tests of sample means when σ is unknown and sample size is small.

t (obtained). The test statistic computed in step 4 of the five-step model. The sample outcome expressed as a t score.

Test statistic. The value computed in step 4 of the five-step model that converts the sample outcome into either a t score or a Z score.

Two-tailed test. A type of hypothesis test used when (1) the direction of the difference cannot be predicted or (2) concern focuses on outcomes in both tails of the sampling distribution.

Type I error (alpha error). The probability of rejecting a null hypothesis that is, in fact, true.

Type II error (beta error). The probability of failing to reject a null hypothesis that is, in fact, false.

Z (critical). The Z score that marks the beginnings of the critical region on a Z distribution.

Z (obtained). The test statistic computed in step 4 of the five-step model. The sample outcomes expressed as a Z score.

PROBLEMS

8.1 a. For each situation, find Z (critical).

Alpha	Form	Z (critical)
.05	One-tailed	
.10	Two-tailed	
.06	Two-tailed	
.01	One-tailed	
.02	Two-tailed	

b. For each situation, find the critical t score.

Alpha	Form	N	t (critical)
.10	Two-tailed	31	
.02	Two-tailed	24	
.01	Two-tailed	121	
.01	One-tailed	31	
.05	One-tailed	61	

c. Compute the appropriate test statistic (Z or t) for each situation:

a. $\mu = 2.40$ $\bar{X} = 2.20$
$\sigma = 0.75$ $N = 200$
b. $\mu = 17.1$ $\bar{X} = 16.8$
$s = 0.9$
$N = 45$
c. $\mu = 10.2$ $\bar{X} = 9.4$
$s = 1.7$
$N = 150$
d. $P_u = .57$ $P_s = 0.60$
$N = 117$
e. $P_u = 0.32$ $P_s = 0.30$
$N = 322$

8.2 [SOC] The student body at St. Algebra College attends an average of 3.3 parties per month. A random sample of 117 sociology majors averages 3.8 parties per month with a standard deviation of 0.53. Do sociology majors have significantly more fun?

8.3 [SW]
a. Nationally, social workers average 10.2 years of experience. In a random sample, 203 social workers in greater metropolitan Shinbone average only 8.7 years with a standard deviation of 0.52. Are social workers in Shinbone significantly less experienced?
b. The same sample of social workers reports an average annual salary of $25,782 with a standard deviation of $622. Is this figure signifi-

cantly higher than the national average of $24,509?

8.4 [SOC] Nationally, the average score on the college entrance exams (verbal test) is 453 with a standard deviation of 95. A random sample of 152 freshmen entering St. Algebra College shows a mean score of 502. Is there a significant difference?

8.5 [SOC] A random sample of 423 Chinese Americans has finished an average of 12.7 years of formal education with a standard deviation of 1.7. Is this significantly different from the national average of 12.2 years?

8.6 [SOC] A sample of 105 workers in the Overkill Division of the Machismo Toy Factory earns an average of $24,375 per year. The average salary for all workers is $24,230 with a standard deviation of $523. Are workers in the Overkill Division overpaid?

8.7 [GER]
a. Nationally, the population as a whole watches 6.2 hours of TV per day. A random sample of 1017 senior citizens report watching an average of 5.9 hours per day with a standard deviation of .7. Is the difference significant?
b. The same sample of senior citizens reports that they belong to an average of 2.1 voluntary organizations and clubs with a standard deviation of .5. Nationally, the average is 1.7. Is the difference significant?

8.8 [SOC] A school system has assigned several hundred "chronic and severe underachievers" to an alternative educational experience. To assess the program, a random sample of 35 has been selected for comparison with all students in the system.
a. In terms of GPA, did the program work?

Systemwide GPA	Program GPA
$\mu = 2.47$	$\bar{X} = 2.55$
	$s = 0.70$
	$N = 35$

b. In terms of absenteeism (number of days missed per year), what can be said about the success of the program?

Systemwide	Program
$\mu = 6.13$	$\bar{X} = 4.78$
	$s = 1.11$
	$N = 35$

c. In terms of standardized test scores in math and reading, was the program a success?

Math Test

Systemwide	Program
$\mu = 103$	$\bar{X} = 106$
	$s = 2.0$
	$N = 35$

Reading Test

Systemwide	Program
$\mu = 110$	$\bar{X} = 113$
	$s = 2.0$
	$N = 35$

8.9 [SOC] A random sample of 26 local sociology graduates scored an average of 458 on the GRE advanced sociology test with a standard deviation of 20. Is this significantly different from the national average? ($\mu = 440$)

8.10 [PA] Nationally, the per capita property tax is $130. A random sample of 36 southeastern cities average $98 with a standard deviation of $5. Is the difference significant? Summarize your conclusions in a sentence or two.

8.11 [GER/CJ] A survey shows that 10% of the population is victimized by property crime each year. A random sample of 527 older citizens (65 years or more of age) shows a victimization rate of 14%. Are older people more likely to be victimized?

8.12 [CJ] A random sample of 113 convicted rapists in a state prison system completed a program designed to change their attitudes towards women, sex, and violence before being released on parole. Fifty-eight eventually became repeat sex offenders. Is this recidivism rate significantly different from the rate for all offenders (57%) in that state? Summarize your conclusions in a sentence or two.

8.13 [PS] In a recent statewide election, 55% of the voters rejected a proposal to institute a state lottery. In a random sample of 150 urban precincts, 49% of the voters rejected the proposal. Is the difference significant? Summarize your conclusions in a sentence or two.

8.14 [CJ] Statewide, the police clear by arrest 35% of the robberies and 42% of the aggravated assaults reported to them. A researcher takes a random sample of all the robberies ($N = 207$) and aggravated assaults ($N = 178$) reported to a metropolitan police department in one year and finds that 83 of the robberies and 80 of the assaults were cleared by arrest. Are the local arrest rates significantly different from the statewide rate? Write a sentence or two interpreting your decision.

8.15 [SOC/SW] A researcher has compiled a file of information on a random sample of 317 families in a city that have chronic, long-term patterns of child abuse. Below are reported some of the characteristics of the sample along with values for the city as a whole. For each trait, test the null hypothesis of "no difference" and summarize your findings.

a. Mothers' educational level (proportion completing high school):

City	Sample
$P_u = 0.63$	$P_s = 0.61$

b. Family size (proportion of families with four or more children):

City	Sample
$P_u = 0.21$	$P_s = 0.26$

c. Mothers' work status (proportion of mothers with jobs outside the home):

City	Sample
$P_u = 0.51$	$P_s = 0.27$

d. Relations with kin (proportion of families that have contact with kin at least once a week):

City	Sample
$P_u = 0.82$	$P_s = 0.43$

e. Fathers' educational achievement (average years of formal schooling):

City	Sample
$\mu = 12.3$	$\bar{X} = 12.5$
	$s = 1.7$

f. Fathers' occupational stability (average years in present job):

City	Sample
$\mu = 5.2$	$\bar{X} = 3.7$
	$s = 0.5$

8.16 SOC A sociologist has developed a file of information on the students graduating from her university for the past several years. She has information on every member of the graduating class from last year and information on a random sample of 213 members of this year's class. On each of the characteristics noted below, is there a significant difference?

a. Average number of major term papers written, all four years:

Last Year's Class	This Year's Class
$\mu = 25.2$	$\bar{X} = 24.7$
	$s = 2.7$

b. Grade-point average:

Last Year's Class	This Year's Class
$\mu = 2.67$	$\bar{X} = 2.71$
	$s = 0.50$

c. Number of job offers received before commencement:

Last Year's Class	This Year's Class
$\mu = 5.4$	$\bar{X} = 3.6$
	$s = 4.0$

d. Proportion majoring in the social sciences:

Last Year's Class	This Year's Class
$P_u = 0.17$	$P_s = 0.21$

e. Proportion attending commencement ceremony:

Last Year's Class	This Year's Class
$P_u = 0.88$	$P_s = 0.80$

8.17 SW You are the head of an agency seeking funding for a program to reduce unemployment among teenage males. Nationally, the unemployment rate for this group is 18%. A random sample of 323 teenage males in your area reveals an unemployment rate of 21.7%. Is the difference significant? Can you demonstrate a need for the program? Explain the result of your test of significance as you would to a funding agency.

8.18 PA The city manager of Shinbone has received a complaint from the local union of firefighters to the effect that they are underpaid. Not having much time, the city manager gathers the records of a random sample of 27 firefighters and finds that their average salary is $38,073 with a standard deviation of $575. If she knows that the average salary nationally is $38,202, how can she respond to the complaint? What would she say in a memo to the union that would respond to the complaint?

8.19 The following essay questions review the basic principles and concepts of inferential statistics. The order of the questions roughly follows the five-step model.

a. Hypothesis testing or significance testing can be conducted only with a random sample. Why?

b. Under what specific conditions can it be assumed that the sampling distribtion is normal in shape?

c. Explain the role of the sampling distribution in a test of hypothesis.

d. The null hypothesis is an assumption about reality that makes it possible to test sample outcomes for their significance. Explain.

e. What is the critical region? How is the size of the critical region determined?

f. Describe a research situation in which a one-tailed test of hypothesis would be appropriate.

g. Thinking about the shape of the sampling distribution, why does use of the t distribution (as opposed to the Z distribution) make it more difficult to reject the null hypothesis?

h. What exactly can be concluded in the one-sample case when the test statistic falls in the critical region?

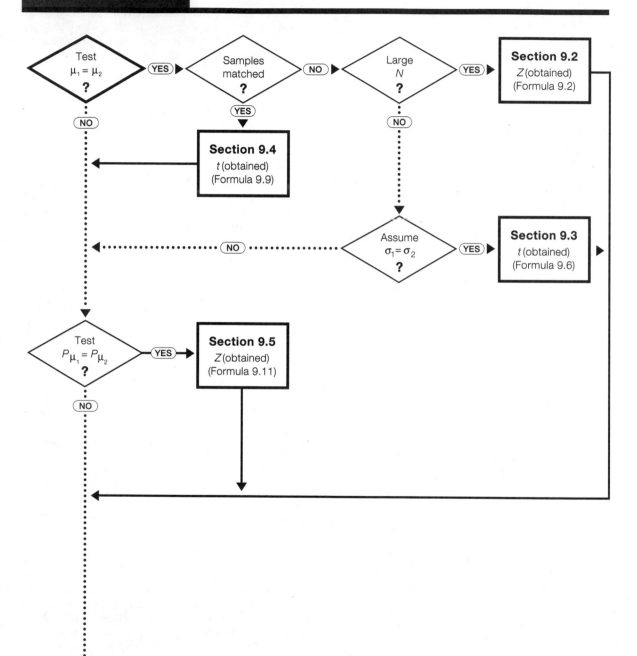

9.1 INTRODUCTION

In Chapter 8, we dealt with hypothesis testing in the one-sample case. In that situation, our concern was with the significance of the difference between a sample value and a population value. In this chapter, we will consider research situations where we will be concerned with the significance of the difference between two separate populations. For example, do men and women in the United States vary in their support for gun control? Obviously, we cannot ask every male and female for their opinions on this issue. Instead, we must draw random samples of both groups and use the information gathered from these samples to infer population patterns.

The central question asked in hypothesis testing in the two-sample case is "Is the difference between the samples large enough to allow us to conclude (with a known probability of error) that the populations represented by the samples are different?" Thus, if we find a large enough difference in support for gun control between random samples of men and women, we can argue that the difference between the samples did not occur by simple random chance but, rather, represents a real difference between men and women in the population.

In this chapter, we will consider tests for the significance of the difference between sample means and sample proportions. In both tests, the five-step model will serve as a framework for organizing our decision making. Although the general flow of the hypothesis-testing process is very similar to the one followed in the one-sample case, some important differences will be identified as we proceed through the material.

9.2 HYPOTHESIS TESTING WITH SAMPLE MEANS (LARGE SAMPLES)

One important difference between the one- and two-sample situations occurs in step 1 of the five-step model. In the one-sample case, we assumed "random sampling" or that the sample was selected from the population following the principle of EPSEM. As you recall, this means that the samples were selected so that each case in the population had an equal chance of being selected for the sample. In the two-sample situation, the samples must be selected independently as well as randomly. Samples meet this requirement if the selection of a case for one sample has no effect on the probability that any particular case will be included in the other sample. In our earlier example, this would mean that the selection of a specific male would have no effect on the probability of selecting any particular female. This new assumption will be stated as **independent random sampling** in step 1.

Drawing EPSEM samples from two separate lists (for example, one for females and one for males) can satisfy the requirement of independent random sampling. It is usually more convenient, however, to draw a single EPSEM sample from a single population list and then subdivide the cases into separate groups (males and females, for example). As long as the original sample is selected randomly, any subsamples created by the researcher will meet the assumption of independent random samples.

The second major difference in the five-step model for the two-sample case is in the form of the null hypothesis. The null is still a statement of "no difference." Now, however, instead of saying that a sample is no different from a population on the trait of interest, it will say that the two populations are no different ("There is no significant difference between men and women in their support of gun control"). This reflects the fact that our interest is centered on the populations per se and not on the samples. If the test statistic falls in the critical region, the null hypothesis of no difference between the populations can be rejected, and the argument that the populations are different on the trait of interest will be supported.

To illustrate the procedure for testing sample means, assume that a researcher has access to a nationally representative random sample and that the individuals in the sample have responded to a scale that measures attitudes toward gun control. The sample is divided by sex, and sample statistics are computed for males and females. Assuming that the scale yields interval-ratio level data, a test for the significance of the difference in sample means can be conducted.

The test statistic will be the difference in sample means. As long as sample size is large (that is, as long as the combined number of cases in the two samples exceeds 100), the sampling distribution of the differences in sample means will be normal in form, and the standard normal curve (Appendix A) can be used to find areas under the sampling distribution and establish the critical regions. The test statistic, Z (obtained), will be computed by the usual formula: sample outcome minus the mean of the sampling distribution, divided by the standard deviation of the sampling distribution. The formula is presented as Formula 9.1. Note that numerical subscripts are used to identify the two samples and the two populations they represent. The subscript $\bar{X} - \bar{X}$ is used to indicate that we are dealing with the sampling distribution of the differences in sample means.

FORMULA 9.1

$$Z\,(\text{obtained}) = \frac{(\bar{X}_1 - \bar{X}_2) - (\mu_1 - \mu_2)}{\sigma_{\bar{X} - \bar{X}}}$$

where $(\bar{X}_1 - \bar{X}_2)$ = the difference in the sample means
$(\mu_1 - \mu_2)$ = the difference in the population means
$\sigma_{\bar{X} - \bar{X}}$ = the standard deviation of the sampling distribution of the differences in sample means

The second term in the numerator, $(\mu_1 - \mu_2)$, might appear to be troublesome since these two values are unknown. Fortunately, this seemingly incalculable term reduces to zero because we assume that the null hypothesis (which will be stated as $H_0: \mu_1 = \mu_2$) is true. Recall that tests of significance are always based on the assumption that the null hypothesis is true. If the means of the two populations are assumed to be equal, then the term $(\mu_1 - \mu_2)$ will be zero and can be dropped from the equation. In effect, then, the formula we will actually use to compute the test statistic in step 4 will be

FORMULA 9.2

$$Z\text{(obtained)} = \frac{(\bar{X}_1 - \bar{X}_2)}{\sigma_{\bar{X}-\bar{X}}}$$

For large samples, the standard deviation of the sampling distribution of the difference in sample means is defined as

FORMULA 9.3

$$\sigma_{\bar{X}-\bar{X}} = \sqrt{\frac{\sigma_1^2}{N_1} + \frac{\sigma_2^2}{N_2}}$$

Since we will rarely be in a position to know the values of the population standard deviations (σ_1 and σ_2), we must use the sample standard deviations, suitably corrected for bias, to estimate them. Formula 9.4 displays the equation used to estimate the standard deviation of the sampling distribution in this situation. This is called a **pooled estimate** since it combines information from both samples.

FORMULA 9.4

$$\sigma_{\bar{X}-\bar{X}} = \sqrt{\frac{s_1^2}{N_1 - 1} + \frac{s_2^2}{N_2 - 1}}$$

The sample outcomes are reported below, and a test for the significance of the difference can now be conducted.

Sample 1 (men)	Sample 2 (women)
$\bar{X}_1 = 6.2$	$\bar{X}_2 = 6.5$
$s_1 = 1.3$	$s_2 = 1.4$
$N_1 = 324$	$N_2 = 317$

We see from the sample statistics that men have a lower average score on the Support for Gun Control Scale, indicating that they are less supportive of gun control than women. The test of hypothesis will tell us if this difference is large enough to justify the conclusion that it did not occur by random chance alone but rather reflects an actual difference between men and women on this issue.

Step 1. **Making Assumptions.** Note that, although we now assume that the random samples are independent, the remaining model assumptions are the same as in the one-sample case.

Model: Independent random samples
Level of measurement is interval-ratio
Sampling distribution is normal

Step 2. **Stating the Null Hypothesis.** The null hypothesis states that the *populations* represented by the samples are not different on this variable. Since no direction for the difference has been predicted, a two-tailed test is called for, as reflected in the research hypothesis.

$$H_0: \mu_1 = \mu_2$$

$$(H_1: \mu_1 \neq \mu_2)$$

Step 3. **Selecting the Sampling Distribution and Establishing the Critical Region.** For large samples, the Z distribution can be used to find areas under the sampling distribution and establish the critical region. Alpha will be set at 0.05.

$$\text{Sampling distribution} = Z\text{ distribution}$$
$$\text{Alpha} = 0.05, \text{two-tailed}$$
$$Z(\text{critical}) = \pm 1.96$$

Step 4. **Computing the Test Statistic.** Since the population standard deviations are unknown, Formula 9.4 will be used to estimate the standard deviation of the sampling distribution. This value will then be substituted into Formula 9.2 and Z (obtained) will be computed.

$$\sigma_{\bar{X}-\bar{X}} = \sqrt{\frac{s_1^2}{N_1 - 1} + \frac{s_2^2}{N_2 - 1}}$$

$$\sigma_{\bar{X}-\bar{X}} = \sqrt{\frac{(1.3)^2}{324 - 1} + \frac{(1.4)^2}{317 - 1}}$$

$$\sigma_{\bar{X}-\bar{X}} = \sqrt{(0.0052) + (0.0062)}$$

$$\sigma_{\bar{X}-\bar{X}} = \sqrt{0.0114}$$

$$\sigma_{\bar{X}-\bar{X}} = 0.107$$

$$Z(\text{obtained}) = \frac{(\bar{X}_1 - \bar{X}_2)}{\sigma_{\bar{X}-\bar{X}}}$$

$$Z(\text{obtained}) = \frac{6.2 - 6.5}{0.107}$$

$$Z(\text{obtained}) = \frac{-0.300}{0.107}$$

$$Z(\text{obtained}) = -2.80$$

Step 5. **Making a Decision.** Comparing the test statistic with the critical region:

$$Z(\text{obtained}) = -2.80$$
$$Z(\text{critical}) = \pm 1.96$$

We see that the Z score clearly falls in the critical region. This outcome indicates that a difference as large as -0.3 ($6.2 - 6.5$) between the sample means is unlikely if the null is true. The null hypothesis of no difference can be rejected, and the notion that men and women are different in terms of their support of gun control is supported. The decision to reject the null hypothesis has only a 0.05 probability (the alpha level) of being incorrect. *(For practice in testing the significance of the difference between sample means for large samples, see problems 9.1 to 9.6 and 9.18d to f.)*

9.3 HYPOTHESIS TESTING WITH SAMPLE MEANS (SMALL SAMPLES)

As with single-sample means, when the population standard deviation is unknown and sample size is small (combined Ns of less than 100), the Z distribution can no longer be used to find areas under the sampling distribution of the differences in sample means. Instead, we will use the t distribution to find the critical region and thus to identify unlikely sample outcomes. To utilize the t distribution for testing two sample means, we need to perform one additional calculation and make one additional assumption. The calculation is for degrees of freedom, a quantity required for proper use of the t distribution table (Appendix B). In the two-sample case, degrees of freedom are equal to $N_1 + N_2 - 2$.

The additional assumption is a more complex matter. With small samples, to justify the assumption of a normal sampling distribution and to form a pooled estimate of the standard deviation of the sampling distribution, we must assume that the variances of the populations of interest are equal ($\sigma_1^2 = \sigma_2^2$). The assumption of equal variance in the population can be tested by an inferential statistical technique known as the analysis of variance or ANOVA (see Chapter 10). For our purposes here, however, we will simply assume equal population variances without formal testing. This assumption can be considered justified as long as sample sizes are approximately equal.

To illustrate this procedure, assume that a researcher believes that center-city families are significantly larger than suburban families, as measured by number of children. Random samples from both areas are gathered and sample statistics computed.

Sample 1 (suburban)	Sample 2 (center-city)
$\bar{X}_1 = 2.37$	$\bar{X}_2 = 2.78$
$s_1 = 0.63$	$s_2 = 0.95$
$N_1 = 42$	$N_2 = 37$

The sample data reveal a difference in the predicted direction. The significance of this observed difference can be tested with the five-step model.

Step 1. Making Assumptions. Sample size is small, and the population standard deviation is unknown. Hence, we must assume equal population variances in the model.

Model: Independent random samples
Level of measurement is interval-ratio
Equal population variances ($\sigma_1^2 = \sigma_2^2$)
Sampling distribution is normal

Step 2. Stating the Null Hypothesis. Since a direction has been predicted (center-city families are larger), a one-tailed test will be used, and the research hypothesis is stated in accordance with this decision.

$$H_0: \mu_1 = \mu_2$$

$$(H_1: \mu_1 < \mu_2)$$

Step 3. Selecting the Sampling Distribution and Establishing the Critical Region. With small samples the t distribution will be used to establish the critical region. Alpha will be set at 0.05, and a one-tailed test will be used.

$$\text{Sampling distribution} = t \text{ distribution}$$
$$\text{Alpha} = 0.05, \text{ one-tailed}$$
$$\text{Degrees of freedom} = N_1 + N_2 - 2 = 42 + 37 - 2 = 77$$
$$t(\text{critical}) = -1.671$$

Note that the critical region is placed in the lower tail of the sampling distribution in accordance with the direction specified in H_1.

Step 4. Computing the Test Statistic. With small samples, a different formula (Formula 9.5) is used for the pooled estimate of the standard deviation of the sampling distribution. This value is then substituted directly into the denominator of the formula for t (obtained) given in Formula 9.6.

FORMULA 9.5

$$\sigma_{\bar{x}-\bar{x}} = \sqrt{\frac{N_1 s_1^2 + N_2 s_2^2}{N_1 + N_2 - 2}} \sqrt{\frac{N_1 + N_2}{N_1 N_2}}$$

$$\sigma_{\bar{x}-\bar{x}} = \sqrt{\frac{(42)(.63)^2 + (37)(.95)^2}{42 + 37 - 2}} \sqrt{\frac{42 + 37}{(42)(37)}}$$

$$\sigma_{\bar{x}-\bar{x}} = \sqrt{\frac{50.06}{77}} \sqrt{\frac{79}{1554}}$$

$$\sigma_{\bar{x}-\bar{x}} = (.81)(.23)$$
$$\sigma_{\bar{x}-\bar{x}} = .19$$

FORMULA 9.6

$$t(\text{obtained}) = \frac{\bar{X}_1 - \bar{X}_2}{\sigma_{\bar{x}-\bar{x}}}$$

$$t(\text{obtained}) = \frac{2.37 - 2.78}{.19}$$

$$t(\text{obtained}) = -\frac{.41}{.19}$$

$$t(\text{obtained}) = -2.16$$

Step 5. Making a Decision. Comparing the test statistic with the critical region,

$$t(\text{obtained}) = -2.16$$
$$t(\text{critical}) = -1.671$$

you can see that the test statistic falls in the critical region. If the null ($\mu_1 = \mu_2$) were true, this would be a very unlikely outcome, so the null can be rejected. There is a statistically significant difference (a difference so large that it is unlikely to be due to random chance) in the sizes of center-city and suburban families. Furthermore, center-city families are significantly larger in size. The test statistic and sampling distribution are depicted in Figure 9.1. *(For practice*

Application 9.1

An attitude scale measuring satisfaction with family life has been administered to a sample of married respondents. On this scale higher scores indicate greater satisfaction. The sample has been divided into respondents with no children and respondents with at least one child, and means and standard deviations have been computed for both groups. Is there a significant difference in satisfaction with family life between these two groups? The information necessary for conducting a test of the null hypothesis ($H_0: \mu_1 = \mu_2$) is

Sample 1 (no children)	Sample 2 (at least one child)
$\bar{X}_1 = 11.3$	$\bar{X}_2 = 10.8$
$s_1 = 0.6$	$s_2 = 0.5$
$N_1 = 78$	$N_2 = 93$

The estimate of the standard deviation of the sampling distribution would be

$$\sigma_{\bar{X}-\bar{X}} = \sqrt{\frac{s_1^2}{N_1 - 1} + \frac{s_2^2}{N_2 - 1}}$$

$$\sigma_{\bar{X}-\bar{X}} = \sqrt{\frac{(0.6)^2}{78 - 1} + \frac{(0.5)^2}{93 - 1}}$$

$$\sigma_{\bar{X}-\bar{X}} = \sqrt{0.008}$$

$$\sigma_{\bar{X}-\bar{X}} = 0.09$$

The test statistic, Z (obtained), would be

$$Z \text{ (obtained)} = \frac{\bar{X}_1 - \bar{X}_2}{\sigma_{\bar{X}-\bar{X}}}$$

$$Z \text{ (obtained)} = \frac{11.3 - 10.8}{0.09}$$

$$Z \text{ (obtained)} = \frac{0.50}{0.09}$$

$$Z \text{ (obtained)} = 5.56$$

If alpha were set at the 0.05 level, the critical region would begin at Z (critical) $= \pm 1.96$. With an obtained Z of 5.56, we would reject the null hypothesis. This test supports the conclusion that parents and childless couples are different with respect to satisfaction with family life.

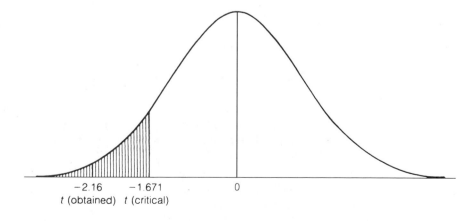

FIGURE 9.1   THE SAMPLING DISTRIBUTION WITH CRITICAL REGION AND TEST STATISTIC DISPLAYED

−2.16   −1.671   0
t (obtained) *t* (critical)

in testing the significance of the difference between sample means for small samples, see problems 9.7 to 9.9.)

9.4 HYPOTHESIS TESTING WITH SAMPLE MEANS (MATCHED SAMPLES)*

Some research designs require samples that are intentionally selected in violation of the assumption of independence. For example, some designs use matched samples in which subjects are selected so as to resemble each other as closely as possible. The matching might be based on any number of traits including age, gender, race, or social class. The strongest form of matching would be a "before/after" design where exactly the same subjects are tested before undergoing some experimental treatment and again after completing the treatment.

Matched-sample designs are frequently found in social psychological experiments, where they are used to eliminate the confounding effects of variables other than those that are relevant to the research project. If the two samples are matched on all dimensions, and one sample receives an experimental treatment and the other does not, the researcher can more confidently attribute any change in the experimental group to the treatment itself. The greater the extent to which the samples are matched, the greater the certainty that changes are not caused by some factor extraneous to the experiment.

In matched designs, the techniques described in the previous two sections can no longer be used because the researcher has intentionally violated the assumption of independent random samples. Instead, an alternative procedure that focuses on the differences between the two sets of scores is used. The null hypothesis is still a statement to the effect that there is "no difference." Specifically, the null states that the average difference between the two sets of scores will be zero. That is, if we found the difference between each pair of scores, added these differences, and divided by the number of pairs of scores, the result would be zero. This would be the case if the sets of scores were exactly the same. The greater the overall differences between the two sets of scores, the greater the average difference, and the more likely we will be to reject the null hypothesis.

To illustrate, suppose that 15 people were tested for their level of prejudice both before and after they viewed a movie that presented minority groups sympathetically. Scores from the scale used to measure prejudice increase as prejudice increases, and the results are presented in Table 9.1. Additional columns for computations have been added.

The difference, or D, column in the table displays the results of subtracting the second score (the posttest) from the first (the pretest). The values in the D^2 column are found by squaring the values in the D column. We will need these quantities to complete our calculations. The mean difference and

---
*This section is optional.

TABLE 9.1   PRETEST AND POSTTEST SCORES ON PREJUDICE, WITH COMPUTATIONAL COLUMNS ADDED

Case	Pretest	Posttest	Difference (D)	D²
1	10	10	0	0
2	11	10	1	1
3	11	11	0	0
4	15	14	1	1
5	16	10	6	36
6	17	10	7	49
7	20	15	5	25
8	20	16	4	16
9	20	19	1	1
10	25	20	5	25
11	26	25	1	1
12	27	23	4	16
13	28	20	8	64
14	30	25	5	25
15	35	30	5	25
	311	258	53	285

the standard deviation of the differences are computed by using formulas 9.7 and 9.8.

FORMULA 9.7
$$\bar{X}_D = \frac{\Sigma D}{N}$$

where $\bar{X}_D$ is the mean of the differences
ΣD is the summation of the D column

FORMULA 9.8
$$s_D = \sqrt{\frac{\Sigma D^2 - (\Sigma D)^2/N}{N}}$$

where s_D is the standard deviation of the differences
ΣD is the summation of the Difference (D) column
ΣD^2 is the summation of the D^2 column

The test statistic computed in step 4 of the five-step model will be a t score:

FORMULA 9.9
$$t\text{(obtained)} = \frac{\bar{X}_D}{s_D/\sqrt{N-1}}$$

We can now apply the five-step model to this situation.

Step 1. Making Assumptions.

Model: Random sampling
Level of measurement is interval-ratio
The population of all differences is normally distributed

Note that we are still treating the 15 cases as a random sample. The assumption of normal population distributions is necessary because of the small sample size.

S

Step 2. Stating the Null Hypothesis.

$$H_0: \mu_D = 0$$

$$(H_1: \mu_D \neq 0)$$

Step 3. Selecting the Sampling Distribution and Establishing the Critical Region.

$$\text{Sampling distribution} = t \text{ distribution}$$
$$\text{Alpha} = .05, \text{two-tailed}$$
$$\text{Degrees of freedom} = N - 1 = 15 - 1 = 14$$
$$t(\text{critical}) = \pm 2.145$$

Note that the degrees of freedom are based on an N of 15, *not* 30, as would be the case if we had independent random samples.

Step 4. Computing the Test Statistic.

$$\bar{X}_D = \frac{\Sigma D}{N}$$

$$X_D = \frac{53}{15} = 3.53$$

$$s_D = \sqrt{\frac{\Sigma D^2 - (\Sigma D)^2/N}{N}}$$

$$s_D = \sqrt{\frac{(285) - (53)^2/15}{15}}$$

$$s_D = \sqrt{\frac{(285) - 187.27}{15}}$$

$$s_D = \sqrt{97.73/15}$$

$$s_D = \sqrt{6.52}$$

$$s_D = 2.55$$

$$t(\text{obtained}) = \frac{X_D}{s_D\sqrt{N-1}}$$

$$t(\text{obtained}) = \frac{3.53}{2.55/\sqrt{14}}$$

$$t(\text{obtained}) = \frac{3.53}{.68}$$

$$t(\text{obtained}) = 5.19$$

Step 5. Making a Decision. The test statistic is in the critical region, so we can reject the null hypothesis. The difference in prejudice scores from the pretest to the posttest is statistically significant. The evidence suggests that the experience of viewing the movie did have an impact on prejudice. *(For practice in significance testing with matched samples, see problems 9.10 to 9.12.)*

9.5 HYPOTHESIS TESTING WITH SAMPLE PROPORTIONS (LARGE SAMPLES)

Testing for the significance of the difference between two sample proportions is analogous to testing sample means. The null hypothesis states that no difference exists between the populations from which the samples are drawn on the trait being tested. The sample proportions form the basis of the test statistic computed in step 4, which is then compared with the critical region. When sample sizes are large (combined Ns of more than 100), the Z distribution may be used to find the critical region. We will not consider tests of significance for proportions based on small samples or matched samples in this text. The formula for finding Z (obtained) in step 4 is

FORMULA 9.10

$$Z\text{(obtained)} = \frac{(P_{s1} - P_{s2}) - (P_{u1} - P_{u2})}{\sigma_{p-p}}$$

where $(P_{s1} - P_{s2})$ = the difference between the sample proportions
$(P_{u1} - P_{u2})$ = the difference between the population proportions
σ_{p-p} = the standard deviation of the sampling distribution of the differences in sample proportions

As was the case with sample means, the second term in the numerator is assumed to be 0 by the null. Therefore, the formula reduces to

FORMULA 9.11

$$Z\text{(obtained)} = \frac{P_{s1} - P_{s2}}{\sigma_{p-p}}$$

The value of the standard deviation of the sampling distribution of the differences in sample proportions is found by solving Formula 9.12:

FORMULA 9.12

$$\sigma_{p-p} = \sqrt{P_u(1 - P_u)} \ \sqrt{(N_1 + N_2)/N_1 N_2}$$

Solving this formula requires that a value for P_u be estimated. This is done by pooling information from both samples as specified in Formula 9.13:

FORMULA 9.13

$$P_u = \frac{N_1 P_{s1} + N_2 P_{s2}}{N_1 + N_2}$$

To find the test statistic, Z (obtained), in step 4, we must begin with Formula 9.13 and work back to Formula 9.11. A sample problem will clarify these procedures.

Random samples of black and white senior citizens have been gathered, and each respondent has been classified as "highly participatory" or "not highly participatory" in terms of the number of memberships he or she holds in voluntary associations. Is there a statistically significant difference in the

---

Application 9.2

Do attitudes toward sex vary by gender? The respondents in a national survey have been asked if they think that premarital sex is "always wrong" or only "sometimes wrong." The proportion of each sex that feels that premarital sex is always wrong is

Sample 1 Females	Sample 2 Males
$P_{s1} = 0.35$	$P_{s2} = 0.32$
$N_1 = 450$	$N_2 = 417$

This is all the information we will need to conduct a test of the null hypothesis (H_0: $P_{u1} = P_{u2}$). The estimate of the population proportion is

$$P_u = \frac{N_1 P_{s1} + N_2 P_{s2}}{N_1 + N_2}$$

$$P_u = \frac{(450)(0.35) + (417)(0.32)}{450 + 417}$$

$$P_u = \frac{290.94}{867}$$

$$P_u = 0.34$$

The standard deviation of the sampling distribution would be

$$\sigma_{p-p} = \sqrt{P_u(1 - P_u)} \sqrt{\frac{N_1 + N_2}{N_1 N_2}}$$

$$\sigma_{p-p} = \sqrt{(0.34)(0.66)} \sqrt{\frac{450 + 417}{(450)(417)}}$$

$$\sigma_{p-p} = \sqrt{0.2244} \sqrt{\frac{867}{187650}}$$

$$\sigma_{p-p} = \sqrt{0.2244} \sqrt{0.0046}$$

$$\sigma_{p-p} = (0.47)(0.068)$$

$$\sigma_{p-p} = 0.032$$

The test statistic, Z (obtained), is

$$Z \text{ (obtained)} = \frac{P_{s1} - P_{s2}}{\sigma_{p-p}}$$

$$Z \text{ (obtained)} = \frac{0.35 - 0.32}{0.032}$$

$$Z \text{ (obtained)} = \frac{0.030}{0.032}$$

$$Z \text{ (obtained)} = 0.94$$

If alpha is set at the 0.05 level, the critical region would begin at Z (critical) = ±1.96. With an obtained Z score of 0.94, we would fail to reject the null hypothesis. There is no statistically significant difference between males and females on attitudes toward premarital sex.

---

participation patterns of black and white elderly? Sample data for the proportions of each group classified as highly participatory are reported below.

Sample 1 (black senior citizens)	Sample 2 (white senior citizens)
$P_{s1} = 0.34$	$P_{s2} = 0.25$
$N_1 = 83$	$N_2 = 103$

Step 1. Making Assumptions.

Model: Independent random samples
Level of measurement is nominal
Sampling distribution is normal

Step 2. **Stating the Null Hypothesis.** Since no direction has been predicted, this will be a two-tailed test.

$$H_0: P_{u1} = P_{u2}$$

$$(H_1: P_{u1} \neq P_{u2})$$

Step 3. **Selecting the Sampling Distribution and Establishing the Critical Region.** Since sample size is large, the Z distribution will be used to establish the critical region. Setting alpha at 0.05, we have

$$\text{Sampling distribution} = Z \text{ distribution}$$
$$\text{Alpha} = 0.05, \text{ two-tailed}$$
$$Z \text{(critical)} = \pm 1.96$$

Step 4. **Computing the Test Statistic.** Remember to begin with the formula for estimating P_u (Formula 9.13), substitute the resultant value in the formula for σ_{p-p} (Formula 9.12), and then solve for Z (obtained) (Formula 9.11).

$$P_u = \frac{N_1 P_{s1} + N_2 P_{s2}}{N_1 + N_2}$$

$$P_u = \frac{(83)(.34) + (103)(.25)}{83 + 103}$$

$$P_u = .29$$

$$\sigma_{p-p} = \sqrt{P_u(1 - P_u)} \sqrt{\frac{(N_1 + N_2)}{N_1 N_2}}$$

$$\sigma_{p-p} = \sqrt{(.29)(.71)} \sqrt{\frac{(83 + 103)}{(83)(103)}}$$

$$\sigma_{p-p} = (.45)(.15)$$

$$\sigma_{p-p} = .07$$

$$Z \text{(obtained)} = \frac{P_{s1} - P_{s2}}{\sigma_{p-p}}$$

$$Z \text{(obtained)} = \frac{.34 - .25}{.07}$$

$$Z \text{(obtained)} = 1.29$$

Step 5. **Making a Decision.** Since the test statistic, Z (obtained) = 1.29, does not fall in the critical region as marked by the Z (critical) of ± 1.96, we fail to reject the null hypothesis. The difference between the sample proportions is no greater than what would be expected if the null hypothesis were true and only random chance were operating. Black and white senior citizens are not significantly different in terms of participation patterns as measured in this test. *(For practice in testing the significance of the difference between sample proportions, see problems 9.13 to 9.18a to c.)*

READING STATISTICS 6: Hypothesis Testing

Professional researchers use a vocabulary that is much terser than ours when presenting the results of tests of significance. Because of space limitations, results are not reported in very much detail. Furthermore, because professional researchers can assume a certain level of statistical literacy in their audiences, they can avoid stating many of the details we have been so careful about. So, you are unlikely to find mention of the null hypothesis, the critical region, or the choice of a sampling distribution in the professional research literature.

Instead, researchers will report only the sample values (for example, means or proportion), the value of the test statistic (for example, a Z or t score), the alpha level, the degrees of freedom (if applicable), and sample size. The results of the example problem in Section 9.3 might be reported in the professional literature as "the difference between the sample means of 2.37 (suburban families) and 2.78 (center-city families) was tested and found to be significant ($t = -2.16$, df $= 77$, $p < 0.05$)." Note that the alpha level is reported as "$p < 0.05$." This is shorthand for "the probability of a difference of this magnitude occurring by chance alone, if the null hypothesis of no difference is true, is less than 0.05" and is a good illustration of how researchers can convey a great deal of information in just a few symbols. In a similar fashion, our somewhat long-winded phrase "the test statistic falls in the critical region and, therefore, the null hypothesis is rejected" is rendered tersely and simply: "the difference . . . was . . . found to be significant."

When researchers need to report the results of many tests of significance, they will often use a summary table in which they report only the sample information and whether the difference is significant at a certain alpha level. If you read the researcher's description and analysis of such tables, you should have little difficulty interpreting and understanding them. As a final note, these comments about how significance tests are re-

ported in the literature apply to all of the tests of hypotheses covered in Part II of this text.

Statistics in the Professional Literature

Professor Jennifer Glass used data from the 1977 Quality of Employment Survey to analyze the effect of sex segregation in occupations on a variety of working conditions. The Quality of Employment Survey was administered to a nationally representative sample of about 1500 Americans in 1972–1973 by the Institute for Social Research. See Reading Statistics 4 in Chapter 7 for a discussion of this type of database. Professor Glass was particularly concerned with the contention that ". . . sex segregation arises from women's preference for jobs that mesh easily with family caregiving . . ." (779). The table on page 217 reports the results of a series of t tests for sex differences on a variety of job characteristics as described or rated by the respondents.

This table reports, in a very compact way, full information for 15 different tests of significance. Asterisks are used to identify level of statistical significance, and the table reports means, standard deviations, and number of cases. Here's what Professor Glass has to say about these results:

Table 2 presents a preliminary breakdown of scores on each job description variable by sex. This table shows that women rate their jobs as slightly more socially rewarding and significantly more conflict-free than men rate their jobs. Women do not indicate that their jobs are any easier or more difficult than men's jobs.

For the rest of the job description variables, men's scores are significantly higher than women's scores. Men report that they are given more opportunity to use their skills in challenging work, that their jobs are more flexible, that promotions are handled more fairly, and that their chances for promotion are better. Men also have significantly higher wages and more benefits.

Professor Glass then goes on to report the results of other statistical tests. If you would like to learn more, bibliographical information is given below.

Means by Sex for Variables Used in the Analysis

	Women $N = 493$	Men $N = 793$
Social Rewards*	3.10 (.65)[a]	3.01 (.66)
Interpersonal Ease***	3.08 (1.01)	2.78 (.99)
Ease of Work	2.08 (.59)	2.10 (.60)
Flexibility**	2.52 (.65)	2.59 (.61)
Total Time[b] for Breaks***	63.53 (52.35)	83.47 (89.74)
Challenge and Interest***	3.16 (.87)	3.35 (.88)
Fairness of Promotion Policies*	2.32 (1.11)	2.46 (1.02)
Chances for Promotion**	2.08 (1.67)	2.39 (1.80)
Wages (in dollars)***	4.66 (2.57)	7.77 (5.07)
Benefits***	2.73 (1.29)	3.10 (1.10)
Age	36.29 (13.09)	37.54 (12.59)
Education (in years)	12.50 (2.77)	12.53 (2.92)
Years at Present Job***	5.68 (6.18)	7.67 (7.41)
Hours Worked per Week***	38.14 (8.67)	44.63 (11.73)
Years of Job Experience***	14.04 (10.53)	20.02 (12.66)

*Difference significant at $p < .05$ **$p < .01$ ***$p < .001$
[a]Standard deviations in parentheses
[b]Time in minutes per day

9.6 THE LIMITATIONS OF HYPOTHESIS TESTING

Given that we are usually interested in rejecting the null hypothesis, we should take a moment to consider systematically the factors that affect our decision in step 5. For all tests of hypothesis, the probability of rejecting the null is a function of four independent factors:

1. The size of the observed difference(s)
2. The alpha level
3. The use of one- or two-tailed tests
4. The size of the sample

Only the first of these four is not under the direct control of the researcher. The size of the difference (either between the sample outcome and the population value or between two sample outcomes) is a function, in part, of the testing procedures (that is, how variables are measured), but for the most part it should reflect the underlying realities we are trying to probe.

The relationship between alpha level and the probability of rejection is straightforward. The higher the alpha level, the larger the critical region, the higher the percentage of all possible sample outcomes that fall in the critical region, and the greater the probability of rejection. Thus, it is easier to reject the H_0 at the 0.05 level than at the 0.01 level, and easier still at the 0.10 level. The danger here, of course, is that higher alpha levels will lead to more frequent Type I errors, and we might find ourselves declaring rather small differences to be statistically significant. In similar fashion, use of the one-tailed test increases the probability of rejection (assuming that the proper direction of the difference has been predicted).

The final factor that affects the probability of rejecting the null hypothesis is sample size. The larger the sample, with all other factors constant, the higher the probability of rejecting H_0. This may appear to be a surprising relationship, but the reasons for it can be appreciated with a brief consideration of the formulas used to compute test statistics in step 4. In all these formulas, sample size (N) is in the "denominator of the denominator." Algebraically, this is equivalent to being in the numerator of the formula and means that the value of the test statistic is directly proportional to N and that the two will increase together. To illustrate, consider Table 9.2, which shows the value of the test statistic for single sample means from samples of various sizes.

The value of the test statistic, Z (obtained), increases as N increases even though none of the other terms in the formula changes. This pattern of higher probabilities for rejecting H_0 with larger samples holds for all tests of significance.

Let me make two points by way of conclusion. On one hand, this relationship between sample size and the probability of rejecting the null should not alarm us unduly. Larger samples are, after all, better approximations of the populations they represent. Thus, decisions based on larger samples, whether to reject or fail to reject the null hypothesis, can be regarded as more trustworthy than decisions based on small samples.

On the other hand, this relationship clearly underlines what is perhaps the most significant limitation of hypothesis testing. Simply because a difference is statistically significant does not guarantee that it is important in any other sense. Particularly with very large samples (say, Ns in excess of 1000),

TABLE 9.2   TEST STATISTICS FOR SINGLE SAMPLE MEANS COMPUTED FROM SAMPLES OF VARIOUS SIZES ($\bar{X} = 80$, $\mu = 79$, $s = 5$ for all problems)

Sample Size	Test Statistic, Z (obtained)
100	1.99
200	2.82
500	4.47

relatively small differences may be statistically significant. Even with small samples, of course, differences that are otherwise trivial or uninteresting may be statistically significant. The point I want to make is that there can be a difference between statistical significance on one hand and theoretical or practical importance on the other. In the context of quantitative research, statistical significance is a necessary but not sufficient condition for theoretical importance. Once the statistical significance of a research result has been demonstrated, the researcher still faces the task of evaluating the results in terms of the theory that guides the inquiry.

SUMMARY

1. A common research situation is to test for the significance of the difference between two populations. Sample statistics are calculated for random samples of each population, and then we test for the significance of the difference between the samples as a way of inferring differences between the specified populations.

2. When sample information is summarized in the form of sample means, and N is large, the Z distribution is used to find the critical region. When N is small, the t distribution is used to establish the critical region. In the latter circumstance, we must also assume equal population variances before forming a pooled estimate of the standard deviation of the sampling distribution.

3. When samples have been matched or paired, an alternative significance test must be used. The t test for matched samples, which is based on the differences between the sets of scores, should be used whenever the research design requires the researcher to violate the assumption of independence.

4. Differences in sample proportions may also be tested for significance. For large samples, the Z distribution is used to find the critical region.

5. In all tests of hypothesis, a number of factors affect the probability of rejecting the null: the size of the difference, the alpha level, the use of one- versus two-tailed tests, and sample size. Statistical significance is not the same thing as theoretical or practical importance. Even after a difference is found to be statistically significant, the researcher must still demonstrate the relevance or importance of his or her findings.

SUMMARY OF FORMULAS

Test statistic for two sample means, large samples:

9.1 $\qquad Z(\text{obtained}) = \dfrac{(\bar{X}_1 - \bar{X}_2) - (\mu_1 - \mu_2)}{\sigma_{\bar{X}-\bar{X}}}$

Test statistic for two sample means, large samples (simplified formula):

9.2 $\qquad Z(\text{obtained}) = \dfrac{(\bar{X}_1 - \bar{X}_2)}{\sigma_{\bar{X}-\bar{X}}}$

Standard deviation of the sampling distribution of the difference in sample means, large samples:

9.3 $\qquad \sigma_{\bar{X}-\bar{X}} = \sqrt{\dfrac{\sigma_1^2}{N_1} + \dfrac{\sigma_2^2}{N_2}}$

Pooled estimate of the standard deviation of the sampling distribution of the difference in sample means, large samples:

9.4 $\qquad \sigma_{\bar{X}-\bar{X}} = \sqrt{\dfrac{s_1^2}{N_1 - 1} + \dfrac{s_2^2}{N_2 - 1}}$

Pooled estimate of the standard deviation of the sampling distribution of the difference in sample means, small samples:

9.5 $\quad \sigma_{\bar{X}-\bar{X}} = \sqrt{\dfrac{N_1 s_1^2 + N_2 s_2^2}{N_1 + N_2 - 2}} \sqrt{\dfrac{N_1 + N_2}{N_1 N_2}}$

Test statistic for two sample means, small samples:

9.6 $\quad t\,(\text{obtained}) = \dfrac{\bar{X}_1 - \bar{X}_2}{\sigma_{\bar{X}-\bar{X}}}$

Mean of the differences, matched samples*:

9.7 $\quad \bar{X}_D = \dfrac{\Sigma D}{N}$

Standard deviation of the differences, matched samples*:

9.8 $\quad s_D = \sqrt{\dfrac{\Sigma D^2 - (\Sigma D)^2/N}{N}}$

Test statistic, matched samples*:

9.9 $\quad t\,(\text{obtained}) = \dfrac{\bar{X}_D}{s_D/\sqrt{N - 1}}$

Test statistic for two sample proportions, large samples:

9.10 $\quad Z\,(\text{obtained}) = \dfrac{(P_{s1} - P_{s2}) - (P_{u1} - P_{u2})}{\sigma_{p-p}}$

Test statistic for two sample proportions, large samples (simplified formula):

9.11 $\quad Z\,(\text{obtained}) = \dfrac{P_{s1} - P_{s2}}{\sigma_{p-p}}$

Standard deviation of the sampling distribution of the difference in sample proportions, large samples:

9.12 $\quad \sigma_{p-p} = \sqrt{P_u(1 - P_u)}\sqrt{\dfrac{(N_1 + N_2)}{N_1 N_2}}$

Pooled estimate of population proportion, large samples:

9.13 $\quad P_u = \dfrac{N_1 P_{s1} + N_2 P_{s2}}{N_1 + N_2}$

GLOSSARY

Independent random samples. Random samples gathered in such a way that the selection of a particular case for one sample has no effect on the probability that any other particular case will be selected for the other samples.

Pooled estimate. An estimate of the standard deviation of the sampling distribution of the difference in

sample means based on the standard deviations of both samples.

σ_{p-p}. Symbol for the standard deviation of the sampling distribution of the differences in sample proportions.

$\sigma_{\bar{X}-\bar{X}}$. Symbol for the standard deviation of the sampling distribution of the differences in sample means.

PROBLEMS

Problems 9.10 to 9.12 are optional (*).

 9.1 For each problem below, test for the significance of the difference in sample statistics using the five-step model. (*HINT: Remember to solve Formula 9.4 for $\sigma_{\bar{X}-\bar{X}}$ before attempting to solve Formula 9.2. Also, in Formula 9.4, perform the mathematical operations in the proper sequence. First*

square each sample standard deviaton, then divide by N − 1, add the resultant values, and then find the square root of the sum.)

a. Sample 1	Sample 2
$\bar{X}_1 = 72.5$	$\bar{X}_2 = 76.0$
$s_1 = 14.3$	$s_2 = 10.2$
$N_1 = 136$	$N_2 = 257$

---

*This formula is optional.

b.

Sample 1	Sample 2
$\bar{X}_1 = 107$	$\bar{X}_2 = 103$
$s_1 = 14$	$s_1 = 17$
$N_1 = 175$	$N_2 = 200$

9.2 [SOC] Gessner and Healey administered questionnaires to samples of undergraduates. Among other things, the questionnaires contained a scale that measured attitudes toward interpersonal violence (higher scores indicate greater approval of interpersonal violence). Test the results as reported below for sexual, racial, and social-class differences.

a.

Males	Females
$\bar{X}_1 = 2.99$	$\bar{X}_2 = 2.29$
$s_1 = 0.88$	$s_2 = 0.81$
$N_1 = 122$	$N_2 = 251$

b.

Blacks	Whites
$\bar{X}_1 = 2.76$	$\bar{X}_2 = 2.49$
$s_1 = 0.68$	$s_2 = 0.91$
$N_1 = 43$	$N_2 = 304$

c.

White Collar	Blue Collar
$\bar{X}_1 = 2.46$	$\bar{X}_2 = 2.67$
$s_1 = 0.91$	$s_2 = 0.87$
$N_1 = 249$	$N_2 = 97$

d. Summarize your results in terms of the significance and the direction of the differences. Which of these three factors seems to make the biggest difference in attitudes toward interpersonal violence?

9.3 [SOC] Do athletes in different sports vary in terms of intelligence? Below are reported College Board scores of random samples of college basketball and football players. Is there a significant difference? Write a sentence or two explaining the difference.

Basketball Players	Football Players
$\bar{X}_1 = 460$	$\bar{X}_2 = 442$
$s_1 = 92$	$s_2 = 57$
$N_1 = 102$	$N_2 = 117$

What about male and female college athletes?

Male	Female
$\bar{X}_1 = 452$	$\bar{X}_2 = 480$
$s_1 = 88$	$s_2 = 75$
$N_1 = 107$	$N_2 = 105$

9.4 [PA] A number of years ago, the fire department in Shinbone, Kansas, began recruiting minority group members through an affirmative action program. In terms of efficiency ratings as compiled by their superiors, how do the affirmative action employees rate? The ratings of random samples of both groups were collected, and the results are reported below (higher ratings indicate greater efficiency).

Affirmative Action	Regular
$\bar{X}_1 = 15.2$	$\bar{X}_2 = 15.5$
$s_1 = 3.9$	$s_2 = 2.0$
$N_1 = 97$	$N_2 = 100$

Write a sentence or two of interpretation.

9.5 [SOC] Are middle-class families more likely than working-class families to maintain contact with kin? Write a paragraph summarizing the results of these tests.

a. A sample of middle-class families reported an average of 7.3 visits per year with close kin while a sample of working-class families averaged 8.2 visits. Is the difference significant?

Middle Class	Working Class
$\bar{X}_1 = 7.3$	$\bar{X}_2 = 8.2$
$s_1 = 0.3$	$s_2 = 0.5$
$N_1 = 89$	$N_2 = 55$

b. The middle-class families averaged 2.3 phone calls and 8.7 e-mail messages per month with close kin. The working-class families averaged 2.7 calls and 5.7 e-mail messages per month. Are these differences significant?

Phone Calls	
Middle Class	Working Class
$\bar{X}_1 = 2.3$	$\bar{X}_2 = 2.7$
$s_1 = 0.5$	$s_2 = 0.8$
$N_1 = 89$	$N_2 = 55$

E-mail Messages	
Middle Class	Working Class
$\bar{X}_1 = 8.7$	$\bar{X}_2 = 5.7$
$s_1 = 0.3$	$s_2 = 1.1$
$N_1 = 89$	$N_2 = 55$

9.6 [SOC] Are college students who live in dormitories significantly more involved in campus life than students who commute to campus? The data below report the average number of hours per week students devote to extracurricular activities. Is the difference between these randomly selected samples of commuter and residential students significant?

Residential	Commuter
$\bar{X}_1 = 12.4$	$\bar{X}_2 = 10.2$
$s_1 = 2.0$	$s_2 = 1.9$
$N_1 = 158$	$N_2 = 173$

9.7 [SOC] Are senior citizens who live in retirement communities more socially active than those who live in age-integrated communities? Write a sentence or two explaining the results of these tests. *(HINT: Remember to use the proper formulas for small sample sizes.)*

a. A random sample of senior citizens living in a retirement village reported that they had an average of 1.42 face-to-face interactions per day with their neighbors. A random sample of those living in age-integrated communities reported 1.58 interactions. Is the difference significant?

Retirement Community	Age-integrated Neighborhood
$\bar{X}_1 = 1.42$	$\bar{X}_2 = 1.58$
$s_1 = 0.10$	$s_2 = 0.78$
$N_1 = 43$	$N_2 = 37$

b. Senior citizens living in the retirement village reported that they had 7.43 telephone calls with friends and relatives each week while those in the age-integrated communities reported 5.50 calls. Is the difference significant?

Retirement Community	Age-integrated Neighborhood
$\bar{X}_1 = 7.43$	$\bar{X}_2 = 5.50$
$s_1 = 0.75$	$s_2 = 0.25$
$N_1 = 43$	$N_2 = 37$

9.8 [SW] As the director of the local Boys Club, you have claimed for years that membership in your club reduces juvenile delinquency. Now, a cynical member of your funding agency has demanded proof of your claim. Fortunately, your local sociology department is on your side and springs to your aid with student assistants, computers, and hand calculators at the ready. Random samples of members and nonmembers are gathered and interviewed with respect to their involvement in delinquent activities. Each respondent is asked to enumerate the number of delinquent acts he has engaged in over the past year. The results are in and reported below (the average number of admitted acts of delinquency). What can you tell the funding agency?

Members	Nonmembers
$\bar{X}_1 = 10.3$	$\bar{X}_2 = 12.3$
$s_1 = 2.7$	$s_2 = 4.2$
$N_1 = 40$	$N_2 = 55$

9.9 [SOC] Are professional basketball teams significantly more oriented toward offense these days? Do they score more points, on the average, than teams in past years? Games are randomly selected from the most recent season and from 10 years ago. Is the difference in average points per game significant?

This Season	10 Years Ago
$\bar{X}_1 = 110$	$\bar{X}_2 = 102$
$s_1 = 10$	$s_2 = 9$
$N_1 = 35$	$N_2 = 37$

*9.10 [SW] A program designed to reduce complaints of sexual harassment in a randomly selected sample of 15 organizations has been implemented. The researchers have information on the average number of complaints per year before and after implementation. What can be said about the effectiveness of the program?

Organization	Average Number of Complaints Before Implementation	Average Number of Complaints After Implementation
A	33	34
B	25	27
C	18	7
D	45	20
E	19	5
F	10	2
G	10	0
H	15	2
I	15	12
J	14	10

Organization	Average Number of Complaints Before Implementation	Average Number of Complaints After Implementation
K	8	2
L	2	8
M	2	0
N	0	0
O	0	1

*9.11 [SOC] A randomly selected sample of entering college freshmen have participated in a special program designed to enhance their academic abilities, and their GPA's at the end of one year have been recorded. A group of 20 students from the same class who did not participate in the program have been selected as a control group, and they have been matched with the experimental group by gender, age, high-school class rank, College Board scores, and declared college major. The results (GPA's) are presented below. Can the program claim that it was successful?

Project Students	Control Group
3.45	3.41
3.10	3.01
3.04	2.41
3.01	2.00
3.00	2.56
3.00	3.76
2.89	2.80
2.76	2.90
2.75	2.01
2.67	1.90
2.50	2.65
2.50	2.56
2.40	2.40
2.33	1.67
2.30	1.50
2.00	1.00
2.00	2.89
2.00	1.79
1.80	1.00
1.50	1.23

*9.12 [SW] As one part of an attempt to deal with the problem of teen pregnancies as well as other sexual issues, the local juvenile court system has begun to refer all female clients to a sex education program. The premise of the program is that, with full and accurate information, people will make better, more responsible decisions about sex. The scores on a test of general knowledge about sex given before and after the course are presented below for a random sample of participants. Are graduates of the course in fact significantly more knowledgeable?

Student	Pretest	Posttest
A	8	10
B	10	12
C	15	16
D	6	18
E	9	12
F	15	14
G	7	14
H	5	7
I	18	18
J	17	18
K	13	15

9.13 For each problem, test the sample statistics for the significance of the difference. (*HINT: In testing proportions, remember to begin with Formula 9.13, then solve Formulas 9.12 and 9.11.*)

a.

Sample 1	Sample 2
$P_{s1} = 0.17$	$P_{s1} = 0.20$
$N_1 = 101$	$N_2 = 114$

b.

Sample 1	Sample 2
$P_{s1} = 0.62$	$P_{s2} = 0.60$
$N_1 = 532$	$N_2 = 478$

9.14 [CJ] About half of the police officers in Shinbone, Kansas, have completed a special course in investigative procedures. Has the course increased their efficiency in clearing crimes by arrest? The proportions of cases cleared by arrest for samples of trained and untrained officers are reported below.

Trained	Untrained
$P_{s1} = 0.47$	$P_{s2} = 0.43$
$N_1 = 157$	$N_2 = 113$

9.15 [SW] A large counseling center needs to evaluate several experimental programs. Write a paragraph summarizing the results of these tests. Did the new programs work?

a. One program is designed for divorce counseling; the key feature of the program is its counselors, who are married couples working in teams. About half of all clients have been randomly assigned to this special program and half to the regular program, and the proportion of cases that eventually ended in divorce was recorded for both. The results for random samples of couples from both programs are reported below. In terms of preventing divorce, did the new program work?

Special Program	Regular Program
$P_{s1} = 0.53$	$P_{s2} = 0.59$
$N_1 = 78$	$N_2 = 82$

b. The agency is also experimenting with peer counseling for depressed children. About half of all clients were randomly assigned to peer counseling. After the program had run for a year, a random sample of children from the new program were compared with a random sample of children who did not receive peer counseling. In terms of the percentage who were judged to be "much improved," did the new program work?

Peer Counseling	No Peer Counseling
$P_{s1} = 0.10$	$P_{s2} = 0.15$
$N_1 = 52$	$N_2 = 56$

9.16 [SOC] At St. Algebra College, the sociology and psychology departments have been feuding for years about the respective quality of their programs. In an attempt to resolve the dispute, you have gathered data about the graduate school experience of random samples of both groups of majors. The results are presented below: the proportion of majors who applied to graduate schools, the proportion of majors accepted in their preferred programs, and the proportion of these who completed their programs. As measured by these data, is there a significant difference in program quality?

a. Proportion of majors who applied to graduate school:

Sociology	Psychology
$P_{s1} = 0.53$	$P_{s2} = 0.40$
$N_1 = 150$	$N_2 = 175$

b. Proportion accepted by program of first choice:

Sociology	Psychology
$P_{s1} = 0.75$	$P_{s2} = 0.85$
$N_1 = 80$	$N_2 = 70$

c. Proportion completing the programs:

Sociology	Psychology
$P_{s1} = 0.75$	$P_{s2} = 0.69$
$N_1 = 60$	$N_2 = 60$

9.17 [CJ] The local police chief started a "crimeline" program some years ago and wonders if it's really working. The program publicizes unsolved violent crimes in the local media and offers cash rewards for information leading to arrests. Are "featured" crimes more likely to be cleared by arrest than other violent crimes? Results from random samples of both types of crimes are reported below.

Crimeline Crimes Cleared by Arrest	Noncrimeline Crimes Cleared by Arrest
$P_{s1} = .35$	$P_{s2} = .25$
$N_1 = 178$	$N_2 = 212$

9.18 [SOC] Some results from the 1996 General Social Survey are reported below in terms of differences by sex. The actual questions are listed in Appendix G. Which of these differences, if any, are significant? Write a sentence or two of interpretation for each test.

a. Proportion favoring gun control (GUNLAW):

Males	Females
$P_{s1} = .73$	$P_{s2} = .85$
$N_1 = 415$	$N_2 = 532$

b. Proportion agreeing that premarital sex is "always wrong" (PREMARSX):

Males	Females
$P_{s1} = .18$	$P_{s2} = .26$
$N_1 = 424$	$N_2 = 568$

c. Proportion voting for President Clinton in 1992 (PRES92):

Males	Females
$P_{s1} = .44$	$P_{s2} = .55$
$N_1 = 404$	$N_2 = 545$

d. Average occupational prestige (PRESTG80):

Males	Females
$\bar{X}_1 = 43.84$	$\bar{X}_2 = 42.95$
$s_1 = 14.27$	$s_2 = 13.46$
$N_1 = 633$	$N_2 = 767$

Ho=No-diff
btwn mean

e. Average rate of church attendance (ATTEND):

Males	Females
$\bar{X}_1 = 3.30$	$\bar{X}_2 = 4.12$
$s_1 = 2.55$	$s_2 = 2.66$
$N_1 = 624$	$N_2 = 795$

f. Number of children (CHILDS):

Males	Females
$\bar{X}_1 = 1.71$	$\bar{X}_2 = 1.98$
$s_1 = 1.65$	$s_2 = 1.69$
$N_1 = 633$	$N_2 = 812$

SPSS for Windows

Using *SPSS for Windows* to Test the Significance of the Difference Between Two Means

DEMONSTRATION 9.1 Do Men or Women Watch More TV?

SPSS for Windows includes several tests for the significance of the difference between means. In this demonstration, I'll explain the use of the **Independent-Samples T Test**, the test we covered in Sections 9.2 and 9.3. As an illustration, we'll use the 1996 GSS to test for differences between men and women in average hours of reported TV watching. If there is a statistically significant difference between the sample mean for men and the sample mean for women, we can reject the null hypothesis and conclude that the populations (men and women) are different on this variable.

Start *SPSS for Windows* and load the 1996 GSS data base. From the main menu bar, click **Statistics**, then **Compare Means**, and then **Independent-Samples T Test**. The **Independent-Samples T Test** dialog box will open with the usual list of variables on the left. Find and move the cursor over *tvhours* in the list (the variable label is HOURS PER DAY WATCHING TV) and click the top arrow in the middle of the window to move *tvhours* to the **Test Variable(s)** box. Next, find and highlight *sex* (the label is RESPONDENTS SEX) and click the bottom arrow in the middle of the window to move *sex* to the **Grouping Variable** box. Two question marks will appear in the **Grouping Variable** box, and the **Define Groups** button will become active. SPSS needs to know which cases go in which groups and, in the case at hand, the instructions we need to supply are straightforward. Males (indicated by a score of 1 on *sex*) go in group 1 and females (a score of 2) will go in group 2.

Click the **Define Groups** button and the **Define Groups** window will appear. The cursor will be blinking in the box beside Group 1—SPSS is asking for the score that will determine which cases go in this group. Type a 1 in this box (for males) and then click the box next to Group 2 and type a 2 (for females). Click **Continue** to return to the **Independent-Samples T Test** window and click **OK**. The following output will be produced:

T-Test

Group Statistics

	RESPONDENTS	N	Mean	Std. Deviation	Std. Error Mean
HOURS PER DAY WATCHING TV	MALE	424	2.88	2.31	.11
	FEMALE	565	3.04	2.47	.10

Independent Samples Test

		Levene's Test for Equality of Variances	
		F	Sig.
HOURS PER DAY WATCHING TV	Equal variances assumed	.362	.547
	Equal variances not assumed		

Independent Samples Test

		t-test for Equality of Means			
		t	df	Sig. (2-tailed)	Mean Difference
HOURS PER DAY WATCHING TV	Equal variances assumed	−1.082	987	.280	−.17
	Equal variances not assumed	−1.092	940.841	.275	−.17

Independent Samples Test

		t-test for Equality of Means		
			95% Confidence Interval of the Difference	
		Std. Error Difference	Lower	Upper
HOURS PER DAY WATCHING TV	Equal variances assumed	.15	−.47	.14
	Equal variances not assumed	.15	−.47	.13

In the first block of output are some descriptive statistics. There were 424 males in the sample and they watched TV an average of 2.88 hours a day with a standard deviation of 2.31. The 565 females watched an average of 3.04 hours per day with a standard deviation of 2.47. We will skip over the next block of output (which reports the results of a test for equality of the population variances).

The results of the test for significance are reported in the next block. *SPSS for Windows* does a separate test for each assumption about the population variance (see Sections 9.2 and 9.3), but we will look only at the 'Equal variances assumed' reported in the top row. This is basically the same model used in this chapter.

SPSS for Windows reports the *t* value (-1.082), the degrees of freedom (df = 987), and the 'Sig. (2-tailed)' (.280). This last piece of information is an alpha level (or a "p" level—see Reading Statistics 6) except that it is the *exact* probability of getting the observed difference in sample means if only chance is operating. Thus, there is no need to look up the test statistic in a *t* table. This value is greater than .05, our usual indicator of significance. We will fail to reject the null hypothesis and conclude that the difference is not statistically significant. On the average, men and women do not have significantly different TV-viewing habits. (The final block of output reports confidence interval estimates to the population mean. See Chapter 7.)

DEMONSTRATION 9.2 Using the COMPUTE Command to Test for Gender Differences in Attitudes About Abortion

I introduced the **Compute** command in Demonstration 4.2. To refresh your memory, I used **Compute** to create a summary scale (*abscale*) for attitudes toward abortion by adding the scores on the two constituent items (a*bhlth* and *abany*). Remember that, once created, a computed variable is added to the active file and can be used like any of the variables actually recorded in the file. If you did not save the data file with *abscale* included, you can quickly recreate the variable by following the instructions in Demonstration 4.2.

Here, we will test *abscale* for the significance of the difference by gender. Our question is: Do men and women have different attitudes toward abortion? If the difference in the sample means is large enough, we can reject the null hypothesis and conclude that the populations are different. Before we conduct the test, I should point out that the abortion scale used in this test is only ordinal in level of measurement. Scales like this are often treated as interval-ratio variables, but we should still be cautious in interpreting our results.

Follow the instructions in Demonstration 9.1 for using the **Independent-Samples T TEST** command, and move *abscale* rather than *tvhours* to the **Test Variable(s)** box. If necessary, repeat the procedure for making *sex* the grouping variable. Your output will be as shown:

`T-Test`

Group Statistics

	RESPONDENTS	N	Mean	Std. Deviation	Std. Error Mean
ABSCALE	MALE	383	2.6031	.6175	3.155E-02
	FEMALE	494	2.6478	.6507	2.927E-02

Independent Samples Test

		Levene's Test for Equality of Variances	
		F	Sig.
ABSCALE	Equal variances assumed	.985	.321
	Equal variances not assumed		

Independent Samples Test

		t-test for Equality of Means			
		t	df	Sig. (2-tailed)	Mean Difference
ABSCALE	Equal variances assumed	−1.030	875	.303	−4.4640E-02
	Equal variances not assumed	−1.037	840.248	.300	−4.4640E-02

Independent Samples Test

		t-test for Equality of Means		
			95% Confidence Interval of the Difference	
		Std. Error Difference	Lower	Upper
ABSCALE	Equal variances assumed	4.333E-02	−.1297	4.040E-02
	Equal variances not assumed	4.304E-02	−.1291	3.984E-02

The sample means are nearly identical in value (2.6031 versus 2.6478). Intuitively we know that this slight difference cannot be statistically significant; sure enough, the 'Sig. (2-tailed)', or alpha (or p), value is .303. As a final point, let me direct your attention to the "Number of Cases" column. This test was based on 383 men and 494 women, for a total of 877 people. The original sample included almost 1500 people. What happened to all those cases?

Recall from Appendix F that many of the items on the GSS are given to only about two-thirds of the sample. Furthermore, as I mentioned at the end of Chapter 4, the **Compute** command is designed to drop a case from the computations if it is missing a score on any of the constituent variables. So, most of the "missing cases" were not asked these questions, and the rest were respondents who did not answer one or both of the constituent abortion items. This phenomenon of diminishing sample size is a common problem in survey research and, at some point, the shrinking sample may jeopardize the integrity of the statistical inquiry.

DEMONSTRATION 9.3 Using the Compute and Recode Commands to Test for Differences in Attitudes on Abortion by Church Attendance

Now things are getting very tricky. In this demonstration, we will combine a computed and a recoded variable to conduct an additional test of significance. I introduced the **Recode** command in Demonstration 2.2. As you may recall, this procedure allows us to change the scores on a variable. In the previous demonstration, I used the **Recode** command to collapse the scores on *age* in order to make that variable more suitable for display in a frequency distribution. Here, we will dichotomize a variable (or recode it so that it has only two scores) so that we can use it in a t test.

In Demonstration 9.2, we tested for the significance of the difference between average *abscale* scores for men and women. We found that the difference was quite trivial and not at all significant. It is common for a researcher, after getting a result like this, to search for some other variable that might have a significant impact on *abscale*. The *attend* variable, a measure of how often the respondents attend church, might be a likely independent variable, but it has eight possible scores, not the two required by the logic of the t test. What can we do?

One possibility would be to collapse the variable into a dichotomy, thus satisfying the requirements for the t test. At what point should we divide the scores to create this dichotomy? If you look at the coding scheme for *attend* (see Appendix G), you will see that there is no obvious cutting point. The scores graduate from "never" to "more than once a week," and there is no point where we could unequivocally divide the sample into "low" and "high." One solution to this problem would be to divide the variable at the median. Although arbitrary, this choice does divide the sample into equal halves, an advantage for many statistical procedures.

Find the median on *attend* by running the **Frequencies** command. Locate the category with the middle case by using the cumulative percent column. This will be 'sevrl times a yr' (several times a year), or a score of 3. Dichotomize the scores at that point with the **Recode** command. Click **Transform** and then click **Recode**, and a window will open, giving us two choices. Remember that if we recode "into same variable," the recoded variable will replace the original variable. Instead, choose "into different variable," and the **Recode into Different Variable** window will open. Find *attend* in the variable list and click the arrow to move the variable name into the **Numeric Variable → Output Variable** box. Click the box below Output Variable and type a name for the new, recoded variable. I suggest *attendr* for 'recoded *attend*'. Click the **Change** button, and the expression

$$attend \rightarrow attendr$$

will appear in the **Numeric Variable → Output Variable** box.

We still have to tell SPSS how we want to recode *attend*. Click the **Old and New Values** button in the middle of the screen, and a new dialog box will open. Click the **Range** button in the left-hand column and type 0 (the lowest value on *attend*) in the left-hand box and 3 (the high score in the first interval for *attendr*) in the right-hand box. In the **New Value** box in the upper-right-hand corner, click **Value** and then type 1 in the **Value** dialog box. Click the **Add** button, and the expression

$$0 - 3 \rightarrow 1$$

will appear in the **Old → New** dialog box.

Click the **Range** button again and specify the lower and upper limits of the second interval in *attendr* as 4 and 8, respectively. Click **Value** in the **New Value** box and then type **2** in the **Value** dialog box. Click the **Add** button, and the expression

$$4 - 8 \rightarrow 2$$

will be added to the **Old → New** dialog box. Click **Continue** and then click **OK**, and *attendr* will be created.

Conduct the *t* test with *attendr* as the independent variable or grouping variable. Assuming that you still have the *abscale* variable available (if not, see Demonstration 9.2), click **Statistics**, **Compare Means**, and then **Independent-Samples T TEST**. Move *abscale* to the **Test Variable(s)** box and *attendr* to the **Grouping Variable** box. Click the **Define Groups** button and type a 1 for Group 1 (this places all respondents with lower rates on church attendance in this group)and a 2 for Group 2 (respondents with higher rates of attendance). Click **Continue** and **OK**, and the following output will be produced:

T-Test

Group Statistics

	ATTENDR	N	Mean	Std. Deviation	Std. Error Mean
ABSCALE	1.00	478	2.4749	.5630	2.575E-02
	2.00	384	2.8151	.6738	3.438E-02

Independent Samples Test

		Levene's Test for Equality of Variances	
		F	Sig.
ABSCALE	Equal variances assumed	.712	.399
	Equal variances not assumed		

Independent Samples Test

		t-test for Equality of Means			
		t	df	Sig. (2-tailed)	Mean Difference
ABSCALE	Equal variances assumed	−8.075	860	.000	−.3402
	Equal variances not assumed	−7.920	744.968	.000	−.3402

Independent Samples Test

		t-test for Equality of Means		
			95% Confidence Interval of the Difference	
		Std. Error Difference	Lower	Upper
ABSCALE	Equal variances assumed	4.213E-02	−.4229	−.2575
	Equal variances not assumed	4.296E-02	−.4245	−.2559

People who are lower in attendance (Group 1) have an average score on *abscale* of 2.4749, and more frequent attendees have an average score of 2.8151. This difference is statistically significant (the Sig. (2-tailed), or alpha, is very low, less than .000) and more frequent attenders are significantly more antiabortion.

Exercises

9.1 Are men significantly different from women in occupational prestige, education, or income? Using Demonstration 9.1 as a guide, substitute *prestg80*, *educ*, and *income91* for *tvhours*. Write a sentence or two summarizing the results of this test. (See Reading Statistics 6 for some ideas on writing up results.)

9.2 Does support for abortion vary by age, education, or income? Using Demonstration 9.3 as a guide, use the **Recode** command to collapse *age, educ*, and *income91* at their median scores. Don't forget to choose the "recode into different variable" option. Substitute the recoded *age, educ*, and *income91* for *attendr*. Write a sentence or two summarizing the results of these tests.

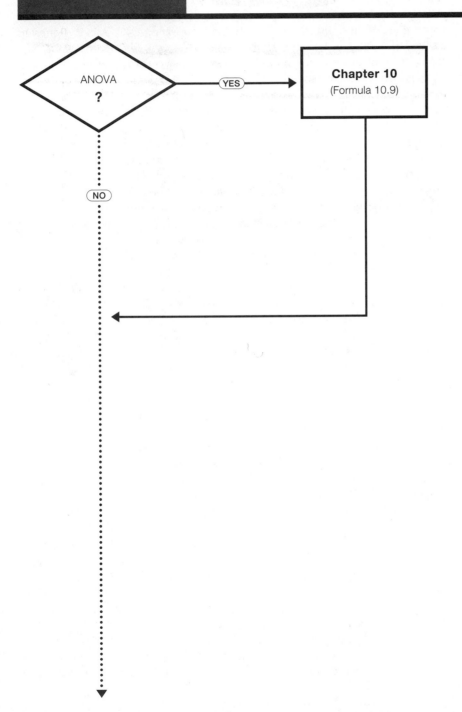

10.1 INTRODUCTION

In this chapter, we will examine a very flexible and widely used test of significance called the **analysis of variance** (often abbreviated as **ANOVA**). This test can be used in a number of situations where previously discussed tests are less than optimum or entirely inappropriate. ANOVA is designed to be used with interval-ratio level variables and is a powerful tool for analyzing the most sophisticated and precise measurements you are likely to encounter.

It is perhaps easiest to think of ANOVA as an extension of the t test for the significance of the difference between two sample means presented in Chapter 9. As you recall, the t test is appropriate when we are interested in the difference between exactly two populations but ANOVA allows us to investigate the significance of differences in a much broader range of situations.

To illustrate, suppose we were interested in examining the social basis of support for capital punishment. Why does support for the death penalty vary from person to person? One possibility would be to explore the relationship between religion and support for capital punishment. The death penalty is a moral issue, and it is certainly reasonable to expect that a person's general morality would be reflected and/or shaped by religion.

Suppose that we decide to conduct an investigation by devising a scale that measures support for capital punishment at the interval-ratio level and administering the scale to a randomly selected sample that includes Protestants, Catholics, Jews, and people with no religious affiliation ("Nones"). Take a couple of seconds to think through the logic of our situation here. We will wind up with four categories of subjects, and we will want to see if a particular attitude or opinion varies significantly by the category (religion) into which a person is classified. We will also want to answer other questions such as, Which religion shows the most support for capital punishment? The least? Are Protestants significantly more supportive than Catholics? Than Jews? How do people with no religious affiliation compare to people in the other categories? ANOVA will not answer all of these questions directly, but—as you will see—it will provide a very useful statistical context in which the questions can be addressed.

In considering this situation, some test options may have occurred to you. One possibility that may occur to you is to run a series of t tests between the various categories. You might begin by testing the mean score for the Protestants against the mean for the Catholics, then against the Jews, and so on, continuing until all possible combinations of religions have been compared. There are two disadvantages to this approach. The first, and more obvious, would simply be the amount of work involved. For the four religious affiliations, we would have to run six different t tests (six different sets of calculations, six trips through the five-step model, and so on) to exhaust all the possibilities.

A second disadvantage, which is less obvious but more important, has to do with Type I error (the probability of rejecting a null hypothesis that is, in fact, true). When conducting multiple tests of significance, the chance of committing at least one alpha error over the series of tests is greater than the selected alpha level for each individual test. Even if each single test is conducted at the .05 level, the combined probability of alpha error over the series of tests will be greater than .05. Thus, not only would multiple *t* tests require a great deal of computational work, but we would also increase our chances of making a rather serious error.

10.2 THE LOGIC OF THE ANALYSIS OF VARIANCE

For ANOVA, the null hypothesis is that the populations from which the samples are drawn are equal on the characteristic of interest. As applied to our problem, the null could be phrased as, "People of the various religious denominations do not vary in their support for the death penalty." The null will be stated symbolically as $\mu_1 = \mu_2 = \mu_3 = \cdots = \mu_k$, and you should see immediately that this is an extended version of the null for the two-sample *t* test. As usual, the researcher will normally be interested in rejecting the null and, in the case at hand, gathering evidence to show that support for capital punishment is related to religion.

If the null hypothesis of "no difference" in the populations is true, then any means calculated from randomly selected samples should be roughly equal in value. The average score for the Protestant sample should be about the same as the average score for the Catholics and the Jews, and so forth. Note that the averages are unlikely to be exactly the same value even if the null really is true, since we will always encounter some error or chance fluctuations in the measurement process. The question we are asking by applying the ANOVA test is not "Are there differences among the samples (or, in our example, religions)?" Rather, we are asking: "Are the differences among the samples large enough to justify a decision to reject the null hypothesis and conclude that the populations represented by the samples are different?"

Now, consider what kinds of outcomes we might encounter if we actually administered a "Support of Capital Punishment Scale" to a sample of respondents and organized the scores by religion. Of the infinite variety of possibilities, let's focus on two extreme outcomes as exemplified by Tables 10.1 and 10.2. In the first set of hypothetical results (Table 10.1), we see that the means and standard deviations of the groups are quite similar. The average scores are about the same for every religious group, and all four groups contain about the same dispersion. These results would be quite consistent with the

TABLE 10.1   SUPPORT FOR CAPITAL PUNISHMENT BY RELIGION (fictitious data)

	Protestant	Catholic	Jew	None
Mean =	10.3	11.0	10.1	9.9
Standard Deviation =	2.4	1.9	2.2	1.7

TABLE 10.2   SUPPORT FOR CAPITAL PUNISHMENT BY RELIGION (fictitious data)

	Protestant	Catholic	Jew	None
Mean =	14.7	11.3	5.7	8.3
Standard Deviation =	.5	1.3	.6	.8

null hypothesis of no difference and would suggest that, on this issue, neither the typical score nor the range of scores changes in any important way by religion.

Now consider another set of fictitious results as displayed in Table 10.2. Here we see substantial differences in average score from category to category, with Jews showing the lowest support and Protestants showing the strongest. Take a minute to observe the standard deviations for each category. The values for this statistic are low and similar from religion to religion, indicating that there is not much variation within the categories. The results in Table 10.2 show marked differences *between* religions combined with homogeneity *within* religions. In other words, each of the four religions is associated with a distinctly different level of support for the death penalty. There are marked differences from religion to religion combined with strong similarities within each religion. These results would certainly support the argument that religion is an important determinant of a person's support for the death penalty and would contradict the null.

In principle, ANOVA proceeds by making the kinds of comparisons outlined above. The test compares the amount of variation *between* categories (for example, from Protestants to Catholics to Jews to "Nones") with the amount of variation *within* categories (among Protestants, among Catholics, and so forth). The greater the differences between categories, relative to the differences within categories, the more likely that the null of "no difference" is false and can be rejected. If support for capital punishment truly varies by religion, then the sample mean for each of the religions should be quite different from the others, and dispersion within the categories should be relatively low.

10.3 THE COMPUTATION OF ANOVA

Even though we have been thinking of ANOVA as a test for the significance of the difference between sample means, the computational routine actually involves developing two separate estimates of the population variance, σ^2 (hence the name *analysis of variance*). Recall from Chapter 4 that the variance of any distribution is the standard deviation squared and that both statistics are measures of dispersion or variability. One estimate of the population variance will be based on the pattern of variation *within* each of the samples separately. The other estimate of the population variance is based on the pattern of variation *between* samples.

Before constructing these estimates, we need to introduce some new concepts and statistics to help us deal with the notion of analyzing the variance of a distribution of scores. The first new concept is the total variation of the scores, which is measured by a quantity called the **total sum of squares**, or **SST**:

total variation

FORMULA 10.1

$$SST = \Sigma(X_i - \bar{X})^2$$

To find this quantity, we would take each score, subtract the mean, square the difference, and then add to get the total of the squared differences. If this formula seems vaguely familiar, it's because the same expression appears in the numerator of the formula for the sample variance and the standard deviation (see Section 4.5). This redundancy is not surprising, considering that the SST is also a way of measuring dispersion.

To construct the two separate estimates of the population variance, the total variation, as measured by SST, is divided into two components. One of these reflects the pattern of variation within the categories and is called the **sum of squares within (SSW)**. The other component is based on the variation between categories and is called the **sum of squares between (SSB)**. The relationship of these three sums of squares is

FORMULA 10.2

$$SST = SSB + SSW$$

This mathematical relationship reflects the fact that SSW and SSB are components of SST.

The sum of squares within is defined as

FORMULA 10.3

$$SSW = \Sigma(X_i - \bar{X}_k)^2$$

where SSW = the sum of the squares within the categories
$\bar{X}_k$ = the mean of a category

This formula directs us to take each score, subtract the mean *of the category* from the score, square the result, and then sum the squared differences. We will do this for each category separately and then summate the squared differences for all categories to find SSW.

The between sum of squares (SSB) reflects the variation between the samples, with the category means serving as summary statistics for each category. Each category mean will be treated as a "case" for purposes of this estimate, and the formula for the sum of squares between (SSB) is

FORMULA 10.4

$$SSB = \Sigma N_k(\bar{X}_k - \bar{X})^2$$

where SSB = the sum of squares between the categories
N_k = the number of cases in a category
$\bar{X}_k$ = the mean of a category

To find SSB, subtract the overall mean of all scores from each category mean, square the difference, multiply by the number of cases in the category, and add the results across all the categories.

Let's pause for a second to remember what we are after here. If the null hypothesis is true, then there should not be much variation from category to category, relative to the variation within categories, and the two estimates to the population variance based on SSW and SSB should be roughly equal. Large differences between the two estimates would indicate important differences between categories relative to the differences within. If the differences between the two estimates are large enough, we will be able to reject the null hypothesis.

The next step in the computational routine is to construct the estimates of the population variance. To do this, we will divide each sum of squares by its respective degrees of freedom. To find the degrees of freedom associated with SSW, subtract the number of categories (k) from the number of cases (N). The degrees of freedom associated with SSB are the number of categories minus one. In summary,

FORMULA 10.5
$$\text{dfw} = N - k$$

where dfw = degrees of freedom associated with SSW
N = number of cases
k = number of categories

FORMULA 10.6
$$\text{dfb} = k - 1$$

where dfb = degrees of freedom associated with SSB
k = number of categories

The estimates, called the **mean square** estimates, are calculated by dividing each sum of squares by its respective degrees of freedom:

FORMULA 10.7
$$\text{Mean square within} = \frac{\text{SSW}}{\text{dfw}}$$

FORMULA 10.8
$$\text{Mean square between} = \frac{\text{SSB}}{\text{dfb}}$$

The actual test statistic, which we will calculate in step 4 of the five-step model, is called the **F ratio**, and its value is determined by the following formula:

FORMULA 10.9
$$F = \frac{\text{Mean square between}}{\text{Mean square within}}$$

As you can see, the value of the F ratio will be a function of the amount of variation between categories to the amount of variation within the categories. The greater the variation between the categories relative to the variation within, the higher the value of the F ratio and the more likely we will be able to reject the null.

**10.4 A COMPUTATIONAL
SHORTCUT**

The computational routine for ANOVA, as summarized in the previous section, requires a number of separate steps and different formulas and will almost certainly seem complicated and time-consuming the first time you see it. As is almost always the case, if you proceed systematically from formula to formula (in the correct order, of course), then the computations won't seem nearly so formidable. Unfortunately, they will still be lengthy and time-consuming. So, at the risk of stretching your patience, let me introduce a way to save some time and computational effort. This will require the introduction of even more formulas, but the eventual savings in time will be worth the effort.

The formulas presented to this point have been definitional formulas and, as we have seen in previous chapters, these are often not very convenient for computation. The inconvenience is particularly a problem for SST and SSW and, fortunately, a computational formula is available for SST:

FORMULA 10.10

$$SST = \Sigma X^2 - N\bar{X}^2$$

To solve this formula, first square the value of the overall mean, multiply the result by the total number of cases in the sample, and subtract that sum from the sum of the squared scores. In finding the latter quantity, ΣX^2, be sure that you use all of the scores, square each one, and then sum the squared scores.

Once we have the values of SST and SSB, we can find SSW by manipulating Formula 10.2 and doing some simple subtraction:

FORMULA 10.11

$$SSW = SST - SSB$$

The computational routine for ANOVA can be summarized as follows:

1. Find SST by Formula 10.10.
2. Find SSB by Formula 10.4.
3. Find SSW by subtraction (Formula 10.11).
4. Calculate the degrees of freedom (Formulas 10.5 and 10.6).
5. Construct the two mean square estimates to the population variance by dividing SSB and SSW by their respective degrees of freedom (Formulas 10.7 and 10.8).
6. Find the obtained F ratio by dividing the between estimate by the within estimate (Formula 10.9).

These computations and the actual test of significance will be illustrated in the next sections.

**10.5 A COMPUTATIONAL
EXAMPLE**

Assume that we have administered our "Support for Capital Punishment Scale" to a sample of 16 individuals who are equally divided into the four religions. (Obviously, this sample is much too small for any serious research and is intended solely for purposes of illustration.) All scores are reported in

TABLE 10.3   SUPPORT FOR CAPITAL PUNISHMENT BY RELIGION FOR 16 SUBJECTS (fictitious data)

Protestant		Catholic		Jew		None	
X	X^2	X	X^2	X	X^2	X	X^2
8	64	12	144	12	144	15	225
12	144	20	400	13	169	16	256
13	169	25	625	18	324	23	529
17	289	27	729	21	441	28	784
50	666	84	1898	64	1078	82	1794
$\bar{X}_k =$ 12.5		21.0		16.0		20.5	

$$\bar{X} = 280/16 = 17.5$$

Table 10.3 along with the squared scores, the category means, and the overall mean.

To organize our computations, we'll follow the routine summarized at the end of Section 10.4. We begin by finding SST by Formula 10.10:

$$\text{SST} = \Sigma X^2 - N\bar{X}^2$$
$$\text{SST} = (666 + 1898 + 1078 + 1794) - (16)(17.5)^2$$
$$\text{SST} = 5436 - (16)(306.25)$$
$$\text{SST} = 5436 - 4900$$
$$\text{SST} = 536$$

The sum of squares between is found by Formula 10.4:

$$\text{SSB} = \Sigma N_k(\bar{X}_k - \bar{X})^2$$
$$\text{SSB} = (4)(12.5 - 17.5)^2 + (4)(21.0 - 17.5)^2$$
$$\qquad\quad + (4)(16.0 - 17.5)^2 + (4)(20.5 - 17.5)^2$$
$$\text{SSB} = 4(-5)^2 + 4(3.5)^2 + 4(-1.5)^2 + 4(3)^2$$
$$\text{SSB} = 100 + 49 + 9 + 36$$
$$\text{SSB} = 194$$

Now SSW can be found by subtraction (Formula 10.11):

$$\text{SSW} = \text{SST} - \text{SSB}$$
$$\text{SSW} = 536 - 194$$
$$\text{SSW} = 342$$

To find the degrees of freedom for the two sums of squares, we use Formulas 10.5 and 10.6:

$$\text{dfw} = N - k = 16 - 4 = 12$$
$$\text{dfb} = k - 1 = 4 - 1 = 3$$

Finally, we are ready to construct the mean square estimates to the population variance. For the estimate based on SSW, we use Formula 10.7:

$$\text{Mean square within} = \frac{\text{SSW}}{\text{dfw}} = \frac{342}{12} = 28.5$$

For the between estimate, we use Formula 10.8:

$$\text{Mean square between} = \frac{\text{SSB}}{\text{dfb}} = \frac{194}{3} = 64.67$$

The test statistic, or F ratio, is found by dividing the between estimate by the within estimate (see Formula 10.9):

$$F(\text{obtained}) = \frac{\text{Mean square between}}{\text{Mean square within}}$$

$$F(\text{obtained}) = \frac{64.67}{28.5}$$

$$F(\text{obtained}) = 2.27$$

This statistic must still be evaluated for its significance. *(For practice in computing these quantities and solving these formulas see problem 10.1.)*

10.6 A TEST OF SIGNIFICANCE FOR ANOVA

In the previous section, we calculated a value for the test statistic F from the data on capital punishment and religion. In this section, we will see how to test that value for its significance and also take a look at some of the assumptions underlying the ANOVA test. As usual, we will follow the five-step model as a convenient way of organizing the decision-making process.

Step 1. **Making Assumptions.**

> Model: Independent random samples
> Level of measurement is interval-ratio
> Populations are normally distributed
> Population variances are equal

The model assumptions are quite stringent and underscore the fact that this technique should be used only with variables that have been carefully and precisely measured. As long as sample sizes are equal (or nearly so), ANOVA can tolerate some violation of the model assumptions, but—in situations where you are uncertain or have samples of very different size—it is probably advisable to use an alternative test. (Chi square in Chapter 12 is one option.)

Step 2. **Stating the Null Hypothesis.** For ANOVA, the null hypothesis always states that the means of the populations from which the samples were drawn are equal. For our example problem, we are concerned with four different populations, so our null hypothesis would be

$$H_0: \mu_1 = \mu_2 = \mu_3 = \mu_4$$

where μ_1 represents the mean for Protestants, μ_2, the mean for Catholics, and so forth.

Application 10.1

An experiment in teaching introductory biology was recently conducted at a large university. One section was taught by the traditional lecture-lab method, a second was taught by an all-lab/demonstration approach with no lectures, and a third was taught entirely by a series of videotaped lectures and demonstrations that the students were free to view at any time and as often as they wanted. Students were randomly assigned to each of the three sections and, at the end of the semester, random samples of final exam scores were collected from each section. Is there a significant difference in student performance by teaching method?

Final exam scores by teaching method

	Lecture		Demonstration		Videotape	
X	X^2	X	X^2	X	X^2
55	3,025	56	3,136	50	2,500
57	3,249	60	3,600	52	2,704
60	3,600	62	3,844	60	3,600
63	3,969	67	4,489	61	3,721
72	5,184	70	4,900	63	3,969
73	5,329	71	5,041	69	4,761
79	6,241	82	6,724	71	5,041
85	7,225	88	7,744	80	6,400
92	8,464	95	9,025	82	6,724

$\Sigma X = 636$ 651   588

$\Sigma X^2 =$ 46,286   48,503   39,420

$\bar{X}_k =$ 70.67   72.33   65.33

$$\bar{X} = 1{,}875/27 = 69.44$$

We can see by inspection that the "Videotape" group had the lowest average score and that the "Demonstration" group had the highest average score. The ANOVA test will tell us if these differ-ences are large enough to justify the conclusion that they did not occur by chance alone. Following the computational routine established at the end of Section 10.4:

$$\text{SST} = \Sigma X^2 - N\bar{X}^2$$
$$\text{SST} = (46286 + 48503 + 39420) - (27)(69.44)^2$$
$$\text{SST} = 134209 - 130191.67$$
$$\text{SST} = 4017.33$$

$$\text{SSB} = \Sigma N_k (\bar{X}_k - \bar{X})^2$$
$$\text{SSB} = 9(70.67 - 69.44)^2 + 9(72.33 - 69.44)^2$$
$$\quad\quad + 9(65.33 - 69.44)^2$$
$$\text{SSB} = 13.62 + 75.17 + 152.03$$
$$\text{SSB} = 240.82$$

$$\text{SSW} = \text{SST} - \text{SSB}$$
$$\text{SSW} = 4017.33 - 240.82$$
$$\text{SSW} = 3776.51$$

$$\text{dfw} = N - k = 27 - 3 = 24$$
$$\text{dfb} = k - 1 = 3 - 1 = 2$$

$$\text{Mean square within} = \frac{\text{SSW}}{\text{dfw}} = \frac{3776.51}{24} = 157.36$$

$$\text{Mean square between} = \frac{\text{SSB}}{\text{dfb}} = \frac{240.82}{2} = 120.41$$

$$F = \frac{\text{Mean square between}}{\text{Mean square within}} = \frac{120.41}{157.36} = 0.77$$

If we set alpha equal to .05, the critical region for 2 and 24 degrees of freedom would begin at 3.40, and our obtained F score is less than 1.00. We would clearly fail to reject the null hypothesis ("the population means are equal") and would conclude that the observed differences among the category means were the results of random chance. Student performance in this course does not vary significantly by teaching method.

The alternative hypothesis for this test states simply that at least one of the population means is different. The wording here is important. If we reject the null, ANOVA does not identify which of the means are significantly different.

(H_1: At least one of the population means is different.)

Step 3. **Selecting the Sampling Distribution and Establishing the Critical Region.** The sampling distribution for ANOVA is the F distribution, which is summarized in Appendix D. Note that there are separate tables for alphas of .05 and .01, respectively. As with the t table, the value of the critical F score will vary by degrees of freedom. For ANOVA, there are two separate degrees of freedom, one for each estimate of the population variance. The numbers across the top of the table are the degrees of freedom associated with the between estimate (dfb), and the numbers down the side of the table are those associated with the within estimate (dfw). In our example, the former is $(k - 1)$, or 3, and the latter is $(N - k)$, or 12 (see Formulas 10.5 and 10.6). So, if we set alpha at .05, our critical F score will be 3.49. Summarizing these considerations:

$$\text{Sampling distribution} = F \text{ distribution}$$
$$\text{Alpha} = .05$$
$$\text{Degrees of freedom (within)} = (N - k) = 12$$
$$\text{Degrees of freedom (between)} = (k - 1) = 3$$
$$F(\text{critical}) = 3.49$$

Taking a moment to inspect the two F tables, you will notice that all the values are greater than 1.00. This is because ANOVA is a one-tailed test, and we are concerned only with outcomes in which there is more variance between categories than within categories. F values of less than 1.00 would indicate that the between estimate was lower in value than the within estimate and, since we would always fail to reject the null in such cases, we simply ignore this class of outcomes.

Step 4. **Computing the Test Statistic.** This was done in the previous section, where we found an obtained F ratio of 2.27.

Step 5. **Making a Decision.** Compare the test statistic with the critical value:

$$F(\text{obtained}) = 2.27$$

$$F(\text{critical}) = 3.49$$

Since the test statistic does not fall in the critical region, our decision would be to fail to reject the null. Support for capital punishment does not differ significantly by religion, and the variation we observed in the sample means is unimportant.

10.7 AN ADDITIONAL EXAMPLE FOR COMPUTING AND TESTING THE ANALYSIS OF VARIANCE

In this section, we will work through an additional example of the computation and interpretation of the ANOVA test. We will first review matters of computation, find the obtained F ratio, and then test the statistic for its significance. In the computational section, we will follow the step-by-step guidelines presented at the end of Section 10.4.

A researcher has been asked to evaluate the efficiency with which each of three social service agencies is administering a particular program. One area of concern is the speed with which the agencies are processing paperwork and determining the eligibility of potential clients. The researcher has gathered information on the number of days required for processing a random sample of 10 cases in each agency. Is there a significant difference? The data are reported in Table 10.4, which also includes some additional information we will need to complete our calculations.

To find SST by the computational formula (10.10):

$$SST = \Sigma \bar{X}^2 - N\bar{X}^2$$
$$SST = (524 + 1816 + 2462) - 30(11.67)^2$$
$$SST = (4802) - 30(136.19)$$
$$SST = 4802 - 4085.70$$
$$SST = 716.30$$

To find SSB by Formula 10.4:

$$SSB = \Sigma N_k(\bar{X}_k - \bar{X})^2$$
$$SSB = 10(7 - 11.67)^2 + 10(13 - 11.67)^2 + 10(15 - 11.67)^2$$
$$SSB = 10(-4.67)^2 + 10(1.33)^2 + 10(3.33)^2$$
$$SSB = 10(21.81) + 10(1.77) + 10(11.09)$$
$$SSB = 218.09 + 17.69 + 110.89$$
$$SSB = 346.7$$

Now we can find SSW by Formula 10.11:

$$SSW = SST - SSB$$
$$SSW = 716.30 - 346.70$$
$$SSW = 369.60$$

The degrees of freedom are found by Formulas 10.5 and 10.6:

$$dfw = N - k = 30 - 3 = 27$$
$$dfb = k - 1 = 3 - 1 = 2$$

The estimates to the population variance are found by Formulas 10.7 and 10.8:

$$\text{Mean square within} = \frac{SSW}{dfw} = \frac{369.60}{27} = 13.69$$

$$\text{Mean square between} = \frac{SSB}{dfb} = \frac{346.7}{2} = 173.35$$

TABLE 10.4   NUMBER OF DAYS REQUIRED TO PROCESS CASES FOR THREE AGENCIES (fictitious data)

Client	Agency A		Agency B		Agency C	
	X	X^2	X	X^2	X	X^2
1	5	25	12	144	9	81
2	7	49	10	100	8	64
3	8	64	19	361	12	144
4	10	100	20	400	15	225
5	4	16	12	144	20	400
6	9	81	11	121	21	441
7	6	36	13	169	20	400
8	9	81	14	196	19	361
9	6	36	10	100	15	225
10	6	36	9	81	11	121
$\Sigma X =$	70		130		150	
$\Sigma X^2 =$		524		1816		2462
$\bar{X}_k =$	7.0		13.0		15.0	

$$\bar{X} = 350/30 = 11.67$$

The F ratio (Formula 10.9) is

$$F = \frac{\text{Mean square between}}{\text{Mean square within}} = \frac{173.35}{13.69}$$

$$F = 12.66$$

And we can now test this value for its significance.

Step 1. Making Assumptions.

Model: Independent random samples
Level of measurement is interval-ratio
Population distributions are normal
Population variances are equal

The researcher will always be in a position to judge the adequacy of the first two assumptions in the model. The second two assumptions are more problematical, but remember that ANOVA will tolerate some deviation from its assumptions as long as sample sizes are roughly equal.

Step 2. Stating the Null Hypothesis.

$$H_0: \mu_1 = \mu_2 = \mu_3$$
(H_1: At least one population mean is different.)

Step 3. Selecting the Sampling Distribution and Establishing the Critical Region.

$$\text{Sampling distribution} = F \text{ distribution}$$
$$\text{Alpha} = .05$$
$$\text{Degrees of freedom (within)} = (N - k) = (30 - 3) = 27$$
$$\text{Degrees of freedom (between)} = (k - 1) = (3 - 1) = 2$$
$$F(\text{critical}) = 3.35$$

Step 4. Computing the Test Statistic. We found an obtained F ratio of 12.66.

Step 5. Making a Decision.

$$F(\text{critical}) = 3.35$$

$$F(\text{obtained}) = 12.66$$

The test statistic is in the critical region, and we would reject the null of no difference. The differences among the three agencies are very unlikely to have occurred by chance alone. The agencies are significantly different in the speed with which they process paperwork and determine eligibility. *(For practice in conducting the ANOVA test, see problems 10.2 to 10.12. Begin with the lower-numbered problems since they have smaller data sets, fewer categories, and, therefore, the simplest calculations.)*

10.8 THE LIMITATIONS OF THE TEST

The test presented in this chapter is appropriate whenever you want to test the significance of a difference across three or more categories of a single variable. This application is called **one-way analysis of variance**, since we are observing the effect of a single variable (for example, religion) on another (for example, support for capital punishment). This is the simplest application of ANOVA, and you should be aware that the technique has numerous more advanced and complex forms. For example, you may encounter research projects in which the effects of two separate variables (for example, religion and gender) on some third variable were observed.

The major limitations of ANOVA are that it requires interval-ratio measurement and roughly equal numbers of cases in each of the categories. The former condition may be difficult to meet with complete confidence for many variables of interest to the social sciences. The latter condition may create problems when the research hypothesis calls for comparisons between groups that are, by their nature, unequal in numbers (for example, white versus black Americans) and may call for some unusual sampling schemes in the data-gathering phase of a research project. Neither of these limitations should be particularly crippling, however, since ANOVA can tolerate some deviation from its model assumptions, but you should be aware of these limitations in planning your own research as well as in judging the adequacy of research conducted by others.

READING STATISTICS 7: Improving the Chances for a Successful Transition from Welfare to Work

One of the most significant recent changes in welfare programs is the push to move clients from the welfare rolls to the work force. Under what conditions is such a transition most likely to be successful? Professors Dale Weaver and Yaheskel Hasenfeld investigated four California programs that had implemented different case management practices. Are these different agency practices associated with different levels of success in moving clients into the work force?

One agency (labeled SC by the researchers) relied primarily on persuasion to motivate clients to comply with program requirements while a second (UC) used inducements or the rewards of completing various educational and vocational programs. A third agency (RC) relied on constraints (denial of benefits) to encourage compliance while a fourth (AC) mixed persuasion and constraints. Which relationship worked best? Some of their results are reported in the table below in the form of mean scores for each agency. The significance of the differences between the means was tested by analysis of variance, and the F ratios and Probabilities (alphas) are also reported.

The researchers note that workers in the SC agency were rated as least bureaucratic and most helpful. This agency also generated the highest level of satisfaction among clients and had the highest evaluation. This reflects the reliance on persuasion as a strategy to increase compliance. The RC agency, in contrast, relied on constraints and was consistently rated lowest. The researchers conclude "Overall, the professional case management approach of SC elicited from participants the most favorable views of their case managers and the program."

Dale Weaver and Yaheskel Hasenfeld: 1997. "Case Management Practices, Participants' Responses, and Compliance in Welfare-to-Work Programs." *Social Work Research.* 21: 92–100.

Mean Scores on Worker-Client Interaction Variables[1]

Case Manager Attribute	SC Persuasion	UC Inducements	AC Persuasion & Inducements	RC Constraints	F Ratio	Probability
Bureaucratic	3.14	3.26	3.63	3.85	5.86	.000
Helper	3.98	3.05	2.54	2.64	11.82	.000
Satisfied	4.52	3.78	3.49	3.31	7.44	.000
Evaluation	4.48	3.80	3.28	3.20	11.49	.000

[1]All scores are based on a scale ranging from 1 = low endorsement to 5 = high endorsement.

A final limitation of this test relates to the research hypothesis. As you recall, when we reject the null hypothesis of no difference, we generate support for the alternative hypothesis, which—in the case of ANOVA—simply asserts that at least one of the population means is different from the others. The next question that will occur to you is, of course, "Which one?" You will sometimes be able to make this determination by inspection of the sample means. In our problem involving social service agencies, for example, it is pretty clear from Table 10.4 that Agency A is the source of most of the differences in the data. There are tests by which you can ascertain exactly which differences are statistically significant, but these tests are beyond the scope of this text. You should be aware that the informal "eyeball" method can be misleading so that, if you reject the null, you should exercise caution in making conclusions about which means are significantly different.

SUMMARY

1. One-way analysis of variance is a powerful test of significance that is commonly used when comparisons across more than two categories or samples are of interest. It is perhaps easiest to conceptualize ANOVA as an extension of the test for the difference in sample means.

2. ANOVA compares the amount of variation within the categories to the amount of variation between categories. If the null of no difference is false, there should be relatively great variation between categories and relatively little variation within categories. The greater the differences from category to category relative to the differences within the categories, the more likely we will be able to reject the null.

3. The computational routine for even simple applications of ANOVA can quickly become quite complex. The basic process is to construct separate estimates to the population variance based on the variation within the categories and the variation between the categories. The test statistic is the F ratio, which is based on a comparison of these two estimates. The basic computational routine is summarized at the end of Section 10.4, and this is probably an appropriate time to mention the widespread availability of statistical packages such as SPSS, the purpose

of which is to perform complex calculations such as these accurately and quickly. If you haven't yet learned how to use such programs, ANOVA may provide you with the necessary incentive.

4. The ANOVA test can be organized into the familiar five-step model for testing the significance of sample outcomes. Although the model assumptions (step 1) require high-quality data, the test can tolerate some deviation as long as sample sizes are roughly equal. The null takes the familiar form of stating that there is no difference of any importance among the population values, while the alternative hypothesis asserts that at least one population mean is different. The sampling distribution is the F distribution, and the test is always one-tailed. The decision to reject or to fail to reject the null is based on a comparison of the obtained F ratio with the critical F ratio as determined for a given alpha level and degrees of freedom. The decision to reject the null indicates only that one or more of the population means is different from the others. We can often determine which sample mean(s) account for the difference by inspecting the sample data, but this informal method should be used with caution.

SUMMARY OF FORMULAS

Total sum of squares:

10.1 $SST = \Sigma(X_i - \bar{X})^2$

The two components of the total sum of squares:

10.2 $SST = SSB + SSW$

Sum of squares within:

10.3 $SSW = \Sigma(X_i - \bar{X}_k)^2$

Sum of squares between:

10.4 $SSB = \Sigma N_k(\bar{X}_k - \bar{X})^2$

Degrees of freedom for SSW:

10.5 $dfw = N - k$

Degrees of freedom for SSB:

10.6 $dfb = k - 1$

Mean square within:

10.7 $\text{Mean square within} = \dfrac{SSW}{dfw}$

Mean square between:

10.8 $\text{Mean square between} = \dfrac{SSB}{dfb}$

F ratio:

10.9 $F = \dfrac{\text{Mean square between}}{\text{Mean square within}}$

Computational formula for SST:

10.10 $SST = \Sigma X^2 - N\bar{X}^2$

Finding SSW by subtraction:

10.11 $SSW = SST - SSB$

GLOSSARY

Analysis of variance. A test of significance appropriate for situations in which we are concerned with the differences among more than two sample means.

ANOVA. See **Analysis of variance**.

F ratio. The test statistic computed in step 4 of the ANOVA test.

Mean square. An estimate of the variance calculated by dividing the sum of squares within (SSW) or the sum of squares between (SSB) by the proper degrees of freedom.

One-way analysis of variance. Applications of ANOVA in which the effect of a single variable on another is observed.

Sum of squares between (SSB). The sum of the squared deviations of the sample means from the overall mean, weighted by sample size.

Sum of squares total (SST). The sum of the squared deviations of the scores from the overall mean.

Sum of squares within (SSW). The sum of the squared deviations of scores from the category means.

PROBLEMS

10.1 Conduct the ANOVA test for each set of scores below. *(HINT: Follow the computational shortcut outlined in Section 10.4 and keep track of all sums and means by constructing computational tables like Table 10.3 or 10.4.)*

a.

Category		
A	B	C
5	10	12
7	12	16
8	14	18
9	15	20

b.

Category		
A	B	C
1	2	3
10	12	10
9	2	7
20	3	14
8	1	1

c.

Category			
A	B	C	D
13	45	23	10
15	40	78	20
10	47	80	25
11	50	34	27
10	45	30	20

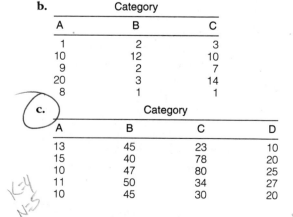

choose one

10.2 [SOC] What type of person is most involved in the neighborhood and community? Who is more likely to volunteer for organizations such as PTA, scouts, or Little League? A random sample of 15 people have been asked for their number of memberships in community voluntary organizations and some other information. Which differences are significant?

a. Membership by education:

Less than High School	High School	College
0	1	0
1	3	3
2	3	4
3	4	4
4	5	4

b. Membership by length of residence in present community:

Less than 2 Years	2–5 Years	More than 5 Years
0	0	1
1	2	3
3	3	3
4	4	4
4	5	4

c. Membership by extent of television watching:

Little or None	Moderate	High
0	3	4
0	3	4
1	3	4
1	3	4
2	4	5

10.3 [SOC] In a local community, a random sample of 25 couples has been assessed on a scale that measures the extent to which power and decision making are shared (lower scores) or monopolized by one party (higher scores). The couples were also classified by type of relationship: traditional (only the husband works outside the home), dual-career (both parties work), and cohabitational (parties living together but not legally married, regardless of work patterns). Does power structure vary significantly by type of relationship?

Traditional	Dual-career	Cohabitational
7	8	2
8	5	1
2	4	3
5	4	4
7	5	1
6	5	2
5	5	1
	8	1
	2	2

10.4 [CJ] Two separate crime-reduction programs have been implemented in the city of Shinbone. One involves a neighborhood watch program with citizens actively involved in crime prevention. The second involves officers patrolling the neighborhoods on foot rather than in patrol cars. In terms of the percentage reduction in crimes reported to the police over a one-year period, were the programs successful? The results are for random samples drawn from all neighborhoods.

Neighborhood Watch	Foot Patrol	No Program
−10	−21	+ 3
− 2	−15	−10
+ 1	− 8	+14
+ 2	−10	+ 8
+ 7	− 5	+ 5
+10	− 1	− 2
+10	+ 1	−10
+12	+ 8	− 2

10.5 [SOC] Are sexually active teenagers any better informed about AIDS and other potential health problems related to sex than teenagers who are sexually inactive? A 15-item test of general knowledge about sex and health was administered to random samples of teens who are sexually inactive, teens who are sexually active but with only a single partner ("going steady"), and teens who are sexually active with more than one partner. Is there any significant difference in the test scores?

Inactive	Active— One Partner	Active—More than One Partner
10	11	12
12	11	12
8	6	10
10	5	4
8	15	3
5	10	15

Inactive	Active—One Partner	Active—More than One Partner
3	11	14
8	9	7
10	14	9
11	11	15

10.6 SW A random sample of counties has been rated as predominantly urban, suburban, or rural. Does infant mortality rate (number of infant deaths per 1000 live births) in these counties vary significantly by this variable?

Rural	Suburban	Urban
15.1	11.0	12.5
14.7	10.9	12.4
14.2	10.1	12.1
13.5	10.0	11.9
12.5	9.9	9.7
11.2	9.8	9.2
10.1	8.5	8.2
9.9	7.1	6.5
8.5	7.0	6.2
7.0	6.9	6.0

(margin notes: $N = 10$, $k = 3$)

10.7 PS Does the rate of voter turnout vary significantly by the type of election? A random sample of voting precincts displays the following pattern of voter turnout by election type. Assess the results for significance.

Local Only	State	National
33	35	42
78	56	40
32	35	52
28	40	66
10	45	78
12	42	62
61	65	57
28	62	75
29	25	72
45	47	51
44	52	69
41	55	59

10.8 SOC Random samples of cities have been drawn from each of four geographical regions. The homicide rates for each city are reported below. Is there a significant difference?

North	South	Midwest	West
5	13	18	34
5	18	25	30

North	South	Midwest	West
10	20	30	8
6	25	20	10
3	12	10	25
4	10	11	26
8	8	10	28
7	18	12	30
12	20	20	28
15	21	21	7

10.9 SOC A random sample of 50 colleges has been taken and, for each college, average growth in student population over the past five years has been calculated. The colleges are divided into four types by size and funding source. Is there a significant difference in growth rate by type of college?

Small Private	Large Private	Small Public	Large Public
1	2	4	2
−2	0	3	3
−10	5	7	2
0	−1	−2	0
−1	0	5	4
−3	−10	0	5
−2	7	8	6
5	4	4	−1
4	2	3	1
1	0	5	2
2	3	−1	4
	2	2	2
		8	−3
		4	

10.10 GFR Do older citizens lose interest in politics and current affairs? A brief quiz on recent headline stories was administered to random samples of respondents from each of four different age groups. Is there a significant difference? The data below represent numbers of correct responses.

High School (15–18)	Young Adult (21–30)	Middle-Aged (40–55)	Retired (65+)
0	0	2	5
1	0	3	6
1	2	3	6
2	2	4	6
2	4	4	7
2	4	5	7
3	4	6	8
5	6	7	10
5	7	7	10

High School (15–18)	Young Adult (21–30)	Middle-Aged (40–55)	Retired (65+)
7	7	8	10
7	7	8	10
9	10	10	10

10.11 SOC A sociologist believes that interest in the American sport institution is almost entirely a phenomenon of white-collar males. A random sample has been given a list of 20 sports and has been asked to check off those they follow on a regular basis. The results for each sex/class group are reported below. Are these results consistent with the hypothesis?

Blue-Collar Males	Blue-Collar Females	White-Collar Males	White-Collar Females
0	0	0	0
1	0	1	0
1	0	3	0
1	0	3	0
1	0	3	0
2	1	3	0
3	3	4	1
3	3	5	1
3	3	6	3
3	3	6	3
5	3	7	4
5	3	10	5
8	4	11	6
10	5	12	9
10	8	15	10

10.12 SOC A small random sample of respondents has been selected from the General Social Survey database. Each respondent has been classified as either a city dweller, a suburbanite, or a rural dweller. Are there statistically significant differences by place of residence for any of the variables listed below?

a. Occupational prestige (PRESTG80)

Urban	Suburban	Rural
32	40	30
45	48	40
42	50	40
47	55	45
48	55	45
50	60	50
51	65	52
55	70	55
60	75	55
65	75	60

b. Number of children (CHILDS)

Urban	Suburban	Rural
1	0	1
1	1	4
0	0	2
2	0	3
1	2	3
0	2	2
2	3	5
2	2	0
1	2	4
0	1	6

c. Family income (INCOME91)

Urban	Suburban	Rural
5	6	5
7	8	5
8	11	11
11	12	10
8	12	9
9	11	6
8	11	10
3	9	7
9	10	9
10	12	8

d. Church attendance (ATTEND)

Urban	Suburban	Rural
0	0	1
7	0	5
0	2	4
4	5	4
5	8	0
8	5	4
7	8	8
5	7	8
7	2	8
4	6	5

e. Hours of TV watching per day (TVHOURS)

Urban	Suburban	Rural
5	5	3
3	7	7
12	10	5
2	2	0
0	3	1
2	0	8
3	1	5
4	3	10
5	4	3
9	1	1

Using *SPSS for Windows* to Conduct Analysis of Variance

DEMONSTRATION 10.1 Does Political Conservatism Increase with Age?

SPSS provides several different ways of conducting the analysis of variance test. The procedure summarized below is the most accessible of these, but it still incorporates options and capabilities that we have not covered in this chapter. If you wish to explore these possibilities, please consult the SPSS manual or use the on-line Help facility.

In designing examples for these sessions, my choices are constrained by the scarcity of interval-ratio variables in the GSS data set. Some variables which 'should be' interval-ratio are actually measured at the ordinal level (e.g., *income* and *childs*) while others are unsuitable dependent variables on logical grounds (e.g., *age*). To have something to discuss, I have taken some liberties with respect to level-of-measurement criteria in several of the examples that follow (a practice that is, in fact, common in social science research).

Let's begin by exploring the idea that political ideology is linked to age in U.S. society. People are often said to become more conservative about a wide range of issues as they age, and this might result in greater political conservatism in older age groups. For a measure of ideology, we will use *polviews*. Higher scores on this variable indicate higher levels of conservatism. We need to collapse *age* into a few categories to fit the ANOVA design. To find reasonable cutting points for *age*, I ran the **Frequencies** command to find the ages that divided the sample into three groups of roughly equal size. Use these scores to recode *age*:

Interval	% of Sample
18–36	34%
37–50	32%
51–89	34%

To use the ANOVA procedure, click **Statistics**, **Compare Means**, and then **One-way Anova**. The **One-way Anova** window appears. Find *polviews* (the label for this variable is THINK OF SELF AS LIBERAL OR CONSERVATIVE) in the variable list on the left and click the arrow to move the variable name into the **Dependent List** box. Note that this procedure will accommodate more than one dependent variable at a time. Next, find the name of the recoded *age* variable (*ager*?) and click the arrow to move the variable name into the **Factor** box. Note that this procedure will accept only one factor or independent variable at a time.

Click **Options** and then click the box next to Descriptive in the Statistics box to request means and standard deviations along with the analysis of variance. Click **Continue** and then click **OK**, and the following output will be produced:

Oneway

Descriptives

THINK OF SELF AS LIBERAL OR CONSERVATIVE

	N	Mean	Std. Deviation	Std. Error	95% Confidence Interval for Mean	
					Lower Bound	Upper Bound
1.00	455	4.07	1.39	6.50E-02	3.94	4.19
2.00	447	4.10	1.41	6.67E-02	3.97	4.23
3.00	466	4.33	1.35	6.27E-02	4.21	4.45
Total	1368	4.17	1.39	3.75E-02	4.09	4.24

Descriptives

THINK OF SELF AS LIBERAL OR CONSERVATIVE

	Minimum	Maximum
1.00	1	7
2.00	1	7
3.00	1	7
Total	1	7

ANOVA

THINK OF SELF AS LIBERAL OR CONSERVATIVE

	Sum of Squares	df	Mean Square	F	Sig.
Between Groups	18.876	2	9.438	4.931	.007
Within Groups	2612.457	1365	1.914		
Total	2631.333	1367			

The output box labeled ANOVA includes the various degrees of freedom, all of the sums of squares, the Mean Squares, the F ratio (4.931), and, at the far right, the exact probability of getting these results if the null is true. This is reported as .007, which is lower than our usual alpha level of .05. The differences in *polviews* for the various age groups are statistically significant.

The report also displays some summary statistics, with the three age groups identified as 1 (this is the youngest group), 2, and 3 (the oldest group). An inspection of the means shows that the average score for the entire sample was 4.17. The oldest age group is the most conservative (remember that higher scores indicate greater conservatism). These results indicate that conservatism is significantly related to age and that older people are more conservative.

DEMONSTRATION 10.2 Does Political Ideology Vary by Social Class?

Let's continue our analysis of *polviews* and see if there are any significant differences in political ideology by social class. The GSS includes several variables that might be used as indicators of class, including *degree, income, class,* and *prestg80*. We should probably investigate all of these, but—to conserve space—we confine our attention to *class* (the label for this variable is SUBJECTIVE CLASS IDENTIFICATION). Follow the instructions for One-way ANOVA in Demonstration 10.1 and specify *polviews* again as the dependent variable and *class* as the factor, or independent variable. The output will look like this:

Oneway

Descriptives

THINK OF SELF AS LIBERAL OR CONSERVATIVE

	N	Mean	Std. Deviation	Std. Error	95% Confidence Interval for Mean	
					Lower Bound	Upper Bound
LOWER CLASS	85	3.94	1.56	.17	3.60	4.28
WORKING CLASS	579	4.09	1.36	5.65E-02	3.98	4.20
MIDDLE CLASS	646	4.25	1.37	5.39E-02	4.15	4.36
UPPER CLASS	54	4.35	1.57	.21	3.92	4.78
Total	1364	4.17	1.39	3.76E-02	4.09	4.24

Descriptives

THINK OF SELF AS LIBERAL OR CONSERVATIVE

	Minimum	Maximum
LOWER CLASS	1	7
WORKING CLASS	1	7
MIDDLE CLASS	1	7
UPPER CLASS	1	7
Total	1	7

ANOVA

THINK OF SELF AS LIBERAL OR CONSERVATIVE

	Sum of Squares	df	Mean Square	F	Sig.
Between Groups	14.489	3	4.830	2.515	.057
Within Groups	2611.400	1360	1.920		
Total	2625.889	1363			

The *F* Ratio (2.515) and Sig. (.057) reported at the far right of the ANOVA output box show that the difference between groups is very close to being statistically significant at the .05 level. Inspect the group means and you will see that lower-class respondents were least conservative (had the lowest average score on *polviews*) and that the level of conservatism increased as score on class increased. The differences among the group means are not quite large enough, however, to reject the null hypothesis of "no difference." These results indicate that, at the .05 level, political conservatism does not vary significantly by social class in the population.

DEMONSTRATION 10.3 Another Test for Differences in Attitudes on Abortion

In Demonstration 9.3, we found that church attendance had a statistically significant impact on *abscale*. In this demonstration, we will investigate the impact of religious affiliation (*relig*). Click **Statistics**, **Compare Means**, and **One-way Anova** and name *abscale* as the dependent variable and *relig* as the factor. If necessary, see Demonstration 9.3 for instructions on computing *abscale*. The output will look like this:

Descriptives

ABSCALE

	N	Mean	Std. Deviation	Std. Error	95% Confidence Interval for Mean Lower Bound	Upper Bound
PROTESTANT	485	2.7093	.6331	2.875E-02	2.6528	2.7658
CATHOLIC	221	2.6380	.6777	4.559E-02	2.5482	2.7279
JEWISH	21	2.1429	.3586	7.825E-02	1.9796	2.3061
NONE	109	2.4312	.5335	5.110E-02	2.3299	2.5325
OTHER	40	2.4000	.5454	8.623E-02	2.2256	2.5744
Total	876	2.6290	.6364	2.150E-02	2.5868	2.6712

Descriptives

ABSCALE

	Minimum	Maximum
PROTESTANT	2.00	4.00
CATHOLIC	2.00	4.00
JEWISH	2.00	3.00
NONE	2.00	4.00
OTHER	2.00	4.00
Total	2.00	4.00

ANOVA

ABSCALE

	Sum of Squares	df	Mean Square	F	Sig.
Between Groups	14.469	4	3.617	9.268	.000
Within Groups	339.954	871	.390		
Total	354.424	875			

The *F* Ratio (9.268) and Sig. (.000) indicate a significant difference at the .05 level. Protestants were the most opposed, and Jews were the most supportive of the right to a legal abortion.

Exercises

10.1 As a follow-up on Demonstration 10.2, test ~~income91~~, prestg80, and pa-pres80 as independent variables (factors) against polviews. Do these measures of social class display the same type of relationship with polviews as class? First, recode each of these independents into three categories. Run **Frequencies** for each and find cutting points that divide the sample into three groups of roughly equal size.

10.2 What other variable might have a significant relationship with abscale? Pick three potential independent variables (factors) and test their relationships with abscale. Some possible independent variables would be sex, race, or any of the measures of social class. If necessary, recode the independent variables into a few (three or four) categories.

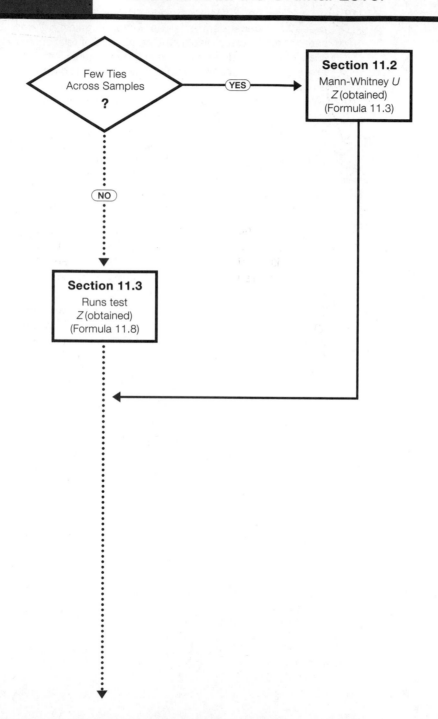

11.1 INTRODUCTION

In this chapter, two **nonparametric tests** of hypothesis, the Mann-Whitney U and the runs test, will be presented. Both tests are appropriate in the two-sample case when the variable of interest is measured at the ordinal level and, under conditions to be specified, may be considered as alternatives to the differences-in-means tests presented in Chapter 9. Before introducing the tests themselves, let us consider the situations where they can or should be used.

The **Mann-Whitney U** and the **runs test** represent a large class of tests of significance called "nonparametric" or "distribution-free" tests. These tests differ from the tests introduced in Chapters 8, 9, and 10 in that they require no particular assumption about the shape of the population distribution. Recall that we needed to assume a normal sampling distribution in order to test proportions and means for significance. This assumption is satisfied only if sample size is large or if the additional assumption of a normally distributed population can be made (see the theorems presented in Chapter 6). Thus, means and proportions based on small samples can be tested for their significance only when we can make assumptions about the particular shape of the population distribution. Needless to say, researchers often find themselves in situations where they are uncomfortable with such precise assumptions about an unknown distribution.

Nonparametric tests, on the other hand, do not require the assumption of a normally shaped population distribution. In fact, with nonparametric tests, we do not need to assume any particular shape for the population distribution. The tests to be presented in this chapter therefore have wide applicability in situations where the researcher is unsure of the form of the population distribution. Since the requirement of a normal population can be relaxed with large samples, these tests are particularly useful when working with small samples.

Besides the assumptions of a normal sampling distribution, the test for means also requires the assumption of interval-ratio level of measurement. The Mann-Whitney U and the runs test may also be used when the researcher is uncertain of the appropriateness of this assumption.

To summarize, you should seriously consider ordinal-level, nonparametric tests of significance when either the assumption of interval-ratio measurement or the assumption of normal sampling distribution is in doubt for a test of sample means. Both of these assumptions are included in the model assumptions (step 1) for tests of means. The model can be thought of as the mathematical foundation for the rest of the test, and the researcher should be very certain of the appropriateness of these assumptions before placing confidence in the test results. If the model assumptions can be satisfied, tests of the differences in means are preferable to the nonparametric alternatives because the assumption of interval-ratio level permits more sophisticated mathematical procedures to be performed on the data. Thus, a greater volume of more precise information can be amassed with respect to the variable of interest. If the model assumptions for means cannot be satisfied, the

Mann-Whitney U and the runs test are very useful alternative tests for investigating the significance of the difference between two samples.

11.2 THE MANN-WHITNEY U

In many ways, this test is similar to the test of significance for the difference in sample means. In both cases we compare random samples as a way of making inferences about the possible differences between two populations. In both cases, the test statistics are computed from the samples and compared with the sampling distribution of all possible sample outcomes. Instead of computing means as the sample statistic, however, the Mann-Whitney test is based on the ranking of the sample scores. This is appropriate since ranking is the most sophisticated mathematical operation that can be performed on ordinal-level data.

The computation of U is straightforward. First, the scores from both samples are pooled and ranked from highest to lowest. Second, the ranks for the two samples are totaled and compared. If the two samples represent very different populations, the cases from each sample should be grouped together. The greater the extent to which the two samples differ, the greater the difference in the sum of the total ranks. If the populations are not significantly different from each other, then the cases from the two samples should be very intermixed, and the sum of the total ranks would have similar values.

An example should clarify the underlying logic as well as the computational routines for this test. Assume that you are concerned with racial differences in the level of satisfaction with the social life available on your campus. You have administered survey instruments to randomly selected samples of black and white students and have devised a scale on which a high score indicates great satisfaction. The scores, by race, are presented in Table 11.1.

TABLE 11.1 SCORES ON SOCIAL SATISFACTION SCALE FOR BLACK AND WHITE STUDENTS

Sample 1 Black Students		Sample 2 White Students	
Case	Score	Case	Score
1	42	13	45
2	35	14	40
3	30	15	32
4	25	16	30
5	19	17	28
6	17	18	27
7	15	19	26
8	14	20	24
9	9	21	20
10	5	22	10
11	4	23	8
12	2	24	7

You may be tempted to compute a mean for each of these samples and test the difference for its significance. If you keep in mind that these scores are only ordinal in level of measurement, you should be able to restrain yourself long enough to compute U.

First, pool the scores and rank them from high to low. If you encounter any tied scores, assign all of them the average of the ranks they would have used up if they had not been tied. For example, cases 3 and 16 are tied with scores of 30. If they had not been tied, they would have used up ranks 6 and 7. Therefore, assign both cases the rank 6.5 [(6 + 7)/2]. Ranked scores are presented in Table 11.2. To compute U, use Formula 11.1:

FORMULA 11.1

$$U = N_1 N_2 + \frac{N_1(N_1 + 1)}{2} - \Sigma R_1$$

where ΣR_1 = the sum of the ranks for sample 1

For our sample problem above:

$$U = (12)(12) + \frac{(12)(13)}{2} - 173.5$$

$$U = 144 + \frac{156}{2} - 173.5$$

$$U = 48.5$$

Note that we could have computed the Mann-Whitney U by using data from sample 2. This alternative solution, which we will label U' (U prime), would have resulted in a larger value for U. The smaller of the two values, U

TABLE 11.2   RANKED SCORES ON SOCIAL SATISFACTION SCALE FOR BLACK AND WHITE STUDENTS

Sample 1 Black Students			Sample 2 White Students		
Case	Score	Rank	Case	Score	Rank
1	42	2	13	45	1
2	35	4	14	40	3
3	30	6.5	15	32	5
4	25	11	16	30	6.5
5	19	14	17	28	8
6	17	15	18	27	9
7	15	16	19	26	10
8	14	17	20	24	12
9	9	19	21	20	13
10	5	22	22	10	18
11	4	23	23	8	20
12	2	24	24	7	21
		$\Sigma R_1 = 173.5$			$\Sigma R_2 = 126.5$

Application 11.1

Do men or women endorse stricter norms of honesty? Samples of 12 men and women have read brief descriptions of various acts that might be considered dishonest (for example, not telling a cashier that he or she has given you too much change) and have been asked to classify each on a scale from "very dishonest" to "not dishonest at all." Summary scores on the scale range from 50 (indicating very strict norms of honesty) to 0 and are reported in the following tables for each subject. Is the difference between men and women statistically significant?

Because the honesty scores are only ordinal, a Mann-Whitney U test of significance will be conducted. The scores have already been ranked and the sum of ranks calculated.

Men			Women		
Case	Score	Rank	Case	Score	Rank
1	47	2	13	48	1
2	44	4	14	45	3
3	40	7	15	43	5
4	35	10	16	42	6
5	32	12	17	39	8
6	31	13	18	36	9
7	30	14	19	33	11
8	29	15	20	28	16
9	25	17	21	23	19
10	24	18	22	21	20
11	20	21	23	15	22
12	12	24	24	14	23
		$\Sigma R_1 = 157$			$\Sigma R_2 = 143$

The value of the Mann-Whitney U is

$$U = N_1 N_2 + \frac{N_1(N_1 + 1)}{2} - \Sigma R_1$$

$$U = (12)(12) + \frac{12(13)}{2} - 157$$

$$U = 144 + 78 - 157$$

$$U = 65$$

The mean of the sampling distribution is

$$\mu_u = \frac{N_1 N_2}{2} = \frac{(12)(12)}{2} = 72$$

The standard deviation of the sampling distribution is

$$\sigma_U = \sqrt{\frac{N_1 N_2 (N_1 + N_2 + 1)}{12}}$$

$$\sigma_U = \sqrt{\frac{(12)(12)(12 + 12 + 1)}{12}}$$

$$\sigma_U = \sqrt{\frac{(144)(25)}{12}}$$

$$\sigma_U = \sqrt{300}$$

$$\sigma_U = 17.32$$

The value of the test statistic is

$$Z \text{ (obtained)} = \frac{U - \mu_u}{\sigma_u}$$

$$Z \text{ (obtained)} = \frac{65 - 72}{17.32}$$

$$Z \text{ (obtained)} = \frac{-7}{17.32}$$

$$Z \text{ (obtained)} = -0.40$$

If alpha is set at the 0.05 level, the critical region would begin at Z (critical) = ± 1.96. With an obtained Z of -0.40, we would fail to reject the null hypothesis (H_0: the populations are identical). The scores of men and women on the honesty scale are not significantly different.

or U', is always taken as the value of U. Once U has been calculated, U' can be quickly determined by means of Formula 11.2:

FORMULA 11.2
$$U' = N_1 N_2 - U$$

Thus, for the sample problem, U' would be $(12)(12) - 48.5$, or 95.5. Remember that the lower of these values is always taken as U.

Once the value of U has been determined, we must still conduct the test of significance. In step 1 of the five-step model, we assume ordinal level of measurement and make no assumption about the shape of the population distribution. The null hypothesis in step 2 is, as always, a statement of "no difference": the two populations represented by these samples are identical. Note that if this assumption is true, then the differences in total ranks should be small.

The alternative or research hypothesis is usually a statement to the effect that the two populations are different. This form for H_1 would direct the use of a two-tailed test. It is perfectly possible to use one-tailed tests with Mann-Whitney U when a direction for the difference can be predicted. But, to conserve space and time, we will consider only the two-tailed case.

In step 3, we will take advantage of the fact that, when both sample N's are greater than 10, the sampling distribution of U approximates normality. This will allow us to use the Z-score tables to find the critical region as marked by Z (critical).

To compute the test statistic (step 4), the necessary formulas are

FORMULA 11.3
$$Z\text{(obtained)} = \frac{U - \mu_u}{\sigma_u}$$

where U = the sample statistic
μ_u = the mean of the sampling distribution of sample U's
σ_u = the standard deviation of the sampling distribution of sample U's

FORMULA 11.4
$$\mu_u = \frac{N_1 N_2}{2}$$

FORMULA 11.5
$$\sigma_u = \sqrt{N_1 N_2 (N_1 + N_2 + 1)/12}$$

For the sample problem, μ_u will be $(12)(12)/2$, or 72. The standard deviation of the sampling distribution is

$$\sigma_u = \sqrt{(12)(12)(12 + 12 + 1)/12}$$
$$\sigma_u = \sqrt{300}$$
$$\sigma_u = 17.32$$

We now have all the information we need to conduct a test of significance for U.

Step 1. Making Assumptions.

Model: Independent random samples
Level of measurement is ordinal

Step 2. Stating the Null Hypothesis.

H_0: The populations from which the samples are drawn
are identical on the variable of interest.
(H_1: The populations from which the samples are drawn
are different on the variable of interest.)

Step 3. Selecting the Sampling Distribution and Establishing the Critical Region.

Sampling distribution = Z distribution
Alpha = 0.05
Z (critical) = ±1.96

Step 4. Calculating the Test Statistic. With U equal to 48.5, μ_u equal to 72, and a σ_u of 17.32,

$$Z(\text{obtained}) = \frac{U - \mu_u}{\sigma_u}$$

$$Z(\text{obtained}) = \frac{48.5 - 72}{17.32}$$

$$Z(\text{obtained}) = -1.36$$

Step 5. Making a Decision. The test statistic, a Z (obtained) of −1.36, does not fall in the critical region as marked by the Z (critical) of ±1.96. Therefore, we fail to reject the null of no difference. Black students are not significantly different from white students in terms of their level of satisfaction with the social life available on campus. Note that if we had used the U' value of 95.5 instead of U in computing the test statistic, the value of Z (obtained) would have been +1.36, and our decision to fail to reject the null would have been exactly the same. *(For practice in conducting the Mann-Whitney U test, see any of the problems at the end of the chapter. Problem 11.1 has the simplest calculations, and it may be a good idea to start with that problem to make sure that you are comfortable with these procedures.)*

11.3 THE RUNS TEST

Far from being a test of physical fitness, the runs test is very similar in logic and form to the Mann-Whitney U test. The null hypothesis is, again, the assumption that there is no significant difference between the populations from which the samples come. To conduct the test, the scores from both samples are pooled and then ranked from high to low as if they were a single sample. If the null is true, then the scores should be very intermixed and there should be many runs. A **run** is defined as *any sequence of one or more scores from the same sample*. If the null is false and the populations are different on the trait being measured, then there should be very few runs.

To illustrate with the data on social satisfaction presented above, if we pooled the two samples and designated a white student with a *W* and a black student with a *B*, we would get the following sequences:

W	B	W	B	W	B	WWWW	B	WW	BBBB	W	B	WW	BBB
1	2	3	4	5	6	7	8	9	10	11	12	13	14

where the *W* at the far left represents the individual from either sample with the highest score (case 13 with a score of 45), the next letter represents the individual with the second highest score (case 1 with a score of 42), and so on. The underlinings represent runs, and by counting we find that there are 14 runs in these data.

Although no prediction of the direction of the difference need be made with the runs test, the smaller the number of runs, the greater the likelihood that the difference between the two samples is significant. The lowest number of runs possible is, of course, two. If there had been racial differences in social satisfaction in the data above, with all white students expressing greater satisfaction than any black student, the data would have looked like this:

WWWWWWWWWW	BBBBBBBBBB
1	2

The fewer the differences between the two populations, the greater the intermixing between the two samples. More intermixing means a higher number of runs and, consequently, a lower probability of being able to reject the null hypothesis.

For situations in which at least one sample is larger than 20, the sampling distribution of all possible sample runs approximates normality. The mean (μ_R) and the standard deviation (σ_R) of the sampling distribution can be found by Formulas 11.6 and 11.7, respectively:

FORMULA 11.6
$$\mu_R = \frac{2N_1 N_2}{N_1 + N_2 + 1}$$

FORMULA 11.7
$$\sigma_R = \sqrt{\frac{2N_1 N_2(2N_1 N_2 - N_1 - N_2)}{(N_1 + N_2)^2(N_1 + N_2 - 1)}}$$

The test statistic, Z (obtained), is computed with Formula 11.8:

FORMULA 11.8
$$Z(\text{obtained}) = \frac{R - \mu_R}{\sigma_R}$$

where $R =$ the number of runs

To illustrate with an example, assume that randomly selected samples of white males and females have been tested for their racial prejudice. Is there

TABLE 11.3   RACIAL PREJUDICE SCORES FOR SAMPLES OF MALES
AND FEMALES

Sample 1 Males		Sample 2 Females	
Case	Score	Case	Score
1	29	23	32
2	29	24	31
3	25	25	30
4	25	26	30
5	25	27	27
6	24	28	26
7	23	29	26
8	23	30	22
9	21	31	22
10	18	32	20
11	18	33	20
12	17	34	19
13	15	35	19
14	13	36	16
15	12	37	14
16	8	38	14
17	8	39	10
18	7	40	9
19	6	41	4
20	5	42	4
21	3	43	2
22	3	44	2

a significant difference in levels of prejudice between the sexes? The scores
are reported in Table 11.3. First, they must be pooled and then ranked; then
we can count the number of runs. Tied scores can be a serious problem with
this particular test because the number of runs can be very much affected by
how the tied cases are arranged. If you encounter tied scores, probably the
best (safest) thing to do is to compute all possible numbers of runs by rear-
ranging the tied cases. If all possible solutions lead to the same decision (reject H_0 or fail to reject H_0), then we are on safe ground in making that decision. If the different arrangements lead to different decisions, then we clearly
must opt for another test of significance. In our example, there are no tied
cases across the samples (ties within samples won't make any difference in
the number of runs).

Designating males with M and females with F, we can array the scores as
follows:

$\overline{FFFF}$	$\overline{MM}$	$\overline{FFF}$	$\overline{MMMMMM}$	$\overline{FF}$	$\overline{M}$	$\overline{FFFF}$	$\overline{MMM}$	$\overline{F}$
1	2	3	4	5	6	7	8	9

$\overline{M}$	$\overline{FF}$	$\overline{MM}$	$\overline{FF}$	$\overline{MMMMM}$	$\overline{FF}$	$\overline{MM}$	$\overline{FF}$
10	11	12	13	14	15	16	17

There are 17 runs in these data, and we can now proceed with the formal test of significance.

Step 1. Making Assumptions.

Model: Independent random sampling
Level of measurement is ordinal

Step 2. Stating the Null Hypothesis.

H_0: The two populations are identical on racial prejudice.
(H_1: The two populations are different on racial prejudice.)

Step 3. Selecting the Sampling Distribution and Establishing the Critical Region.

Sampling distribution = Z distribution
Alpha = 0.05
Z(critical) = ±1.96

Step 4. Calculating the Test Statistic. Before solving for Z (obtained), both μ_R (the mean of the sampling distribution) and σ_R (the standard deviation of the sampling distribution) must be calculated. The necessary formulas were presented above as Formulas 11.6 and 11.7.

$$\mu_R = \frac{2N_1N_2}{N_1 + N_2} + 1$$

$$\mu_R = \frac{(2)(22)(22)}{22 + 22} + 1$$

$$\mu_R = 23$$

$$\sigma_R = \sqrt{\frac{2N_1N_2(2N_1N_2 - N_1 - N_2)}{(N_1 + N_2)^2 (N_1 + N_2 - 1)}}$$

$$\sigma_R = \sqrt{\frac{(2)(22)(22) [(2)(22)(22) - 22 - 22]}{(22 + 22)^2 (22 + 22 - 1)}}$$

$$\sigma_R = 3.28$$

$$Z\text{(obtained)} = \frac{R - \mu_R}{\sigma_R}$$

$$Z\text{(obtained)} = \frac{17 - 23}{3.28}$$

$$Z\text{(obtained)} = -1.83$$

Step 5. Making a Decision. The test statistic, a Z (obtained) of -1.83, does not fall in the critical region as marked by the Z (critical) of ±1.96. Therefore, we must fail to reject the null hypothesis. Males and females do not differ significantly on the trait of racial prejudice. *(For practice in conducting the runs test, see problems 11.1 to 11.5. Problems 11.6 to 11.12 have too many ties to justify use of this test. Problem 11.1 has the simplest calculations, and it may be a good idea to start with that problem to make sure that you are comfortable with these procedures.)*

11.4 CHOOSING AN ORDINAL TEST OF SIGNIFICANCE

Several other ordinal tests of significance have been invented by statisticians. Most are based on logic similar to that presented in Sections 11.2 and 11.3, although some are appropriate only in specific circumstances. For example, a test statistic called Wilcoxon's *T* is designed for the situation where the two samples are not independent (that is, when the samples have been matched on some criterion as in a "before-after" test of the same subjects). There is even a way of conducting a kind of analysis of variance on ordinal data for situations where the researcher has more than two samples.

The two tests presented above are among the more popular general tests of significance for ordinal data, and choosing between them may be something of a problem. In most cases, which test is selected makes little difference, because the decision made in step 5 will be the same. One criterion you can use in making the choice is the number of ties across the samples. As was pointed out above, a large number of such ties makes the runs test rather troublesome and, in such a case, you should opt for the Mann-Whitney test. Also, remember that, when the more restrictive model assumptions can be satisfied, the test for the significance of the difference in sample means is preferred over either of the tests presented here.

SUMMARY

1. When testing for the significance of the difference between two sample means, researchers sometimes must make assumptions (step 1) about which they are uncertain. If sample size is small and/or if it is not clear that interval-ratio level of measurement can be assumed, serious consideration should be given to using nonparametric tests. These tests require only ordinal level of measurement and make no particular assumptions about the shape of the population distribution. Two of these tests are the Mann-Whitney *U* and the runs test.

2. In the Mann-Whitney *U* test, the scores from both samples are pooled and then ranked from high to low. The ranks are summed for each sample, and the test is based on a comparison of the summed ranks. If the null is true (if the samples come from the same population), the summed ranks should be approxi-

mately equal. The greater the difference between the summed ranks, the greater the probability of rejecting the null.

3. The runs test also involves pooling and ranking the scores from both samples. For this test, we count the number of runs (sequences of one or more scores from the same sample) in the pooled scores. If the null hypothesis is true, there should be many runs (the scores should be very intermixed). The fewer the runs, the greater the probability of rejecting the null.

4. If there are many ties across samples, the Mann-Whitney *U* is preferred to the runs test. If ties across samples are few in number, both tests will result in the same decision (reject or fail to reject) in the huge majority of cases.

SUMMARY OF FORMULAS

Mann-Whitney U:

11.1 $\quad U = N_1 N_2 + \dfrac{N_1(N_1 + 1)}{2} - \Sigma R_1$

U':

11.2 $\quad U' = N_1 N_2 - U$

Z (obtained) for Mann-Whitney U:

11.3 $\quad Z\text{(obtained)} = \dfrac{U - \mu_u}{\sigma_u}$

Mean of the sampling distribution for sample U's:

11.4 $\quad \mu_u = \dfrac{N_1 N_2}{2}$

Standard deviation of the sampling distribution for sample U's:

11.5 $\quad \sigma_u = \sqrt{N_1 N_2 (N_1 + N_2 + 1)/12}$

Mean of the sampling distribution for sample runs:

11.6 $\quad \mu_R = \dfrac{2 N_1 N_2}{N_1 + N_2 + 1}$

Standard deviation of the sampling distribution for sample runs:

11.7 $\quad \sigma_R = \sqrt{\dfrac{2 N_1 N_2 (2 N_1 N_2 - N_1 - N_2)}{(N_1 + N_2)^2 (N_1 + N_2 - 1)}}$

Z (obtained) for runs test:

11.8 $\quad Z\text{(obtained)} = \dfrac{R - \mu_R}{\sigma_R}$

GLOSSARY

Mann-Whitney U. A nonparametric test of significance for the two-sample case when the variable of interest has been measured at the ordinal level.

μ_R. The symbol for the mean of a sampling distribution of all possible sample runs.

μ_u. The symbol for the mean of the sampling distribution of all possible sample U's.

Nonparametric test. A type of significance test where no assumptions about the precise shape of the population need be made.

Run. A sequence of one or more scores from the same sample. The basis of the test statistic used in the runs test.

Runs test. A nonparametric test of significance for the two-sample case when the variable of interest has been measured at the ordinal level.

σ_R. The symbol for the standard deviation of the sampling distribution of all possible sample runs.

σ_u. The symbol for the standard deviation of the sampling distribution of all possible sample U's.

PROBLEMS

For problems 11.1 to 11.5, conduct both the Mann-Whitney U and the runs test. Conduct only the Mann-Whitney U for problems 11.6 to 11.12 as they have too many ties for the runs test. Use the normal approximations for the sampling distributions throughout.

11.1 SOC What are the sources of job satisfaction? One possibility is the nature of the work: Are jobs that involve working with people (e.g., retail clerk) more or less satisfying than those that involve working with things (e.g., computer repair)? Are jobs where the supervision is relatively loose and the worker has a lot of autonomy more or less satisfying than those where workers are closely controlled? Or, is it just a matter of money with better paying jobs being more satisfying? Results of a survey of job satisfaction are presented below. Scores on the job satisfaction scale range from 1–30 with higher scores indicating greater satisfaction. Which differences are significant? (Obviously, these samples are much too small for serious research. Problems have been simplified for ease of computation.) Rank order the scores from both samples as the first step in computations.

a.

Working More with People	Working More with "Things"
3	2
5	4
7	6
8	6
11	7
15	16
17	20
18	21
24	25
26	27

b.

Greater Autonomy	Lesser Autonomy
7	1
10	2
11	5
15	6
20	9
21	14
25	17
26	18
27	19
28	24

c.

Higher Salary	Lower Salary
10	1
11	4
12	5
15	7
16	8
19	10
20	14
25	17
27	18
28	21

11.2 SOC Random samples of freshmen and seniors were tested for their level of racial prejudice, and the scores are reported below (higher scores indicate greater prejudice). Is there a significant difference between the two samples? Write a sentence or two interpreting your results.

Freshmen		Seniors	
Case	Score	Case	Score
1	59	21	44
2	58	22	39
3	57	23	39
4	56	24	38
5	54	25	38
6	52	26	37
7	52	27	37
8	45	28	36
9	40	29	35
10	30	30	34
11	30	31	31
12	29	32	10
13	25	33	9
14	25	34	8
15	24	35	8
16	23	36	7
17	22	37	5
18	20	38	1
19	15	39	1
20	12	40	0

11.3 SW/CJ Your local juvenile probation department sponsors summer camping trips for some of its clients and wants to know if these wilderness experiences have any effect on the attitudes of the campers. You have gathered random samples of participants and nonparticipants and have administered a scale that measures respect for authority (higher scores indicate greater respect). What can you tell the agency?

Campers		Noncampers	
Case	Score	Case	Score
1	35	21	27
2	35	22	26
3	34	23	23
4	30	24	22
5	30	25	22
6	29	26	17
7	28	27	17
8	25	28	12
9	24	29	11
10	23	30	10
11	23	31	10
12	23	32	7
13	19	33	7
14	18	34	3
15	18	35	3
16	18	36	3
17	16	37	2
18	15	38	2
19	15	39	1
20	14	40	1

11.4 GER A nursing home has an unusually large number of patients who are uncommunicative and socially isolated. In an effort to alleviate this condition, the staff began a limited program of art and music therapy for some of the patients. After six months of operation, the staff rated the patients in terms of responsiveness to social interaction. Random samples of both therapy and nontherapy patients were assembled. Is there a significant difference between the two groups? (Higher scores indicate greater responsiveness.)

Therapy		Nontherapy	
Case	Score	Case	Score
1	102	22	100
2	98	23	99
3	95	24	96
4	85	25	70
5	77	26	65
6	70	27	65
7	69	28	50
8	68	29	30
9	68	30	30
10	66	31	29
11	51	32	28
12	49	33	21
13	45	34	15
14	42	35	10
15	40	36	9
16	39	37	3
17	35	38	1
18	35		
19	34		
20	33		
21	32		

11.5 SOC As is the case with most institutions of higher education, St. Algebra College provides sociology majors with numerous ways of avoiding courses in mathematics. This arrangement is generally satisfactory for all concerned until, of course, the majors encounter their required sociology statistics course (sound familiar?). The instructor of this course has been keeping careful track of the mathematical background of the students and is now prepared to do a little hypothesis testing. The question is, Do students with college mathematics courses perform at higher levels in the sociology statistics course than students who have taken no college mathematics courses? Random samples are collected and final course averages are used as a measure

of student performance. Being extremely conservative, the instructor decides that grades must be considered ordinal-level data. What can be concluded?

Students with at Least One College Math Course		Students with No College Math Courses	
Case	Final Average in Statistics	Case	Final Average in Statistics
1	95.6	26	95.4
2	94.3	27	93.7
3	92.1	28	93.2
4	90.1	29	90.9
5	89.9	30	87.6
6	88.0	31	83.2
7	85.3	32	80.9
8	82.1	33	74.2
9	81.1	34	74.1
10	79.2	35	69.5
11	78.0	36	68.4
12	76.3	37	66.3
13	76.2	38	66.0
14	75.7	39	62.0
15	75.1	40	61.0
16	74.9	41	60.0
17	73.5	42	59.2
18	72.0	43	58.7
19	71.3		
20	70.5		
21	69.3		
22	65.0		
23	64.0		
24	60.0		
25	53.4		

11.6 SOC A random sample of students in a large introductory sociology class has rated the effectiveness of the instructor on a scale that ranges from 10 (excellent) to 1 (poor). Is there a difference in the ratings for sociology majors as opposed to nonmajors? Use the Mann-Whitney U test for significance and explain your conclusions. Could the runs test be used for these data? Why or why not?

Majors		Nonmajors	
Case	Rating	Case	Rating
1	10	21	10
2	9	22	9
3	10	23	7
4	10	24	10
5	7	25	8

Major	s	Nonmaj	ors
Case	Rating	Case	Rating
6	10	26	6
7	6	27	10
8	10	28	9
9	8	29	6
10	6	30	4
11	10	31	7
12	9	32	6
13	5	33	10
14	3	34	2
15	9	35	9
16	5	36	3
17	4	37	7
18	9	38	5
19	3	39	3
20	8	40	6

Greek	s	Independe	nts
Case	Score	Case	Score
1	74	16	85
2	53	17	84
3	53	18	84
4	52	19	75
5	52	20	75
6	52	21	74
7	51	22	50
8	50	23	49
9	47	24	48
10	47	25	48
11	40	26	29
12	35	27	22
13	34	28	21
14	33	29	15
15	32	30	11

11.7 SW A family therapist used a new approach with half of her case load for a year and then had an associate rate each family for level of conflict (higher scores indicate greater conflict). Is there a significant difference between the samples?

New The	rapy	Old The	rapy
Case	Score	Case	Score
1	53	16	50
2	53	17	49
3	40	18	49
4	35	19	45
5	33	20	44
6	32	21	44
7	30	22	41
8	20	23	39
9	17	24	39
10	16	25	32
11	10	26	27
12	9	27	22
13	4	28	21
14	4	29	15
15	3	30	11

11.8 SOC College fraternities and sororities are sometimes accused of being anti-intellectual in that they encourage and reward behaviors and lifestyles incompatible with scholarly pursuits. To test the accuracy of this charge, samples of "Greeks" and independents have been given a survey that measures support for scholarship, and the scores are reported below. Is there a significant difference between the samples?

11.9 SOC A scale measuring support for legalized abortion has been administered to samples of men and women. Is there a significant difference?

Men		Women	
Case	Score	Case	Score
1	13	12	13
2	12	13	13
3	12	14	11
4	11	15	11
5	10	16	9
6	10	17	8
7	10	18	8
8	9	19	6
9	9	20	4
10	8	21	3
11	5	22	1

11.10 PS Samples of Republican and Democratic legislators have been rated on their support for a new missile system being requested by the Pentagon. Is there a significant difference between the parties?

Democr	ats	Republic	ans
Case	Score	Case	Score
1	25	13	35
2	24	14	34
3	23	15	32
4	22	16	31
5	14	17	30
6	10	18	30
7	10	19	29

Democrats		Republicans	
Case	Score	Case	Score
8	10	20	26
9	9	21	21
10	4	22	15
11	2	23	9
12	2	24	8

11.11 SW For a number of months, some surgical patients in a large hospital have been visited shortly after release by a social worker and a public health nurse working as a team. This new program is designed to ease the recovery period and ensure that all patients have access to all necessary services. Samples of participants and nonparticipants have been asked to rate the difficulty of their recovery period. As assessed by these scores, did the program work?

Participants		Nonparticipants	
Case	Score	Case	Score
1	23	12	25
2	22	13	25
3	19	14	25
4	16	15	24
5	9	16	24
6	5	17	20
7	5	18	20
8	5	19	17
9	4	20	14
10	1	21	10
11	1	22	7

11.12 SOC Twenty males and 20 females have been randomly selected from the General Social Survey data file. Their scores on each of five variables are reported below. Conduct the Mann-Whitney U test for each variable. There are too many tied scores to justify use of the runs test. See Appendix G for the exact wording of each question. Be sure to rank order the scores before conducting the tests.

a. Occupational prestige (*Prestg80*).

Males		Females	
Case	Score	Case	Score
1	39	21	26
2	32	22	47
3	36	23	60
4	52	24	12
5	32	25	58
6	33	26	50
7	15	27	29
8	43	28	63
9	78	29	32
10	42	30	34
11	17	31	55
12	19	32	60
13	50	33	26
14	32	34	60
15	40	35	39
16	51	36	25
17	47	37	62
18	55	38	51
19	49	39	34
20	37	40	62

b. Father's occupational prestige (*Papres*).

Males		Females	
Case	Score	Case	Score
1	45	21	29
2	28	22	30
3	47	23	40
4	34	24	23
5	50	25	50
6	70	26	35
7	17	27	41
8	45	28	25
9	51	29	45
10	31	30	21
11	41	31	43
12	52	32	47
13	54	33	52
14	58	34	62
15	51	35	41
16	39	36	61
17	61	37	45
18	56	38	47
19	27	39	41
20	29	40	51

c. Number of children (*Childs*).

Males		Females	
Case	Score	Case	Score
1	1	21	2
2	2	22	0
3	2	23	2
4	0	24	3
5	3	25	0
6	1	26	0
7	0	27	2

Males		Females	
Case	Score	Case	Score
8	2	28	2
9	0	29	3
10	0	30	1
11	0	31	0
12	0	32	0
13	2	33	1
14	0	34	3
15	7	35	1
16	4	36	4
17	1	37	0
18	2	38	2
19	3	39	2
20	2	40	3

d. Income (*Income91*).

Males		Females	
Case	Score	Case	Score
1	12	21	6
2	9	22	12
3	12	23	11
4	12	24	11
5	12	25	12
6	10	26	9
7	8	27	8
8	9	28	12
9	10	29	7
10	11	30	5
11	12	31	12
12	7	32	4
13	12	33	12

Males		Females	
Case	Score	Case	Score
14	9	34	12
15	8	35	10
16	10	36	10
17	10	37	9
18	8	38	8
19	7	39	7
20	6	40	12

e. Church attendance (*Attend*).

Males		Females	
Case	Score	Case	Score
1	0	21	3
2	1	22	3
3	7	23	3
4	2	24	8
5	3	25	3
6	3	26	0
7	0	27	4
8	7	28	5
9	7	29	8
10	2	30	4
11	7	31	6
12	0	32	5
13	5	33	7
14	0	34	6
15	6	35	2
16	2	36	4
17	7	37	7
18	0	38	4
19	8	39	4
20	4	40	2

SPSS for Windows

Using *SPSS for Windows* to Produce Nonparametric Tests of Significance

DEMONSTRATION 11.1 Are People Who Are Afraid More Likely to Attend Religious Services?

SPSS for Windows includes a number of nonparametric tests for a variety of situations. In this section, we will use only the Mann-Whitney *U* test and the runs test, but you should take some time to familiarize yourself with the variety of alternative tests available. As an example, we'll see if attendance at religious services (*attend*) is significantly related to a measure of how fearful people are for their own safety (*fear*). Are people who are more fearful more likely to attend religious services? If you inspect *attend* in Appendix G, you will see that the categories are unequal in size, and the variable can be considered ordinal in level of measurement, so the tests introduced in this chapter are apppropriate.

To run the Mann-Whitney *U* test, click **Statistics**, then click **Nonparametric tests**, and **2 independent samples**. The **Two Independent Samples** dialog box will open with a list of variables on the left and a choice of four tests at the bottom. Make sure the boxes beside the Mann-Whitney *U* test and the Wald-Wolfowitz runs test are checked. Select *attend* and click the arrow to move the variable name into the **Test Variable List** box and move *fear* (the label for this variable is AFRAID TO WALK AT NIGHT IN NEIGHBORHOOD) into the **Grouping Variable** box. The **Define Groups** button will activate. Click this button and type 1 (fearful) into the box beside Group 1 and 2 into the box next to Group 2. This instruction tells SPSS to divide the sample into two groups based on their scores on *fear*. Click **Continue** and **OK**, and the following output will be produced:

NPar Tests
Mann-Whitney Test

Ranks

	AFRAID TO WALK	N	Mean Rank	Sum of Ranks
HOW OFTEN R ATTENDS RELIGIOUS SERVICES	YES	385	494.62	190429.50
	NO	535	435.94	233230.50
	Total	920		

Test Statistics[a]

	HOW OFTEN R ATTENDS RELIGIOUS SERVICES
Mann-Whitney U	89850.500
Wilcoxon W	233230.500
Z	−3.335
Asymp. Sig. (2-tailed)	.001

a. Grouping Variable: AFRAID TO WALK AT NIGHT IN NEIGHBORHOOD

Wald-Wolfowitz Test

Frequencies

	AFRAID TO WALK	N
HOW OFTEN R ATTENDS RELIGIOUS SERVICES	YES	385
	NO	535
	Total	920

Test Statistics[b,c]

		Number of Runs	Z	Asymp. Sig. (1-tailed)
HOW OFTEN R ATTENDS RELIGIOUS SERVICES	Minimum Possible	10[a]	−29.739	.000
	Maximum Possible	744[a]	20.010	1.000

a. There are 9 inter-group ties involving 920 cases.
b. Wald-Wolfowitz Test
c. Grouping Variable: AFRAID TO WALK AT NIGHT IN NEIGHBORHOOD

Looking at the results of the Mann-Whitney U test first, the mean rank is the sum of ranks divided by N. People who are fearful have the higher mean rank (494.62 versus 435.94), which indicates that, on the average, they have higher rates of attendance at religious services than people who are not fearful. The output also shows the number of cases included, the U statistic (89850.5), the obtained Z score (−3.335), and the exact probability that the Z score occurred by random chance alone. The significance is reported as an "Asymp. Sig. (2-tailed)" of .001. The difference in attendance rates is statistically significant.

Turning to the runs test, with only 9 possible scores for *attend*, there will be many ties across the samples, and this will make the results of the runs test hard to interpret. When there are ties, the SPSS output will include the minimum and maximum number of runs possible in the data and the Z scores for each value. If both these Z scores are significant, or if both are not significant, we are safe in coming to a conclusion. In this case, the minimum number of runs (10) is significant, but the maximum (744) is not. This inconsistency makes any conclusion based on the runs test difficult, and we should probably disregard these results.

DEMONSTRATION 11.2 Can Money Buy Happiness?

Is there any empirical support for the widely held view that money can't buy happiness? A variable called *happy*, available in the GSS data set, records the respondent's level of happiness (see Appendix G). We can use *income91* as an indicator of "money." If the adage is true, there should be no significant relationship between these two variables, and people with more income should report about the same levels of happiness as people with less income.

We need to collapse *income91* into just two categories in order to conduct the test. There is no obvious place to split this variable, so we need to run a **Frequencies** for the variable, find the median, and split the variable at that point. To save you the trouble, I'll just let you know that the median for this variable is a score of 16 (or an income of $30,000 to $34,999). Remember to recode into a different variable and assign a new name (say, *rinc*) to the recoded variable. The **recode** instructions in the **Old → New** values box should look like this:

1 thru 16 → 1
17 thru 21 → 2

Follow the instructions in Demonstration 11.1 to generate the two tests with *happy* as the dependent, or test, variable and *rinc* (or whatever name you gave to recoded *income91*) as the grouping variable. Define the groups on *rinc* by assigning 1 to Group 1 and 2 to Group 2. The output for this test will be:

NPar Tests
Mann-Whitney Test

Ranks

	RINC	N	Mean Rank	Sum of Ranks
GENERAL HAPPINESS	1.00	658	675.53	444495.50
	2.00	599	577.89	346157.50
	Total	1257		

Test Statistics[a]

	GENERAL HAPPINESS
Mann-Whitney U	166457.500
Wilcoxon W	346157.500
Z	−5.406
Asymp. Sig. (2-tailed)	.000

a. Grouping Variable: RINC

Wald-Wolfowitz Test

Frequencies

	RINC	N
GENERAL HAPPINESS	1.00	658
	2.00	599
	Total	1257

Test Statistics[b,c]

		Number of Runs	Z	Asymp. Sig. (1-tailed)
GENERAL HAPPINESS	Minimum Possible	4[a]	−35.299	.000
	Maximum Possible	1092[a]	26.236	1.000

a. There are 3 inter-group ties involving 1257 cases.
b. Wald-Wolfowitz Test
c. Grouping Variable: RINC

The Mann-Whitney *U* test shows that the higher-income category (Group 2) has a lower mean rank on *happy*, and the difference is significant at less than .001. Because *happy* is coded so that lower scores indicate higher levels of reported happiness, these results indicate that higher-income respondents are significantly happier than lower-income respondents. The results of the runs test are again ambiguous because, with only three possible scores for *happy*, there are many ties. Overall, we could conclude that the evidence does not support the old adage that money can't buy happiness.

Exercises

11.1 As long as we have *income91* recoded and in a suitable form for use as an independent variable for these tests, take a look at the relationship between income and political ideology. Run both the Mann-Whitney *U* and the runs test with *polviews* as the test variable and recoded *income91* as the independent. It is commonly argued that a conservative political ideology is associated with higher levels of material wealth. Is there any support for this view in these data? Write a few sentences of interpretation.

11.2 Is church attendance (*attend*) related to gender? Use both ordinal tests of significance and write a few sentences summarizing your conclusions.

11.3 Remember *abscale*? If not, see Demonstration 4.2 and use the **Compute** command to recreate this scale. Use the scale as the dependent variable and run separate tests with *sex* and recoded *income91* as independent variables. Recode *degree* into two categories (1 = high school and less, and 2 = at least some college and more) and conduct a third test with recoded *degree* as an independent variable. Write a brief report summarizing your results, and indicate which of the three independent variables had the strongest effect on *abscale*.

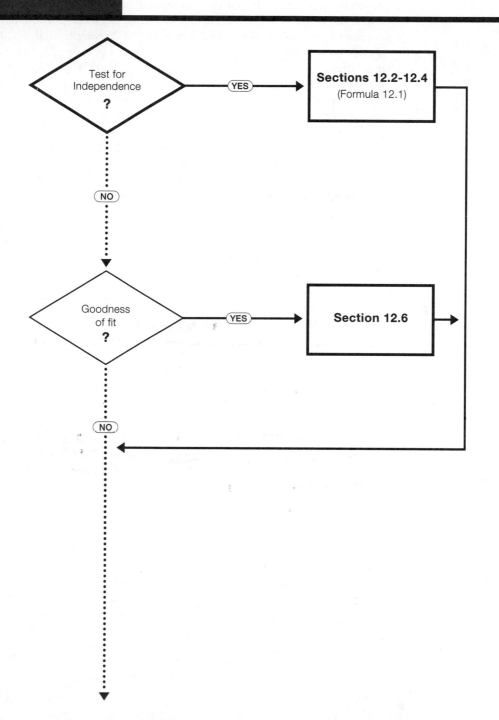

12.1 INTRODUCTION

The **chi square test** is probably the most frequently used test of hypothesis in the social sciences. In part, this popularity is due to the relative ease with which the model assumptions in step 1 can be satisfied. Chi square is a non-parametric test of hypothesis and requires no assumption about the exact shape of the population distribution. Since this test is appropriate for nominally measured variables, we make only the minimum possible assumption with respect to level of measurement.

The popularity of chi square is also related to its usefulness in a wider variety of research situations than the other tests of significance we have covered to this point. The chi square test can be used in the familiar two-sample case but, like the ANOVA test, it is also applicable to research situations where the variables of interest have more than two categories and in situations where we have more than two samples. For example, in Chapter 9, we compared a sample of black senior citizens with a sample of white senior citizens in terms of participation patterns (see Section 9.5). What if the researcher desired to expand the test to include samples of Spanish-speaking and Asian-American senior citizens in the comparison? Two-sample tests are not readily adaptable to research designs with many samples, but chi square handles these situations easily and in a single set of calculations.

12.2 BIVARIATE TABLES

Chi square is computed from **bivariate tables**, so called because they display the scores of cases on two different variables at the same time. Bivariate tables are used to ascertain if there is a significant relationship between the variables and for other purposes that we will investigate in later chapters. In fact, these tables are very commonly used in social research, and a detailed examination of them is in order.

First of all, bivariate tables have (of course) two dimensions. The horizontal (across) dimension is referred to in terms as **rows**, and the vertical dimension (up and down) is referred to in terms of **columns**. Each column or row represents a score on a variable, and the intersections of the row and columns (called **cells**) represent the various combined scores on both variables.

Let's use an example to clarify. Suppose a researcher is interested in the relationship between racial group membership and participation in voluntary groups, community service organizations, and so forth. Is there a difference in the level of involvement in volunteer groups between the races? We have two variables here (race and membership) and, for the sake of simplicity, assume that both are simple dichotomies. That is, people have been classified as either black or white and as either high or low in their level of involvement in voluntary associations.

So far, we have two variables and each has two possible scores. One variable will be placed in the columns and the other in the rows. By convention, the independent (or causal) variable is placed in the columns and the dependent variable in the rows. In the example at hand, race would be the

TABLE 12.1 RATES OF PARTICIPATION IN VOLUNTARY ASSOCIATIONS BY
RACIAL GROUP FOR 100 SENIOR CITIZENS

Participation Rates	Racial Group		
	Black	White	
High			50
Low			50
	50	50	100

causal variable, and each column will represent a score on this variable. Each row will represent a score on level of membership (high or low), and the outline of the bivariate table, for a sample of 100, will look like Table 12.1.

Note that Table 12.1 has a descriptive title that includes the names of the variables contained in the table with the dependent variable listed first. Clear, concise titles should be included in all tables, graphs, and charts.

To finish Table 12.1, we would classify people in terms of both their race and their level of participation and keep a count of how often each combination of scores occurred. Note that with just two scores per variable, there are four possible combinations, each represented by a cell. Blacks with high levels of participation would be counted in the upper-left-hand cell, whites with low levels of participation would be counted in the lower-right-hand cell, and so forth. When we had finished sorting cases into cells, each cell would display the number of times each combination of scores occurred.

Finally, most tables will also include the row and column subtotals or **marginals.** The total number of cases in the table (*N*) is reported at the intersection of the row and column marginals.

12.3 THE LOGIC OF CHI SQUARE

The chi square test has several different uses. We will look first at (and, in fact, spend most of this chapter on) an application called the chi square test for independence. We have encountered the term *independence* in connection with the requirements for random sampling in the two-sample case. In that context, we noted that independent random samples are gathered such that the selection of a particular case for one sample has no effect on the probability that any particular case will be selected for the other sample (see Section 9.2).

In the context of chi square, the meaning of **independence** is closely related to its earlier meaning. Two variables are independent if, for all cases, the classification of a case into a particular category of one variable has no effect on the probability that the case will fall into any particular category of the second variable. For example, race and rates of participation in voluntary association would be independent of each other if the classification of the cases on one variable (as black or white) has no effect on the classification of

TABLE 12.2   THE CELL FREQUENCIES THAT WOULD BE EXPECTED IF RATES
OF PARTICIPATION AND RACIAL GROUP WERE INDEPENDENT

Participation Rates	Racial Group		Totals
	Black	White	
High	25	25	50
Low	25	25	50
Totals	50	50	100

the cases on the other variable (as high or low on participation). In other words, the variables would be independent if level of participation and race were completely unrelated to each other.

Consider Table 12.1 again. If these two variables are independent, then the cell frequencies will be determined solely by random chance. For example, we would find that, just as an honest coin will show heads about 50% of the time when flipped, about half of the black respondents will rank high on participation and half will rank low. If the variables are independent, the same pattern would hold for the 50 white respondents and, therefore, each of the four cells should have about 25 cases in it, as illustrated in Table 12.2. This pattern of cell frequencies indicates that the racial classification of the subjects has no effect on the probability that they would be either high or low in participation. The probability of being classified as high or low would be 0.5 for both blacks and whites, and the variables would therefore be independent.

In the case of chi square, the null hypothesis is that the variables are independent. Under the assumption that the null hypothesis is true, the cell frequencies we would expect to find if only random chance were operating are computed. These frequencies, called **expected frequencies** (symbolized f_e), are then compared, cell by cell, with the frequencies actually observed in the table ("**observed frequencies**," symbolized f_o). If the null hypothesis is true and the variables are independent, then there should be little difference between the expected and observed frequencies. If the null is false, however, there should be large differences between the two. The greater the differences between expected (f_e) and observed (f_o) frequencies, the less likely that the variables are, in fact, independent and the more likely that we will be able to reject the null hypothesis.

12.4 THE COMPUTATION OF CHI SQUARE

As is the case with all tests of hypothesis, the test with chi square consists of computing a test statistic, χ^2 **(obtained)**, from the sample data and placing that value on the sampling distribution of all possible sample outcomes. Specifically, the χ^2 (obtained) will be compared with the value of χ^2 **(critical)** that will be determined by consulting a chi square table (Appendix C) for a particular alpha level and degrees of freedom. Prior to conducting the formal test of hypothesis, let us take a moment to consider the calculation of chi square:

FORMULA 12.1

$$\chi^2 \text{ (obtained)} = \sum \frac{(f_o - f_e)^2}{f_e}$$

where f_o = the cell frequencies observed in the bivariate table
f_e = the cell frequencies that would be expected
if the variables were independent

Solving this formula requires that an expected frequency be computed for each cell in the table. In Table 12.2, because the marginals are the same value for all rows and columns, the expected frequencies are obvious by intuition and will be the same value (f_e = 25) for all four cells. In the more usual case, the expected frequencies will not be obvious, marginals will be unequal, and we must use Formula 12.2 to find the expected frequency for each cell:

FORMULA 12.2

$$f_e = \frac{\text{Row marginal} \times \text{Column marginal}}{N}$$

That is, the expected frequency for any cell is equal to the total number of cases in the row (the row marginal) times the total number of cases in the column (the column marginal) divided by the total number of cases in the table (N).

Once the expected frequencies have been calculated, Formula 12.1 directs us to subtract the expected frequency from the observed frequency for each cell, square this difference, divide by the expected frequency for that cell, and then sum the resultant values for all cells.

An example using Table 12.3 should clarify these procedures. In a random sample, 100 social work majors have been classified in terms of whether the Council on Social Work Education has accredited their undergraduate programs (the column variable) and whether they were hired in social work positions within three months of graduation (the row variable).

TABLE 12.3 EMPLOYMENT OF 100 SOCIAL WORK MAJORS BY ACCREDITATION STATUS OF UNDERGRADUATE PROGRAM

Employment Status	Accreditation Status		Totals
	Accredited	Not Accredited	
Working as a social worker	30	10	40
Not working as a social worker	25	35	60
Totals	55	45	100

TABLE 12.4 EXPECTED FREQUENCIES FOR TABLE 12.3

	Accreditation Status		
Employment Status	Accredited	Not Accredited	Totals
Working as a social worker	22	18	40
Not working as a social worker	33	27	60
Totals	55	45	100

Beginning with the upper-left-hand cell (graduates of accredited programs who are working as social workers), the expected frequency for this cell, using Formula 12.2, is (40)(55)/100, or 22. For the other cell in this row (graduates of nonaccredited programs who are working as social workers), the expected frequency is (40)(45)/100, or 18. For the two cells in the bottom row, the expected frequencies are (60)(55)/100, or 33, and (60)(45)/100, or 27, respectively. The expected frequencies for all four cells are displayed in Table 12.4.

Note that the row and column marginals as well as the total number of cases are all exactly the same as they were in Table 12.3. The row and column marginals for the expected frequencies must always equal those of the original observed frequencies. This relationship provides a convenient way of checking your arithmetic to this point.

The value for chi square for these data can now be found by making the appropriate substitutions into Formula 12.1.

$$\chi^2 \text{ (obtained)} = \sum \frac{(f_o - f_e)^2}{f_e}$$

$$\chi^2 \text{ (obtained)} = \frac{(30 - 22)^2}{22} + \frac{(10 - 18)^2}{18} + \frac{(25 - 33)^2}{33} + \frac{(35 - 27)^2}{27}$$

$$\chi^2 \text{ (obtained)} = \frac{64}{22} + \frac{64}{18} + \frac{64}{33} + \frac{64}{27}$$

$$\chi^2 \text{ (obtained)} = 2.91 + 3.56 + 1.94 + 2.37$$

$$\chi^2 \text{ (obtained)} = 10.78$$

This sample value for chi square must still be tested for its significance. *(For practice in computing chi square, see problem 12.1.)*

12.5 THE CHI SQUARE TEST FOR INDEPENDENCE

As always, the five-step model for significance testing will provide the framework for organizing our decision making. The data presented in Table 12.3 will serve as our example.

Step 1. Making Assumptions.

> Model: Independent random samples
> Level of measurement is nominal

Step 2. Stating the Null Hypothesis. As stated previously, the null hypothesis in the case of chi square states that the two variables are independent. If the null is true, the differences between the observed and expected frequencies should be small. As usual, the research hypothesis directly contradicts the null. Thus, if we reject H_0, the research hypothesis will be supported.

> H_0: The two variables are independent.
> (H_1: The two variables are dependent.)

Step 3. Choosing a Sampling Distribution and Establishing the Critical Region. The sampling distribution of sample chi squares, unlike the Z and t distributions, is positively skewed, with higher values of sample chi squares in the upper tail of the distribution (to the right). Thus, with the chi square test, the critical region is established in the upper tail of the sampling distribution.

Values for χ^2 (critical) are given in Appendix C. This table is similar to the t table, with alpha levels arrayed across the top and degrees of freedom down the side. A major difference from the table of critical t scores is that, for chi square, degrees of freedom (df) are found by the following formula:

FORMULA 12.3

$$df = (r - 1)(c - 1)$$

where df = degrees of freedom
$(r - 1)$ = number of rows minus 1
$(c - 1)$ = number of columns minus 1

A table with two rows and two columns (a 2×2 table) has one degree of freedom regardless of the number of cases in the sample.* A table with two rows and three columns would have $(2 - 1)(3 - 1)$, or two, degrees of freedom. Since our sample problem involves a 2×2 table (df = 1), if we set

---

*Degrees of freedom are the number of values in a distribution that are free to vary for any particular statistic. A 2×2 table has one degree of freedom because, for a given set of marginals, once one cell frequency is determined, all other cell frequencies are fixed (that is, they are no longer free to vary). In Table 12.3, for example, if any cell frequency is known, all others are determined. If the upper-left-hand cell is known to be 30, the remaining cell in that row must be 10, since there are 40 cases total in the row and $40 - 30 = 10$. Once the frequencies of the cells in the top row are established, cell frequencies for the bottom row are determined by subtraction from the column marginals. Incidentally, this relationship can be used to good advantage when computing expected frequencies. For example, in a 2×2 table, only one expected frequency needs to be computed. The f_e's for all other cells can then be found by subtraction.

FIGURE 12.1 THE CHI SQUARE DISTRIBUTION WITH $\alpha = 0.05$, df = 1

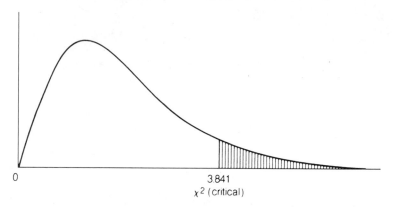

χ^2 (critical)

alpha at 0.05, the critical chi square score would be 3.841. Summarizing these decisions, we have

$$\text{Sampling distribution} = \chi^2 \text{ distribution}$$
$$\text{Alpha} = 0.05$$
$$\text{Degrees of freedom} = 1$$
$$\chi^2 \text{ (critical)} = 3.841$$

The sampling distribution of χ^2 is depicted in Figure 12.1.

Step 4. Computing the Test Statistic. The mechanics of these computations were introduced in Section 12.4. As you recall, we had

$$\chi^2 \text{ (obtained)} = \sum \frac{(f_o - f_e)^2}{f_e}$$

$$\chi^2 \text{ (obtained)} = 10.78$$

Step 5. Making a Decision. Comparing the test statistic with the critical region,

$$\chi^2 \text{ (obtained)} = 10.78$$
$$\chi^2 \text{ (critical)} = 3.841$$

we see that the test statistic falls in the critical region and, therefore, we reject the null hypothesis of independence. The pattern of cell frequencies observed in Table 12.3 is unlikely to have occurred by chance alone. The variables are dependent. Specifically, based on these sample data, the probability of securing employment in the field of social work seems to be dependent on the accreditation status of the program. *(For practice in conducting and interpreting the chi square test for independence, see problems 12.2 to 12.15 and problem 12.19.)*

12.6 THE CHI SQUARE TEST: AN EXAMPLE

To this point, we have confined our attention to 2 × 2 tables. For purposes of illustration, we will work through the computational routines and decision-making process for a larger table. As you will see, larger tables require more computations (because they have more cells); but, in all other essentials, they are dealt with in the same way as the 2 × 2 table.

A researcher is concerned with the possible effects of marital status on the academic progress of college students. Do married students, with their extra burden of family responsibilities, suffer academically as compared to unmarried students? (Or: Is academic performance dependent on marital status?) A random sample of 453 students is gathered, and each student is classified as either married or unmarried, and—using grade-point average (GPA) as a measure—as a good, average, or poor student. Results are presented in Table 12.5.

For the top left-hand cell (married students with good GPA's) the expected frequency would be (160)(175)/453, or 61.8. For the other cell in this row, expected frequency is (160)(278)/453, or 98.2. In similar fashion, all expected frequencies are computed (being very careful to use the correct row and column marginals) and displayed in Table 12.6.

TABLE 12.5 GRADE-POINT AVERAGE (GPA) BY MARITAL STATUS FOR 453 COLLEGE STUDENTS

GPA	Married	Not Married	Totals
Good	70	90	160
Average	60	110	170
Poor	45	78	123
Totals	175	278	453

TABLE 12.6 EXPECTED FREQUENCIES FOR TABLE 12.5

GPA	Married	Not Married	Totals
Good	61.8	98.2	160
Average	65.7	104.3	170
Poor	47.5	75.5	123
Totals	175	278	453

With all expected frequencies computed, the formula for χ^2 (obtained) can be solved. In working through this formula, double-check to be certain that you are using the proper f_o's and f_e's for each cell.

$$\chi^2 \text{ (obtained)} = \sum \frac{(f_o - f_e)^2}{f_e}$$

$$\chi^2 \text{ (obtained)} = \frac{(70 - 61.8)^2}{61.8} + \frac{(90 - 98.2)^2}{98.2} + \frac{(60 - 65.7)^2}{65.7} + \frac{(110 - 104.3)^2}{104.3}$$

$$+ \frac{(45 - 47.5)^2}{47.5} + \frac{(78 - 75.5)^2}{75.5}$$

$$\chi^2 \text{ (obtained)} = \frac{(8.2)^2}{61.8} + \frac{(-8.2)^2}{98.2} + \frac{(-5.7)^2}{65.7} + \frac{(5.7)}{104.3} + \frac{(-2.5)^2}{47.5} + \frac{(2.5)^2}{75.5}$$

$$\chi^2 \text{ (obtained)} = 1.09 + 0.69 + 0.49 + 0.31 + 0.13 + 0.08$$

$$\chi^2 \text{ (obtained)} = 2.79$$

This value for the test statistic can now be tested for its significance.

Step 1. Making Assumptions.

Model: Independent random samples
Level of measurement is nominal

Step 2. Stating the Null Hypothesis.

H_0: The variables are independent.
(H_1: The variables are dependent.)

Step 3. Selecting the Sampling Distribution and Establishing the Critical Region.

Sampling distribution = χ^2 distribution
Alpha = 0.05
Degrees of freedom = $(r - 1)(c - 1) = (3 - 1)(2 - 1) = 2$
χ^2 (critical) = 5.991

Step 4. Computing the Test Statistic.

$$\chi^2 \text{ (obtained)} = \sum \frac{(f_o - f_e)^2}{f_e}$$

$$\chi^2 \text{ (obtained)} = 2.79$$

Step 5. Making a Decision. The test statistic, χ^2 (obtained) = 2.79, does not fall in the critical region, which, for alpha = 0.05, df = 2, begins at a χ^2 (critical) of 5.991. Therefore, we fail to reject the null. The observed

Application 12.1

Do members of different groups have different levels of "narrow-mindedness"? A random sample of 47 white and black Americans have been rated as high or low on a scale that measures intolerance of viewpoints or belief systems different from their own. The results are

	Group		
Intolerance	White	Black	Totals
High	15	5	20
Low	10	17	27
Totals	25	22	47

The frequencies we would expect to find if the null hypothesis (H_0: the variables are independent) were true are

	Group		
Intolerance	White	Black	Totals
High	10.64	9.36	20.00
Low	14.36	12.64	27.00
Totals	25.00	22.00	47.00

Expected frequencies are found on a cell-by-cell basis by the formula

$$f_e = \frac{(\text{Row marginal})(\text{Column marginal})}{N}$$

for "White-High": $f_e = \dfrac{(20)(25)}{47} = 10.64$

for "Black-High": $f_e = \dfrac{(20)(22)}{47} = 9.36$

for "White-Low": $f_e = \dfrac{(27)(25)}{47} = 14.36$

for "Black-Low": $f_e = \dfrac{(27)(22)}{47} = 12.64$

The value of χ^2 (obtained) would be

$$\chi^2 \text{ (obtained)} = \sum \frac{(f_o - f_e)^2}{f_e}$$

$$\chi^2 \text{ (obtained)} = \frac{(15 - 10.64)^2}{10.64} + \frac{(5 - 9.36)^2}{9.36}$$
$$+ \frac{(10 - 14.36)^2}{14.36} + \frac{(17 - 12.64)^2}{12.64}$$

$$\chi^2 \text{ (obtained)} = \frac{19.01}{10.64} + \frac{19.01}{9.36} + \frac{19.01}{14.36} + \frac{19.01}{12.64}$$

$$\chi^2 \text{ (obtained)} = 1.79 + 2.03 + 1.32 + 1.50$$

$$\chi^2 \text{ (obtained)} = 6.64$$

If alpha is set at the 0.05 level, the critical region, with 1 degree of freedom, would begin at χ^2 (critical) = 3.841. With an obtained χ^2 of 6.64, we would reject the null hypothesis of independence. For this sample there is a statistically significant relationship between group membership and intolerance.

frequencies are not significantly different from the frequencies we would expect to find if the variables were independent and only random chance were operating. Based on these sample results, we can conclude that the academic performance of college students is not dependent on their marital status.

12.7 AN ADDITIONAL APPLICATION OF THE CHI SQUARE TEST: THE GOODNESS-OF-FIT TEST*

To this point we have dealt with the chi square test for independence for situations involving two variables, each of which has two or more categories. Another situation in which the chi square statistic is useful, called the **goodness-of-fit test,** is one in which the distribution of scores on a single variable must be tested for significance. The logic underlying this second research situation is quite similar to that of the test for independence. The chi square statistic will be computed by comparing the actual distribution of a variable against a set of expected frequencies. The greater the difference between the observed distribution of scores and the expected distribution, the more likely that the observed pattern did not occur by random chance alone. If the observed and expected frequencies are similar, it is said that there is a "good fit" (hence the name of this application), and we would conclude that the two distributions were not significantly different.

A major difference in this new application lies in the way in which the expected frequencies are ascertained. Instead of computing these scores, the null hypothesis is used to figure out what the expected frequencies should be. An example may make this process clear. Suppose you were gambling on coin tosses and suspected that a particular coin was biased toward heads. Over a series of tosses, what percentage of heads would you expect to observe from an unbiased coin? In an actual test of significance for this problem, the null hypothesis would be that the coin was unbiased and that half of all tosses should be heads and half tails. Notice what we have just done: we figured out the expected frequencies (half of the flips should be heads) from the null hypothesis (the coin is unbiased), rather than by computing them.

Let's consider another example of a situation in which the goodness-of-fit test is appropriate. Is there a seasonal rhythm to the crime rate? If we gathered crime statistics on a monthly basis for a given jurisdiction, what would we expect to find? Given the way in which the problem has been stated, there is only one variable (crime rate), and the focus is on the distribution of that variable over a given set of categories (by month). If the crime rate does not vary by month, we would expect that about 1/12 of all crimes committed in a year would be committed in each and every month. Our null hypothesis would be that the crime rate does not vary across time, and our expected frequencies would be calculated by dividing the actual (observed) number of crimes equally by the number of months. (To conserve time and space, we will ignore the slight complexities that would be introduced by taking account of the varying number of days per month.)

In a particular jurisdiction, a total of 2172 crimes were committed last year. The actual distribution of crimes by month is displayed in Table 12.7. The expected distribution of crimes per month would be found by dividing the total number of crimes by 12: $2172/12 = 181$. Note that the expected

---

*This section is optional.

TABLE 12.7 NUMBER OF CRIMES PER MONTH

Month	Number of Crimes
January	190
February	152
March	121
April	110
May	147
June	199
July	250
August	247
September	201
October	150
November	193
December	212
	2172

frequency will be the same for every month. With these values determined, we can calculate chi square by using Formula 12.1.

$$\chi^2 \text{ (obtained)} = \frac{(f_o - f_e)^2}{f_e}$$

$$\chi^2 \text{ (obtained)} = \frac{(190 - 181)^2}{181} + \frac{(152 - 181)^2}{181} + \frac{(121 - 181)^2}{181}$$

$$+ \frac{(110 - 181)^2}{181} + \frac{(147 - 181)^2}{181} + \frac{(199 - 181)^2}{181}$$

$$+ \frac{(250 - 181)^2}{181} + \frac{(247 - 181)^2}{181} + \frac{(201 - 181)^2}{181}$$

$$+ \frac{(150 - 181)^2}{181} + \frac{(193 - 181)^2}{181} + \frac{(212 - 181)^2}{181}$$

$$\chi^2 \text{ (obtained)} = \frac{81}{181} + \frac{841}{181} + \frac{3600}{181} + \frac{5041}{181} + \frac{1156}{181} + \frac{324}{181}$$

$$+ \frac{4761}{181} + \frac{4356}{181} + \frac{400}{181} + \frac{961}{181} + \frac{144}{181} + \frac{961}{181}$$

$$\chi^2 \text{ (obtained)} = .45 + 4.65 + 19.89 + 27.85 + 6.39 + 1.79$$

$$+ 26.30 + 24.07 + 2.21 + 5.31 + .80 + 5.31$$

$$\chi^2 \text{ (obtained)} = 125.02$$

The χ^2 of 125.02 can now be tested for significance in the usual manner.

Step 1. Making Assumptions.

Model: Random sampling
Level of measurement is nominal

In this application of significance testing, we are assuming that the observed frequencies represent a random sample of the theoretical frequency distribution.

Step 2. Stating the Null Hypothesis.

H_0: There is no difference in the crime rate by month.
(H_1: There is a difference in crime rate by month.)

Step 3. Selecting the Sampling Distribution and Establishing the Critical Region. In the goodness-of-fit test, degrees of freedom are equal to the number of categories minus 1, or df $= k - 1$. In the problem under consideration, there are 12 months or categories and, therefore, 11 degrees of freedom.

$$\text{Sampling distribution} = \chi^2 \text{ distribution}$$
$$\text{Alpha} = .05$$
$$\text{Degrees of freedom} = k - 1 = 12 - 1 = 11$$
$$\chi^2 \text{ (critical)} = 19.675$$

Step 4. Computing the Test Statistic.

$$\chi^2 \text{ (obtained)} = 125.02$$

Step 5. Making a Decision. The test statistic (125.02) clearly falls into the critical region (which begins at 19.675), so we may reject the null. These data suggest that, at least for this jurisdiction, crime rate does vary by month in a nonrandom fashion. *(For practice in conducting and interpreting the chi square goodness-of-fit test, see problems 12.16 and 12.17.)*

12.8 THE LIMITATIONS OF THE CHI SQUARE TEST

Two potential difficulties occur with the chi square test. The first occurs with small samples and the second, interestingly enough, occurs with large samples. When sample size is small, you can no longer assume that the sampling distribution of all possible sample test statistics is accurately described by the chi square distribution. In the case of the chi square test, a small sample is defined as one where a high percentage of the cells have expected frequencies (f_e) of 5 or less. Various rules of thumb have been developed to help the researcher decide what constitutes a "high percentage of cells." Probably the safest course is to take corrective action whenever any of the cells have expected frequencies of 5 or less.

In the case of 2×2 tables, the value of χ^2 (obtained) can be adjusted by applying Yate's correction for continuity, the formula for which is

FORMULA 12.4

$$\chi_c^2 = \sum \frac{(|f_o - f_e| - .5)^2}{f_e}$$

where χ_c^2 = corrected chi square
$|f_o - f_e|$ = the absolute value of the difference between the observed and expected frequency for each cell

---

READING STATISTICS 8: What *Really* Goes On at Mardi Gras?

In a somewhat unusual research project, sociologists Wesley Schrum and John Kilburn observed instances of public nudity during the 1991 Mardi Gras celebration in the French Quarter of New Orleans. They note that nudity is common in this locale in the days preceding Lent and almost always involves brief, exhibitionistic acts of "flashing" by both males and females. Commonly, people disrobe in competition for beads—a form of symbolic wealth—thrown by the riders on the floats in the parades that are at the center of the ceremony. Why does this behavior—deviant and criminal in other contexts and officially illegal in New Orleans as well—occur, and what does it signify?

Schrum and Kilburn present an elaborate and interesting analysis that sees disrobing in this context as a type of ritual that reflects and reinforces the market, power, and gender relations in the larger society. For example, the float riders are elaborately costumed and represent "aristocrats" who, by tossing trinkets to the crowd of "commoners" reflect patterns of economic inequalities. The competition for the otherwise worthless trinkets reflects the competition for wealth and success in everyday life.

The following table presents some of the results of the study and shows that disrobement is a very localized behavior. Bourbon Street is the center of Mardi Gras, and virtually all of the public nudity occurs in that location. As the chi square shows,

the differences between Bourbon Street and Royal Street (only a short block away and running parallel to Bourbon) are statistically significant.

Type of Action by Street Location

	Bourbon Street	Royal Street
Disrobement	55.4 (504)	4.7 (14)
Refusal	6.0 (55)	0 (0)
Beads without Disrobement	38.6 (351)	95.3 (281)
Totals	100% (910)	100% (295)

Of the 910 observed episodes on Bourbon Street, over half involved disrobement in exchange for beads or other trinkets. In almost 40% of the cases, however, beads were obtained without disrobement. Does this mean that nudity is unnecessary and excessive? The researchers say

Of the thousands of revelers who strolled down Bourbon Street . . . only 351 received . . . beads without disrobing. . . . Disrobing increases the chances of receiving beads by at least an order of magnitude. . . . No significant accumulation of symbolic wealth is likely unless participants (1) are children, (2) purchase beads in shops, or (3) disrobe.

Table and excerpts reprinted from *Social Forces* 75(1996). "Ritual Disrobement at Mardi Gras: Ceremonial Exchange and Moral Order" by Wesley Schrum and John Kilburn. Copyright © The University of North Carolina Press.

The correction factor is applied by reducing the value of the term $(f_o - f_e)$ by .5 before squaring the difference and dividing by the expected frequency for the cell.

For tables larger than 2 × 2, there is no correction formula for computing χ^2 (obtained) for small samples. It may be possible to combine some of the categories of the variables and thereby increase cell sizes. Obviously, however, this course of action should be taken only when it is sensible to do so.

In other words, distinctions that have clear theoretical justifications should not be erased merely to conform to the requirements of a statistical test. In any case where you feel that categories cannot be combined to build up cell frequencies and the percentage of cells with expected frequencies of 5 or less is small, it is probably justifiable to continue with the uncorrected chi square test as long as the results are regarded with a suitable amount of caution.

The second potential problem with chi square occurs with large samples. I pointed out at the end of Chapter 9 that all tests of hypothesis are sensitive to sample size. That is, the probability of rejecting the null hypothesis increases with sample size regardless of the size of the difference and the selected alpha level. It turns out that chi square, too, is sensitive to sample size and that larger samples may lead to the decision to reject the null when the actual relationship is trivial. In fact, chi square is more responsive to changes in sample size than the other test statistics we have encountered, since the value of χ^2 (obtained) will increase at the same rate as sample size. That is, if sample size is doubled, the value of χ^2 (obtained) will be doubled. *(For an illustration of this principle, see problem 12.18.)*

In a sense, this general relationship between sample size and the value of the test statistic is more a theoretical, mathematical problem than a realistic or practical one. My major purpose in stressing this relationship is really to emphasize the distinction between statistical significance and theoretical importance. On one hand, tests of significance play a crucial role in research. As long as we are working with random samples, we must know if our research results could have been produced by mere random chance.

On the other hand, tests of hypothesis, like any other statistical techniques, are limited in the range of questions they can answer. Specifically, these tests will tell us whether our results are statistically significant or not. They will not necessarily tell us if the results are important in any other sense. To deal more directly with questions of importance, we must apply an additional set of statistical techniques called measures of association. These techniques will be the subject of Part III.

SUMMARY

1. The chi square test for independence is appropriate for situations in which the variables of interest have been organized into table format. The null hypothesis is that the variables are independent or that the classification of a case into a particular category on one variable has no effect on the probability that the case will be classified into any particular category of the second variable.

2. Since chi square is nonparametric and requires only nominally measured variables, its model assump-

tions are easily satisfied. Furthermore, since it is computed from bivariate tables, in which the number of rows and columns can be easily expanded, the chi square test can be used in many situations in which other tests are inapplicable.

3. In the chi square test, we first find the frequencies that would appear in the cells if the variables were independent (f_e) and then compare those frequencies, cell by cell, with the frequencies actually observed in the cells (f_o). If the null is true, expected and observed frequencies should be quite close in value. The greater the difference between the observed and expected frequencies, the greater the possibility of rejecting the null.

4. In the chi square goodness-of-fit test, the observed frequencies are compared against a set of expected frequencies as derived from the null hypothesis. This application is used to test for the randomness of the distribution of a single variable across a series of categories.

5. As sample size (N) decreases, the chi square test becomes less trustworthy, and corrective action may be required. As sample size increases, the value of the obtained chi square also increases, and with very large samples, we may declare relatively trivial relationships to be statistically significant. As is the case with all tests of hypothesis, statistical significance is not the same thing as "importance" in any other sense. As a general rule, statistical significance is a necessary but not sufficient condition for theoretical or practical importance.

SUMMARY OF FORMULAS

Chi square (obtained):

12.1 $\qquad \chi^2 \text{ (obtained)} = \sum \dfrac{(f_o - f_e)^2}{f_e}$

Expected frequencies:

12.2 $\qquad f_e = \dfrac{\text{(Row marginal)(Column marginal)}}{N}$

Degrees of freedom, bivariate tables:

12.3 $\qquad df = (r - 1)(c - 1)$

Yate's correction for continuity:

12.4 $\qquad \chi_c^2 = \sum \dfrac{(|f_o - f_e| - .5)^2}{f_e}$

GLOSSARY

Bivariate table. A table that displays the joint frequency distributions of two variables

Cells. The cross-classification categories of the variables in a bivariate table.

χ^2 **(critical).** The score on the sampling distribution of all possible sample chi squares that marks the beginning of the critical region.

χ^2 **(obtained).** The test statistic as computed from sample results.

Chi square test. A nonparametric test of hypothesis for variables that have been organized into a bivariate table.

Column. The vertical dimension of a bivariate table. By convention, each column represents a score on the independent variable.

Expected frequency (f_e). The cell frequencies that would be expected in a bivariate table if the variables were independent.

Goodness-of-fit test. A chi square test to see if a variable is randomly distributed across a series of categories.

Independence. The null hypothesis in the chi square test. Two variables are independent if, for all cases, the classification of a case on one variable has no effect on the probability that the case will be classified in any particular category of the second variable.

Marginals. The row and column subtotals in a bivariate table.

Observed frequency (f_o). The cell frequencies actually observed in a bivariate table.

Row. The horizontal dimension of a bivariate table. By convention, each row represents a score on the dependent variable.

PROBLEMS

Asterisk indicates an optional problem.

12.1 For each table below, calculate the obtained chi square. (*HINT: Calculate the expected frequencies for each cell with Formula 12.2. Double-check to make sure you are using the correct row and column marginals for each cell. It may be helpful to record the expected frequencies in table format as well—see Tables 12.4 and 12.3, 12.6 and 12.5. Next, use Formula 12.1 to calculate chi square. For each cell, subtract expected frequency from obtained frequency, square the result, and divide by expected frequency. Sum the results to determine the value of chi square. Double-check to make sure that you are using the correct values for each cell.*)

a.

20	25	45
25	20	45
45	45	90

b.

10	15	25
20	30	50
30	45	75

c.

25	15	40
30	30	60
55	45	100

d.

20	45	65
15	20	35
35	65	100

12.2 SOC A sample of 25 cities have been classified as high or low on their homicide rates and on the number of handguns sold within the city limits. Is there a relationship between these two variables? Explain your results in a sentence or two.

Volume of Gun Sales	Homicide Rate Low	High	Totals
High	8	5	13
Low	4	8	12
Totals	12	13	25

12.3 SW A local politician is concerned that a program for the homeless in her city is discriminating against blacks and other minorities. The data below were taken from a random sample of black and white homeless people. Is there a statistically significant relationship between race and whether or not the person has received services from the program?

Received Services?	Race Black	White	Totals
Yes	6	7	13
No	4	9	13
Totals	10	16	26

12.4 PS Many analysts have noted a "gender gap" in elections for the U.S. presidency with women more likely to vote for the Democratic candidate. A sample of university faculty has been asked about their political party preference. Do their responses indicate a significant relationship between gender and party preference?

Party Preference	Gender Male	Female	Totals
Democratic	10	15	25
Republican	15	10	25
Totals	25	25	50

12.5 PA Is there a relationship between salary levels and unionization for public employees? The data below represent this relationship for fire departments in a random sample of 100 cities of roughly the same size. Salary data have been dichotomized at the median. Summarize your findings.

Salary	Union	Nonunion	Totals
High	21	29	50
Low	14	36	50
Totals	35	65	100

12.6 GER Are retired citizens who maintain close relations with their kin happier than those who do not? The table below relates reported "happiness" to kin relations for a random sample of 123 retired citizens. Is there a significant relationship?

Happiness	Relations with Kin Close	Not Close	Totals
"Very happy"	26	44	70
"Not very happy"	32	21	53
Totals	58	65	123

12.7 SOC A program of pet therapy has been running at a local nursing home. Are the participants in the program more alert and responsive than nonparticipants? The results, drawn from a random sample of residents, are reported below.

Alertness	Participants	Nonparticipants	Totals
High	23 19.28	15 18.81	38
Low	11 14.62	18 14.19	29
Totals	34	33	67

12.8 |SOC| The state Department of Education has rated a sample of local school systems for compliance with state-mandated guidelines for quality. Is the quality of a school system significantly related to the affluence of the community as measured by per capita income?

| | Per Capita Income | | |
Quality	Low	High	Totals
Low	16	8	24
High	9	17	26
Totals	25	25	50

12.9 |CJ| A local judge has been allowing some individuals convicted of "driving under the influence" to work in a hospital emergency room as an alternative to fines, suspensions, and other penalties. A random sample of offenders has been drawn. Do participants in this program have lower rates of recidivism for this offense?

Recidivist?	Participants	Nonparticipants	Totals
Yes	60	123	183
No	55	108	163
Totals	115	231	346

12.10 |SOC| Is there a relationship between length of marriage and satisfaction with marriage? The necessary information has been collected from a random sample of 100 respondents drawn from a local community. Write a sentence or two explaining your decision.

| | Length of Marriage (in years) | | | |
Satisfaction	More than 10	5–10	Less than 5	Totals
Low	20	20	10	50
High	10	20	20	50
Totals	30	40	30	100

12.11 |PS| Is there a relationship between political ideology and class standing? Are upperclass students significantly different from underclass students on this variable? The table below reports the relationship between these two variables for a random sample of 267 college students.

| Political Ideology | Class Standing | | |
	Underclass	Upperclass	Totals
Liberal	43	40	83
Moderate	50	50	100
Conservative	40	44	84
Totals	133	134	267

12.12 |SOC| The new police chief in Shinbone is concerned about the image of the department and wants to know which sectors of the community have the least favorable attitudes toward the police. The results of a police opinion survey conducted on a random sample of 520 citizens are reported below. Are attitudes related to social class, ethnicity, or contact with members of the force? Summarize your findings. Does the police force have an "image problem"? Where?

a.

| Attitude | Social Class | | | |
	Lower	Middle	Upper	Totals
Positive	40	120	140	300
Negative	110	80	30	220
Totals	150	200	170	520

b.

| Attitude | Ethnicity | | | |
	Black	White	Other	Totals
Positive	70	190	40	300
Negative	60	150	10	220
Totals	130	340	50	520

c. Have you had any contacts with the police during the past six months?

Attitude	Yes, Many	Yes, Few	No	Totals
Positive	15	100	185	300
Negative	55	100	65	220
Totals	70	200	250	520

12.13 |SOC| At a large urban college, about half of the students live off campus in various arrangements, and the other half live in dormitories on campus. Is academic performance dependent on living arrangements? The results based on a random sample of 300 students are presented below.

| | Living Arrangement Off Campus | | | |
GPA	With Roommates	With Parent	On Campus	Totals
High	32 *(24.3)*	10 *(18.9)*	38 *(37.8)*	80
Moderate	36 *(38.7)*	40 *(30.1)*	54 *(60.7)*	130
Low	22 *(27)*	20 *(21)*	48 *(42)*	90
Totals	90	70	140	300

12.14 SOC An urban sociologist has built up a database describing a sample of the neighborhoods in her city and has developed a scale by which each area can be rated for the "quality of life" (this includes measures of pollution, noise, open space, services available, and so on). She has also asked samples of residents of these areas about their level of satisfaction with their neighborhoods. Is there significant agreement between the sociologist's objective ratings of quality and the respondent's self-reports of satisfaction?

| Reported Satisfaction | Quality of Life | | | |
	Low	Moderate	High	Totals
High	8 *(14.76)*	17 *(20.5)*	32 *(21.29)*	57
Moderate	12 *(15.17)*	25 *(21.09)*	21 *(21.83)*	58
Low	21 *(11.07)*	15 *(15.39)*	6 *(15.93)*	42
Totals	41	57	59	157

12.15 SOC Does support for the legalization of marijuana vary by region of the country? The table displays the relationship between the two variables for a random sample of 1020 adult citizens. Is the relationship significant?

| Legalize Marijuana? | Region | | | | |
	North	Midwest	South	West	Totals
Yes	60	65	42	78	245
No	245	200	180	150	775
Totals	305	265	222	228	1020

*12.16 The director of athletics at the local high school wonders if the sports program is getting a proportional amount of support from each of the four classes. If there are roughly equal numbers of students in each of the classes, what does the following breakdown of attendance figures from a random sample of students in attendance at a recent basketball game suggest?

Class	Frequency
Freshman	200
Sophomore	150
Junior	120
Senior	110
	580

*12.17 SOC A small western town has roughly equal numbers of Hispanic, Asian, and Anglo-American residents. Are the three groups equally represented at town meetings? The attendance figures for a random sample drawn from those attending a meeting were

Group	Frequencies
Hispanic	74
Asian	55
Anglo	53

Is there a statistically significant pattern here?

12.18 SOC A researcher is concerned with the relationship between attitudes toward violence and violent behavior. If attitudes "cause" behavior (a very debatable proposition), then people who have positive attitudes toward violence should have high rates of violent behavior. A pretest was conducted on 70 respondents and, among other things, the respondents were asked, "Have you been involved in a violent incident of any kind over the past six months?" The researcher established the following relationship:

| Involvement | Attitude Toward Violence | | Totals |
	Favorable	Unfavorable	
Yes	16	19	35
No	14	21	35
Totals	30	40	70

The chi square calculated on these data is .23, which is not significant at the .05 level (confirm this conclusion with your own calculations).

Undeterred by this result, the researcher proceeded with the project and gathered a random sample of 7000. In terms of percentage distributions, the results for the full sample were exactly the same as for the pretest:

Attitude Toward Violence

Involvement	Favorable	Unfavorable	Totals
Yes	1600	1900	3500
No	1400	2100	3500
Totals	3000	4000	7000

However, the chi square obtained is a very healthy 23.4 (confirm with your own calculations). Why is the full-sample chi square significant when the pretest was not? What happened? Do you think that the second result is important?

12.19 SOC Some results from the General Social Survey are presented below. The survey is administered every year to a nationally representative sample. See Appendix G for the complete questions and other details about the survey. For each table, conduct the chi square test of significance. Write a sentence or two of interpretation for each test.

a. Support for the legal right to an abortion for any reason (ABANY) by age:

Support for Abortion	40 and Younger	Older than 40	Totals
Yes	195	213	408
No	216	279	495
Totals	411	492	903

b. Support for the death penalty by age:

Support for Capital Punishment:	40 and Younger	Older than 40	Totals
Favor	461	557	1018
Oppose	130	181	311
Totals	591	738	1329

c. Fear of walking alone at night by age:

Fear	40 and Younger	Older than 40	Totals
Yes	163	226	389
No	263	284	547
Totals	426	510	936

d. Support for legalizing marijuana by age:

Legalize?	40 and Younger	Older than 40	Totals
Legal	128	113	241
Not Legal	262	403	665
Totals	390	516	906

e. Support for legalizing suicide when a person has an incurable disease by age:

Suicide?	40 and Younger	Older than 40	Totals
Yes	287	326	613
Not Legal	137	191	328
Totals	424	517	941

SPSS for Windows

Using *SPSS for Windows* to Produce the CHI Square Test for Independence

DEMONSTRATION 12.1 Do Childcare Practices Vary by Social Class?

The **Crosstabs** procedure in SPSS produces bivariate tables and a wide variety of statistics. This procedure is very commonly used in social science research at all levels, and you will see many references to **Crosstabs** in chapters to come. I will introduce the command here, and we will return to it often in later sessions.

Do people in different social classes raise their children differently? Which social class would be most likely to approve of spanking as a disciplinary technique? We can answer these questions, at least for the 1996 GSS sample, by constructing a bivariate table to display the relationship between *class* and *spanking*. We will also request a chi square test for the table.

Begin by clicking **Statistics**, then click **Summarize** and **Crosstabs**. The **Crosstabs** dialog box will appear with the variables listed in a box on the left. Highlight *spanking* (the label for this variable is FAVOR SPANKING TO DISCIPLINE CHILD) and click the arrow to move the variable name into the **Rows** box and then highlight *class* (the label for this variable is SUBJECTIVE CLASS IDENTIFICATION) and move it into the **Columns** box. Click the **Statistics** button at the bottom of the window and click the box next to Chi-square. Click **Continue** and **OK**, and the following output will be produced:

Crosstabs

Case Processing Summary

	Cases					
	Valid		Missing		Total	
	N	Percent	N	Percent	N	Percent
FAVOR SPANKING TO DISCIPLINE CHILD * SUBJECTIVE CLASS IDENTIFICATION	969	66.7%	483	33.3%	1452	100.0%

FAVOR SPANKING TO DISCIPLINE CHILD* SUBJECTIVE CLASS IDENTIFICATION
Crosstabulation

Count

		SUBJECTIVE CLASS IDENTIFICATION		
		LOWER CLASS	WORKING CLASS	MIDDLE CLASS
FAVOR SPANKING TO DISCIPLINE CHILD	STRONGLY AGREE	26	122	115
	AGREE	18	218	187
	DISAGREE	13	51	95
	STRONGLY DISAGREE	3	25	55
Total		60	416	452

FAVOR SPANKING TO DISCIPLINE CHILD* SUBJECTIVE CLASS IDENTIFICATION
Crosstabulation

Count

		SUBJECTIVE	
		UPPER CLASS	Total
FAVOR SPANKING TO DISCIPLINE CHILD	STRONGLY AGREE	11	274
	AGREE	16	439
	DISAGREE	9	168
	STRONGLY DISAGREE	5	88
Total		41	969

Chi-Square Tests

	Value	df	Asymp. Sig. (2-sided)
Pearson Chi-Square	37.390[a]	9	.000
Likelihood Ratio	37.719	9	.000
Linear-by-Linear Association	17.136	1	.000
N of Valid Cases	969		

a. 1 cells (6.3%) have expected count less than 5. The minimum expected count is 3.72.

The table displays the number of cases in each cell and the row and column totals. The results of the chi square test are reported in the output block that follows the table. The value of chi square is 37.390, the degrees of freedom are 9, and the exact significance of the chi square is less than .001. This is well below the standard indicator of a significant result (alpha = .05), so we may conclude that there is a statistically significant relationship between *class* and *spanking*. Support for spanking does vary by social class.

Remember that the chi square test tells us only that the frequencies in the table did not occur by random chance. Naturally, we would also like to know *which* social class is most and least supportive of spanking. We will take this question up in Demonstration 13.1.

DEMONSTRATION 12.2 Who Watches X-Rated Movies?

Is there a relationship between social class and the viewing of X-rated movies? Run the **Crosstabs** procedure again with *xmovie* (the label for this variable is SEEN X-RATED MOVIE IN LAST YEAR) as the row variable (in place of *spanking*) and class as the column variable. Don't forget to request chi square. The output will be as shown:

Crosstabs

Case Processing Summary

	Cases					
	Valid		Missing		Total	
	N	Percent	N	Percent	N	Percent
SEEN X-RATED MOVIE IN LAST YEAR* SUBJECTIVE CLASS IDENTIFICATION	947	65.2%	505	34.8%	1452	100.0%

SEEN X-RATED MOVIE IN LAST YEAR* SUBJECTIVE CLASS IDENTIFICATION Crosstabulation

Count

		SUBJECTIVE CLASS IDENTIFICATION		
		LOWER CLASS	WORKING CLASS	MIDDLE CLASS
SEEN X-RATED MOVIE IN LAST YEAR	YES	14	115	90
	NO	52	302	336
Total		66	417	426

SEEN X-RATED MOVIE IN LAST YEAR* SUBJECTIVE CLASS IDENTIFICATION Crosstabulation

Count

		SUBJECTIVE	
		UPPER CLASS	Total
SEEN X-RATED MOVIE IN LAST YEAR	YES	11	230
	NO	27	717
Total		38	947

Chi-Square Tests

	Value	df	Asymp. Sig. (2-sided)
Pearson Chi-Square	5.558[a]	3	.135
Likelihood Ratio	5.555	3	.135
Linear-by-Linear Association	.763	1	.382
N of Valid Cases	947		

a. 0 cells (.0%) have expected count less than 5. The minimum expected count is 9.23.

In this case, there is no significant relationship between the variables. The exact probability of getting this pattern of cell frequencies by random chance alone is .135, well above the standard alpha level of .05. What other variables might be significantly related to *xmovie*? See exercise 12.2 for an opportunity to investigate further.

DEMONSTRATION 12.3 What Are the Sources of Happiness? Is Ignorance Bliss?

In Demonstration 11.2, we tested the old adage that money can't buy happiness. The Mann-Whitney *U* test indicated that the adage was false and that people with higher levels of income also tended to report higher levels of happiness. What effect would education have on happiness? Would more-educated people be more or less happy? Is there any truth to the saying that "ignorance is bliss"?

To test these relationships, we will use *degree* as our measure of education. As originally coded, the variable has five categories (see Appendix G), but we will simplify the analysis and use only two categories, distinguishing between people with a high school education or less and people with at least some education beyond high school. Remember to choose "Recode into Different variable" and give the recoded variable a new name (say, *rdeg*). The recode instruction in the **Old →** **New** box of the **Recode into Different Variable** window should look like this:

$$0 \text{ thru } 1 \rightarrow 1$$
$$2 \text{ thru } 4 \rightarrow 2$$

Follow the instructions in Demonstration 12.1 for the **Crosstabs** command. Specify *happy* as the row variable (the label for this variable is GENERAL HAPPINESS) and recoded *degree* as the column variable. Don't forget to request chi square. The output should look like this:

Crosstabs

Case Processing Summary

	Cases					
	Valid		Missing		Total	
	N	Percent	N	Percent	N	Percent
GENERAL HAPPINESS * RDEG	1441	99.2%	11	.8%	1452	100.0%

GENERAL HAPPINESS * RDEG Crosstabulation

Count

		RDEG		
		1.00	2.00	Total
GENERAL HAPPINESS	VERY HAPPY	294	145	439
	PRETTY HAPPY	587	245	832
	NOT TOO HAPPY	128	42	170
Total		1009	432	1441

.05 ∠ rej Ho
.05 > fail to reject Ho

Chi-Square Tests

	Value	df	Asymp. Sig. (2-sided)
Pearson Chi-Square	4.310[a]	2	.116
Likelihood Ratio	4.360	2	.113
Linear-by-Linear Association	4.258	1	.039
N of Valid Cases	1441		

a. 0 cells (.0%) have expected count less than 5. The minimum expected count is 50.96.

The table displays the observed frequencies for the cells, along with row and column marginals and total number of cases. The value of chi square (4.310), the degrees of freedom (2), and the exact probability that this pattern of cell frequencies occurred by chance (.116) are reported in the output block below the table. The reported significance is greater than .05, the standard value for alpha, so we would fail to reject the null hypothesis of independence. There is no relationship between level of education and degree of happiness. The idea that "ignorance is bliss" is neither supported nor refuted by these results.

Exercises

income only

12.1 For a follow-up on Demonstration 12.1, pick two variables that measure social class. Recode these variables so that they have only two or three categories. Run the **Crosstabs** procedure with *spanking* as the row variable and your new measures of social class as the column variables. How do these relationships compare with the table generated in Demonstration 12.1? Are these tables consistent with the idea that approval of spanking varies by social class?

12.2 Find three more variables that might be related to *xmovie* (*sex* and *relig* seem like possibilities). Use the **Crosstabs** procedure, with chi square, to see if any of your variables have a significant relationship with *xmovie*. Which of the variables had the most significant relationship?

12.3 Since we already recoded *degree* for Demonstration 12.3, let's see if education has any significant relationship with *grass, gunlaw,* or *cappun*. Run the **Crosstabs** procedure with these three variables in the **Rows** box and recoded *degree* in the **Columns** box. Write a paragraph summarizing the results of these three tests. Which relationships are significant? At what levels?

PART II CUMULATIVE EXERCISES

1. Conduct the appropriate test of significance for each research situation. Problems are stated in no particular order and include research situations from each chapter in Part II *except* Chapter 11.

   a. Is there a gender gap in use of the Internet? Random samples of men and women have been questioned about the average number of minutes they spend each week on the Internet for any purpse. Is the difference significant?

Women	Men
$\bar{X}_1 = 55.0$	$\bar{X}_2 = 60.0$
$s_1 = 2.5$	$s_1 = 2.0$
$N_1 = 520$	$N_1 = 515$

   b. For high school students, is there a relationship between social class and involvement in activities such as clubs and sports? Data have been gathered for a random sample of students. Is the relationship significant?

	Class		
Involvement	Middle	Lower	Totals
High	11	19	30
Moderate	19	21	40
Low	8	12	20
Totals	38	52	90

   c. The General Social Survey asks respondents how many total sex partners they have had over their lifetimes. For a small subsample of 23 respondents, does the number vary significantly by educational level?

Less Than High School	High School	At Least Some College
1	1	2
1	2	1
2	3	5
1	8	9
9	10	11
2	5	2
9	4	1
	3	1

   d. A random sample of the U.S. population was asked how many times they had moved since they were 18 years of age. The results are presented below. On the average, how many times do adult Americans move?

   $$\bar{X} = 3.5$$
   $$s = 0.4$$
   $$N = 1450$$

e. On the average, school districts in a state receive budget support from the state government of $623 per student. A random sample of 107 rural schools reports that they received an average of $605 per pupil. Is the difference significant?

$$\mu = 623 \quad \bar{X} = 605$$
$$s = 74$$
$$N = 107$$

2. Below are a number of research questions that can be answered by the techniques presented in Chapters 7–10 and 12. For each question, select the most appropriate test, compute the necessary statistics, and state your conclusions.

In order to complete some problems, you must first calculate the sample statistics (for example, means or proportions) that are used in the test of hypothesis. Use alpha = 0.05 throughout. The questions are presented in random order. There is at least one problem for each chapter, but I have not included a research situation for every statistical procedure covered in the chapters.

In selecting tests and procedures, you need to consider the question, the number of samples or categories being compared, and the level of measurement of the variables. The flowcharts at the beginning of each chapter may be helpful.

The situations refer to the database below, which is based on the General Social Survey (GSS). The actual questions asked and the complete response codes for the GSS are presented in Appendix G. Abbreviated codes are listed below. The scores of some variables have been simplified, collapsed, or recoded for this exercise. These problems are based on a small sample, and you may have to violate some assumptions about sample size in order to complete this exercise.

a. Is there a statistically significant difference in average hours of TV watching by income level? By race?
b. Is there a statistically significant relationship between age and happiness?
c. Estimate the average number of hours spent watching TV for the entire population.
d. If Americans currently average 2.3 children, is this sample representative of the population?
e. Are the educational levels of Catholics and Protestants significantly different?
f. Does average hours of TV watching vary by level of happiness?
g. Based on the sample data, estimate the proportion of Black Americans in the population.

SURVEY ITEMS

1. How many children have you ever had? (CHILDS) Scores are actual numbers.
2. Respondent's educational level (DEGREE)
   0. Less than HS
   1. HS
   2. At least some college

3. Race (RACE)
   1. White
   2. Black
4. Age (AGE)
   1. Younger than 35
   2. 35 and older
5. Number of hours of TV watched per day. (TVHOURS) Values are actual number of hours.
6. What is your religious preference? (RELIG)
   1. Protestant
   2. Catholic
7. Respondent's income (INCOME91)
   1. 24999 or less
   2. 25000 or more
8. Respondent's overall level of happiness (HAPPY)
   1. Very happy
   2. Pretty happy
   3. Not too happy

Case	No. of Children	Educational Level	Race	Age	TV Hours	Religious Preference	Income	Happiness
1	3	1	1	1	3	1	1	2
2	2	0	1	1	1	1	2	3
3	4	2	1	2	3	1	1	1
4	0	2	1	1	2	1	1	1
5	5	1	1	1	2	1	2	2
6	1	1	1	2	3	1	2	1
7	9	0	1	1	6	1	1	1
8	6	1	1	2	4	1	1	2
9	4	2	1	1	2	2	2	2
10	2	1	1	1	1	1	2	3
11	2	0	1	2	4	1	1	3
12	4	1	2	1	5	2	1	2
13	0	1	1	2	2	2	1	2
14	2	1	1	2	2	1	1	2
15	3	1	2	2	4	1	1	1
16	2	0	1	2	2	1	2	1
17	2	1	1	2	2	1	2	3
18	0	2	1	2	2	1	2	1
19	3	0	1	2	5	2	1	1
20	2	1	2	1	10	1	1	3
21	2	1	1	2	4	1	2	1
22	1	0	1	2	5	1	2	1
23	0	2	1	1	2	2	1	1
24	0	1	1	2	0	2	2	1
25	2	2	1	1	1	2	2	1
26	1	0	1	1	10	2	1	1
27	4	2	1	2	3	2	2	2

Case	No. of Children	Educational Level	Race	Age	TV Hours	Religious Preference	Income	Happiness
28	0	2	1	2	3	2	2	2
29	2	2	1	2	1	1	2	1
30	3	0	1	2	9	1	1	1
31	4	1	1	2	3	1	2	1
32	0	2	2	2	4	1	1	2
33	2	1	2	1	1	1	2	2
34	2	1	1	2	2	1	2	1
35	1	1	1	1	4	1	1	1
36	4	0	1	1	1	1	1	1
37	3	1	2	1	0	1	2	1
38	0	1	2	2	2	2	2	3
39	3	2	1	2	3	2	2	3
40	0	1	1	1	0	1	2	2
41	4	1	1	1	2	1	1	3
42	0	2	1	1	0	2	1	2
43	2	1	1	1	6	2	1	1
44	3	0	1	1	3	1	2	2
45	6	1	1	1	7	2	1	3

Part III | Measures of Association

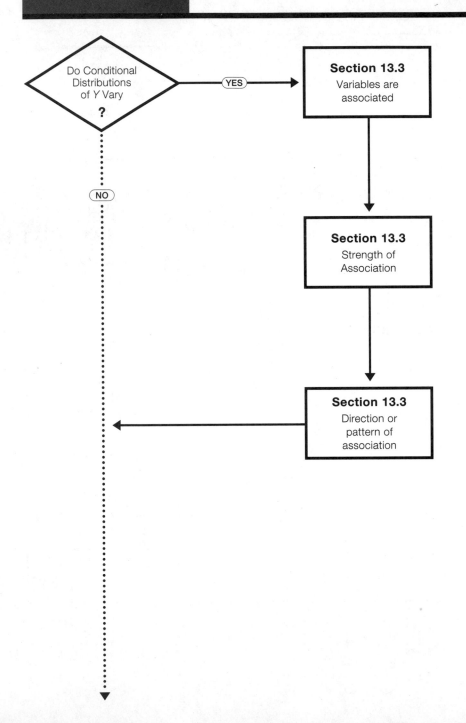

13.1 STATISTICAL SIGNIFICANCE AND THEORETICAL IMPORTANCE

As we have seen over the past several chapters, tests of statistical significance are extremely important in social science research. As long as social scientists must work with random samples rather than populations, these tests are indispensable for dealing with the possibility that our research results are the products of mere random chance. However, tests of significance are often merely the first step in the analysis of research results. We have seen that these tests do have limitations. Statistical significance is not necessarily the same thing as relevance or importance. Furthermore, all tests of significance are affected by sample size: tests performed on large samples may result in decisions to reject the null hypothesis when, in fact, the observed differences are quite minor.

Beginning with this chapter, we will be working with **measures of association**, a new type of statistic that will provide additional information about our research results. Whereas tests of significance are designed to detect nonrandom relationships, measures of association are designed to give us an indication of the strength and direction of relationships. The two statistical techniques perform complementary functions and yield different types of information about our research results.

Measures of association are useful because they can help us reach some important scientific goals. First, they can help us trace causal relationships among variables. The theories that guide scientific research are often stated in cause-and-effect terms: Variable X causes variable Y to occur or to change values. Recall our discussion of the contact hypothesis in Chapter 1. In that theory, the causal (or independent) variable was equal status intergroup contacts, and the "effect" (or dependent) variable was level of individual prejudice. The theory asserts that involvement in such contact *causes* prejudice to decline. Measures of association are our most important and powerful statistical tools for documenting, measuring, and analyzing causal relationships between variables.

Of course, measures of association, as useful and important as they are, have their limitations. Most importantly, these statistics, by themselves, cannot *prove* that two variables are causally related. That is, even if we find a strong statistical association between two variables, we cannot necessarily conclude that one variable causes the other. We will explore causation in more detail in Part IV, but for now you should keep in mind that causation and association are two different things. We can use a statistical association between variables as evidence for a causal relationship, but association by itself is not proof of a causal relationship between variables.

Prediction is a second important use for measures of association. If two variables are associated, we will be able to predict the score of a case on one variable from the score of that case on the other variable. For example, if equal status contacts and prejudice are associated, we can predict that people who have experienced many such contacts will be less prejudiced than those who have had few or no contacts. Note that prediction and causation can be two separate matters. If variables are associated, we can predict from one to the other even if the variables are not causally related.

This chapter will introduce the concept of **association** and illustrate what an association between two variables "looks like" when the variables are presented in table format. We will also see how to use percentages to learn more about the bivariate association. In the chapters that follow, we will concentrate on the logic, calculation, and interpretation of the various measures of association. Finally, in Part IV, we will extend some of these ideas to the multivariate (more than two variables) case.

13.2 ASSOCIATION BETWEEN VARIABLES AND THE BIVARIATE TABLE

Most generally, two variables are said to be associated if the distribution of the categories or scores of one of them changes under the various categories or scores of the other. For example, suppose that an industrial sociologist was concerned with the relationship between job satisfaction and productivity for assembly-line workers. If these two variables are associated, then the distribution of scores on productivity will change under the different conditions of satisfaction. Highly satisfied workers will have different patterns of productivity than workers who are low on satisfaction and levels of productivity will vary by levels of satisfaction.

This relationship will become clearer with the use of bivariate tables. As you recall (see Chapter 12), bivariate tables are devices for displaying the scores of cases on two different variables. By convention, the **independent**, or **X**, **variable** (that is, the variable taken as causal) is arrayed in the columns, and the **dependent**, or **Y**, **variable** in the rows.* That is, each column of the table (the vertical dimension) represents a score or category of the independent variable (X), and each row (the horizontal dimension) represents a score or category of the dependent variable (Y). Table 13.1 illustrates, with fictitious data, the basic features of a bivariate table.

This table displays the relationship between productivity and job satisfaction for a sample of 173 factory workers. To investigate the association between variables displayed in table format, we focus our attention on the columns. Each column shows the pattern of scores on the dependent variable for each score on the independent variable. For example, the first column indicates that, of the 60 workers who were low on job satisfaction, 30 were low on productivity, 20 were moderately productive, and 10 were high on productivity. The second column shows that, of the 61 moderately satisfied workers, 21 were low, 25 were moderate, and 15 were high on productivity. What is the pattern for the 52 highly satisfied workers?

Inspecting the table from column to column allows us to observe the effects of the independent variable on the dependent variable (provided, of

---

*In the material that follows, we will often, for the sake of brevity, refer to the independent variable as X and the dependent variable as Y.

TABLE 13.1   PRODUCTIVITY BY JOB SATISFACTION

Productivity (Y)	Job Satisfaction (X)			Totals
	Low	Moderate	High	
Low	30	21	7	58
Moderate	20	25	18	63
High	10	15	27	52
Totals	60	61	52	173

course, that the table is constructed with the independent variable in the columns). These "within-column" frequency distributions are called the **conditional distributions of Y**, since they display the distribution of scores on the dependent variable for each condition (or score) of the independent variable.

Table 13.1 indicates that productivity and satisfaction are associated: the distribution of scores on Y (productivity) changes across the various conditions of X (satisfaction). For example, half of the workers who were low on satisfaction were also low on productivity. On the other hand, over half of the workers who were high on satisfaction were high on productivity. Since the conditional distributions of Y change, these two variables are associated in some way.

There is, as you may have realized, another way to detect the existence of an association between two variables that have been organized into table format. In Chapter 12, we introduced the chi square statistic. Even though this statistic is designed as a test of significance, it can also be used as an indicator of association. Just as any change in the conditional distributions of the dependent variable indicates the existence of an association, so too does any nonzero value for the obtained chi square. For example, the obtained chi square for Table 13.1 would be 24.2, and this nonzero value affirms our previous conclusion, based on the observation of the conditional distributions of Y, that an association of some sort exists between job satisfaction and productivity.

Often, the researcher will have already conducted a chi square test for independence before considering matters of association. In such cases, an extended inspection of the conditional distributions of Y will not be necessary to ascertain whether or not the two variables are associated. If the obtained chi square is zero, the two variables are independent and not mutually associated. Any value other than zero indicates some association between the variables. Remember, however, that statistical significance and association are two different things. It is perfectly possible for two variables to be associated (as indicated by a nonzero chi square) but still independent (if we fail to reject the null hypothesis).

In this section, we have defined, in a general way, the concept of association between two variables. We have also shown two different ways to detect the presence of an association. In the next section, we will extend the analysis beyond questions of the mere presence or absence of an association and, in a systematic way, show how additional very useful information about the relationship between two variables can be developed.

13.3 THREE CHARACTERISTICS OF BIVARIATE ASSOCIATIONS

Bivariate associations possess three different characteristics, and each must be analyzed for a full investigation of the relationship. Investigating these characteristics may be thought of as a process of finding answers to three questions:

1. Does an association exist?
2. If an association does exist, how strong is it?
3. What is the pattern and/or the direction of the association?

We will consider each of these questions in the rest of this section.

Does an Association Exist?   We have already discussed the general definition of association, and we have seen how a bivariate association would manifest itself in changing conditional distributions of Y in a table or in a nonzero value for chi square. In Table 13.1, since the conditional distributions of productivity (Y) are different across the various categories of satisfaction (X) and chi square is a nonzero value, we conclude that these two variables are associated to some extent.

In Table 13.1, comparisons from column to column are relatively easy to make because the column totals are roughly equal. This will not always be the case, and it is usually helpful to use percentages to control for varying column totals. Calculating percentages also helps to make the pattern of association more visible.

The general rule for calculating percentages with bivariate tables is to compute them in the direction of the independent variable and then compare in the opposite direction. When the independent variable has been arrayed in the columns, compute the percentages within each column separately. Table 13.2 presents percentages calculated in the direction of the independent variable based on the data in Table 13.1.

Note that Table 13.2 omits the row marginals and reports the column marginals in parentheses. Besides controlling for any differences in column totals, tables in percentage form are usually easier to read because changes in the conditional distributions of Y are easier to detect.

In Table 13.2, we can readily see that the largest cell changes position from column to column. For workers who were low on satisfaction, the single largest cell is in the top row (low on productivity). For the middle column (moderate on satisfaction), the largest cell is in the middle row (moderate on

TABLE 13.2 PRODUCTIVITY BY JOB SATISFACTION (in percentages)

Productivity (Y)	Job Satisfaction (X)		
	Low	Moderate	High
Low	50.00%	34.43%	13.46%
Moderate	33.33%	40.98%	34.62%
High	16.67%	24.59%	51.92%
	100%	100%	100%
	(60)	(61)	(52)

positive assoc go together

TABLE 13.3 PRODUCTIVITY BY HAIR COLOR (an illustration of no association)

Productivity (Y)	Hair Color (X)		
	Brown	Blonde	Red
Low	33.33%	33.33%	33.33%
Moderate	33.33%	33.33%	33.33%
High	33.33%	33.33%	33.33%
	100%	100%	100%

negative assoc

productivity), and, for the right-hand column (high on satisfaction), it is in the bottom row (high on productivity). Even a cursory glance at the conditional distributions of Y in Table 13.2 reinforces our conclusion that an association does exist between these two variables.

If two variables are not associated, then the conditional distributions of Y will not change across the columns. The distribution of Y would be the same for each condition of X. Table 13.3 illustrates a "perfect nonassociation" between the variables hair color and productivity for assembly-line workers. Table 13.3 is only one of many patterns that indicate "no association"; but the important point is that, comparing across the table from column to column, the conditional distributions of Y are the same. Levels of productivity do not change at all for the various hair colors and, therefore, no association exists between these variables. Also, the obtained chi square computed from this table would have a value of zero, again indicating no association between these two variables.

How Strong Is the Association? Once we establish the existence of the association, we need to develop some idea of how strong the association is. This is essentially a matter of determining the amount of change in the conditional distributions of Y. At one extreme, of course, there is the case of "no association" where the conditional distributions of Y do not change at all (see Table 13.3). At the other extreme is the case of "perfect association"—the

TABLE 13.4   PRODUCTIVITY BY HAIR COLOR (an illustration of perfect association)

Productivity (Y)	Hair Color (X)		
	Brown	Blonde	Red
Low	0%	0%	100%
Moderate	0%	100%	0%
High	100%	0%	0%
	100%	100%	100%

strongest possible relationship between X and Y. In general, a perfect association exists between two variables if each value of Y is associated with one and only one value of X.* In a bivariate table, all cases in each column would be located in a single cell. There would be no variation in Y for a given value of X, as illustrated in Table 13.4.

A perfect relationship would be taken as very strong evidence of a causal relationship between the variables, at least for the particular sample at hand. In fact, the results presented in Table 13.4 would indicate that, for this sample, hair color is the sole cause of productivity. Also, in the case of a perfect relationship, predictions from one variable to the other could be made without error. If we knew that a particular worker had brown hair, for example, we could precisely predict that that worker would be highly productive.

Of course, the huge majority of relationships will fall somewhere between these two extremes of no association and perfect association. We need to develop some way of describing these intermediate relationships consistently and meaningfully. For example, Tables 13.1 and 13.2 show that there is an association between productivity and job satisfaction. How could this relationship be described in terms of strength? How close is the relationship to perfect? How far away from no association?

Although you might be able to describe this association impressionistically, the task would certainly be easier if you had a set of objective indices to consistently describe the strength of the association (for example, as strong, moderate, or weak). As we shall see in Chapters 14–16, these objective indices of the strength of a relationship are supplied by the various measures of association. Virtually all of these statistics are designed so that they

---

*Each measure of association that will be introduced in the following chapters incorporates its own definition of a "perfect association," and these definitions vary somewhat depending on the specific logic and mathematics of the statistic. That is, for different measures computed from the same table, some measures will possibly indicate perfect relationships when others will not. We will note these variations in the mathematical definitions of a perfect association at the appropriate time.

Application 13.1

Why are many Americans attracted to movies that emphasize graphic displays of violence? One idea is that "slash" movie fans feel threatened by violence in their daily lives and use these movies as a means of coping with their fears. In the safety of the theater, violence can be vicariously experienced, and feelings and fears can be expressed privately. Also, highly violent movies almost always, as a necessary plot element, provide a role model of one character who does deal with violence successfully (usually, of course, with more violence).

Is fear of violence associated with frequent attendance at high-violence movies? The following table reports the joint frequency distributions of "fear" and "attendance" in percentages for a sample of 600.

	Fear		
Attendance	Low	Moderate	High
Rare	50%	20%	30%
Occasional	30	60	30
Frequent	20	20	40
	100%	100%	100%
	(200)	(200)	(200)

The conditional distributions of attendance (Y) do change across the values of fear (X), so these variables are associated. The clustering of cases in the diagonal from upper left to lower right suggests a substantial relationship in the predicted direction. People who are low on fear are infrequent attenders; people who are high on fear are frequent attenders. These results do suggest an important relationship between fear and attendance. Notice that the results, as presented here, pose an interesting causal problem. The table supports the idea that fearful and threatened people attend violent movies as a coping mechanism (X causes Y). However, the table is also consistent with the reverse causal argument: attendance at violent movies increases fears for one's personal safety (Y causes X). The results support both causal arguments, and I remind you that association is not the same thing as causation.

have a lower limit of 0 and an upper limit of 1 (±1 for ordinal and interval-ratio measures of association). A measure that equals 0 indicates no association between the variables (the conditional distributions of Y do not vary), and a measure of 1 (±1 in the case of ordinal and interval-ratio measures) indicates a perfect relationship. The exact meaning of values between 0 and 1 varies from measure to measure, but, for all measures, the closer the value is to 1.00, the stronger the relationship (the greater the change in the conditional distributions of Y).

You may be tempted to regard the chi square test as an indication of the strength of a relationship. Remember that significance and association are two separate matters and chi square, by itself, is not a measure of association. While nonzero values for chi square indicate the existence of an association, the actual numerical value for obtained chi square bears no particular relationship to the strength of the association. That is, the obtained chi square

READING STATISTICS 9: Bivariate Tables

The conventions for constructing and interpreting bivariate tables presented in this text are commonly but not universally followed in the professional literature. Tables will usually be constructed with the independent variable in the columns, the dependent variable in the rows, and percentages calculated in the columns. However, you should be careful to check the format of every table you attempt to read to see if these conventions have been observed. If the table is presented with the independent variable in the rows, for example, you will have to reorient your analysis (or redraw the table) to account for this. Above all, you should convince yourself that the percentages have been calculated in the correct direction. It is possible for even skilled professionals to calculate percentages incorrectly and potentially misinterpret the table as a consequence.

Once you have assured yourself that the table is properly presented, you can apply the analytical techniques developed in this chapter. By comparing the conditional distributions of the dependent variable, you can ascertain for yourself if the variables are associated and check the strength and pattern of the association. You may then compare your conclusions with those of the researchers. As an aid in the interpretation of bivariate tables, researchers will usually compute and report other statistics in addition to percentages. We'll talk about these statistics in Reading Statistics 11 in Chapter 15.

Statistics in the Professional Literature

In 1989, a group of sociologists and lawyers began a study of community standards regarding pornography and obscenity. The results of the study were used in a court case involving charges of pornography in Charlotte, North Carolina. One of the legal criteria currently used to define pornography across the nation is that a judge or jury must determine that the " . . . average person, ap-

plying contemporary community standards, would find that the work, taken as a whole, appeals to . . . prurient interests." This standard is one of three established by the U.S. Supreme Court in *Miller vs. California* (1973). The researchers gathered an EPSEM sample of 129 residents of the Charlotte area, and each respondent viewed either one of the sexually explicit films that were the subjects of the litigation or a nonexplicit control film. Each subject filled out a survey measuring their views of community standards before and after viewing the films. The table below presents some of the results of this study for the subjects who viewed the explicit film.

Note that the table is constructed with the independent variable in the columns and the dependent variable in the rows. Percentages are calculated within the columns, and the cell frequencies are reported as well. The conditional distributions change so we may conclude that viewing explicit films (the independent variable) changed opinions about whether or not such films were unhealthy (the dependent variable).

"Are explicit movies unhealthy?" Before and After Viewing an Explicit, Sexually-Oriented Film[1]

	Before	After
Definitely Yes	43.6%	16.4%
(N)	48	18
Definitely No	56.4%	83.6%
(N)	62	92
	110	110

[1]Adapted from Table 4, page 96. The actual question was: "Some adult movies, video cassettes, and magazines show actual sex acts in great detail and with close-ups of sexual organs. Would viewing this type of material appeal to an unhealthy, shameful, or morbid interest in sex?"

What do these results suggest about community standards? What do they suggest about people's attitudes towards sexually explicit material after actually viewing the films? Here's part of how the authors interpreted this table and other results:

> . . . a majority . . . believe that . . . (adult movies, video cassettes, and magazines are) . . . tolerated in their community. . . . Seventy-five percent believe they personally should be able to obtain and see any such showing of (these materials). . . . (P)re- and post-film . . . responses on the belief that these materials are tolerated showed no significant change (the test of significance used was chi square). Likewise, there was no statistically significant change in subjects' opinions about whether they should be allowed to see such materials. There was, however, a statistically significant change in . . . perception that the materials appealed to unhealthy, shameful, or morbid interest in sex [where the percentage drops from 43.6% to 16.4%].

Quoted material from *Public Opinion Quarterly* 55(1991). "Estimating Community Standards: The Use of Social Science Evidence in an Obscenity Prosecution" by Daniel Linz, et al. Published by the University of Chicago Press. Used by permission.

might be quite large in value while the actual association is weak. Chapter 14 will introduce some ways to transform chi square into other statistics that do measure the strength of the association between two variables. *(For practice in computing percentages and judging the existence and strength of an association, see any of the problems at the end of this chapter.)*

What Is the Pattern and/or the Direction of the Association?   Investigating the pattern of the association requires that we ascertain which values or categories of one variable are associated with which values or categories of the other. We have already remarked on the pattern of the relationship between productivity and satisfaction. Table 13.2 indicates that low scores on satisfaction are associated with low scores on productivity, moderate satisfaction with moderate productivity, and high satisfaction with high productivity.

When both variables are at least ordinal in level of measurement, the pattern of association may also have a direction to it.* The direction of the association can be either positive or negative.

An association is positive if the variables vary in the same direction. That is, in a **positive association**, high scores on one variable are associated with high scores on the other variable, and low scores on one variable are associated with low scores on the other. Still another way to express this relationship would be to say that, in a positive association, as one variable increases in value, the other also increases; and, as one variable decreases in value, the other also decreases. Table 13.5 displays, with fictitious data, a positive

---

*Variables measured at the nominal level have no numerical order to them (by definition). Therefore, associations including nominal-level variables, while they may have a pattern, cannot be said to have a direction.

TABLE 13.5   LIBRARY USE BY EDUCATION (an illustration of a positive relationship)

	Education		
Library Use	Low	Moderate	High
Low	60%	20%	10%
Moderate	30	60	30
High	10	20	60
	100%	100%	100%

positive assoc.

TABLE 13.6   AMOUNT OF TELEVISION VIEWING BY EDUCATION (an illustration of a negative relationship)

	Education		
Television Viewing	Low	Moderate	High
Low	10%	20%	60%
Moderate	30	60	30
High	60	20	10
	100%	100%	100%

negative assoc. (opposites)

relationship between education and use of public libraries. As education increases, library use also increases. The association between job satisfaction and productivity, as displayed in Tables 13.1 and 13.2, is also a positive association.

In a **negative association**, the variables vary in opposite directions. High scores on one variable are associated with low scores on the other, and increases in one variable are accompanied by decreases in the other. Table 13.6 displays a negative relationship, again with fictitious data, between education and television viewership. The amount of television viewing decreases as education increases.

Measures of association for ordinal and interval-ratio variables are designed so that they will take on positive values for positive associations and negative values for negative associations. Thus, a measure of association preceded by a plus sign indicates a positive relationship between the two variables, with the value $+1.00$ indicating a perfect positive relationship. A negative sign indicates a negative relationship, with -1.00 indicating a perfect negative relationship. *(For practice in determining the pattern of an association, see any of the end-of-chapter problems. For practice in determining the direction of a relationship, see problems 13.1, 13.6, 13.8, 13.9, 13.10, 13.12, and 13.15.)*

SUMMARY

1. The analysis of associations between variables provides information that is complementary to tests of significance. Whereas the latter are designed to detect nonrandom relationships, measures of association are designed to quantify the importance or strength of a relationship.
2. Relationships between variables have three characteristics: the existence of an association, the strength of the association, and the direction or pattern of the association. These three characteristics can be investigated by calculating percentages for a bivariate table in the direction of the independent variable (vertically) and then comparing in the opposite direction (horizontally).
3. Tables 13.1 and 13.2 can be analyzed in terms of these three characteristics. Clearly, a relationship does exist between job satisfaction and productivity, since the conditional distributions of the dependent variable (productivity) are different for the three different conditions of the independent variable (job satisfaction). Even though we cannot quantify the strength of the relationship yet, we can see that the association is substantial in that the change in Y (productivity) across the three categories of X (satisfaction) is marked. Furthermore, the relationship is positive in direction. Productivity increases as job satisfaction rises, and workers who report high job satisfaction tend to be also high on productivity. Workers with little job satisfaction tend to be low on productivity.
4. Given the nature and strength of the relationship, it could be predicted with fair accuracy that highly satisfied workers would tend to be highly productive ("happy workers are busy workers"), and these results might be taken as evidence of a causal relationship between these two variables. However, these results cannot, by themselves, prove that a causal relationship exists. Tables 13.1 and 13.2 show only that the variables are associated, and association is not the same thing as causation. In fact, although we have presumed that job satisfaction is the independent variable in the relationship, we could have argued the reverse causal sequence ("busy workers are happy workers"). The results presented in Tables 13.1 and 13.2 are consistent with both causal arguments.

Thus, the analysis of the association between variables may be seen as a very important process by which evidence for (or against) a causal relationship is systematically gathered. Ultimate proof for causal relationships depends more on logical, theoretical, and methodological grounds (actually proving causation is a rather difficult task). As we shall see in Part IV, some of the multivariate techniques are quite useful for probing possible causal relationships, and we will return to some of these concerns at that point.

GLOSSARY

Association. The relationship between two (or more) variables. Two variables are said to be associated if the distribution of one variable changes for the various categories or scores of the other variable.

Conditional distribution of Y. The distribution of scores on the dependent variable for a specific score or category of the independent variable when the variables have been organized into table format.

Dependent variable. In a bivariate relationship, the variable that is taken as the effect.

Independent variable. In a bivariate relationship, the variable that is taken as the cause.

Measures of association. Statistics that quantify the strength of the association between variables.

Negative association. A bivariate relationship where the variables vary in opposite directions. As one variable increases, the other decreases, and high scores on one variable are associated with low scores on the other.

Positive association. A bivariate relationship where the variables vary in the same direction. As one variable increases, the other also increases, and high scores on one variable are associated with high scores on the other.

X. Symbol used for any independent variable.
Y. Symbol used for any dependent variable.

PROBLEMS

13.1 | PA | Various supervisors in the city government of Shinbone, Kansas, have been rated on the extent to which they practice authoritarian styles of leadership and decision making. The efficiency of each department has also been rated, and the results are summarized below. Figure percentages for the table so that it shows the effect of leadership style on efficiency. Is there an association between these two variables? What is the strength and direction of the relationship?

Efficiency	Authoritarianism	
	Low	High
Low	10	12
High	17	5
Totals	27	17

13.2 | SOC | The administration of a local college campus has proposed an increase in the mandatory student fee in order to finance an upgrading of the intercollegiate football program. A sample of the faculty has completed a survey on the issues. Is there any association between support for raising fees and the gender, discipline, or tenured status of the faculty? Describe the strength and direction of the association.

a. Support for raising fees by gender:

Support	Males	Females
For	12	8
Against	15	12
Totals	27	20

b. Support for raising fees by discipline:

Support	Liberal Arts	Science & Business
For	6	13
Against	14	14
Totals	20	27

c. Support for raising fees by tenured status:

Support	Tenured	Nontenured
For	15	4
Against	18	10
Totals	33	14

13.3 | PS | How consistent are people in their voting habits? Do people vote for the same party from election to election? Below are the results of a poll in which people were asked if they had voted Democrat or Republican in each of the last two presidential elections. Assess the strength of this relationship.

1996 Election	1992 Election	
	Democrat	Republican
Democrat	117	23
Republican	17	178
Totals	134	201

13.4 | SOC | A needs assessment survey has been distributed in a large retirement community. Residents were asked to check off the services or programs they thought should be added. Is there any association between gender and the perception of a need for more social occasions? Write a few sentences describing the relationship in terms of pattern and strength of the association.

More Parties?	Male	Female
Yes	321	426
No	175	251
Totals	496	677

13.5 For problems 12.2 to 12.9 in Chapter 12, calculate percentages in the direction of the independent variable. Is there an association between the variables? Characterize each association in terms of the strength and direction of the relationship.

13.6 SW As the state director of mental health programs, you note that some local mental health facilities have very high rates of staff turnover. You believe that part of this problem is a result of the fact that some of the local directors have very little training in administration and poorly developed leadership skills. Before implementing a program to address this problem, you collect some data to make sure that your beliefs are supported by the facts. Is there a relationship between staff turnover and the administrative experience of the directors? Describe the relationship in terms of pattern and strength of the association.

Turnover	Director Experienced?	
	No	Yes
Low	4	9
Moderate	9	8
High	15	5
Totals	28	22

13.7 CJ About half the neighborhoods in a large city have instituted programs to increase citizen involvement in crime prevention. Do these areas experience less crime? Write a few sentences describing the relationship in terms of pattern and strength of the association.

Crime Rate	Program	
	No	Yes
Low	29	15
Moderate	33	27
High	52	45
Totals	114	87

13.8 PS Are people who think of themselves as political liberals really more liberal on social issues? The table below summarizes the relationship between political ideology and the respondent's position on the legalization of marijuana. Are the variables associated? Write a few sentences describing the relationship in terms of pattern and strength of the association.

Should Marijuana Be Legal?	Liberals	Moderates	Conservatives
Yes	132	78	52
No	101	87	109
Totals	233	165	161

13.9 SOC A researcher hypothesizes that physical attractiveness is related to academic achievement (that is, the more attractive the student, the higher the grade). One hundred students are rated on physical attractiveness and asked for their grade-point averages. The bivariate table below summarizes the data. Calculate percentages for the table in the proper direction and compare the conditional distributions. Is there an association between these variables? What is your impression of the strength of the association? What is the direction of the association? Is the hypothesis supported? Why or why not?

GPA	Attractiveness		
	Low	Moderate	High
Low	7	8	15
Moderate	10	10	16
High	8	12	14
Totals	25	30	45

13.10 SOC Is interest in sports associated with social class? Write a few sentences describing the relationship in terms of pattern and strength of the association.

Interest	Working Class	Middle Class	Upper Class
Low	21	14	21
Moderate	24	40	23
High	12	45	7
Totals	57	99	51

13.11 SOC The table below shows the relationship between place of birth and racial prejudice for a sample of white Americans. Calculate percentages for the table in the direction of the independent variable. Characterize the relationship in terms of strength. Which region seems to be associated with high levels of prejudice? Which with low levels of prejudice?

	Place of Birth			
Prejudice	North	South	Midwest	West
Low	10	5	30	25
Moderate	20	10	10	15
High	10	20	10	5
Totals	40	35	50	45

13.12 [SOC] The latest fad to sweep college campuses is streaking to panty raids while swallowing live goldfish. A researcher is interested in how closely the spread of this bizarre form of behavior is linked to the amount of coverage and publicity provided by local campus newspapers. For a sample of 25 universities, the researcher has rated the amount of press coverage (as extensive, moderate, or no coverage) and how much the student body was involved in this new fad. The data for each campus are reported below. Organize the data into a properly labeled table in percentage form. Does the table indicate an association between press coverage and fad behavior?

Campus	Amount of Press Coverage	Extent of Student Involvement
1	Extensive	Extensive
2	Extensive	Some
3	Moderate	Some
4	Moderate	Some
5	Moderate	Extensive
6	Extensive	Some
7	Extensive	Extensive
8	Moderate	Some
9	None	Some
10	Moderate	None
11	None	None
12	Extensive	Extensive
13	None	Some
14	Extensive	Extensive
15	Moderate	None
16	Moderate	Some
17	Moderate	Extensive
18	Moderate	None
19	None	None
20	Extensive	Some
21	None	Extensive
22	Moderate	None
23	None	None
24	Extensive	Extensive
25	Moderate	Extensive

13.13 [SOC] Twenty-seven men and women were asked if they favored (F) or opposed (O) the death penalty. Is there any association between gender and support for capital punishment?

Subject	Gender	Support for Capital Punishment
1	M	F
2	M	F
3	M	F
4	M	F
5	M	O
6	M	O
7	M	O
8	M	F
9	M	O
10	M	F
11	M	O
12	M	O
13	M	F
14	F	F
15	F	O
16	F	F
17	F	F
18	F	O
19	F	F
20	F	F
21	F	O
22	F	O
23	F	O
24	F	F
25	F	O
26	F	F
27	F	O

13.14 In any social science journal, find an article that includes a bivariate table. Inspect the table and the related text carefully and answer the following questions:

a. Identify the variables in the table. What values (categories) does each possess? What is the level of measurement for each variable?

b. Is the table in percentage form? In what direction? Are comparisons made between columns or rows?

c. Is one of the variables identified by the author as independent? Are the percentages in the direction of the independent variable?

d. How is the relationship characterized by the author in terms of the strength of the association? In terms of the direction (if any) of the association?

e. Find the measure of association (if any) calculated for the table. What is the numerical value of the measure? What is the sign (if any) of the measure?

13.15 Let's see if political ideology is associated with a variety of different issues and concerns. The tables below are taken from the General Social Survey. Political ideology (POLVIEWS) has been collapsed into three categories and is used as the "column," or independent, variable. If liberals are generally progressive and conservatives are generally traditional (with moderates in between), what relationships would you expect to find between political ideology and these issues?

a. Support for the legal right to an abortion
b. The death penalty
c. The legal right to commit suicide for people with incurable disease
d. Sex education in schools
e. Legalization of marijuana

For each table, compute percentages and describe the pattern and strength of the relationship. Are the patterns consistent with your predictions? Overall, is political ideology an important factor in shaping people's views and opinions? See Appendix G for the exact wording of each item.*

a. Support for the legal right to an abortion (ABANY) by political ideology:

| | Political Ideology | | |
Right to Abortion?	Liberal	Moderate	Conservative
Yes	135	155	101
No	81	177	203
Totals	216	332	304

b. Support for capital punishment (CAPPUN) by political ideology:

| | Political Ideology | | |
Capital Punishment	Liberal	Moderate	Conservative
Favor	213	370	388
Oppose	121	107	68
Totals	334	477	456

c. Approve of suicide for people with incurable disease (SUICIDE1) by political ideology:

| | Political Ideology | | |
Right to Suicide?	Liberal	Moderate	Conservative
Yes	192	214	186
No	61	107	139
Totals	253	321	325

d. Support for sex education in public schools (SEXEDUC) by political ideology:

| | Political Ideology | | |
Sex Education?	Liberal	Moderate	Conservative
Favor	246	305	258
Oppose	11	33	66
Totals	257	338	324

e. Support for legalization of marijuana (GRASS) by political ideology:

| | Political Ideology | | |
Legalize Marijuana?	Liberal	Moderate	Conservative
Legal	92	80	62
Not Legal	124	262	242
Totals	216	342	304

---

*Sample size (*N*) varies from table to table because not all respondents were asked every question.

Using *SPSS for Windows* to Analyze Bivariate Association

DEMONSTRATION 13.1 Do Childcare Practices Vary by Social Class? Another Look

In Demonstration 12.1, the **Crosstabs** procedure and chi square indicated that there is a statistically significant relationship between social class and child rearing practices. Now, we can add to our information and investigate the strength and direction of the relationship. Which class is most supportive of spanking and which is least supportive?

Click **Statistics**, **Summarize**, and **Crosstabs** and name *spanking* as the row variable and *class* as the column variable. Click the **Cells** button and request column percentages by clicking the box next to Column in the **Percentages** box. Click **Continue** and **OK**, and the following output will be produced:

Crosstabs

Case Processing Summary

	Cases					
	Valid		Missing		Total	
	N	Percent	N	Percent	N	Percent
FAVOR SPANKING TO DISCIPLINE CHILD* SUBJECTIVE CLASS IDENTIFICATION	969	66.7%	483	33.3%	1452	100.0%

FAVOR SPANKING TO DISCIPLINE CHILD* SUBJECTIVE CLASS IDENTIFICATION
Crosstabulation

			SUBJECTIVE CLASS IDENTIFICATION		
			LOWER CLASS	WORKING CLASS	MIDDLE CLASS
FAVOR SPANKING TO DISCIPLINE CHILD	STRONGLY AGREE	Count	26	122	115
		% within SUBJECTIVE CLASS IDENTIFICATION	43.3%	29.3%	25.4%
	AGREE	Count	18	218	187
		% within SUBJECTIVE CLASS IDENTIFICATION	30.0%	52.4%	41.4%

			13	51	95
DISAGREE	Count		13	51	95
	% within SUBJECTIVE CLASS IDENTIFICATION		21.7%	12.3%	21.0%
STRONGLY DISAGREE	Count		3	25	55
	% within SUBJECTIVE CLASS IDENTIFICATION		5.0%	6.0%	12.2%
Total	Count		60	416	452
	% within SUBJECTIVE CLASS IDENTIFICATION		100.0%	100.0%	100.0%

FAVOR SPANKING TO DISCIPLINE CHILD* SUBJECTIVE CLASS IDENTIFICATION Crosstabulation

			SUBJECTIVE	
			UPPER CLASS	Total
FAVOR SPANKING TO DISCIPLINE CHILD	STRONGLY AGREE	Count	11	274
		% within SUBJECTIVE CLASS IDENTIFICATION	26.8%	28.3%
	AGREE	Count	16	439
		% within SUBJECTIVE CLASS IDENTIFICATION	39.0%	45.3%
	DISAGREE	Count	9	168
		% within SUBJECTIVE CLASS IDENTIFICATION	22.0%	17.3%
	STRONGLY DISAGREE	Count	5	88
		% within SUBJECTIVE CLASS IDENTIFICATION	12.2%	9.1%
Total		Count	41	969
		% within SUBJECTIVE CLASS IDENTIFICATION	100.0%	100.0%

Chi-Square Tests

	Value	df	Asymp. Sig. (2-sided)
Pearson Chi-Square	37.390[a]	9	.000
Likelihood Ratio	37.719	9	.000
Linear-by-Linear Association	17.136	1	.000
N of Valid Cases	969		

a. 1 cells (6.3%) have expected count less than 5. The minimum expected count is 3.72.

We already knew that the relationship was statistically significant from Demonstration 12.1. To this information, we can add that there is an association between the variables (the conditional distribution of Y or *spanking* changes from column to column), and—judging by the amount of change—the relationship appears to be moderate in strength. What makes this relationship "moderate"? Basically, the conditional distributions change dramatically in some instances but not in all.

For example, observe the top row of the table ("strongly agree") and you will see a change of 14% from column 1 (lower class) to column 2 (working class). However, the percentage of cases in this row stays about the same for the middle and upper classes. In a strong relationship, we would expect to find more pronounced changes throughout the table.

As for the direction of the relationship, we have already identified it. Observe the top row again and you will see that the approval of spanking decreases as social class increases. In other words, the higher the social class, the lower the approval of spanking.

DEMONSTRATION 13.2 Does Support for Gun Control Vary by Political Ideology?

Gun control is a highly politicized issue in our society and, given the way in which this issue is usually debated, we would expect liberals to support gun control, conservatives to oppose it, and moderates to be intermediate. In the 1996 GSS, support for gun control is measured by the variable *gunlaw*, and political ideology is measured by *polviews*. The latter has too many categories (7) to be used in a bivariate table, so we must first recode the variable. Click **Transform** from the main menu and remember to choose "Recode into Different Variable." In this demonstration, I named the recoded version of the variable *pol*. The original scores for this variable are given in Appendix G and the recoding instructions that should appear in the **Old → New** box of the **Recode into Different Variable** window should be

1 thru 3 → 1   *Lib*
4 → 2   *mod*
5 thru 7 → 3   *cons*

This scheme groups all liberals (a score of 1 on *pol*) and conservatives (a score of 3 on *pol*) together and changes the score associated with "moderates" (a score of 2 on *pol*) so that they remain intermediate between the other two.

The **Crosstabs** procedure is accessed by following the steps explained in Demonstration 12.1. Click **Statistics**, **Summarize** and **Crosstabs**. Specify *gunlaw* as the row variable, or the dependent variable (the label for this variable is FAVOR OR OPPOSE GUN PERMITS), and *pol* (recoded *polviews*) as the column, or independent, variable. On the **Crosstabs** window, click **Cells** and, under Percentages, click the **radio** button next to Columns. With the dependent variable in the rows and the independent variable in the columns and with percentages calculated within columns, we will be able to read the table by following the rules developed in this chapter. You may also, if you wish, click the **Statistics** button and request chi square. The bivariate table for *gunlaw* and *pol* is shown below.

Crosstabs

Case Processing Summary

	Cases					
	Valid		Missing		Total	
	N	Percent	N	Percent	N	Percent
FAVOR OR OPPOSE GUN PERMITS* POL	870	59.9%	582	40.1%	1452	100.0%

FAVOR OR OPPOSE GUN PERMITS* POL Crosstabulation

			POL			Total
			1.00	2.00	3.00	
FAVOR OR OPPOSE GUN PERMITS	FAVOR	Count	190	283	236	709
		% within POL	87.6%	83.7%	74.9%	81.5%
	OPPOSE	Count	27	55	79	161
		% within POL	12.4%	16.3%	25.1%	18.5%
Total		Count	217	338	315	870
		% within POL	100.0%	100.0%	100.0%	100.0%

Chi-Square Tests

	Value	df	Asymp. Sig. (2-sided)
Pearson Chi-Square	15.434[a]	2	.000
Likelihood Ratio	15.319	2	.000
Linear-by-Linear Association	14.580	1	.000
N of Valid Cases	870		

a. 0 cells (.0%) have expected count less than 5. The minimum expected count is 40.16.

In each cell of the table, the count is reported first. For example, there are 190 cases in the upper-left-hand cell. These are people who are liberal (1 on *pol*) and are in favor of gun control. Below the count is the percentage of all the cases in the column in that cell. So, 87.6% of the liberals favor gun control, as compared with 83.7% of the moderates and 74.9% of the conservatives. These results are similar to what we expected and suggest that the variables are associated. This conclusion is reinforced by the fact that chi square is a nonzero value and, in fact, is significant at less than .05 ("Significance" is reported as .000, to be exact). The differences in column percentages aren't very great, and the sample as a whole is fairly homogeneous on the issue. The majority of cases in each column are in favor of gun control, and there is not much variation in support from column to column. The relationship between support for gun control and political ideology is, at best, moderate in strength.

DEMONSTRATION 13.3 The Direction of Relationships: Do Attitudes Towards Sex or the Frequency Vary by Age?

Let's take a look at a relationship between two ordinal-level variables so that you can develop some experience in describing the direction as well as the strength of bivariate relationships. Both our variables have to be at least ordinal in level of measurement, so let's start with *age* as an independent variable and **Recode** it into, for example, three categories, as in Demonstration 10.1. The recoding instructions were

$$18 \text{ thru } 36 \rightarrow 1$$
$$37 \text{ thru } 50 \rightarrow 2$$
$$51 \text{ thru } 89 \rightarrow 3$$

If you've forgotten, I used these particular cutting points to divide the sample into three categories of roughly equal size.

Without looking at the data, I'm willing to bet that *age* will have a negative relationship with approval of premarital sex (*premarsx*). Run the **Crosstabs** procedure (see Demonstration 13.1) with *premarsx* as the row variable and *ager* (recoded *age*) as the column variable. Click the **Cells** button and make sure that the **radio** button next to Columns under Percentages is checked. If you wish, click the **Statistics** button and request a chi square test. The bivariate table for these two variables is shown.

Crosstabs

Case Processing Summary

	Cases					
	Valid		Missing		Total	
	N	Percent	N	Percent	N	Percent
SEX BEFORE MARRIAGE *AGER	951	65.5%	501	34.5%	1452	100.0%

SEX BEFORE MARRIAGE* AGER Crosstabulation

			AGER			
			1.00	2.00	3.00	Total
SEX BEFORE MARRIAGE	ALWAYS WRONG	Count	51	68	107	226
		% within AGER	16.1%	22.0%	32.8%	23.8%
	ALMOST ALWAYS	Count	28	20	44	92
		% within AGER	8.9%	6.5%	13.5%	9.7%
	SOMETIMES WRONG	Count	73	70	77	220
		% within AGER	23.1%	22.7%	23.6%	23.1%
	NOT WRONG AT ALL	Count	164	151	98	413
		% within AGER	51.9%	48.9%	30.1%	43.4%
Total		Count	316	309	326	951
		% within AGER	100.0%	100.0%	100.0%	100.0%

Chi-Square Tests

	Value	df	Asymp. Sig. (2-sided)
Pearson Chi-Square	48.605[a]	6	.000
Likelihood Ratio	49.543	6	.000
Linear-by-Linear Association	39.479	1	.000
N of Valid Cases	951		

a. 0 cells (.0%) have expected count less than 5. The minimum expected count is 29.89.

Is there a relationship between these two variables? Do the conditional distributions change? Inspect the table column by column; I think you will agree that there is a relationship. Is this a positive or negative relationship? Remember that in a positive association, high scores on one variable will be associated with high scores on the other and low scores will be associated with low. In a negative relationship, high scores on one variable are associated with low scores on the other.

Now go back and look at the table. Find the single largest cell in each column and see if you can detect the pattern. The lowest score (or youngest age group) on *ager* is in the left-hand column. For this age group, the most common score (51.9% of this age group) on *premarsx* is 4 (not wrong at all). (To view the numerical codes associated with each response on a variable, see Appendix G or click **Utilities** and **Variables**.) In other words, a slight majority of the youngest respondents supports the most permissive position on premarital sex. This score is also the most common response for the other two age groups but the percentage of cases taking this most permissive view of premarital sex declines as age increases. So, we could say that permissiveness *declines* as age *increases*—the relationship is *negative* in direction.

[handwritten margin notes: "row → dep / column → indep" and "0–10 weak / 11–30 mod / 31+ strong"]

Now look at the top row of the table. This is the least permissive position on premarital sex, and the percentage of cases in this row increases as age increases. So, we could say that opposition to premarital sex *increases* as age *increases*—the variables change in the same direction, so the relationship is *positive*.

Which of these two characterizations of the relationship is correct? Both: the relationship is positive if you think of *premarsx* as measuring permissiveness and negative if you think of the variable as measuring opposition to premarital sex. The point, of course, is to call your attention to the fact that the direction of a relationship can be ambiguous when we are dealing with ordinal-level variables like *premarsx*. The coding for these variables is always arbitrary, and higher scores may indicate an increase in the quantity being measured or a decrease. Always exercise additional caution when analyzing the direction of relationships that involve ordinal-level variables.

Exercises

13.1 As long as *polviews* has already been recoded, pick a few more "social issues" (such as *grass, cappun,* or *busing*—see problem 13.15) and, with *polviews* as the independent variable, see if the patterns of association conform to expectations. Are "liberals" really more liberal on these issues? For each table, be sure to request column percentage in the cells and the chi square test. For each table, write a sentence or two of interpretation.

13.2 See if you can assess the strongest determinant of attitudes on an issue such as *cappun, grass, gunlaw,* or *busing*. You already have results for *polviews* as an independent variable. Run the same dependent or row variables against *sex, relig, class,* and *race* and compare the strength and direction or pattern of the relationships with each other and with *polviews*. Which independent variable has the strongest effect on the "social issue" you choose to analyze?

14 Association Between Variables Measured at the Nominal Level

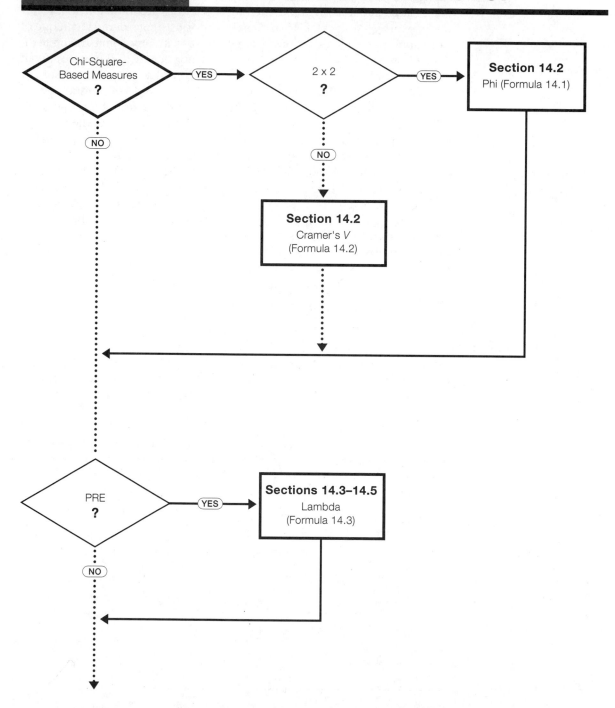

14.1 INTRODUCTION

Measures of association are descriptive statistics that summarize the overall strength of the association between two variables. Because they represent that relationship in a single number, these statistics are more efficient methods of expressing an association than the previously considered technique of calculating percentages for bivariate tables. As with any summarizing technique, however, a certain amount of information (detail and nuance) about the relationship will be lost if the researcher considers only the measure of association. Make a habit of inspecting the patterns of cell frequencies or percentages in the table along with the summary measure of association in order to maximize the amount of information you have about the relationship. This practice should be followed regardless of the level of measurement of the data or the specific measure that has been calculated.

As we shall see, there are many measures of association. In this text, these statistics have been organized according to the level of measurement for which they are most appropriate. In this chapter, we will consider measures appropriate for nominally measured variables. Several of the research situations used as examples involve variables measured at different levels (for example, one nominal-level variable and one ordinal-level variable). The general procedure in the situation of "mixed levels" is to be conservative and select measures of association appropriate for the lower of the two levels of measurement.

14.2 CHI SQUARE-BASED MEASURES OF ASSOCIATION

Over the years, statisticians have relied heavily on a group of measures of association, all based on the value of chi square. Since, ordinarily, the value of chi square will have already been obtained, these measures are quite easy to calculate and are, perhaps, most useful as "quick and dirty" indicators of the strength of a relationship.

To illustrate, let us reconsider Table 12.3, which displayed, with fictitious data, a relationship between accreditation status of undergraduate programs and employment patterns of social work majors. For the sake of convenience, this table is reproduced here as Table 14.1.

TABLE 14.1 EMPLOYMENT STATUS OF SOCIAL WORK MAJORS BY ACCREDITATION STATUS OF UNDERGRADUATE PROGRAM (Fictitious data)

| Employment | Accreditation Status | | Totals |
	Accredited	Not Accredited	
Employed as a social worker	30	10	40
Not employed as a social worker	25	35	60
Totals	55	45	100

TABLE 14.2   EMPLOYMENT STATUS BY ACCREDITATION STATUS (percentages)

Employment	Accreditation Status	
	Accredited	Not Accredited
Employed as a social worker	54.55%	22.22%
Not employed as a social worker	45.45%	77.78%
	100.00%	100.00%

As we saw in Chapter 12, this relationship was statistically significant. The chi square test statistic was 10.78, which was significant at an alpha of .05. The question now concerns the strength of the association. A brief glance at Table 14.1 shows that the conditional distribution of employment status does change, so the variables are associated. To emphasize this point, it is always helpful to calculate the percentages, as in Table 14.2.

So far, we know that the relationship between these two variables is statistically significant and that there is an association of some kind between accreditation and employment. To assess the strength of the association, we will compute a **phi (ϕ)**. This statistic is a frequently used chi square-based measure of association appropriate for 2 × 2 tables (that is, tables with two rows and two columns). The formula for phi is

FORMULA 14.1

$$\phi = \sqrt{\frac{\chi^2}{N}}$$

One of the attractions of phi is that it is easy to calculate. Simply divide the value of the obtained chi square by N and take the square root of the result. For the data displayed in Table 14.1, the chi square was 10.78. Therefore, phi is

$$\phi = \sqrt{\frac{\chi^2}{N}}$$

$$\phi = \sqrt{\frac{10.78}{100}}$$

$$\phi = .33$$

For a 2 × 2 table, phi ranges in value from 0 (no association) to 1.00 (perfect association). The phi of .33 for Table 14.1 indicates a moderate relationship between accreditation and employment status. The relationship between these two variables is statistically significant but not particularly strong. As for the pattern of the association, Table 14.2 shows that graduates of accredited programs were more often employed as social workers.

For tables larger than 2 × 2 (specifically, for tables with more than two columns *and* more than two rows), the upper limit of phi can exceed 1.00. This makes phi difficult to interpret, and a more general form of the statistic called **Cramer's V** must be used for larger tables. The formula for Cramer's *V* is

FORMULA 14.2

$$V = \sqrt{\frac{\chi^2}{(N)(\text{minimum of } r-1, c-1)}}$$

where (minimum of $r-1$, $c-1$) = the minimum
value of $r-1$ (number of rows minus 1)
or $c-1$ (number of columns minus 1)

To calculate *V*, find the square root of chi square divided by the quantity *N* multiplied by the number of rows minus 1 or the number of columns minus 1, whichever is the lower (or "minimum") value. Cramer's *V* has an upper limit of 1.00 for any size table and will be the same value as phi if the table has either two rows or two columns. Like phi, Cramer's *V* can be interpreted as an index that measures the strength of the association between two variables.

To illustrate the computation of *V*, suppose you had gathered the data displayed in Table 14.3, which shows the relationship between membership in student organizations and academic achievement for a sample of college students. The obtained chi square for this table is 31.5, a value that is significant at the .05 level (see Chapter 12 for a detailed account of the computation of chi square). Cramer's *V* is

$$V = \sqrt{\frac{\chi^2}{(N)(\text{minimum of } r-1, c-1)}}$$

$$V = \sqrt{\frac{31.5}{(75)(2)}}$$

$$V = \sqrt{\frac{31.5}{150}}$$

$$V = \sqrt{.21}$$

$$V = .46$$

TABLE 14.3 ACADEMIC ACHIEVEMENT BY CLUB MEMBERSHIP

Academic Achievement	Fraternity or Sorority	Other Organization	No Memberships	Totals
Low	4	4	17	25
Moderate	15	6	4	25
High	4	16	5	25
Totals	23	26	26	75

TABLE 14.4   ACADEMIC ACHIEVEMENT BY CLUB MEMBERSHIP (percentages)

Academic Achievement	Membership		
	Fraternity or Sorority	Other Organization	No Memberships
Low	17.4	15.4	65.4
Moderate	65.2	23.1	15.4
High	17.4	61.5	19.2
	100%	100%	100%
	(23)	(26)	(26)

Since Table 14.3 has the same number of rows and columns, we may use either $(r - 1)$ or $(c - 1)$ in the denominator. In either case, the value of the denominator is N multiplied by $(3 - 1)$, or 2. The computed value of V of .46 suggests a moderate to strong association between club membership and academic achievement.

Calculating percentages as in Table 14.4 will help identify the pattern of the relationship. Fraternity and sorority members tend to be moderate, members of other organizations tend to be high, and nonmembers tend to be low in academic achievement.

A limitation of phi and Cramer's V, as you may have noticed, is the absence of a direct or meaningful interpretation for values between the extremes of 0.00 and 1.00. Both measures are indices of the strength of the association; for example, a value of .75 indicates a stronger relationship than a value of .50. The problem is that the values between 0.00 and 1.00 cannot be interpreted as anything other than an index of the relative strength of the association. Because they are easy to calculate (once the value of chi square has been obtained), perhaps the most appropriate use for these measures is as first indicators of the importance of an association.* *(For practice in computing phi and Cramer's V, see any of the problems at the end of this chapter or at the end of Chapter 13. To minimize computations, however, you could use problems from the end of Chapter 12 for which you already have a value for chi square. Otherwise, use the problems based on 2 × 2 tables, especially problems 13.1, 13.2, and 14.1. Remember that for tables that have either two rows or two columns, phi and Cramer's V will have the same value. Cramer's V would be computed only for problems 13.9 to 13.12, and 14.14.)*

---

*Two other chi square-based measures of association, T^2 and C (the contingency coefficient), are sometimes reported in the literature. Both of these measures have serious limitations. T^2 has an upper limit of 1.00 only for tables with an equal number of rows and columns, and the upper limit of C varies depending on the dimensions of the table. These characteristics make these measures more difficult to interpret and thus less useful than phi or Cramer's V.

READING STATISTICS 10: The Importance of Percentages

In this chapter and the next, our primary concern is with measures of association for bivariate tables. These statistics are extremely useful for summarizing the strength and—as we will see in Chapter 15—the direction of relationships. Nonetheless, the first step in analyzing bivariate tables should always be to apply the techniques introduced in Chapter 13. The column percentages and conditional distributions will give you more detail about the relationship than measures of association. As useful as they are, the latter should be regarded as summary statements rather than analysis-in-depth.

Of course, even statistics as humble as percentages can be miscalculated and misunderstood. Sociologists Wilbert Leonard and John Phillips illustrate the importance of calculating percentages in the correct direction in a research note

concerning the phenomenon of racial stacking in professional sports. This is the practice of reserving certain positions for white players—usually the positions that require leadership skills or decision making (e.g., the quarterback in football). Racial stacking has been widely documented but sometimes misinterpreted because of incorrectly calculated percentages.

Leonard and Phillips illustrate their point with the tables reproduced below. Note that the first table has the independent variable (race) in the columns—consistent with the convention followed in this text—but incorrectly computes the percentages within the rows. The second table has the independent variable in the rows and correctly computes the percentages in the direction of the independent variable.

TABLE A RACE BY POSITION IN MAJOR LEAGUE BASEBALL (IN PERCENTAGES)

| | Race | | | | |
Position	White	Black	Latin American	N	%
Catcher	87	1	12	98	100
1st Base	69	19	11	36	100
2nd Base	59	15	26	34	100
3rd Base	76	12	12	33	100
Shortstop	42	8	50	36	100
Outfield	34	49	17	194	100
Utility	54	21	25	111	100

TABLE B RACE BY POSITION IN MAJOR LEAGUE BASEBALL (IN PERCENTAGES)

Race	Catcher	1B	2B	3B	SS	OF	Util	N	%
White	29	8	7	8	5	22	20	296	100
Black	1	5	4	3	2	69	17	138	100
Latin American	11	4	8	4	17	31	26	108	100

Data are from R. E. Lapchick and J. R. Benedict. 1993. *Racial Report Card*. Boston: Center for the Study of Sport in Society.

Here's what the authors have to say about these tables:

Table [A] does not provide information about the proportion of players from each ethnic group. . . . Utility infielder (player capable of "filling in" at most infield positions and in the outfield when needed) appears to be a "white" job. One might be tempted to construct a hypothesis to explain the absence of minority players from this unglamorous but important role. In fact, Table [B] shows that Black, White, and Latin American players are about equally likely to be utility infielders. The differences in Table [A] say more about the proportions of White, Black, and Latin American players in the major leagues than they do about whether oneracial/ethnic group is better or more competent to fill the job of utility infielder.

In other words, white players appear to dominate certain positions in Table A because they outnumber players from the other groups. What does Table B show? Although the groups are about evenly distributed across most positions, whites dominate the position of catcher—a primary decision maker on the field, and blacks are overrepresented among outfielders—the position most distant from decision-making responsibilities.

Table and excerpts reprinted by permission from W. M. Leonard II and J. Phillips, 1997, "The Cause and Effect Rule for Percentaging Tables: An Overdue Statistical Correction for 'Stacking' Studies,"; *Sociology of Sport Journal*, 14(5): 283-289.

14.3 PROPORTIONAL REDUCTION IN ERROR (PRE)

In recent years, chi square-based measures of association—once very popular—have been supplanted by a group of measures that are based on a logic known as **proportional reduction in error (PRE)**. For nominal-level variables, the logic of PRE involves first attempting to guess or predict the category into which each case will fall on the dependent variable (Y) while ignoring the independent variable (X). Since we would in effect be predicting blindly in this case, we would make many errors (that is, incorrectly predict the category of many cases on the dependent variable).

The second step would be to predict again the category of each case on the dependent variable, this time taking the independent variable into account. If the two variables are associated, the additional information supplied by the independent variable should reduce our errors of prediction (that is, we should misclassify fewer cases). The stronger the association between the variables, the greater the reduction in errors. In the case of a perfect association, we would make no errors at all when predicting score on Y from score on X. When there is no association between the variables, on the other hand, knowledge of the independent will not improve the accuracy of our predictions. We would make just as many errors of prediction with knowledge of the independent variable as we did without knowledge of the independent variable.

To illustrate these principles with an example, suppose you were placed in the rather unusual position of having to predict whether each of the next 100 people you meet will be less than 5 feet 9 inches in height ("short") or 5 feet 9 inches or taller ("tall"), under the condition that you would have no knowledge of these people at all (for example, the condition that you are in a sensory deprivation tank and can use none of your senses to help you with your predictions). Clearly, your predictions will be wrong quite often (you will frequently misclassify a tall person as short and vice versa).

Now assume that you must go through this ordeal twice; but, on the second round, just prior to your prediction, you are supplied with information about the sex of the person whose height you must predict. Since height is associated with sex and females are, on the average, shorter than males, the optimal strategy would be to predict that all females are short and all males are tall. Of course, you will still make errors on this second round; but, given an association between the variables, the number of errors on the second round will be less than the number of errors on the first. That is, by taking information from an associated variable into account, you will have reduced your errors of prediction. How can these unusual thoughts be translated into a useful statistic?

14.4 A PRE MEASURE FOR NOMINAL-LEVEL VARIABLES: LAMBDA

One hundred individuals have been categorized by gender and height, and the data are displayed in Table 14.5. It is clear, even without percentages, that the two variables are associated. To measure the strength of this association, a PRE measure called **lambda** (symbolized by the Greek letter λ) will be calculated. Following the logic introduced in the previous section, we must find two quantities. First, the number of prediction errors made while ignoring the independent variable (gender) must be found. Then, we will find the number of prediction errors made while taking gender into account. These two sums will then be compared to derive the statistic.

First, the information given by the independent variable (gender) can be ignored, in effect, by working only with the row marginals. Two different predictions can be made by using these marginals. We can predict either that all subjects are tall or that all subjects are short.* For the first prediction (all subjects are tall), 48 errors will be made. That is, for this prediction, all 100 cases would be placed in the first row. Since only 52 of the cases actually belong in this row, this prediction would result in $(100 - 52)$, or 48, errors. If we had predicted that all subjects were short, on the other hand, we would have made 52 errors $(100 - 48 = 52)$. We will take the lesser of these two numbers and refer to this quantity as E_1 for the number of errors made while ignoring the independent variable. So, $E_1 = 48$.

The second step in the computation of lambda is again to predict score on Y (height), this time taking X (gender) into account. Follow the same procedure as in the first step, but this time move from column to column. Since each column is a category of X, we thus take X into account in making our predictions. For the left-hand column (males), we predict that all 50 cases will be tall and make six errors $(50 - 44 = 6)$. For the second column (females), our prediction is that all females are short, and eight errors will be

---

*Other predictions are, of course, possible, but these are the only two permitted by lambda.

TABLE 14.5 HEIGHT BY GENDER

	Gender		
Height	Male	Female	Totals
Tall	44	8	52
Short	6	42	48
Totals	50	50	100

made. By moving from column to column, we have taken X into account and have made a total of 14 errors of prediction, a quantity we will label E_2 ($E_2 = 6 + 8 = 14$).

Now, the logic of lambda is that, if the variables are associated, fewer errors will be made under the second procedure than under the first. Clearly, gender and height are associated, since we made fewer errors of prediction while taking gender into account ($E_2 = 14$) than while ignoring gender ($E_1 = 48$). Our errors have been reduced from 48 to only 14. To find the proportional reduction in error, use Formula 14.3:

FORMULA 14.3

$$\lambda = \frac{E_1 - E_2}{E_1}$$

For the sample problem, the value of lambda would be

$$\lambda = \frac{E_1 - E_2}{E_1}$$

$$\lambda = \frac{48 - 14}{48}$$

$$\lambda = \frac{34}{48}$$

$$\lambda = .71$$

The value of lambda ranges from 0.00 to 1.00. A lambda of 0.00 would indicate that the information supplied by the independent variable does not improve our ability to predict the dependent and, therefore, that there is no association between the variables. A lambda of 1.00 would mean that we could predict Y without error from X.

Part of what has made PRE measures popular is that values between 0.00 and 1.00, such as the lambda of .71 calculated above, have a direct and meaningful interpretation. These values can be seen as indices of the extent to which X helps us to predict (or, more loosely, understand) Y. When multiplied by 100, the value of the lambda indicates the percentage reduction in error and, therefore, the strength of the association. Thus, the lambda above would be interpreted by concluding that knowledge of gender improves our ability to predict height by 71%. Or, we are 71% better off knowing gender when attempting to predict height than we are not knowing gender.

14.5 THE COMPUTATION OF LAMBDA

Let us work through another example in order to state the computational routine for lambda in more general terms. Suppose a researcher was concerned with the relationship between religious denomination and attitude toward capital punishment and had collected the data presented in Table 14.6 from a sample of 130 respondents.

Step 1. To find E_1, the number of errors made while ignoring X (religion, in this case), subtract the largest row total from N. For Table 14.6, E_1 will be

$$E_1 = N - \text{(Largest row total)}$$
$$E_1 = 130 - 50$$
$$E_1 = 80$$

Thus, we will misclassify 80 cases on attitude toward capital punishment while ignoring religion.

Step 2. Next, E_2—the number of errors made when taking the independent variable into account—must be found. For each column, subtract the largest cell frequency from the column total and then add the subtotals together. For the data presented in Table 14.6:

$$\text{For Catholics: } 35 - 14 = 21$$
$$\text{For Protestants: } 25 - 12 = 13$$
$$\text{For "Others": } 40 - 25 = 15$$
$$\text{For "None": } 30 - 14 = \underline{16}$$
$$E_2 = 65$$

A total of 65 errors are made when predicting attitude on capital punishment while taking religion into account.

Step 3. In step 1, 80 errors of prediction were made as compared to 65 errors in step 2. Since the number of errors has been reduced, the variables

TABLE 14.6 ATTITUDE TOWARD CAPITAL PUNISHMENT BY RELIGIOUS DENOMINATION (fictitious data)

Attitude	Religion				Totals
	Catholic	Protestant	Other	None	
Favors	10	9	5	14	38
Neutral	14	12	10	6	42
Opposed	11	4	25	10	50
Totals	35	25	40	30	130

Application 14.1

A random sample of students at a large urban university have been classified as either "traditional" (18–23 years of age and unmarried) or "nontraditional" (24 or older or married). Subjects have also been classified as "vocational," if their primary motivation for college attendance is career or job oriented, or "academic," if their motivation is to pursue knowledge for its own sake. Are these two variables associated?

Motivation	Traditional	Non-traditional	Totals
	Type		
Vocational	25	60	85
Academic	75	15	90
Totals	100	75	175

A lambda will be computed to measure the association between these two variables.

$E_1 = 175 - 90 = 85$
$E_2 = (100 - 75) + (75 - 60)$
$E_2 = 25 + 15 = 40$
$$\lambda = \frac{E_1 - E_2}{E_1}$$
$$\lambda = \frac{85 - 40}{85}$$
$$\lambda = \frac{45}{85}$$
$\lambda = 0.53$

A lambda of 0.53 indicates that we would make 53% fewer errors in predicting motivation from student type as opposed to predicting motivation while ignoring student type. The association is moderately strong, and, by inspection of the table, we can see that traditional students are more likely to have academic motivations, and nontraditional types are more likely to be vocational in motivation.

are associated. To find the proportional reduction in error, the values for E_1 and E_2 can be directly substituted into Formula 14.3:

$$\lambda = \frac{E_1 - E_2}{E_1}$$

$$\lambda = \frac{80 - 65}{80}$$

$$\lambda = \frac{15}{80}$$

$$\lambda = .19$$

When attempting to predict attitude toward capital punishment, we would make 19% fewer errors by taking religion into account. Knowledge of a respondent's religious denomination does improve the accuracy of our predictions (the variables are associated) but not by a very large margin. The relatively low value of this lambda indicates that factors other than religion are associated with the dependent variable.

As a measure of association, lambda has two characteristics that should be stressed. First, lambda is asymmetric. This means that the value of the statistic will vary depending on which variable is taken as independent. For example, in Table 14.6, the value of lambda would be .14 if attitude toward capital punishment had been taken as the independent variable (verify this with your own computation). Thus, you should exercise some caution in the designation of an independent variable. If you consistently follow the convention of arraying the independent variable in the columns and compute lambda as outlined above, the asymmetry of the statistic should not be confusing.

Second, when one of the row totals is much larger than the others, lambda can take on a value of 0.00 even when other measures of association would not be 0.00, and calculating percentages for the table indicates some association between the variables. This anomaly is a function of the way lambda is calculated and suggests that great caution should be exercised in the interpretation of lambda when the row marginals are very unequal. In fact, in the case of very unequal row marginals, a chi square-based measure of association might well be preferable to lambda. *(For practice in computing lambda, see any of the problems at the end of this chapter, Chapter 13, or Chapter 12. As with phi and Cramer's V, it's probably a good idea to start with small samples and 2 × 2 table, especially problems 13.1, 13.2, and 14.1.)*

SUMMARY

1. Three measures of association—phi, Cramer's V, and lambda—were introduced. Each is used to summarize the overall strength of the association between two variables that have been organized into a bivariate table.

2. Phi and Cramer's V are chi square-based measures of association and have the advantage of being easy to compute (once the value of chi square is found). Phi is used for 2 × 2 tables while Cramer's V can be used for any size table. Both indicate the strength of the relationship, but values between 0.00 and 1.00 have no direct interpretation.

3. Lambda is a PRE-based measure and provides a more direct interpretation for values between the extremes of 0.00 and 1.00. Lambda indicates the improvement in predicting the dependent variable with knowledge of the independent, compared to predicting the dependent without knowledge of the independent. Because of the meaningfulness of values between the extremes, lambda is often preferred over the more traditional chi square-based measures.

SUMMARY OF FORMULAS

Phi	14.1	$\phi = \sqrt{\dfrac{\chi^2}{N}}$
Cramer's V	14.2	$V = \sqrt{\dfrac{\chi^2}{(N)(\text{minimum of } r-1, c-1)}}$
Lambda	14.3	$\lambda = \dfrac{E_1 - E_2}{E_1}$

GLOSSARY

Cramer's V. A chi square-based measure of association. Appropriate for nominally measured variables that have been organized into a bivariate table of any number of rows and columns.

E_1. For lambda, the number of errors of prediction made when predicting which category of the dependent variable cases will fall into while ignoring the independent variable.

E_2. For lambda, the number of errors of prediction made when predicting which category of the dependent variable cases will fall into while taking account of the independent variable.

Lambda (λ). A measure of association appropriate for nominally measured variables that have been organized into a bivariate table. Lambda is based on the logic of proportional reduction in error (PRE).

Phi (Φ). A chi square-based measure of association. Appropriate for nominally measured variables that have been organized into a 2 × 2 bivariate table.

Proportional reduction in error (PRE). A logic that underlies the definition and computation of lambda (and several other measures of association). Basically, the logic involves comparing the number of errors made in predicting the dependent variable while ignoring the independent variable (E_1) with the number of errors made in predicting the dependent variable while taking account of the independent (E_2). These two figures, E_1 and E_2, are then manipulated to produce the final statistic.

PROBLEMS

14.1 SOC Who is most likely to be victimized by crime? A small sample of city residents has been asked if they were the victims of burglary or robbery over the past year. The tables below report relationships between several variables and victimization. Compute phi and lambda for each table. Which relationship is strongest?

a. Victimization by sex of respondent:

	Sex		
Victimized?	Male	Female	Totals
Yes	10	12	22
No	15	18	33
Totals	25	30	55

b. Victimization by age of respondent:

	Age		
Victimized?	21 or Younger	22 or Older	Totals
Yes	12	10	22
No	15	18	33
Totals	27	28	55

c. Victimization by race of respondent:

	Race		
Victimized?	Black	White	Totals
Yes	9	13	22
No	6	27	33
Totals	15	40	55

14.2 Compute a phi and a lambda for problems 13.1 to 13.8. Compare the value of the measure of association with your impressions of the strength of the relationships based on the percentages.

14.3 SOC There is concern that suicides are motivated, in part, by imitation. Especially among young people, it may be that "epidemics" of self-destructive behaviors follow publication of suicides in local media. A number of cities have been classified by rate of suicide and by whether or not they experienced a publicized suicide within the past year. Is there an association between these two variables? Summarize your conclusions in a sentence or two.

	Publicized Suicide?		
Suicide Rate	Yes	No	Totals
Low	15	20	35
High	15	10	25
Totals	30	30	60

14.4 [SW] The director of a shelter for battered women has noticed that many of the women who are referred to the shelter eventually return to their violent husbands even when there is every indication that the husband will continue the pattern of abuse. The director suspects that the women who return to their husbands do so because they have no place else to go—for example, no close relatives in the area with whom the women could reside. Do the data below support the director's suspicion? (Data are from the case files of former clients.)

Return to Husband?	Relatives Nearby	No Relatives Nearby	Totals
Yes	10 19.8	23 13.2	33
No	50 40.2	17 26.8	67
Totals	60	40	100

14.5 [PA] Traditionally, bus ridership in your town has been confined to lower-income and blue-collar patrons. As head of transportation planning for the city, you believe that ridership from white-collar, middle-income neighborhoods can be increased if bus routes linking these neighborhoods to the downtown area (where most people work) are increased. A survey is conducted, and the results are displayed below. Is willingness to ride the bus related to job location? What is the pattern of the relationship (if any)?

Potential Ridership	Job Location Downtown	Other	Totals
Would use bus	55	20	75
Would not use bus	15	21	36
Totals	70	41	111

14.6 [GER] A survey of senior citizens who live in either a housing development specifically designed for retirees or an age-integrated neighborhood has been conducted. Is type of living arrangement related to sense of social isolation?

Sense of Isolation	Housing Development	Integrated Neighborhood	Totals
Low	80	30	110
High	20	120	140
Totals	100	150	250

Living Arrangement

14.7 [SOC] Is there an association between the gender of college instructors and the teaching effectiveness ratings they receive from students? Write a few sentences summarizing your findings.

Teaching Effectiveness	Gender Female	Male	Totals
High	115	241	356
Low	54	113	167
Totals	169	354	523

14.8 [SOC] A researcher has conducted a survey on sexual attitudes for a sample of 317 teenagers. The respondents were asked whether they considered premarital sex to be "always wrong" or "OK under certain circumstances." The tables below summarize the relationship between responses to this item and several other variables. For each table, assess the strength and pattern of the relationship and write a paragraph interpreting these results.

a. Attitudes toward premarital sex by gender:

Premarital Sex	Sex Male	Female	Totals
Always wrong	90	105	195
Not always wrong	65	57	122
Totals	155	162	317

b. Attitudes toward premarital sex by courtship status:

Premarital Sex	Ever "Gone Steady"? No	Yes	Totals
Always wrong	148	47	195
Not always wrong	42	80	122
Totals	190	127	317

c. Attitudes toward premarital sex by social class:

Premarital Sex	Social Class Blue Collar	White Collar	Totals
Always wrong	72	123	195
Not always wrong	47	75	122
Totals	119	198	317

14.9 [SOC] St. Algebra College has a problem with attrition. A sizable number of nongraduating students do not return to classes each semester. Is attrition importantly related to race, status, or age? For each table below, how strong is the association (if any) between each of the independent variables and attrition? Does the relationship have a pattern? (Calculating percentages for the table will help to answer the second question.) Write a paragraph summarizing the results presented in these three tables.

a. Attrition by race for 532 students enrolled in fall semester:

Attrition	Race		Totals
	White	Black	
Returned spring semester	280	100	380
Did not return	105	47	152
Totals	385	147	532

b. Attrition by status for 532 students enrolled in fall semester:

Attrition	Status		Totals
	Part-time	Full-time	
Returned spring semester	42	353	395
Did not return	87	50	137
Totals	129	403	532

c. Attrition by age for 532 students enrolled in fall semester:

Attrition	Age		Totals
	18–24	25 or More	
Returned spring semester	322	73	395
Did not return	57	80	137
Totals	379	153	532

14.10 SOC/CJ A sociologist is researching public attitudes toward crime and has asked a sample of residents of his city if they think that the crime rate in their neighborhoods is rising. Is there a relationship between sex and perception of the crime rate? Between race and perception of the crime rate? What is the pattern of the relationship? Write a paragraph summarizing the information presented in these tables.

a. Perception of crime rate by sex:

Crime Rate Is	Sex		Totals
	Male	Female	
Rising	200	225	425
Stable	175	150	325
Falling	125	125	250
Totals	500	500	1000

b. Perception of crime rate by race:

Crime Rate Is	Race		Totals
	White	Black	
Rising	300	150	450
Stable	230	85	315
Falling	170	65	235
Totals	700	300	1000

14.11 PS You are running for mayor of Shinbone, Kansas, and realize that, if you are to win, you must win the support of blue-collar voters. (You already have strong support in the white-collar neighborhoods.) You have a very limited advertising budget and wonder how best to reach your intended audience. An aide has found the data below, which show the relationship between social class and "main source of news" for a sample of Shinboneites. Will this information help you make a decision?

Main Source of News	Blue Collar	White Collar	Totals
Television	140	200	340
Radio	25	40	65
Newspapers	85	100	185
Totals	250	340	590

14.12 PS Does the American electorate really have a tendency, as is sometimes alleged, to vote for physically attractive candidates regardless of competence? A sample of adult voters were shown photographs of a physically attractive male or an unattractive male and asked if the individual pictured "seems like he would be a competent elected official." The results are presented below. Is there an association between attractiveness and electability?

Would He Be Competent?	Attractiveness		Totals
	Attractive	Unattractive	
Yes	15	12	27
No	17	15	32
Can't say	18	23	41
Totals	50	50	100

14.13 GER Community leaders are wondering if there is any relationship between the age of city residents and their attendance at city-sponsored festivities like the annual Fourth of July celebration. Does the table below indicate an association between senior citizen status and attendance at last year's celebration? Summarize your conclusions in a sentence or two.

Did You Attend last Year's Fourth of July Celebration?	Senior Citizen		Totals
	Yes	No	
Yes	45	178	223
Wanted to but couldn't	13	42	55
No	145	332	477
Totals	203	552	755

14.14 SW Ninety male first-graders have been rated by their teachers for levels of aggressiveness (for example, hitting other children, "acting out" in the classroom), and a school social worker has rated the families of each of these children for internal conflict and stress. Is there a relationship between these two variables? Do more aggressive children tend to come from high-conflict families? Summarize your conclusions in a sentence or two.

Aggressiveness	Level of Family Conflict			Totals
	Low	Moderate	High	
Low	10	8	17	35
Moderate	15	10	10	35
High	5	12	3	20
Totals	30	30	30	90

14.15 Problem 13.15 analyzed some bivariate relationships taken from the General Social Survey data set. Political ideology was used as the independent variable for five different dependent variables and, using only percentages, you were asked to characterize the relationships in terms of strength and direction. Now, with the aid of measures of association, these characterizations should be easier to develop. Compute a Cramer's V and a lambda for each table in problem 13.15. Compare the measures of association with your characterizations based on the percentages. Below are the same five dependent variables cross-tabulated against sex as an independent variable. Compare the strength of these relationships with those for political ideology. Which independent variable has the stronger associations?

a. Support for the legal right to an abortion (ABANY) by sex:

Right to Abortion?	Sex		Totals
	Male	Female	
Yes	185	224	409
No	214	281	495
Totals	399	505	904

b. Support for capital punishment (CAPPUN) by sex:

Capital Punishment?	Sex		Totals
	Male	Female	
Favor	497	523	1020
Oppose	104	208	312
Totals	601	731	1332

c. Approve of suicide for people with incurable disease (SUICIDE1) by sex:

Right to Suicide?	Sex		Totals
	Male	Female	
Yes	287	328	615
No	116	212	328
Totals	403	540	943

d. Support for sex education in public schools (SEXEDUC) by sex:

Sex Education?	Sex		Totals
	Male	Female	
Favor	356	489	845
Oppose	58	63	121
Totals	414	552	966

e. Support for legalization of marijuana (GRASS) by sex:

Legalize Marijuana?	Sex		Totals
	Male	Female	
Legal	133	108	241
Not legal	281	386	667
Totals	414	494	908

Using *SPSS for Windows* to Produce Nominal-Level Measures of Association

DEMONSTRATION 14.1 Does Support for Gun Control Vary by Political Ideology? Another Look

In Demonstration 13.2, we used **Crosstabs** to look for an association between *cappun* and *polviews*. We saw that these variables were associated and that liberals were the most supportive of gun control and conservatives were the least supportive. The sample was fairly homogeneous on this issue, and the majority of people from each of the three political ideologies were in favor of gun control. The differences in conditional distributions were not great, and the relationship seemed to be, at best, "moderate" in strength. In this demonstration, we will re-examine the relationship and have SPSS compute some measures of association. The commands should repeat those in Demonstration 13.2 (remember to recode *polviews* into three categories). Click **Statistics**, then **Summarize**, and then **Crosstabs**. Move *gunlaw* into the Row(s) box and recoded *polviews* into the Column(s) box. Click the **Cells** button and request column percentages. Click the **Statistics** button and request chi square, phi, Cramer's *V*, and lambda. Click **Continue** and **OK**, and the following output will be produced:

Crosstabs

Case Processing Summary

	Cases					
	Valid		Missing		Total	
	N	Percent	N	Percent	N	Percent
FAVOR OR OPPOSE GUN PERMITS* POL	870	59.9%	582	40.1%	1452	100.0%

FAVOR OR OPPOSE GUN PERMITS* POL Crosstabulation

			POL			
			1.00	2.00	3.00	Total
FAVOR OR OPPOSE GUN PERMITS	FAVOR	Count	190	283	236	709
		% within POL	87.6%	83.7%	74.9%	81.5%
	OPPOSE	Count	27	55	79	161
		% within POL	12.4%	16.3%	25.1%	18.5%
Total		Count	217	338	315	870
		% within POL	100.0%	100.0%	100.0%	100.0%

Chi-Square Tests

	Value	df	Asymp. Sig. (2-sided)
Pearson Chi-Square	15.434[a]	2	.000
Likelihood Ratio	15.319	2	.000
Linear-by-Linear Association	14.580	1	.000
N of Valid Cases	870		

a. 0 cells (.0%) have expected count less than 5. The minimum expected count is 40.16.

Directional Measures

			Value	Asymp. Std. Error[a]	Approx. T[b]	Approx. Sig.
Nominal by Nominal	Lambda	Symmetric	.035	.016	2.078	.038
		FAVOR OR OPPOSE GUN PERMITS Dependent	.000	.000	.[c]	.[c]
		POL Dependent	.045	.021	2.078	.038
	Goodman and Kruskal tau	FAVOR OR OPPOSE GUN PERMITS Dependent	.018	.009		.000[d]
		POL Dependent	.009	.005		.000[d]

a. Not assuming the null hypothesis.
b. Using the asymptotic standard error assuming the null hypothesis.
c. Cannot be computed because the asympotic standard error equals zero.
d. Based on chi-square approximation.

Symmetric Measures

		Value	Approx. Sig.
Nominal by Nominal	Phi	.133	.000
	Cramer's V	.133	.000
N of Valid Cases		870	

a. Not assuming the null hypothesis.
b. Using the asympotic standard error assuming the null hypothesis.

The measures of association are reported below the output for the chi square tests. Three values for lambda are reported in the Directional Measures output block. Remember that lambda is asymmetric and will change value depending on which variable is taken as dependent. In this case, *gunlaw* (FAVOR OR OPPOSE GUN PERMITS) is the dependent variable and the proper lambda is .000, a value that would indicate that there is no relationship between the variables. We saw previously, however, that the conditional distributions in the table do change, indicating that there *is* a relationship. The problem here is that the row totals are very unequal and lambda is misleading and should be disregarded (see Section 14.5).

Phi and Cramer's *V* are reported in the Symmetric Measures output block. The statistics are identical in value (.133), as they will be whenever the table has either two rows or two columns. This indicates, as we suspected, a weak association between the variables. Political ideology does not seem to be an especially important cause of support for gun control.

DEMONSTRATION 14.2 What Variables Affect Support for Gun Control?

Let's see if we can do any better in "explaining" support of gun control with some other commonly used independent variables. As you are no doubt aware, attitudes are often associated with various measures of social location. In keeping with our focus on the nominal level of measurement, let's investigate relationships between *gunlaw* and *sex, marital* (marital status), *race*, and religious preference, or *relig*. Incorporating a number of potential independent variables, although quite common in social science research, generates a large volume of output, and we will abbreviate the actual output from *SPSS* in this demonstration.

	Percent in Favor	Significance of Chi Square	Phi or *V*	Lambda
SEX				
Male	73.8			
Female	88.0			
		.000	.18	.00
RELIGION				
Protestant	82.3			
Catholic	83.4			
Jew	90.5			
None	70.5			
Other	92.7			
		.006	.13	.00
RACE				
White	80.2			
Black	85.7			
Other	92.7			
		.031	.09	.00
MARITAL				
Married	80.2			
Widowed	83.7			
Divorced	83.9			
Separated	85.7			
Never Married	81.6	.782	.04	.00

Basically, these results indicate that *gunlaw* is not particularly related to any of these demographic variables. The strongest relationship is with *sex*, with males less supportive than females. Since the rows are very unequal, lambda should be disregarded.

Exercises

14.1 See if you can do any better finding a variable with a strong association with *gunlaw*. If necessary, use the **Recode** command to reduce the number of categories in the column variable. As a suggestion, try *degree* and *income91* as independent variables.

14.2 Run some of the tables you did for Exercises 13.1 and 13.2 again with a request for phi, Cramer's *V*, and lambda. Do the measures of association help you interpret the relationships?

Association Between Variables Measured at the Ordinal Level

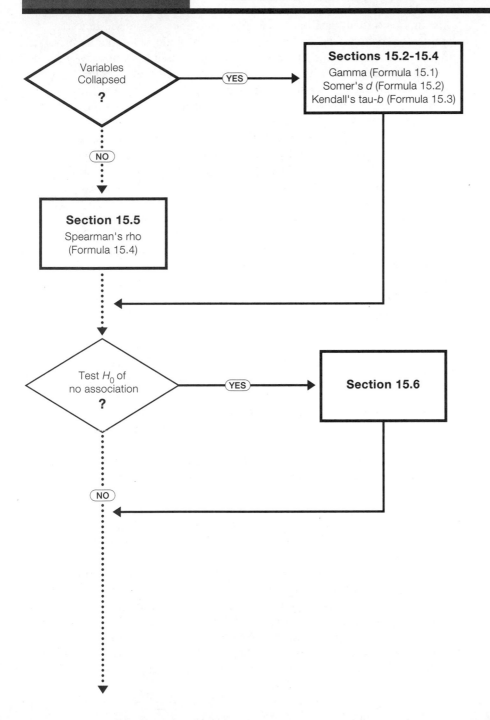

15.1 INTRODUCTION

The measurement process for ordinal-level variables can produce data of two different forms. The first form, which we shall call "continuous ordinal data," is easily recognizable by the wide range of many possible values per variable. Continuous ordinal data bear a close resemblance to interval-ratio data. Examples of such data might be found in any attitude scale that incorporated many different items and, therefore, had many possible values.

The second form, which we shall call "collapsed ordinal data," has only a few (no more than five or six) values per variable and can arise either by collecting data in collapsed form or by collapsing a continuous scale. Collapsed ordinal data would be produced, for example, by measuring the variable "social class" with only a few categories (such as upper, middle, and lower). Collapsed ordinal data could also be produced by dividing the scores on an attitude scale into just a few categories (such as high, moderate, and low).

For situations involving "continuous" ordinal variables, a statistic called **Spearman's rho** (r_s) is commonly used, and we will cover this measure of association toward the end of this chapter. A number of measures of association have been invented for use with collapsed ordinal-level variables. Rather than attempt a comprehensive coverage of all of these statistics, we will concentrate on **gamma (G)**. Although widely used, gamma is not ideal for all possible situations, and we will briefly cover two other measures (**Somer's d** and **Kendall's tau-b**) so that you will have access to some alternative measures of association for collapsed ordinal scales.

15.2 PROPORTIONAL REDUCTION IN ERROR (PRE)

For nominal-level variables, the logic of PRE was based on two prediction rules. First, we predicted Y while ignoring X and, second, we predicted Y while taking X into account. Lambda measured the proportional reduction in the errors of prediction made with the second rule as compared to the first.

For variables measured at the ordinal level, we have two similar prediction rules, and gamma, like lambda, measures the proportional reduction in error gained by predicting one variable while taking the other into account. The major difference in the logic of gamma and lambda lies in the way predictions are made. In the case of gamma, we predict the order of pairs of cases. That is, we predict whether one case will have a higher or lower score than the other on the variable in question. The first prediction rule is to predict the order of a pair of cases on one variable while ignoring their order on the other. The second rule is to predict the order of a pair of cases on one variable while taking their order on the other variable into account.

As an illustration, assume that a researcher is concerned about the causes of "burnout" (that is, demoralization and loss of commitment) among elementary school teachers and wonders if levels of burnout might be related to the number of years a teacher has been employed in the profession. Another way to state the problem would be to ask if teachers who rank higher on years of service would also rank higher on burnout. If we knew that teacher

A had more years of service than teacher B, would we be able to predict accurately that teacher A is also higher than teacher B on burnout? That is, would knowledge of the order of this pair of cases on one variable help us predict their order on the other?

If the two variables are associated, we will reduce our errors when our predictions about one of the variables are based on knowledge of the other. Furthermore, the stronger the association, the fewer the errors we will make. With no association between the variables, gamma will be 0, and knowledge of the order of a pair of cases on one variable will not improve our ability to predict their order on the other. A gamma of ± 1.00 denotes a perfect relationship: the order of all pairs of cases on one variable would be predictable without error from their order on the other variable.

Unlike variables measured at the nominal level, the scores of ordinal-level variables vary from "high" to "low." Thus, relationships between ordinal-level variables have a direction: they can be either positive or negative. In a positive relationship, cases tend to be ranked in the same order on both variables. For example, if Case A is ranked above Case B on one variable, it would also be ranked above Case B on the second variable. The relationship suggested above between years of service and burnout would be a positive relationship. In a negative relationship, the order of the cases would be reversed between the two variables. If Case A ranked above Case B on one variable, it would tend to rank below Case B on the second variable.

To this point, the logic of ordinal measures of association has been explained in the language most appropriate for an understanding of gamma. Thus, we have discussed the order of pairs of cases and how that order is predicted from variable to variable. Note, however, that we are still trying to ascertain information about the same three characteristics of an association between two variables that we introduced in Chapter 13. We are still concerned with the existence of the association, its strength, and its pattern or direction. Associations between collapsed ordinal-level variables can be (and should be) investigated by calculating percentages for the table and observing the conditional distributions of the dependent variable. By computing gamma (or any measure of association), we get a single number that summarizes the existence, strength, and direction of the relationship.

15.3 THE COMPUTATION OF GAMMA

Table 15.1 summarizes the relationship between "length of service" and "burnout" for a fictitious sample of 100 teachers. To compute gamma, two sums are needed. First, we must find the number of pairs of cases that are ranked the same on both variables (we will label this N_s) and then the number of pairs of cases ranked differently on the variables (N_d). We do so by working with the cell frequencies cell by cell.

To find the number of pairs of cases ranked the same (N_s), begin with the cell containing the cases that were ranked the lowest on both variables.

TABLE 15.1 BURNOUT BY LENGTH OF SERVICE (fictitious data)

Burnout	Length of Service			Totals
	Low	Moderate	High	
Low	20	6	4	30
Moderate	10	15	5	30
High	8	11	21	40
Totals	38	32	30	100

In Table 15.1, this would be the upper-left-hand cell. *(NOTE: Not all tables are constructed with values increasing from left to right across the columns and from top to bottom across the rows. When using other tables, always be certain that you have located the proper cell.)* The 20 cases in the upper-left-hand cell all rank low on both burnout and length of service, and we will refer to these cases as "low-lows," or "LLs."

Now, form a pair of cases by selecting one case from this cell and one from any other cell—for example, the middle cell in the table. All 15 cases in this cell are moderate on both variables and, following our practice above, can be labeled moderate-moderates or MMs. Any pair of cases formed between these two cells will be ranked the same on both variables. That is, all LLs are lower than all MMs on both variables (on X, low is less than moderate, and on Y, low is less than moderate). The total number of pairs of cases is given by multiplying the cell frequencies. So, the contribution of these two cells to the total N_s is (20)(15), or 300.

Gamma ignores all pairs of cases that are tied on either variable. For example, any pair of cases formed between the LLs and any other cell in the top row (low on burnout) or the left-hand column (low on length of service) will be tied on one variable. Also, any pair of cases formed within any cell will be tied on both X and Y. Gamma does not take into account any pair of cases formed within the same row, column, or cell. Practically, this means that, in computing N_s, we will take account only of the pairs of cases that can be formed between each cell and the cells below and to the right of it.

The computational routine for finding N_s can by summarized. *To find the total number of pairs of cases ranked the same on both variables, multiply the frequency in each cell by the total of all frequencies below and to the right of that cell.* Repeat for each cell and add the resultant products. The total of these products is N_s. This procedure is displayed below for each cell in Table 15.1. Note that none of the cells in the bottom row or the right-hand column can contribute to N_s because they have no cells below and to the right of them. Figure 15.1 shows the direction of multiplication for each of the four cells that, in a 3 × 3 table, can contribute to N_s. Computing N_s for Table 15.1:

	Contribution to N_s	
For LLs, $20(15 + 5 + 11 + 21)$	=	1040
For MLs, $6(21 + 5)$	=	156
For HLs, $4(0)$	=	0
For LMs, $10(11 + 21)$	=	320
For MMs, $15(21)$	=	315
For HMs, $5(0)$	=	0
For LHs, $8(0)$	=	0
For MHs, $11(0)$	=	0
For HHs, $21(0)$	=	0
	$N_s =$	1831

The computations above indicate that a total of 1831 pairs of cases are ranked the same on both variables. Our next step is to find the number of pairs of cases ranked differently (N_d) on both variables. *To find the total number of pairs of cases ranked in different order on the variables, multiply the frequency in each cell by the total of all frequencies below and to the left of that cell.* Note that the pattern for computing N_d is the reverse of the pattern for N_s. This time, we begin with the upper-right-hand cell (high-lows, or HLs)

FIGURE 15.1   COMPUTING N_s IN A 3 × 3 TABLE

and multiply the number of cases in the cell by the total frequency of cases below and to the left. The four cases in the upper-right-hand cell are low on Y and high on X; and, if a pair is formed with any case from this cell and any cell below and to the left, the cases will be ranked differently on the two variables. For example, if a pair is formed between any HL case and any case from the middle cell (moderate-moderates, or MMs), the HL case would be less than the MM case on Y ("low" is less than "moderate") but more than the MM case on X ("high" is greater than "moderate"). The computation of N_d is detailed below and shown graphically in Figure 15.2. In the computations, we have omitted cells that cannot contribute to N_d because they have no cells below and to the left of them.

Computing N_d for Table 15.1:

		Contribution to N_d
For HLs, $4(10 + 15 + 8 + 11)$	=	176
For MLs, $6(10 + 8)$	=	108
For HMs, $5(8 + 11)$	=	95
For MMs, $15(8)$	=	120
	$N_d =$	499

FIGURE 15.2 COMPUTING N_d IN A 3 × 3 TABLE

Application 15.1

For local political elections, do voters turn out at higher rates when the candidates for office spend more money on media advertising? Each of 177 localities has been rated as high or low on voter turnout. Also, the total advertising budgets for all candidates in each locality have been ascertained and classified as high, moderate, or low. Are these variables associated?

Voter	Expenditure			
Turnout	Low	Moderate	High	Totals
Low	35	32	17	84
High	23	27	43	93
Totals	58	59	60	177

Because both variables are ordinal in level of measurement, we will compute a gamma to summarize the strength and direction of the association. The number of pairs of cases ranked in the same order on both variables (N_s) would be

$N_s = 35(27 + 43) + 32(43) = 2450 + 1376 = 3826$

The number of pairs of cases ranked in different order on both variables (N_d) would be

$N_d = 17(27 + 23) + 32(23) = 850 + 736 = 1586$

Gamma is

$$G = \frac{N_s - N_d}{N_s + N_d}$$

$$G = \frac{3826 - 1586}{3826 + 1586}$$

$$G = \frac{2240}{5412}$$

$$G = 0.41$$

A gamma of 0.41 means that, when predicting the order of pairs of cases on voter turnout, we would make 41% fewer errors by taking candidates' advertising expenditures into account, as opposed to ignoring the latter variable. There is a moderate, positive association between these two variables.

Thus, Table 15.1 has 499 pairs of cases ranked in different order and 1831 pairs of cases ranked in the same order. The formula for computing gamma is

FORMULA 15.1

$$G = \frac{N_s - N_d}{N_s + N_d}$$

where N_s = the number of pairs of cases ranked the same as both variables
N_d = the number of pairs of cases ranked differently on the two variables

For Table 15.1, the value of gamma would be

$$G = \frac{1831 - 499}{1831 + 499}$$

$$G = \frac{1332}{2330}$$

$$G = .57$$

A gamma of .57 indicates that we would make 57% fewer errors if we predicted the order of pairs of cases on one variable from the order of pairs

READING STATISTICS 11: Bivariate Tables and Associated Statistics

The statistics associated with any bivariate table will usually be reported directly below the table itself. This information would include the name of the measure of association, its value, and (if relevant) its sign. Also, if the research involves a random sample, the results of the chi square test, including the value of the obtained chi square, the degrees of freedom in the table, and the significance (alpha level), will be reported. So the information will look something like this:

$$\lambda = 0.47$$
$$\chi^2 = 13.23 \ (2df, \ p < 0.05)$$

Note that the alpha level is reported in the "$p <$" format, which we discussed in Reading Statistics 5.

Besides simply reporting the value of these statistics, the researcher will also interpret them in the text of the article. The preceding statistics might be characterized by the statement: "The association between the variables is statistically significant and moderately strong." Again we see that researchers will avoid our rather wordy (but more careful) style of stating results. Where we might say that "a lambda of 0.47 indicates that we will reduce our errors of prediction by 47% when predicting the dependent variable from the independent variable, as opposed to predicting the dependent while ignoring the independent," the professional researcher will simply report the value of lambda and characterize that value in

a word or two (for example, "The association is moderately strong"). The researcher assumes that his or her audience is statistically literate and can make the more detailed interpretations for themselves.

Statistics in the Professional Literature

What is the relationship between exposure to TV violence and violent behavior? A group of researchers proposed the hypothesis that ". . . exposure to TV violence will be positively related to violent behavior." (p. 343) More than 2000 high school juniors and seniors were asked to name their favorite TV show and also asked to indicate their involvement in violent behavior. The violence level of the TV shows was rated, and the table below displays the relationship between the two variables.

Analyze this table carefully. Remember that there are no absolute conventions for presenting tables in the literature. In this case, the researchers have used a format that differs from the one used in this text. The independent variable (Average Violence Rating of the TV show) is in the rows, not the columns. Percentages are still calculated in the direction of the independent variable but, for this table, that means that they total 100% in the rows.

We can see that the conditional distributions do change (from row to row), so the variables are related. The chi square is not zero and is reported to

TABLE 1 VIOLENT BEHAVIOR BY AVERAGE VIOLENCE RATING OF FAVORITE TELEVISION SHOWS

Average Violence Rating	Violent Behavior						
	Low		Medium		High		
Low	409	(63%)	179	(28%)	62	(10%)	650
Medium	518	(56%)	255	(28%)	147	(16%)	920
High	280	(50%)	175	(31%)	103	(19%)	558
	1,207		609		312		2,128

Chi square = 28.85   Significant at .001   Gamma = .16

be significant at the .001 level. Gamma is .16, indicating a weak to moderate relationship between the variables. Here's what the authors have to say about this table:

A positive but low correlation (gamma = .16) was found between the average violence rating and violent behavior. (W)hile only 10 percent of the respondents who were low on the average violence rating were high on violent behavior, only 19 percent of those who were high on average violence rating were high on violent behavior. . . . Therefore, only minimal support for the . . . hypothesis was obtained.

Make sure you understand how the percentages in the table support this conclusion.

To help you interpret the direction of this relationship, remember that, in a positive relationship, low values on one variable occur together with low values on the other, and high values also occur together. For the table above, a positive relationship would mean that the bulk of the cases will lie along the diagonal from the upper left-hand cell ("low-lows") through the middle cell ("medium-mediums") to the lower right-hand cell ("high-highs").

Table and excerpts reprinted from *Social Forces* 54(2)1975. "Television Violence and Violent Behavior" by T. F. Hartnagel, J.T. Teevan, and J.J. McIntyre. Copyright © The University of North Carolina Press.

of cases on the other—as opposed to predicting order while ignoring the other variable. Length of service is associated with degree of burnout, and the relationship is positive. Knowing the respective rankings of two teachers on length of service (Case A is higher on length of service than Case B) will help us predict their ranking on burnout (we would predict that Case A will also be higher than Case B on burnout).

To use the computational routine for gamma presented above, you must arrange the table in the manner of Table 15.1, with the column variable increasing in value as you move from left to right and the row variable increasing from top to bottom. Be careful to construct your tables according to this format; if you are working with data already in table format, you may have to rearrange the table or rethink the direction of patterns. Gamma is a symmetrical measure of association; that is, the value of gamma will be the same regardless of which variable is taken as independent. *(To practice computing and interpreting gamma, see problems 15.1 to 15.12 and 15.18. Begin with some of the smaller, 2 × 2 tables such as in problems 15.1, 15.2, and 15.3 until you are comfortable with these procedures.)*

15.4 SOMER'S *d* AND KENDALL'S TAU-*b*

This section will present two new measures of association that are similar to gamma in many ways. Both are appropriate for collapsed ordinal-level variables that have been arrayed in a bivariate table. Both can be interpreted by the logic of PRE and, computationally, they are both based on a comparison of the number of pairs of cases ranked in the same order on both variables (N_s) with the number of pairs ranked differently (N_d).

The difference between these statistics and gamma lies in the way they treat pairs of cases that are tied. Gamma ignores all ties and thus represents the proportional reduction in error only for pairs of cases that are not tied. Somer's d and Kendall's tau-b, on the other hand, incorporate pairs of cases that are tied. Specifically, Somer's d includes pairs that are tied on the dependent variable but not the independent, and Kendall's tau-b includes pairs that are tied on either the dependent or the independent variable. Before addressing the statistics themselves, let us see how to compute the number of pairs of cases that are tied.

The number of pairs of cases tied on the dependent variable (T_y) is determined by finding the number of pairs of cases within each row, since all cases in the same row are, by definition, tied on the dependent variable. To find the contribution of each row to T_y, multiply each cell frequency by all frequencies to the right. For example, in the bottom row of Table 15.1, there are 8 cases in the far left-hand cell (LHs) and 32 cases (11 + 21) to the right. Hence, (8)(32) or 256 pairs of cases formed from this cell are tied on the dependent variable. Similarly, from the 11 cases in the next cell to the right (MHs), we get (11)(21), or 231, pairs of cases tied on "burnout" (not counting the pairs of cases from the first cell). The third cell in this row cannot contribute to T_y because there are no cells to the right of it. Thus, for the bottom row, the total number of pairs of cases tied on the dependent variable is 256 + 231, or 487.

Moving to the middle row, we see that there are 10 cases in the first cell and a total of 15 + 5 = 20 cases to the right. Thus, the total number of pairs of cases tied on the dependent variable that can be formed from this row is 10(20) + 15(5), or 275. In similar fashion, the contribution of the top row to T_y is 20(6 + 4) + 6(4), or 224. Adding these subtotals from all three rows, we find that T_y equals 487 + 275 + 224, or 986. In Table 15.1, a total of 986 pairs of cases are tied on the dependent variable ($T_y = 986$).

The number of pairs of cases tied on the independent variable (T_x) is found in an analogous manner by working within columns instead of rows. The contribution of each cell to the total of T_x would be found by multiplying the frequency in the cell by all frequencies below the cell. Using Table 15.1 as our example, the number of pairs of cases tied on the independent variable for the left-hand column would be 20(10 + 8) + 10(8), or 440. The contribution of the middle column to T_x would be 6(15 + 11) + 15(11), or 321. Finally, the contribution of the right-hand column to T_x would be 4(5 + 21) + 5(21), or 209. The total number of pairs of cases tied on the independent variable would be 440 + 321 + 209, or 970. In Table 15.1, 970 pairs of cases are tied on the independent variable ($T_x = 970$). As we will see below, the quantity T_y is used in the computation of Somer's d, while Kendall's tau-b uses both T_y and T_x.

The formula for Somer's d is

FORMULA 15.2

$$d = \frac{N_s - N_d}{N_s + N_d + T_y}$$

Note that this formula is the same as that for gamma except for the addition of the quantity T_y in the denominator. For all practical purposes, the inclusion of this quantity in the denominator of the equation means that the value of Somer's d will always be less than the value of gamma for the same table. For example, the gamma previously computed for Table 15.1 was .57, while Somer's d is .40:

$$d = \frac{N_s - N_d}{N_s + N_d + T_y}$$

$$d = \frac{1831 - 499}{1831 + 499 + 986}$$

$$d = \frac{1332}{3316}$$

$$d = .40$$

A Somer's d of .40 indicates a positive relationship of moderate strength. In predicting the order of pairs of cases on "burnout," we would make 40% fewer errors when predicting from the order of the pairs on the independent variable (length of service). Somer's d is a lower value than gamma because it includes pairs tied on the dependent variable in the pool of pairs of cases for which predictions will be made.

Somer's d is an asymmetric measure of association and will change values depending on which variable is taken as independent. If the row variable is taken as independent, then ties would be computed within columns instead of rows (that is, T_x becomes T_y). For Table 15.1, the values of T_x and T_y are nearly identical, and the value of d would not have changed appreciably had "burnout" been taken as the independent variable. But T_x and T_y will not always be similar, and you should exercise caution when deciding which variable to take as independent, for the value of this measure can be appreciably affected by your decision.

Kendall's tau-b differs from gamma and Somer's d in the way that ties are handled. This feature is reflected in the denominator, which incorporates not only pairs of cases tied on the dependent variable (like Somer's d) but also pairs of cases tied on the independent (T_x). The formula for Kendall's tau-b is

FORMULA 15.3

$$\text{tau-}b = \frac{N_s - N_d}{\sqrt{(N_s + N_d + T_y)(N_s + N_d + T_x)}}$$

Using Table 15.1 as an illustration once again, we would have

$$\text{tau-}b = \frac{1831 - 499}{\sqrt{(1831 + 499 + 986)(1831 + 499 + 970)}}$$

$$\text{tau-}b = \frac{1332}{\sqrt{(3316)(3300)}}$$

$$\text{tau-}b = \frac{1332}{\sqrt{10,942,800}}$$

$$\text{tau-}b = \frac{1332}{3307.99}$$

$$\text{tau-}b = .40$$

A value of .40 for Kendall's tau-b indicates that we will make 40% fewer errors in predicting the order of pairs on one variable from their order on the other variable (as opposed to predicting the order of pairs while ignoring their order on the other variable). This indicates a positive association of moderate strength.

Kendall's tau-b is a symmetrical measure of association. Its major peculiarity is that it can achieve a value of ±1.00 only in a square table (that is, a table where the number of rows equals the number of columns) and, therefore, tau-b would probably not be the measure of first choice in any rectangular table. Since this measure incorporates tied pairs of cases in the denominator, it will tend to be lower in value than gamma for the same table.

In summary, for the situation where we have two ordinally measured variables arrayed in table format, we have three different ways of measuring the strength of the association. The three measures differ from each other primarily in the way they treat tied pairs of cases. Gamma ignores all ties; Somer's d includes only pairs tied on the dependent variable; and tau-b includes pairs tied on either the dependent or the independent variable.

Which of these three is "best"? Unfortunately, there is no simple answer to this question. Gamma is probably the most commonly used of the three but is certainly not equally preferable in all possible situations. When a sharp distinction can be made between independent and dependent variables, Somer's d is in many ways preferable to gamma, since it allows for pairs of cases that are different on the independent variable but tied on the dependent and thus includes more information (that is, more pairs of cases). Kendall's tau-b includes the most information about the relationship (pairs tied on either X or Y) but should not be used for tables that are not square. At any rate, perhaps the most important point to be made is that, in terms of the existence, strength, and direction of an association, the three measures will generally result in the same conclusion. For Table 15.1, for example, all three measures indicate a moderate, positive association between burnout and length of service. *(For practice in computing and interpreting Somer's d and Kendall's tau-b, see problems 15.1 to 15.12 and 15.18. If possible, choose problems for which you already have a value for gamma—this will give you values for N_s and N_d. All you will have to add is the appropriate number of tied cases.)*

15.5 SPEARMAN'S RHO (r_s)   To this point, we have considered ordinal variables that have a limited number of categories (possible values) and are presented in tables. However, many ordinal-level variables are designed with a broad range of scores and many distinct values. Such data may be collapsed into a few broad categories (such as high, moderate, low), organized into a bivariate table, and analyzed with the measures introduced in the previous sections. Collapsing scores in this manner may be beneficial and desirable in many instances, but some important distinctions between cases may be obscured or lost as a consequence.

For example, suppose a researcher wished to test the claim that jogging is beneficial not only physically but also psychologically. Specifically, suppose that the researcher wanted to know if people who are deeply involved in jogging experience an enhanced sense of self-esteem. To this end, 10 joggers are measured on two scales, the first measuring involvement in jogging and the other measuring self-esteem. The scores of the 10 joggers on the two scales are reported in Table 15.2.

These data could be collapsed and a bivariate table produced. We could, for example, dichotomize both variables to create only two values (high and low) for both variables. Although collapsing scores in this way is certainly legitimate and often necessary,* two difficulties with this practice must be noted. First, the scores seem continuous and there are no obvious or natural division points in the distribution that would allow us to distinguish, in a non-arbitrary fashion, between high scores and low ones. Second, and more important, grouping these cases into broader categories will cause us to lose information. That is, if both Wendy and Debbie are placed in the category "high" on involvement, we would lose sight of the fact that they had different scores on this variable. If differences like this are important and meaningful, then we should opt for a measure of association that permits the retention of as much detail and precision in the scores as possible.

Spearman's rho (r_s) is a measure of association for ordinal-level variables that have a broad range of many different scores and few ties between cases on either variable. Scores on ordinal-level variables cannot, of course, be manipulated mathematically except for judgments of "greater than" or "less than." To compute Spearman's rho, cases are first ranked from high to low

---

*For example, collapsing scores may be advisable when the researcher is not certain that fine distinctions between scores are meaningful.

TABLE 15.2 THE SCORES OF 10 SUBJECTS ON INVOLVEMENT IN JOGGING AND A MEASURE OF SELF-ESTEEM

Joggers	Involvment in Jogging (X)	Self-Esteem (Y)
Wendy	18	15
Debbie	17	18
Phyllis	15	12
Frank	12	16
Evelyn	10	6
Tricia	9	10
Christy	8	8
Patsy	8	7
Marsha	5	5
Siegfried	1	2

TABLE 15.3 COMPUTING SPEARMAN'S RHO

	Involvement (X)	Rank	Self-Image (Y)	Rank	D	D^2
Wendy	18	1	15	3	−2	4
Debbie	17	2	18	1	1	1
Phyllis	15	3	12	4	−1	1
Frank	12	4	16	2	2	4
Evelyn	10	5	6	8	−3	9
Tricia	9	6	10	5	1	1
Christy	8	7.5	8	6	1.5	2.25
Patsy	8	7.5	7	7	.5	.25
Marsha	5	9	5	9	0	0
Siegfried	1	10	2	10	0	0
					$\Sigma D = 0$	$\Sigma D^2 = 22.5$

on each variable and then the ranks (not the scores) are manipulated to produce the final measure. Table 15.3 displays the original scores and the rankings of the cases on both variables.

To rank the cases, first find the highest score on each variable and assign it rank 1. Wendy has the high score on X (18) and is thus ranked number 1. Debbie, on the other hand, is highest on Y and is ranked first on that variable. All other cases are then ranked in descending order of scores. If any cases have the same score on a variable, assign them the average of the ranks they would have used up had they not been tied. Christy and Patsy have identical scores of 8 on involvement. Had they not been tied, they would have used up ranks 7 and 8. The average of these two ranks is 7.5, and this average of

---

Application 15.2

Five cities have been rated on an index that measures the quality of life. Also, the percentage of the population that has moved into each city over the past year has been determined. Have cities with higher quality-of-life scores attracted more new residents? The table below summarizes the scores, ranks, and differences in ranks for each of the five cities.

City	Quality of Life	Rank	Percentage of New Residents	Rank	D	D²
A	30	1	17	1	0	0
B	25	2	14	3	−1	1
C	20	3	15	2	1	1
D	10	4	3	5	−1	1
E	2	5	5	4	1	1
					0	4

Spearman's rho for these variables is

$$r_s = 1 - \frac{6\Sigma D^2}{N(N^2 - 1)}$$

$$r_s = 1 - \frac{(6)(4)}{5(25 - 1)}$$

$$r_s = 1 - \frac{24}{120}$$

$$r_s = 1 - 0.20$$

$$r_s = 0.80$$

These variables have a strong, positive association. The higher the quality-of-life score, the greater the percentage of new residents. The value of r_s^2 is 0.64 ($0.80^2 = 0.64$), which indicates that we will make 64% fewer errors when predicting rank on one variable from rank on the other, as opposed to ignoring rank on the other variable.

---

used ranks is assigned to all tied cases. (For example, had Marsha also had a score of 8, three ranks—7, 8, and 9—would have been used, and all three tied cases would have been ranked eighth.)

The formula for Spearman's rho is

FORMULA 15.4

$$r_s = 1 - \frac{6\Sigma D^2}{N(N^2 - 1)}$$

where D^2 = the sum of the differences in ranks, the quantity squared

To compute ΣD^2, the rank of each case on Y is subtracted from its rank on X (D is the difference between rank on Y and rank on X). A column has been provided in Table 15.3 so that these differences may be recorded on a case-by-case basis. Note that the sum of this column (ΣD) is 0. That is, the negative differences in rank are equal to the positive differences, as will always be the case. Thus, you should find the total of this column as a check on your computations to this point. If the ΣD is not equal to 0, you have made a mistake either in ranking the cases or in subtracting the differences.

In the column headed D^2, each difference is squared to eliminate negative signs. The sum of this column is ΣD^2, and this quantity is entered directly into the formula. For our sample problem:

$$r_s = 1 - \frac{6\Sigma D^2}{N(N^2 - 1)}$$

$$r_s = 1 - \frac{(6)(22.5)}{10(100 - 1)}$$

$$r_s = 1 - \frac{135}{990}$$

$$r_s = 1 - .14$$

$$r_s = .86$$

Spearman's rho is an index of the strength of association between the variables; it ranges from 0 (no association) to ±1.00 (perfect association). A perfect positive association ($r_s = +1.00$) would exist if there were no disagreements in ranks between the two variables (if cases were ranked in exactly the same order on both variables). A perfect negative relationship ($r_s = -1.00$) would exist if the ranks were in perfect disagreement (if the case ranked highest on one variable were lowest on the other, and so forth).

The Spearman's rho above indicates a strong, positive relationship between these two variables. The respondents who were highly involved in jogging also ranked high on self-image. These results would be supportive of claims regarding the psychological benefits of jogging.

Spearman's rho is an index of the relative strength of a relationship, and values between 0 and ±1.00 have no direct interpretation. However, if the value of rho is squared, a PRE interpretation is possible. Rho squared (r_s^2) represents the proportional reduction in errors of prediction when predicting rank on one variable from rank on the other variable, as compared to predicting rank while ignoring the other variable. In the example above, r_s was .86 and r_s^2 would be .74. Thus, our errors of prediction would be reduced by 74% if, when predicting the rank of a subject on self-image, the rank of the subject on involvement in jogging were taken into account. *(For practice in computing and interpreting Spearman's rho, see problems 15.13 to 15.17. Problem 15.13 has the fewest number of cases and is probably a good choice for a first attempt at these procedures.)*

15.6 TESTING THE NULL HYPOTHESIS OF "NO ASSOCIATION" WITH GAMMA AND SPEARMAN'S RHO

Whenever a researcher is working with EPSEM, or random, samples, he or she will need to ascertain if the sample findings can be generalized to the population. In Part II of this text, we considered various ways that information taken from samples—for example, the difference between two sample means—could be generalized to the populations from which the samples were drawn. A test of the null hypothesis, regardless of the form or specific test used, asks essentially if the patterns (or differences, or relationships) that have been observed in the samples can be assumed to exist in the population.

The various measures of association can also be tested for their statistical significance. Like means and proportions, measures of association are descriptive statistics that supply the researcher with certain kinds of information about the nature of the data at hand. When data concerning the relationship between two variables have been collected from a random sample, we will not only need to measure the existence, strength, and direction of the association, but we will also want to know if we can assume that the variables are related in the population.

For nominal-level variables, the statistical significance of a relationship is usually judged by the chi square test. Chi square tests could also be conducted on tables displaying the relationship between ordinal-level variables. However, chi square tests deal with the probability that the observed cell frequencies occurred by chance alone and is therefore not a direct test of the measure of association itself.

Below, we will consider tests of significance for gamma and for Spearman's rho. In both cases, the null hypothesis will state that there is no association between the variables in the population and that, therefore, the population values for both sample measures are 0. The population values will be denoted by the Greek letters gamma (γ) and rho (ρ_s). For both measures, the test procedures will be organized around the familiar five-step model (see Chapter 8).

To illustrate the test of significance for gamma, we will use Table 15.1, where gamma was .57.

Step 1. Making Assumptions. When sample size is greater than 10, the sampling distribution of all possible sample gammas can be assumed to be normal in shape.

Model: Random sampling
Level of measurement is ordinal
Sampling distribution is normal

Step 2. Stating the Null Hypothesis.

$$H_0: \gamma = 0.0$$
$$(H_1: \gamma \neq 0.0)$$

Step 3. Selecting the Sampling Distribution and Establishing the Critical Region. For samples of 10 or more, the Z distribution (Appendix A) can be used to find areas under the sampling distribution:

Sampling distribution = Z distribution
Alpha = .05
Z(critical) = ± 1.96

Step 4. Computing the Test Statistic.

$$Z(\text{obtained}) = G\sqrt{\frac{N_s + N_d}{N(1 - G^2)}}$$

$$Z(\text{obtained}) = .57\sqrt{\frac{1831 + 499}{100(1 - .33)}}$$

$$Z(\text{obtained}) = .57\sqrt{\frac{2330}{100(.67)}}$$

$$Z(\text{obtained}) = (.57)\sqrt{34.78}$$

$$Z(\text{obtained}) = (.57)(5.90)$$

$$Z(\text{obtained}) = 3.36$$

Step 5. **Making a Decision.** Comparing the Z (obtained) with the Z (critical):

$$Z(\text{obtained}) = 3.36$$
$$Z(\text{critical}) = \pm 1.96$$

We see that the null hypothesis can be rejected. The sample gamma is unlikely to have occurred by chance alone, and we may conclude that these variables are related in the population from which the sample was drawn. *(For practice in conducting and interpreting the test of significance for gamma, see problems 15.2, 15.4, 15.7, 15.10 to 15.12, and 15.18.)*

Spearman's rho can be tested in similar fashion. The null hypothesis is again a statement that the population value (ρ_s) is actually 0 and, therefore, that the value of the sample Spearman's rho (r_s) is the result of mere random chance. When the number of cases in the sample is 10 or more, the sampling distribution of Spearman's rho approximates the t distribution, and we will use this distribution to conduct the test. To illustrate, the Spearman's rho computed in Section 15.5 will be used.

Step 1. **Making Assumptions.**

Model: Random sampling
Level of measurement is ordinal
Sampling distribution is normal

Step 2. **Stating the Null Hypothesis.**

$$H_0: \rho_s = 0.00$$
$$(H_1: \rho_s \neq 0.00)$$

Step 3. Selecting the Sampling Distribution and Establishing the Critical Region.

$$\text{Sampling distribution} = t \text{ distribution}$$
$$\text{Alpha} = .05$$
$$\text{Degrees of freedom} = N - 2 = 8$$
$$t(\text{critical}) = \pm 2.306$$

Step 4. Computing the Test Statistic.

$$t(\text{obtained}) = r_s \sqrt{\frac{N-2}{1-r_s^2}}$$

$$t(\text{obtained}) = .86 \sqrt{\frac{8}{1-.74}}$$

$$t(\text{obtained}) = .86 \sqrt{\frac{8}{.26}}$$

$$t(\text{obtained}) = .86\sqrt{30.77}$$

$$t(\text{obtained}) = (.86)(5.55)$$

$$t(\text{obtained}) = 4.77$$

Step 5. Making a Decision. Comparing the test statistic with the critical region:

$$t(\text{obtained}) = 4.77$$
$$t(\text{critical}) = \pm 2.306$$

We see that the null hypothesis can be rejected. We may conclude, with a .05 chance of making an error, that the variables are related in the population from which the samples were drawn. *(For practice in conducting and interpreting the test of significance for Spearman's rho, see problems 15.13, 15.16, and 15.17.)*

SUMMARY

1. Three measures of association for variables with collapsed ordinal scales were covered: gamma, Somer's d, and Kendall's tau-b. Spearman's rho is appropriate for measuring association between "continuous" ordinal variables. All four measures summarize the overall strength and direction of the association between the variables.

2. Gamma is a PRE-based measure that shows the improvement in our ability to predict the order of pairs of cases on one variable from the order of pairs of cases on the other variable, as opposed to ignoring the order of the pairs of cases on the other variable.

3. Gamma ignores all pairs of cases that are tied on either or both variables. Somer's d incorporates pairs of cases that are tied on the dependent but not on the independent. Kendall's tau-*b* incorporates all cases tied on either variable but should be computed only for "square" tables. All three measures can be interpreted by the logic of PRE.

4. Spearman's rho is computed from the ranks of the scores of the cases on two "continuous" ordinal variables and, when squared, can be interpreted by the logic of PRE.

5. Both gamma and Spearman's rho should be tested for their statistical significance when computed for a random sample drawn from a defined population. The null hypothesis is that the variables are not related in the population, and the test can be organized by using the familiar five-step model.

SUMMARY OF FORMULAS

Gamma   15.1 $$G = \frac{N_s - N_d}{N_s + N_d}$$

Somer's *d*   15.2 $$d = \frac{N_s - N_d}{N_s + N_d + T_y}$$

Kendall's tau-*b*   15.3

$$\text{tau-}b = \frac{N_s - N_d}{\sqrt{(N_s + N_d + T_y)(N_s + N_d + T_x)}}$$

Spearman's rho   15.4 $$r_s = 1 - \frac{6\Sigma D^2}{N(N^2 - 1)}$$

GLOSSARY

Gamma. A measure of association appropriate for variables measured with "collapsed" ordinal scales that have been organized into table format; G is the symbol for any sample gamma; γ is the symbol for any population gamma.

Kendall's tau-*b*. A measure of association appropriate for variables measured at the ordinal level that have been organized into table format. Especially appropriate for square tables.

N_s. The number of pairs of cases ranked in the same order on two variables.

N_d. The number of pairs of cases ranked in different order on two variables.

Somer's *d*. A measure of association appropriate for variables measured at the ordinal level that have been organized into table format.

Spearman's rho. A measure of association appropriate for ordinally measured variables that are "continuous" in form; r_s is the symbol for any sample Spearman's rho; ρ_s is the symbol for any population Spearman's rho.

T_x. The number of pairs of cases tied on the independent variable.

T_y. The number of pairs of cases tied on the dependent variable.

PROBLEMS

For problems 15.1 to 15.14 and 15.18 calculate percentages for the bivariate tables as described in Chapter 13. Use the percentages to help analyze the strength and direction of the association. As an alternative to gamma, Somer's *d* and Kendall's tau-*b* can be computed for problems 15.1 to 15.14 and 15.18. Remember that both of these measures will usually

be lower in value than gamma because of the way they handle ties.

15.1 SOC A small sample of non-English-speaking immigrants to the United States has been interviewed about their levels of assimilation. Is the pattern of adjustment affected by length of residence in the United States? For each table compute gamma and summarize the relationship in terms of strength and direction. *(HINT: In 2 × 2 tables, only two cells can contribute to N_s or N_d. To compute N_s, multiply the number of cases in the upper-left-hand cell by the number of cases in the lower-right-hand cell. For N_d, multiply the number of cases in the upper-right-hand cell by the number of cases in the lower-left-hand cell. In other words, multiply the diagonal cells in the table.)*

a. Facility in English:

English Facility	Length of Residence		Totals
	Less than Five Years (1)	More than Five Years (2)	
Low (1)	20	10	30
High (2)	5	15	20
Totals	25	25	50

b. Total family income:

Income	Length of Residence		Totals
	Less than Five Years (1)	More than Five Years (2)	
Below national average (1)	20	10	30
Above national average (2)	5	15	20
Totals	25	25	50

c. Extent of contact with country of origin:

Contact	Length of Residence		Totals
	Less than Five Years (1)	More than Five Years (2)	
Rare (1)	5	20	25
Frequent (2)	20	5	25
Totals	25	25	50

15.2 Compute gamma for the tables presented in problems 12.2, 12.6, and 12.8. Since these tables are based on random samples, test the gammas you computed for significance.

15.3 [PA] Compute gamma for the table presented in problem 13.1. If you computed a nominal measure of association for this table (problem

14.2), compare the measures of association. Are they similar in value? Do they characterize the strength for the association in the same way? What information about the relationship does gamma provide that is not available from nominal measures of association?

15.4 [CJ] A random sample of 150 cities has been classified as small, medium, or large by population and as high or low on crime rate. Is there a relationship between city size and crime rate?

Crime Rate	Size			Totals
	Small	Medium	Large	
Low	21	17	8	46
High	29	33	42	104
Totals	50	50	50	150

a. Describe the strength and direction of the relationship.

b. Is the relationship significant?

15.5 [SOC] Some research has shown that families vary by how they socialize their children to sports, games, and other leisure-time activities. In middle-class families, such activities are carefully monitored by parents and are, in general, dominated by adults (for example, Little League baseball). In working-class families, children more often organize and initiate such activities themselves, and parents are much less involved (for example, sandlot or playground baseball games). Are the data below consistent with these findings? Summarize your conclusions in a few sentences.

As a Child, Did You Play Mostly Organized or Sandlot Sports?	Social Class Background	
	White-Collar	Blue-Collar
Organized	155	123
Sandlot	101	138
Totals	256	261

15.6 [PA] Do municipal employees become more or less efficient as a function of longevity? Are the "old guard" better at their jobs than the younger, less experienced workers? A city manager has gathered the yearly efficiency ratings for all municipal employees along with data on longevity. Does the table below suggest any relationship? Summarize your conclusions in a few sentences.

Efficiency	Longevity		Totals
	Low	High	
Low	54	25	79
Moderate	149	254	403
High	257	387	644
Totals	460	666	1126

15.7 [PA] All applicants for municipal jobs in Shinbone, Kansas, are given an aptitude test, but the test has never been evaluated to see if test scores are in any way related to job performance. The following table reports aptitude test scores and job performance ratings for a random sample of 75 city employees.

Efficiency Ratings	Test Scores			Totals
	Low	Moderate	High	
Low	11	6	7	24
Moderate	9	10	9	28
High	5	9	9	23
Totals	25	25	25	75

a. Are these two variables associated? Describe the strength and direction of the relationship in a sentence or two.
b. Is gamma statistically significant?
c. Should the aptitude test continue to be administered? Why or why not?

15.8 [SW] A sample of children has been observed and rated for symptoms of depression. Their parents have been rated for authoritarianism. Is there any relationship between these variables? Write a few sentences stating your conclusions.

Symptoms of Depression	Authoritarianism			Totals
	Low	Moderate	High	
Few	7	8	9	24
Some	15	10	18	43
Many	8	12	3	23
Totals	30	30	30	90

15.9 [SOC] Are prejudice and level of education related? State your conclusion in a few sentences.

Prejudice	Level of Education				Totals
	Elementary School	High School	Some College	College Graduate	
Low	48	50	61	42	201
High	45	43	33	27	148
Totals	93	93	94	69	349

15.10 [SOC] In a recent survey, a random sample of respondents was asked to indicate how happy they were with their situations in life. Are their responses related to income level?

Happiness	Income			Totals
	Low	Moderate	High	
Not happy	101	82	36	219
Pretty happy	40	227	100	367
Very happy	216	198	203	617
Totals	357	507	339	1203

a. Describe the strength and direction of the relationship.
b. Is the relationship significant?

15.11 [SW] Over the past few years, a large number of Indochinese refugees have settled in and around greater metropolitan Shinbone, Kansas. To better identify which of these immigrants might benefit from special programs designed to help them adjust to American society, the Social Service Bureau has surveyed a random sample of 178 refugees. As a part of the survey, each respondent was asked to judge subjectively his or her overall degree of success in coping with the new environment. The table below reports the relationship between the "subjective success score" and the number of members in the household of the respondent. Is "living arrangement" predictive of "success in adjustment?"

Problems of Adjustment	Size of Household				Totals
	1	2	3–5	More than 5	
Few	2	12	15	7	36
Moderate	4	27	27	15	73
Difficult	7	25	10	10	52
Severe	10	4	2	1	17
Totals	23	68	54	33	178

a. Is the relationship significant?

b. What is the strength and direction of the relationship?

c. Do these results suggest a client population needing special programs?

15.12 [SOC] The table below reports the scores of a random sample of 300 undergraduates on a measure of political alienation and a measure of the extent of their use of illegal drugs.

Drug Use	Political Alienation				
	Low	Moderately Low	Moderately High	High	Totals
Nonusers	30	15	8	10	63
Rare	25	25	12	20	82
Occasional	20	20	20	25	85
Frequent	10	10	15	35	70
Totals	85	70	55	90	300

a. Write a sentence describing this relationship in terms of strength and direction.

b. Conduct and interpret a test of significance on gamma.

15.13 [SOC] A random sample of eleven neighborhoods in Shinbone, Kansas, have been rated by an urban sociologist on a "quality-of-life" scale (which includes measures of affluence, availability of medical care, and recreational facilities) and a social cohesion scale. The results are presented below in scores. Higher scores indicate higher "quality of life" and greater social cohesion.

Neighborhood	Quality of Life	Social Cohesion
Happy Acres	17	8.8
North End	40	3.9
Brentwood	47	4.0
Rolling Meadows	90	3.1
Midtown	35	7.5
Bliss Park	52	3.5
Church View	23	6.3
Hidden Valley	67	1.7
College Park	65	9.2
Beaconsdale	63	3.0
Riverview	100	5.3

a. Are the two variables associated? What is the strength and direction of the association? Summarize the relationship in a sentence or two. (*HINT: Don't forget to square the value*

of Spearman's rho for a PRE interpretation.)

b. Conduct a test of significance for this relationship. Summarize your findings.

15.14 [SW] Several years ago, a job training program began, and a team of social workers screened the candidates for suitability for employment. Now the screening process is being evaluated, and the actual work performance of these same individuals has been rated. Did the screening process work? Is there a relationship between the original scores and performance evaluation on the job?

Case	Original Score	Performance Evaluation
A	17	78
B	17	85
C	15	82
D	13	92
E	13	75
F	13	72
G	11	70
H	10	75
I	10	92
J	10	70
K	9	32
L	8	55
M	7	21
N	5	45
O	2	25

15.15 [SOC] Below are the scores of 15 nations on a measure of ethnic diversity and a measure of civil strife (such as riots, demonstrations, and assassinations). Are these variables related? Do ethnically diverse nations experience more strife?

Nation	Ethnic Diversity	Strife
A	93	126
B	85	125
C	82	478
D	80	502
E	75	100
F	60	7
G	45	78
H	43	75
I	42	70
J	39	126
K	29	7
L	25	52
M	15	48
N	5	48
O	0	33

15.16 SOC Twenty ethnic, racial, or national groups were rated by a random sample of white and black students on a Social Distance Scale. Lower scores represent less social distance and less prejudice. How similar are these rankings? Is the relationship statistically significant?

	Average Social Distance Scale Scores	
Group	White Students	Black Students
1. White Americans	1.2	2.6
2. English	1.4	2.9
3. Canadians	1.5	3.6
4. Irish	1.6	3.6
5. Germans	1.8	3.9
6. Italians	1.9	3.3
7. Norwegians	2.0	3.8
8. American Indians	2.1	2.7
9. Spanish	2.2	3.0
10. Jews	2.3	3.3
11. Poles	2.4	4.2
12. Black Americans	2.4	1.3
13. Japanese	2.8	3.5
14. Mexicans	2.9	3.4
15. Koreans	3.4	3.7
16. Russians	3.7	5.1
17. Arabs	3.9	3.9
18. Vietnamese	3.9	4.1
19. Turks	4.2	4.4
20. Iranians	5.3	5.4

15.17 PA For years the city of Shinbone, Kansas, has been appropriating funds to help police officers defray the cost of college courses. Now the city council wants to know if this college experience has produced better police officers. Below, the number of college credits and efficiency ratings are reported for a random sample of 20 police officers. What can you tell the city council? (Efficiency ratings can range from a high score of 20 to a low of 0.)

Officer	Number of College Credits	Efficiency Ratings
A	124	17
B	120	19
C	100	18
D	90	16
E	89	15
F	75	20
G	70	14
H	60	12

Officer	Number of College Credits	Efficiency Ratings
I	60	13
J	59	11
K	45	6
L	30	5
M	24	8
N	21	9
O	18	2
P	12	0
Q	9	10
R	6	3
S	0	4
T	0	1

15.18 SOC In problems 13.15 and 14.15, we looked at the relationships between five dependent variables and, respectively, political ideology and sex. In this exercise, we'll use income (INCOME91) as an independent variable and assess its relationship with this set of variables. Income has been collapsed into three categories to make it suitable for tabular analysis. For each table, calculate percentages and gamma. Describe the strength, direction, and statistical significance of each relationship in a few sentences. *Be careful in interpreting direction.* Data are from the 1996 General Social Survey, which is summarized in Appendix G.

a. Support for the legal right to an abortion (ABANY) by income:

Right to Abortion?	Less than $22,500	$22,500 to $50,000	More than $50,000	Totals
Yes	108	126	134	368
No	159	169	113	441
Totals	267	295	247	809

b. Support for capital punishment (CAPPUN) by income:

Capital Punishment?	Less than $22,500	$22,500 to $50,000	More than $50,000	Totals
Favor	270	338	280	888
Oppose	97	101	83	281
Totals	367	439	363	1169

c. Approval of suicide for people with an incurable disease (SUICIDE1) by income:

Right to Suicide?	Less than $22,500	$22,500 to $50,000	More than $50,000	Totals
Yes	149	198	185	532
No	112	108	69	289
Totals	261	306	254	821

d. Support for sex education in public schools (SEXEDUC) by income:

Sex Education?	Less than $22,500	$22,500 to $50,000	More than $50,000	Totals
Favor	228	278	229	735
Oppose	41	31	26	98
Totals	269	309	255	833

e. Support for legalization of marijuana (GRASS) by income:

Legalize Marijuana?	Less than $22,500	$22,500 to $50,000	More than $50,000	Totals
Legal	64	91	66	221
Not legal	180	219	168	567
Totals	244	310	234	788

SPSS for Windows

Using *SPSS for Windows* to Produce Ordinal-Level Measures of Association

DEMONSTRATION 15.1 Do Sexual Attitudes Vary by Age? Another Look

In Demonstration 13.3, we used percentages to look at the relationships between *premarsx* (attitude toward premarital sex) and recoded *age*. Let's reexamine this relationship and see if the ordinal measures of association introduced in this chapter can add any new information. We will once again use the **Crosstabs** program to calculate the statistics. The independent variable was *ager* (recoded age), and it will be placed in the columns. If you no longer have access to the recoded version of this variable, follow the directions in Demonstration 13.3. The dependent variable *premarsx* will go in the rows. On the **Crosstabs** dialog window, click the **Statistics** button and request gamma, Somer's *d*, and Kendall's tau-*b*. Don't forget to click the **Cells** button and request column percentages. Click **OK**, and the output should look like this:

Crosstabs

Case Processing Summary

	Cases					
	Valid		Missing		Total	
	N	Percent	N	Percent	N	Percent
SEX BEFORE MARRIAGE *AGER	951	65.5%	501	34.5%	1452	100.0%

SEX BEFORE MARRIAGE* AGER Crosstabulation

			AGER			
			1.00	2.00	3.00	Total
SEX BEFORE MARRIAGE	ALWAYS WRONG	Count	51	68	107	226
		Column %	16.1%	22.0%	32.8%	23.8%
	ALMOST ALWAYS	Count	28	20	44	92
		Column %	8.9%	6.5%	13.5%	9.7%
	SOMETIMES WRONG	Count	73	70	77	220
		Column %	23.1%	22.7%	23.6%	23.1%
	NOT WRONG AT ALL	Count	164	151	98	413
		Column %	51.9%	48.9%	30.1%	43.4%
Total		Count	316	309	326	951
		Column %	100.0%	100.0%	100.0%	100.0%

Directional Measures

			Value	Asymp. Std. Error[a]	Approx. T[b]	Approx. Sig.
Ordinal by Ordinal	Somers' d	Symmetric	−.178	.027	−6.544	.000
		SEX BEFORE MARRIAGE Dependent	−.182	.028	−6.544	.000
		AGER Dependent	−.175	.027	−6.544	.000

a. Not assuming the null hypothesis.
b. Using the asymptotic standard error assuming the null hypothesis.

Symmetric Measures

		Value	Asymp. Std. Error[a]	Approx. T[b]	Approx. Sig.
Ordinal by Ordinal	Kendall's tau-b	−.178	.027	−6.544	.000
	Gamma	−.260	.039	−6.544	.000
N of Valid Cases		951			

a. Not assuming the null hypothesis.
b. Using the asymptotic standard error assuming the null hypothesis.

Somer's *d* is reported in the Directional Measures output block and gamma and Kendall's tau-*b* are reported in the Symmetric Measures block. The three measures range in value from about −.18 for Kendall's tau-*b* to −.26 for gamma. Gamma has the highest value—as it usually will—but the three measures are comparable and consistent with each other. All three indicate that the relationship between *premarsx* and *ager* is weak to moderate in strength and negative in direction. Older respondents (a score of 3 on *ager*) were most opposed to premarital sex (they had the highest percentage who felt it was "always wrong"), and younger respondents were least opposed (they had the highest percentage who felt it was "not wrong at all"). As age increases, support of premarital sex (the percentage who say it's "not wrong at all") decreases.

Which of the three measures should we choose? The table is not "square," so we should not use tau-*b*. The other two measures are about equally logical and appropriate. Whichever you choose to report, the conclusion will be that the association is weak to moderate in strength and negative.

DEMONSTRATION 15.2 A Note about the Direction of Ordinal Relationships

Let's pause to briefly consider the direction of relationships between ordinal-level variables. Direction for ordinal variables is a surprisingly tricky matter, mostly because of the arbitrary nature of coding at the ordinal level. We usually think of *larger* scores as indicating *more* of the quantity being measured and, for every interval-ratio level variable I can think of, this pattern will be true. For ordinal variables, however, larger scores may indicate *less* of the quantity because the codes are arbitrary. Depending on how you code the values, a high score on a scale measuring, say, prejudice might indicate great prejudice or its complete absence. Looking at the table for *premarsx* and *ager* in Demonstration 15.1, remember that a negative gamma means that *high* values on one variable are associated with *low* values on the other. This means, for example, that those who scored a 1 (the lowest possible value) on *ager* tended to score a 4 (the highest possible value) on *premarsx*. You may think of a 4 on *premarsx* as representing "high support for premarital sex" or "low opposition to premarital sex." A negative gamma *always* means that the scores of the variables are inversely related, but this does *not* necessarily mean that the underlying relationship is truly negative. Always inspect tables carefully to make sure that you are interpreting the direction of the relationship properly.

DEMONSTRATION 15.3 Interpreting the Direction of Relationships: Are Prestigious Jobs More Satisfying?

What's the relationship between prestige of occupation and job satisfaction? Some would argue that high-prestige jobs are rewarding in a variety of ways (besides simply income) and that people in such jobs would express great satisfaction with their work. Others might argue an opposing point of view: because people in lower-prestige jobs have less pressure and fewer responsibilities, they will experience high levels of satisfaction. In an attempt to resolve the debate, we'll run a **Crosstabs** on *prestg80* and *satjob*. I first found the median on *prestg80* with the **Frequencies** command and recoded the variable into a dichotomy. I also recoded *satjob* into two approximately equal categories. Don't forget to recode "into different variable" and give the recoded variables new names (I used *rprest* and *rsat*). The recoding instructions for *prestg80* are

$$0 \text{ thru } 42 \rightarrow 1$$
$$43 \text{ thru } 86 \rightarrow 2$$

For *satjob*, I left the score of 1 alone and collapsed scores 2–4:

$$1 = 1$$
$$2 \text{ thru } 4 \rightarrow 2$$

Use **Crosstabs** once again with recoded *prestg80* as the independent (column) variable and recoded *satjob* as the dependent variable. Request all three measures of association, chi square, and column percentages. The output should look like this:

Crosstabs

Case Processing Summary

	Cases					
	Valid		Missing		Total	
	N	Percent	N	Percent	N	Percent
RSAT* RPREST	1151	79.3%	301	20.7%	1452	100.0%

RSAT* RPREST Crosstabulation

			RPREST		
			1.00	2.00	Total
RSAT	1.00	Count	235	288	523
		Column %	39.8%	51.4%	45.4%
	2.00	Count	356	272	628
		Column %	60.2%	48.6%	54.6%
Total		Count	591	560	1151
		Column %	100.0%	100.0%	100.0%

Chi-Square Tests

	Value	df	Asymp. Sig. (2-sided)	Exact Sig. (2-sided)	Exact Sig. (1-sided)
Pearson Chi-Square	15.783[b]	1	.000		
Continuity Correction[a]	15.316	1	.000		
Likelihood Ratio	15.815	1	.000		
Fisher's Exact Test				.000	.000
Linear-by-Linear Association	15.769	1	.000		
N of Valid Cases	1151				

a. Computed only for a 2 × 2 table.
b. 0 cells (.0%) have expected count less than 5. The minimum expected count is 254.46.

Directional Measures

			Value	Asymp. Std. Error[a]	Approx. T[b]	Approx. Sig.
Ordinal by Ordinal	Somers' d	Symmetric	−.117	.029	−3.998	.000
		RSAT Dependent	−.117	.029	−3.998	.000
		RPREST Dependent	−.118	.029	−3.998	.000

a. Not assuming the null hypothesis.
b. Using the asymptotic standard error assuming the null hypothesis.

Symmetric Measures

		Value	Asymp. Std. Error[a]	Approx. T[b]	Approx. Sig.
Ordinal by Ordinal	Kendall's tau-b	-.117	.029	-3.998	.000
	Gamma	-.232	.056	-3.998	.000
N of Valid Cases		1151			

a. Not assuming the null hypothesis.
b. Using the asymptotic standard error assuming the null hypothesis.

This relationship is statistically significant (the significance of the chi square is less than .05), weak to moderate in strength (the measures range from −.12 to −.23), and all three measures are negative. As prestige increases, job satisfaction decreases. Right? Wrong. Look at the codes for *satjob*. A *higher* score indicates a *lower* level of satisfaction. So, "high" (or a score of 2) on *prestg80* is associated with "high" (or a score of 1) on *satjob*. In spite of the negative signs for all three measures, this is a *positive* relationship, and job satisfaction increases with prestige. I know—it shouldn't be this complicated. Reread Demonstration 15.2 and try to make the best of it.

Exercises

15.1 Follow up on Demonstration 15.1 with some new ordinal-level independent variables. Among other possibilities, consider *degree* or *class* as independent variables. Do sexual attitudes vary by education or social class? Other possibilities include *income91*, *prestg80*, or *educ* (these would have to be recoded), *polviews*, or *attend*.

15.2 Following the procedure in Demonstration 15.3, use other measures of social class besides prestige (*class* or *degree*) as independent variables in the relationship with *satjob*. Summarize the strength and direction of the relationships in a few sentences. Are these relationships stronger or weaker than those with recoded *prestg80*? Be careful in interpreting the direction of the relationships.

15.3 Examine the relationship between recoded *polviews* and recoded *age*. Does conservatism increase with age? Interpret the strength and direction of the relationship.

16

Association Between Variables Measured at the Interval-Ratio Level

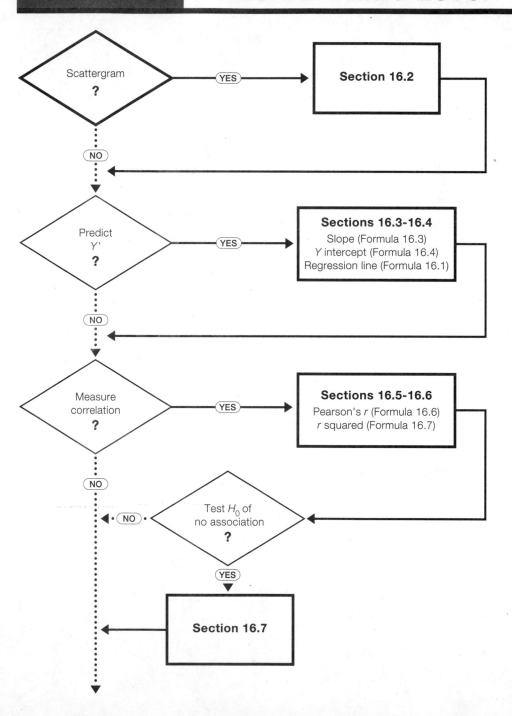

16.1 INTRODUCTION

This chapter presents a set of statistical techniques for analyzing the association (or correlation) between variables measured at the interval-ratio level.* As we shall see, these techniques are rather different in their logic and computation from those covered in Chapters 14 and 15. Let me stress at the outset, therefore, that we are still asking the same three questions about bivariate relationships: Is there a relationship? How strong is it? What is the direction of the relationship? You might tend to become preoccupied by the technical details of the underlying logic and the computational routines necessary for these new techniques, but remind yourself occasionally that the goals of the analysis are identical to those of other measures of association. Ultimately, we seek to understand the nature of a bivariate relationship as a way of probing possible causal relationships and improving our ability to predict scores of a case on one variable from that case's score on another variable.

16.2 SCATTERGRAMS

Over the past several chapters, we have repeatedly noted that the researcher can gain important information about a bivariate association by properly calculating percentages for bivariate tables. When analyzing nominal- or ordinal-level variables, you will almost always be rewarded for taking the time to examine the pattern of cell frequencies in the table in addition to computing a summary measure of association.

By the same token, the usual first step in analyzing a relationship between interval-ratio variables is to construct and examine a **scattergram**. Scattergrams are graphic display devices that are functionally analogous to bivariate tables, since they permit the researcher to quickly perceive several important features of the relationship.

An example will illustrate the construction and use of scattergrams. Suppose a researcher is interested in analyzing how dual-wage-earner families (that is, families where both husband and wife have jobs outside the home) cope with housework. Specifically, the researcher wonders if the number of children in the family is related to the amount of time the husband contributes to housekeeping chores. The relevant data for a sample of 12 dual-wage-earner families are displayed in Table 16.1.

To construct a scattergram by hand, begin by drawing two axes of about equal length and at right angles to each other. Array the independent (X) variable along the horizontal axis (the abscissa) and the dependent (Y) variable along the vertical axis (the ordinate). Calibrate both axes in convenient units of the original scale. Then, for each case, locate the point along the abscissa that corresponds to the score of that case on the X variable. Draw a straight line up from that point and at right angles to the axis. Then, locate the point along the ordinate that corresponds to the score of that same case

---

*The term *correlation* is commonly used instead of *association* when discussing the relationship between interval-ratio variables. We will use the two terms interchangeably.

TABLE 16.1 NUMBER OF CHILDREN AND HUSBAND'S CONTRIBUTION
TO HOUSEWORK (fictitious data)

Family	Number of Children	Hours per Week Husband Spends on Housework
A	1	1
B	1	2
C	1	3
D	1	5
E	2	3
F	2	1
G	3	5
H	3	0
I	4	6
J	4	3
K	5	7
L	5	4

on the *Y* variable. Draw a straight line out from this point and perpendicular to the ordinate. Where the line from the ordinate crosses the line from the abscissa, place a dot to represent the case. Repeat with all cases.

Figure 16.1 shows a scattergram displaying the relationship between "number of children" and "husband's housework" for the sample of 12 families presented in Table 16.1. Note that each family (case) is represented by a dot lying at the intersection of the two scores for that family. Also note that, as always, the scattergram is clearly titled and both axes are labeled.

The overall pattern of the dots (or observation points) succinctly summarizes the nature of the relationship between the two variables. The clarity of the pattern formed by the dots can be enhanced by drawing a straight line through the cluster such that the line touches every dot or comes as close to doing so as possible. In Section 16.3, a technique for fitting this line to the pattern of the dots will be explained. For now, an "eyeball" approximation will suffice. This summarizing line is called the **regression line** and has already been added to the scattergram.

Even a crudely drawn scattergram with a freehand regression line can be used for a variety of purposes. Scattergrams provide at least impressionistic information about the existence, strength, and direction of the relationship and can also be used to check the relationship for linearity (that is, how well the pattern of dots can be approximated with a straight line). Finally, the scattergram can be used to predict the score of a case on one variable from the score of that case on the other variable. We will briefly examine each of these uses.

To ascertain the existence of a relationship, we can return to the basic definition of an association stated in Chapter 13. Two variables are associated if the distributions of *Y* (the dependent variable) change for the various conditions of *X* (the independent variable). In Figure 16.1, the conditions or val-

FIGURE 16.1   HUSBAND'S HOUSEWORK BY NUMBER OF CHILDREN

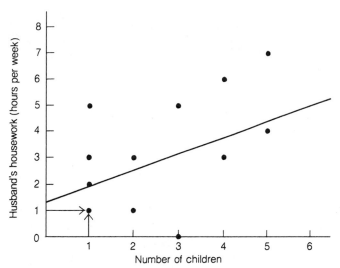

ues of X are arrayed along the abscissa (number of children). The dots above each score on X are the conditional distributions of Y (that is, the dots represent scores on Y for each value of X). Figure 16.1 shows that these conditional distributions of Y change as X changes in that the Y scores vary across the scores of X. The existence of an association is further reinforced by the fact that the regression line lies at an angle to the X axis (the abscissa). If these two variables had not been associated, the conditional distributions of Y would not have changed, and the regression line would have been parallel to the abscissa.

The strength of the association can be judged by observing the spread of the dots around the regression line. In a perfect association, all dots would lie on the regression line. The less the scattering of the dots around the regression line, the stronger the association.

The direction of the relationship can be detected by observing the angle of the regression line with respect to the abscissa. Figure 16.1 displays a positive relationship because families with more children (higher scores on X) tend to have higher scores on "husband's housework." If the relationship had been negative, the regression line would have sloped in the opposite direction to indicate that high scores on one variable were associated with low scores on the other.

To summarize these points about the existence, strength, and direction of the relationship, Figure 16.2 shows a perfect positive and a perfect negative relationship and a "zero relationship," or "nonrelationship" between two variables.

One key assumption underlying the statistical techniques to be introduced later in this chapter is that the two variables have an essentially **linear**

FIGURE 16.2   POSITIVE, NEGATIVE, AND ZERO RELATIONSHIPS

A.

B.

C.

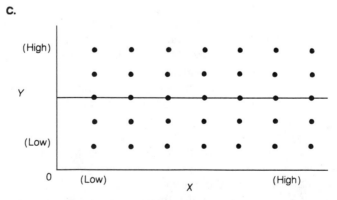

FIGURE 16.3   SOME NONLINEAR RELATIONSHIPS

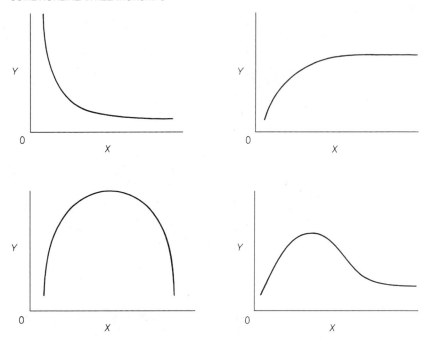

relationship. In other words, the observation points or dots in the scattergram must form a pattern that can be approximated with a straight line. Significant departures from linearity would require the use of statistical techniques beyond the scope of this text. Examples of some common curvilinear relationships are presented in Figure 16.3.

If the scattergram indicates that the variables are associated in a nonlinear fashion, the techniques described in this chapter should be used with great caution or, more circumspectly, not at all. Checking for the linearity of the relationship is perhaps the most important reason for constructing at least a crude, hand-drawn scattergram before proceeding with the statistical analysis. If the relationship is nonlinear, you might need to treat the variables as if they were ordinal rather than interval-ratio in level of measurement. *(For practice in constructing and interpreting scattergrams, see problems 16.1 to 16.4.)*

16.3 REGRESSION AND PREDICTION

A final use of the scattergram is to predict scores of cases on one variable from their score on the other. To illustrate, suppose that, based on the relationship between number of children and husband's housework displayed in Figure 16.1, we wish to predict the number of hours of housework a husband

FIGURE 16.4 PREDICTING HUSBAND'S HOUSEWORK

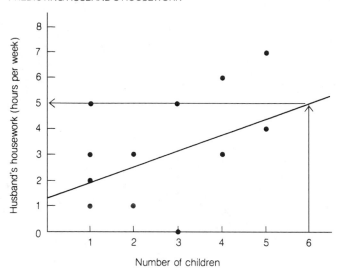

with a family of six children would do each week. The sample of 12 families had no cases of six-children families; but, if the axes and regression line in Figure 16.1 are extended to incorporate this score, a prediction is possible. Figure 16.4 reproduces the scattergram and illustrates how the prediction would be made.

The predicted score on Y—which is symbolized as Y' to distinguish predictions of Y from actual Y scores—is found by first locating the relevant score on X ($X = 6$ in this case) and then drawing a straight line from that point on the abscissa to the regression line. From the regression line, another straight line parallel to the abscissa is drawn across to the Y-axis, or ordinate. The predicted Y score (Y') is found at the point where the line from the regression line crosses the Y-axis. In our example, we would predict that, in a dual-wage-earner family with six children, the husband would devote about five hours per week to housework.

Of course, this technique for predicting Y' is crude and impressionistic. The most serious limitation of this informal prediction technique is that Y' can change in value depending on how accurately the freehand regression line is drawn. One way to eliminate this source of error would be to find the straight line that most accurately summarizes the pattern of the observation points and so best describes the relationship between the two variables. Is there such a "best-fitting" straight line? If there is, how is it defined?

Recall that our criterion for the freehand regression line was that it touch all the dots or come as close to doing so as possible. Also, recall that the dots above any given value of X can be thought of as conditional distributions of Y, the dependent variable. Within each conditional distribution of Y, there is a point around which the variation of the scores is at a minimum: the mean of the conditional distributions of Y.

In Chapter 3, we noted that the mean of any distribution of scores is the point around which the variation of the scores, as measured by squared deviations, is minimized:

$$\Sigma(X_i - \bar{X})^2 = \text{minimum}$$

Thus, if the regression line is fitted so that it touches the mean of each conditional distribution of Y, we would have a line coming as close to all the scores as possible. Such a line would minimize the deviations of the Y scores because it would contain all **conditional means of Y**; and the mean of any distribution is the point of minimized variation.

Conditional means are found by summing all Y values for each value of X and then dividing by the number of cases. For example, four families had one child ($X = 1$), and the husbands of these four families devoted 1, 2, 3, and 5 hours per week to housework. Thus, for $X = 1$, $Y = 1, 2, 3,$ and 5, and the conditional mean of Y for $X = 1$ is 2.75 (11/4 = 2.75). Husbands in families with one child worked an average of 2.75 hours per week doing housekeeping chores. Conditional means of Y are computed in the same way for each value of X displayed in Table 16.2 and plotted on a scattergram in Figure 16.5.

Let us quickly remind ourselves of the reason for these calculations. We are seeking the single best-fitting regression line for summarizing the relationship between X and Y, and we have seen that a line drawn through the conditional means of Y will minimize the spread of the observation points around the line. It will come as close to all the observation points as possible and will therefore be the single best-fitting regression line for these data.

Now, a line drawn through the points on Figure 16.5 (the conditional means of Y) will be the best-fitting line we are seeking. But, you can see from the scattergram that such a line will not be straight. In fact, only rarely will

TABLE 16.2 CONDITIONAL MEANS OF Y (husband's housework) FOR VARIOUS VALUES OF X (number of children)

Number of Children (X)	Husband's Housework (Y)	Conditional Means of Y
1	1,2,3,5	2.75
2	3,1	2.00
3	5,0	2.50
4	6,3	4.50
5	7,4	5.50

FIGURE 16.5   CONDITIONAL MEANS OF *Y*

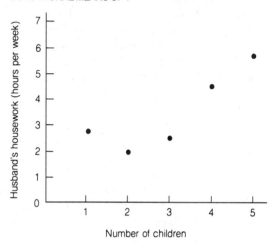

conditional means fall in a perfectly straight line (that is, in a perfect relationship between *X* and *Y*). Since we still must meet the condition of linearity, let us revise our criterion and define the regression line as the unique straight line that touches all conditional means of *Y* or comes as close to doing so as possible. It can be shown that the straight line described in Formula 16.1 will conform to this criterion:

FORMULA 16.1

$$Y = a + bX$$

where Y = score on the dependent variable
 a = the **Y intercept**, or the point where the
   regression line crosses the *Y* axis
 b = the **slope** of the regression line, or
   the amount of change produced in *Y* by a
   unit change in *X*
 X = score on the independent variable

This formula describes the "least-squares" regression line, or the regression line that best fits the pattern of the data points. The formula introduces two new concepts. First, the *Y* intercept (a) is the point at which the regression line crosses the vertical, or *Y*, axis. Second, the slope (b) of the least-squares regression line is the amount of change produced in the dependent variable (Y) by a unit change in the independent variable (X). Think of the slope of the regression line as a measure of the effect of the *X* variable on the *Y* variable. If the variables have a strong association, then changes in the value of *X* will be accompanied by substantial changes in the value of *Y*, and the slope (b) will have a high value. The weaker the effect of *X* on *Y* (the weaker the association between the variables), the lower the value of the slope (b). If the two variables are unrelated, the least-squares regression line would be parallel to the abscissa, and b would be 0 (the line would have no slope).

With the least-squares formula (Formula 16.1), we can predict values of Y in a much less arbitrary and impressionistic way than through mere eye-balling. This will be so, remember, because the least-squares regression line as defined by Formula 16.1 touches or comes as close to the conditional means of Y as possible and is, therefore, the line that best fits the data. Before seeing how predictions of Y can be made, however, we must first calculate a and b. *(For practice in using the regression line to predict scores on Y from scores on X, see problems 16.1 to 16.3 and 16.5.)*

16.4 THE COMPUTATION OF *a* AND *b*

Since the value of b is needed to solve for a, we will begin with the computation of the slope of the least-squares regression line. The definitional formula for the slope is

FORMULA 16.2

$$b = \frac{\Sigma(X - \bar{X})(Y - \bar{Y})}{\Sigma(X - \bar{X})^2}$$

The numerator of this formula is called the *covariation* of X and Y. It is a measure of how X and Y vary together, and its value will reflect both the direction and strength of the relationship. It is tedious to find the slope with this formula, and the following computational formula, which can be derived from Formula 16.2, should be used instead:

FORMULA 16.3

$$b = \frac{N\Sigma XY - (\Sigma X)(\Sigma Y)}{N\Sigma X^2 - (\Sigma X)^2}$$

where b = the slope
N = the number of cases
ΣXY = the summation of the crossproducts of the scores
ΣX = the summation of the scores on X
ΣY = the summation of the scores on Y
ΣX^2 = the summation of the squared scores on X

Admittedly, this formula appears formidable at first glance; but it can be solved without too much difficulty if computations are organized into table format. The computing table displayed in Table 16.3 has a column for each of the four quantities needed to solve the formula. The data are from the dual-wage-earner family sample (see Table 16.1).

In Table 16.3, the first two columns list the original X and Y scores for each case. The third column contains the squared scores on X, and the fourth

TABLE 16.3 COMPUTATION OF THE SLOPE (*b*)

X	Y	X^2	Y^{2*}	XY
1	1	1	1	1
1	2	1	4	2
1	3	1	9	3
1	5	1	25	5
2	3	4	9	6
2	1	4	1	2
3	5	9	25	15
3	0	9	0	0
4	6	16	36	24
4	3	16	9	12
5	7	25	49	35
5	4	25	16	20
Total = 32	40	112	184	125

$$\Sigma X = 32$$
$$\Sigma Y = 40$$
$$\Sigma X^2 = 112$$
$$\Sigma Y^2 = 184$$
$$\Sigma XY = 125$$

*The quantity ΣY^2 is not used in the computation of *b*. We will need it later, however, when we compute Pearson's *r* (see Section 16.5).

lists the squared scores on *Y*. The fifth column lists the crossproducts of the scores for each case. In other words, the entries in the last column are determined by multiplying both scores for each case. We can now replace the symbols in Formula 16.3 with the proper sums:

$$b = \frac{N\Sigma XY - (\Sigma X)(\Sigma Y)}{N\Sigma X^2 - (\Sigma X)^2}$$

$$b = \frac{(12)(125) - (32)(40)}{(12)(112) - (32)^2}$$

$$b = \frac{1500 - 1280}{1344 - 1024}$$

$$b = \frac{220}{320}$$

$$b = .69$$

A slope of .69 indicates that, for each unit change in *X*, there is an increase of .69 units in *Y*. For the specific example under consideration, the addition of each child (an increase of one unit in *X*) results in an increase of .69 hours of housework being done by the husband (an increase of .69 units—or hours—in *Y*).

Once the slope has been calculated, finding the intercept (a) is relatively easy. To compute the mean of X ($\bar{X}$) and the mean of Y ($\bar{Y}$), divide the sums of columns 1 and 2 of Table 16.3 by N and enter these figures in Formula 16.4:

FORMULA 16.4
$$a = \bar{Y} - b\bar{X}$$

For our sample problem, the value of a would be

$$a = \bar{Y} - b\bar{X}$$
$$a = 3.33 - (.69)(2.67)$$
$$a = 3.33 - 1.84$$
$$a = 1.49$$

Thus, the least-squares regression line will cross the Y axis at the point where Y equals 1.49.

The full least-squares regression line for our sample data can now be specified:

$$Y = a + bX$$
$$Y = (1.49) + (.69)X$$

This formula can be used to estimate or predict scores on Y for any value of X. In Section 16.3, we used the freehand regression line to predict a score on Y (husband's housework) for a family with six children ($X = 6$). Our prediction was that, in families of six children, husbands would contribute about five hours per week to housekeeping chores. By using the least-squares regression line, we can see how close our impressionistic, eyeball prediction was.

$$Y' = a + bX$$
$$Y' = (1.49) + (.69)(6)$$
$$Y' = (1.49) + (4.14)$$
$$Y' = 5.63$$

Based on the least-squares regression line, we would predict that in a dual-wage-earner family with six children, husbands would devote 5.63 hours a week to housework. What would our prediction of husband's housework be for a family of seven children ($X = 7$)?

Note that our predictions of Y scores are basically "educated guesses." We will be unlikely to predict values of Y exactly except in the (relatively rare) case where the bivariate relationship is perfect and perfectly linear. Note also, however, that the accuracy of our predictions will increase as relationships become stronger. That is, the less the spread of the observation points around the least-squares regression line (the stronger the relationship), the more accurate will be our predictions of scores of the dependent variable from scores of the independent variable. *(The slope and Y intercept may be computed for any problem at the end of this chapter, but see problems 16.1 to 16.5 in particular. These problems have smaller data sets and will provide good practice until you are comfortable with these calculations.)*

16.5 THE CORRELATION COEFFICIENT (PEARSON'S *r*)

I pointed out in Section 16.4 that the slope of the least-squares regression line (*b*) is a measure of the effect of *X* on *Y*. Since the slope is the amount of change produced in *Y* by a unit change in *X*, *b* will increase in value as the relationship increases in strength. Thus, the value of *b* is a function of the strength of the relationship; but, *b* does not vary between 0 and 1 and is therefore awkward to use as a measure of association per se. For a measure of association for two interval-ratio variables, researchers rely heavily (almost exclusively) on a statistic called **Pearson's *r*,** or the correlation coefficient.

Like the ordinal measures of association discussed in Chapter 15, Pearson's *r* varies from 0.00 to ±1.00, with 0.00 indicating no association and +1.00 and −1.00 indicating perfect positive and perfect negative relationships, respectively. The definitional formula for Pearson's *r* is

FORMULA 16.5

$$r = \frac{\Sigma(X - \bar{X})(Y - \bar{Y})}{\sqrt{[\Sigma(X - X)^2][\Sigma(Y - Y)^2]}}$$

Note that the numerator of this formula is the covariation of *X* and *Y*, as was the case with Formula 16.2. This formula is awkward to use, and the computational Formula 16.6 is usually preferred.

FORMULA 16.6

$$r = \frac{N\Sigma XY - (\Sigma X)(\Sigma Y)}{\sqrt{[N\Sigma X^2 - (\Sigma X)^2][N\Sigma Y^2 - (\Sigma Y)^2]}}$$

A computing table such as Table 16.3 is strongly recommended as a way of organizing the quantities needed to solve this equation. For our sample problem involving dual-wage-earner families, the quantities displayed in Table 16.3 can be substituted directly into Formula 16.6:

$$r = \frac{(12)(125) - (32)(40)}{\sqrt{[(12)(112) - (32)^2][(12)(184) - (40)^2]}}$$

$$r = \frac{1500 - 1280}{\sqrt{(1344 - 1024)(2208 - 1600)}}$$

$$r = \frac{220}{\sqrt{194{,}560}}$$

$$r = \frac{220}{441.09}$$

$$r = .50$$

An *r* value of .50 indicates a moderately strong, positive linear relationship between the variables. As the number of children in the family increases, the hourly contribution of husbands to housekeeping duties also increases. (*Every problem at the end of this chapter requires the computation of Pearson's r. It is probably a good idea to practice with smaller data sets and easier computations first—see problem 16.1 in particular.*)

16.6 INTERPRETING THE CORRELATION COEFFICIENT: r^2

Pearson's r is an index of the strength of the linear relationship between two variables. While a value of 0.00 indicates no linear relationship and a value of ± 1.00 indicates a perfect linear relationship, values between these extremes have no direct interpretation. We can, of course, describe relationships in terms of how closely they approach the extremes (for example, coefficients approaching 0.00 can be described as "weak" and those approaching ± 1.00 as "strong"), but this description is somewhat subjective and less than desirable.

Fortunately, a more direct interpretation is provided by calculating an additional statistic called the **coefficient of determination**. This statistic, which is simply the square of Pearson's r (r^2), can be interpreted with a logic akin to proportional reduction in error (PRE). As you recall, the logic of PRE measures of association is to predict the value of the dependent variable under two different conditions. First, Y is predicted while ignoring the information supplied by X and, second, the independent is taken into account when predicting the dependent. In previous applications of the logic of PRE, we actually counted the number of errors made under each prediction rule and manipulated the sums to derive a final measure. In the case of r^2, both the method of prediction and the construction of the final statistic are somewhat different and require the introduction of some new concepts.

When working with variables measured at the interval-ratio level, the predictions of Y under the first condition (while ignoring X) will be the mean of the Y scores ($\bar{Y}$) for every case. Given no information on X, this prediction strategy will be optimal because we know that the mean of any distribution is closer than any other point to all the scores in the distribution. I remind you of the principle of minimized variation introduced in Chapter 3 and expressed as

$$\Sigma(Y - \bar{Y})^2 = \text{minimum}$$

This statement expresses the fact that the scores of any variable vary around the mean less than they vary around any other point. If we predict the mean of Y for every case, we will make fewer errors of prediction than if we predict any other value for Y.

Of course, we will still make many errors in predicting Y even if we faithfully follow this strategy. The amount of error is represented in Figure 16.6, which displays the relationship between number of children and husband's housework with the mean of Y ($\bar{Y}$) noted. The vertical lines from the actual scores to the predicted score (Y') represent the amount of error we would make when predicting Y while ignoring X.

We can find a more precise measure of the extent of our prediction error under the first condition by taking each actual Y score, subtracting the mean of Y from the score, and squaring and summing these deviations. The resultant figure, which can be noted as $\Sigma(Y - \bar{Y})^2$, is called the **total variation** in Y. We now have a visual representation (Figure 16.6) and a method for

FIGURE 16.6   PREDICTING Y WITHOUT X (dual-career families)

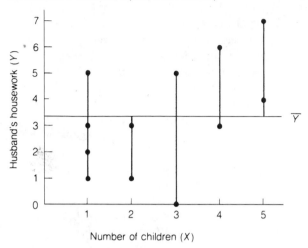

calculating the error we incur by predicting Y without knowledge of X. As we shall see below, we do not need to actually calculate the total variation to find the value of the coefficient of determination, r^2.

Our next step will be to determine the extent to which knowledge of X improves our ability to predict Y. If the two variables have a linear relationship, then predicting scores on Y from the least-squares regression equation will incorporate knowledge of X and reduce our errors of prediction. So, under the second condition, our predicted Y score for each value of X will be

$$Y' = a + bX$$

Figure 16.7 displays the data from the dual-career families with the regression line, as determined by the above formula, drawn in. The vertical lines from each data point to the regression line represent the amount of error in predicting Y that remains even after X has been taken into account.

As was the case under the first condition, we can develop precise ways of describing the reduction in error that results from taking X into account. Specifically, two different sums can be found and then compared with the total variation of Y to construct a statistic that will indicate the improvement in prediction.

The first sum, called the **explained variation**, represents the improvement in our ability to predict Y when taking X into account. This sum is found by subtracting $\bar{Y}$ (our predicted Y score without X) from the score predicted by the regression equation (Y', or the Y score predicted with knowledge of X) for each case and then squaring and summing these differences. These operations can be summarized as $\Sigma(Y' - \bar{Y})^2$, and the resultant figure could

FIGURE 16.7   PREDICTING Y WITH X (dual-career families)

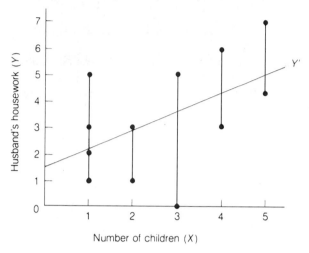

then be compared with the total variation in Y to ascertain to what extent our knowledge of X improves our ability to predict Y. Specifically, it can be shown mathematically that

FORMULA 16.7

$$r^2 = \frac{\Sigma(Y' - \bar{Y})^2}{\Sigma(Y - \bar{Y})^2} = \frac{\text{Explained variation}}{\text{Total variation}}$$

Thus, the coefficient of determination, or r^2, is the proportion of the total variation in Y attributable to or explained by X. Like other PRE measures, r^2 indicates, in a precise way, to what extent knowledge of X helps us to predict, or understand, or explain Y.

Above, we refer to the improvement in predicting Y with X as the ex-plained variation. The use of this term suggests that some of the variation in Y will be "unexplained" or not attributable to the influence of X. In fact, the vertical lines in Figure 16.7 represent the **unexplained variation**, or the difference between our best prediction of Y with X and the actual scores. The unexplained variation is thus the scattering of the actual scores around the regression line and can be found by subtracting the predicted Y scores from the actual Y scores for each case and then squaring and summing these differences. These operations can be summarized as $\Sigma(Y - Y')^2$, and the resultant sum would measure the amount of error in predicting Y that remains even after X has been taken into account. The proportion of the total varia-tion in Y unexplained by X can be found by subtracting the value of r^2 from 1.00. Unexplained variation is usually attributed to the influence of some combination of other variables, measurement error, and random chance.

As you may have recognized by this time, the explained and unexplained variations bear a reciprocal relationship with each other. As one of these

Application 16.1

For five cities, information has been collected on number of civil disturbances (riots, strikes, and so forth) over the past year and on unemployment rate. Are these variables associated?

The data are presented in the table below. Columns have been added for all necessary sums.

The slope (b) is

$$b = \frac{N\Sigma XY - (\Sigma X)(\Sigma Y)}{N\Sigma X^2 - (\Sigma X)^2}$$

$$b = \frac{(5)(985) - (76)(53)}{(5)(1290) - (76)^2}$$

$$b = \frac{897}{674}$$

$$b = 1.33$$

A slope of 1.33 means that for every unit change in X (for every increase of 1 in the unemployment rate) there was a change of 1.33 units in Y (the number of civil disturbances increased by 1.33).

The Y intercept (a) is

$$a = \bar{Y} - b\bar{X}$$

$$a = \frac{53}{5} - (1.33)\frac{76}{5}$$

$$a = 10.6 - (1.33)(15.20)$$

$$a = 10.6 - 20.2$$

$$a = -9.6$$

The least-squares regression equation is

$$Y = a + bX = -9.6 + (1.33)X$$

The correlation coefficient is

$$r = \frac{N\Sigma XY - (\Sigma X)(\Sigma Y)}{\sqrt{[N\Sigma X^2 - (\Sigma X)^2][N\Sigma Y^2 - (\Sigma Y)^2]}}$$

$$r = \frac{(5)(985) - (76)(53)}{\sqrt{[(5)(1290) - (76)^2][(5)(919) - (53)^2]}}$$

$$r = \frac{897}{\sqrt{(674)(1786)}}$$

$$r = \frac{897}{\sqrt{1203764}}$$

$$r = \frac{897}{1097.16}$$

$$r = 0.82$$

These variables have a strong, positive association. The number of civil disturbances increases as the unemployment rate increases. The coefficient of determination, r^2, is $(0.82)^2$, or 0.67. This indicates that 67% of the variance in civil disturbances is explained by the unemployment rate.

City	Unemployment Rate (X)	Civil Disturbances (Y)	X^2	Y^2	XY
A	22	25	484	625	550
B	20	13	400	169	260
C	10	10	100	100	100
D	15	5	225	25	75
E	9	0	81	0	0
	76	53	1290	919	985

sums increases in value, the other decreases. Furthermore, the stronger the linear relationship between X and Y, the greater the value of the explained variation and the lower the unexplained variation. In the case of a perfect relationship ($r = \pm 1.00$), the unexplained variation would be 0 and r^2 would be 1.00. This would indicate that X explains or accounts for all the variation in Y and that we could predict Y from X without error. On the other hand, when X and Y are not linearly related ($r = 0.00$), the explained variation would be 0 and r^2 would be 0.00. In such a case, we would conclude that X explains none of the variation in Y and does not improve our ability to predict Y.

Relationships intermediate between these two extremes can be interpreted in terms of how much X increases our ability to predict or explain Y. For the dual-career families, we calculated an r of 0.50. Squaring this value yields a coefficient of determination of 0.25 ($r^2 = 0.25$), which indicates that number of children (X) explains 25% of the total variation in husband's housework (Y). When predicting the number of hours per week that husbands in such families would devote to housework, we will make 25% fewer errors by basing the predictions on number of children and predicting from the regression line, as opposed to ignoring this variable and predicting the mean of Y for every case. Also, 75% of the variation in Y is unexplained by X and presumably due to some combination of the influence of other variables, measurement error, and random chance. *(For practice in the interpretation of r^2, see any of the problems at the end of this chapter.)*

16.7 TESTING PEARSON'S *r* FOR SIGNIFICANCE

When the relationship measured by Pearson's r is based on data from a random sample, it will usually be necessary to test r for its statistical significance. That is, we will need to know if a relationship between the variables can be assumed to exist in the population from which the sample was drawn. In the example below, the r of .50 from the dual-wage-earner family sample is tested for its statistical significance. As was the case when testing gamma and Spearman's rho, the null hypothesis states that there is no linear association between the two variables in the population from which the sample was drawn. The population parameter is symbolized as ρ (rho), and the appropriate sampling distribution is the t distribution.

To conduct this test, we need to make a number of assumptions in step 1. Several of these assumptions should be quite familiar by now, but several others are new. First, we must assume that both variables are normal in distribution (**bivariate normal distributions**). Second, we must assume that the relationship between the two variables is roughly linear in form.

The third assumption involves a new concept: **homoscedasticity**. Basically, a homoscedastistic relationship is one where the variance of the Y scores is uniform for all values of X. That is, if the Y scores are evenly spread above and below the regression line for the entire length of the line, the relationship is homoscedastistic.

A visual inspection of the scattergram will usually be sufficient to appraise the extent to which the relationship conforms to the assumptions of linearity and homoscedasticity. As a rule of thumb, if the data points fall in a roughly symmetrical, cigar-shaped pattern, whose shape can be approximated with a straight line, then it is appropriate to proceed with this test of significance. Any significant evidence of nonlinearity or marked departures from homoscedasticity may indicate the need for an alternative measure of association and thus a different test of significance. As always, of course, when not certain that the assumptions of the technique are satisfied, the researcher must demonstrate that the conclusions are reasonable and sensible.

Step 1. Making Assumptions.

> Model: Random sampling
> Level of measurement is interval-ratio
> Bivariate normal distributions
> Linear relationship
> Homoscedasticity
> Sampling distribution is normal

Step 2. Stating the Null Hypothesis.

$$H_0: \rho = 0.0$$
$$(H_1: \rho \neq 0.0)$$

Step 3. Selecting the Sampling Distribution and Establishing the Critical Region. With the null of "no relationship" in the population, the sampling distribution of all possible sample r's is approximated by the t distribution. Degrees of freedom are equal to $(N - 2)$.

$$\text{Sampling distribution} = t \text{ distribution}$$
$$\text{Alpha} = .05$$
$$\text{Degrees of freedom} = N - 2 = 10$$
$$t \text{ (critical)} = \pm 2.228$$

Step 4. Computing the Test Statistic.

$$t \text{ (obtained)} = r\sqrt{\frac{N-2}{1-r^2}}$$

$$t \text{ (obtained)} = (.50)\sqrt{\frac{12-2}{1-(.50)^2}}$$

$$t \text{ (obtained)} = (.50)\sqrt{\frac{10}{.75}}$$

$$t \text{ (obtained)} = \frac{(.50)}{\sqrt{13.33}}$$

$$t \text{ (obtained)} = (.50)(3.65)$$

$$t \text{ (obtained)} = 1.83$$

Step 5. Making a Decision. Since the test statistic does not fall in the critical region as marked by t (critical), we fail to reject the null hypothesis. Even though the variables are substantially related in the sample, we do not have sufficient evidence to conclude that the variables are also related in the population. The test indicates that the sample value of $r = .50$ could have occurred by chance alone if the null hypothesis is true and the variables are unrelated in the population. *(For practice in conducting and interpreting tests of significance with Pearson's r, see problems 16.1, 16.2, 16.4, 16.5, 16.8, and 16.12.)*

SUMMARY

1. This summary is based on the example used throughout the chapter. We began with a question: Is the number of children in dual-wage-earner families related to the number of hours per week husbands devote to housework? We presented the observations in a scattergram (Figure 16.1), and our visual impression was that the variables were associated in a positive direction. The pattern formed by the observation points in the scattergram could be approximated with a straight line; thus, the relationship was roughly linear.

2. Values of Y can be predicted with the freehand regression line, but predictions are more accurate if the least-squares regression line is used. The least-squares regression line is the line that best fits the data by minimizing the variation in Y. Using the formula that defines the least-squares regression line ($Y = a + bX$), we found a slope (b) of .69, which indicates that each additional child (a unit change in X) is accompanied by an increase of .69 hours of housework per week for the husbands. We also predicted, based on this formula, that in a dual-wage-earner family with six children ($X = 6$), husbands would contribute 5.63 hours of housework a week ($Y' = 5.63$ for $X = 6$).

3. Pearson's r is a statistic that measures the overall lin-ear association between X and Y. Our impression from the scattergram of a substantial positive relationship was confirmed by the computed r of .50. We also saw that this relationship yields an r^2 of .25, which indicates that 25% of the total variation in Y (husband's housework) is accounted for or explained by X (number of children).

4. Assuming that the 12 families represented a random sample, we tested the Pearson's r for its statistical significance and found that, at the .05 level, we could not assume that these two variables were also related in the population.

5. We acquired a great deal of information about this bivariate relationship. We not only know the strength and direction of the relationship, but have also identified the regression line that best summarizes the effect of X on Y. We know the amount of change we can expect in Y for a unit change in X. In short, we have a greater volume of more precise information about this association between interval-ratio variables than we ever did about associations between ordinal or nominal variables. This is possible, of course, because the data generated by interval-ratio measurement are more precise and flexible than those produced by ordinal or nominal measurement techniques.

SUMMARY OF FORMULAS

Least-squares regression line:

16.1 $Y = a + bX$

Definitional formula for the slope:

16.2 $b = \dfrac{\Sigma(X - \bar{X})(Y - \bar{Y})}{\Sigma(X - \bar{X})^2}$

Computational formula for the slope:

16.3 $b = \dfrac{N\Sigma XY - (\Sigma X)(\Sigma Y)}{N\Sigma X^2 - (\Sigma X)^2}$

Y intercept:

16.4 $a = \bar{Y} - b\bar{X}$

Definitional formula for Pearson's r:

16.5 $$r = \frac{\Sigma(X - \bar{X})(Y - \bar{Y})}{\sqrt{[\Sigma(X - \bar{X})^2][\Sigma(Y - \bar{Y})^2]}}$$

Computational formula for Pearson's r:

16.6 $$r = \frac{N\Sigma XY - (\Sigma X)(\Sigma Y)}{\sqrt{[N\Sigma X^2 - (\Sigma X)^2][N\Sigma Y^2 - (\Sigma Y)^2]}}$$

Coefficient of determination:

16.7 $$r^2 = \frac{\Sigma(Y' - \bar{Y})^2}{\Sigma(Y - \bar{Y})^2} = \frac{\text{explained variation}}{\text{total variation}}$$

GLOSSARY

Bivariate normal distributions. The model assumption in the test of significance for Pearson's r that both variables are normally distributed.

Coefficient of determination (r^2). The proportion of all variation in Y that is explained by X. Found by squaring the value of Pearson's r.

Conditional means of Y. The mean of all scores on Y for each value of X.

Explained variation. The proportion of all variation in Y that is attributed to the effect of X. Equal to $\Sigma(Y' - \bar{Y})^2$.

Homoscedasticity. The model assumption in the test of significance for Pearson's r that the variance of the Y scores is uniform across all values of X.

Linear relationship. A relationship between two variables in which the observation points (dots) in the scattergram can be approximated with a straight line.

Pearson's r (r). A measure of association for variables that have been measured at the interval-ratio level; ρ (Greek letter rho) is the symbol for the population value of Pearson's r.

Regression line. The single, best-fitting straight line that summarizes the relationship between two variables. Regression lines are fitted to the data points by the least-squares criterion whereby the line touches all conditional means of Y or comes as close to doing so as possible.

Scattergram. Graphic display device that depicts the relationship between two variables.

Slope. The amount of change in one variable per unit change in the other; b is the symbol for slope of a regression line.

Total variation. The spread of the Y scores around the mean of Y. Equal to $\Sigma(Y - \bar{Y})^2$.

Unexplained variation. The proportion of the total variation in Y that is not accounted for by X. Equal to $\Sigma(Y - Y')$.

Y intercept (a). The point where the regression line crosses the Y axis.

Y'. Symbol for predicted score on Y.

PROBLEMS

16.1   [PS] Why does voter turnout vary from election to election? For municipal elections in five different cities, information has been gathered on the percent of registered voters who actually voted, unemployment rate, average years of education for the city, and the percentage of all political ads that used "negative campaigning" (personal attacks, negative portrayals of the opponent's record, etc.). For each relationship:

a. Draw a scattergram and a freehand regression line.

b. Compute the slope (b) and find the Y intercept (a). (*HINT: Remember to compute b before computing a. A computing table such as Table 16.3 is highly recommended.*)

c. State the least-squares regression line and predict the voter turnout for a city in which the unemployment rate was 12, a city in which the average years of schooling was 11, and an election in which 90% of the ads were negative.

d. Compute r and r^2. *(HINT: A computing table such as Table 16.3 is highly recommended. If you constructed one for computing b, you already have most of the quantities you will need to solve for r.)*

e. Assume these cities are a random sample and conduct a test of significance for each relationship.

f. Describe the strength and direction of the relationships in a sentence or two. Which (if any) relationships were significant? Which factor had the strongest effect on turnout?

Turnout and Unemployment:

City	Turnout	Unemployment Rate
A	55	5
B	60	8
C	65	9
D	68	9
E	70	10

Turnout and Level of Education:

City	Turnout	Average Years of School
A	55	11.9
B	60	12.1
C	65	12.7
D	68	12.8
E	70	13.0

Turnout and Negative Campaigning:

City	Turnout	% of Negative Advertisements
A	55	60
B	60	63
C	65	55
D	68	53
E	70	48

16.2 SOC Occupational prestige scores for a sample of fathers and their oldest son and oldest daughter are shown in the table.

Family	Father's Prestige	Son's Prestige	Daughter's Prestige
A	80	85	82
B	78	80	77
C	75	70	68
D	70	75	77
E	69	72	60
F	66	60	52
G	64	48	48
H	52	55	57

Analyze the relationship between father's and son's prestige and the relationship between father's and daughter's prestige. For each relationship:

a. Draw a scattergram and a freehand regression line.

b. Compute the slope (b) and find the Y intercept (a).

c. State the least-squares regression line. What prestige score would you predict for a son whose father had a prestige score of 72? What prestige score would you predict for a daughter whose father had a prestige score of 72?

d. Compute r and r^2.

e. Assume these families are a random sample and conduct a test of significance for both relationships.

f. Describe the strength and direction of the relationships in a sentence or two. Does the occupational prestige of the father have an impact on his children? Does it have the same impact for daughters as it does for sons?

16.3 GER The residents of a housing development for senior citizens have completed a survey whereon they indicated how physically active they are and how many visitors they receive each week. Are these two variables related for the 10 cases reported below? Draw a scattergram and compute r and r^2. Find the least-squares regression line. What would be the predicted number of visitors for a person whose level of activity was a 5? How about a person who scored 18 on level of activity?

Case	Level of Activity	Number of Visitors
A	10	14
B	11	12

Case	Level of Activity	Number of Visitors
C	12	10
D	10	9
E	15	8
F	9	7
G	7	10
H	3	15
I	10	12
J	9	2

16.4 [PS] The variables below were collected for a random sample of 10 precincts during the last national election. Draw scattergrams and compute r and r^2 for each combination of variables and test the correlations for their significance. Write a paragraph interpreting the relationship between these variables.

Precinct	Percent Democrat	Percent Minority	Voter Turnout
A	50	10	56
B	45	12	55
C	56	8	52
D	78	15	60
E	13	5	89
F	85	20	25
G	62	18	64
H	33	9	88
I	25	0	42
J	49	9	36

16.5 [SOC/CJ] The table presents the scores of 10 states on each of six variables: three measures of criminal activity and three measures of population structure. Crime rates are number of incidents per 100,000 population.

State	Homicide	Robbery	Auto Theft	Population Growth[1]	Density[2]	Urbanization[3]
Maine	2	22	143	1	40	36
New York	11	477	709	1	384	92
Ohio	6	188	427	3	272	81
Iowa	3	47	180	2	51	44
Virginia	9	133	279	7	167	78
Kentucky	6	94	224	5	97	48
Texas	11	205	603	10	72	84
Arizona	11	162	1067	15	37	87
Washington	6	140	501	12	82	83
California	12	257	981	6	203	97

[1]Percentage change in population from 1990 to 1995.
[2]Population per square mile of land area, 1995.
[3]Percent of population living in metropolitan areas, 1994.

Source: United States Bureau of the Census, *Statistical Abstracts of the United States: 1996* (116th edition). Washington, D.C., 1996.

For each combination of crime rate and population characteristic:
a. Draw a scattergram and a freehand regression line.
b. Compute the slope (b) and find the Y intercept (a).
c. State the least-squares regression line. What homicide rate would you predict for a state with a growth rate of -1? What robbery rate would you predict for a state with a population density of 250? What auto theft rate would you predict for a state in which 50% of the population lived in urban areas?
d. Compute r and r^2.
e. Assume these states are a random sample and conduct a test of significance for both relationships.
f. Describe the strength and direction of the relationships in a sentence or two.

16.6 [PS] The city of Shinbone recently had an election for city council. For the 15 randomly selected precincts listed below, was voter turnout related to the percentage of minority-group population?

Precinct	Registered Voters Who Voted (%)	Minority Group Voters (%)
A	23	0
B	47	15
C	65	75
D	17	10

Precinct	Registered Voters Who Voted (%)	Minority Group Voters (%)
E	9	5
F	64	25
G	50	97
H	49	45
I	45	7
J	43	16
K	27	17
L	25	8
M	33	23
N	34	24
O	25	25

a. Compute r and r^2.
b. Is r statistically significant?
c. Summarize the relationship in terms of its strength, direction, and statistical significance.

16.7 SOC The basketball coach at a small local college believes that his team plays better and scores more points in front of larger crowds. The number of points scored and attendance for all home games last season are reported below. Do these data support the coach's argument?

Game	Points Scored	Attendance
1	54	378
2	57	350
3	59	320
4	80	478
5	82	451
6	75	250
7	73	489
8	53	451
9	67	410
10	78	215
11	67	113
12	56	250
13	85	450
14	101	489
15	99	472

16.8 The table below presents the scores of 15 states on three variables. Compute r and r^2 for each combination of variables. Assume that these 15 states are a random sample of all states and test the correlations for their significance. Write a paragraph interpreting the relationship among these three variables.

State	Per Capita Expenditures on Education	School Dropout Rate[1]	Income per Capita
Arkansas	840	11	17429
Colorado	1067	10	23449
Connecticut	1363	9	30303
Florida	985	14	22916
Illinois	912	11	24763
Kansas	1048	9	21825
Louisiana	865	13	18827
Maryland	1090	11	25927
Michigan	1264	10	23551
Mississippi	819	12	16531
Nebraska	981	7	21703
New Hampshire	1012	9	25151
North Carolina	815	13	20604
Pennsylvania	1025	9	23279
Wyoming	1320	7	21321

[1] Percentage of people ages 16–19 not in school and not graduated.

Source: United States Bureau of the Census, *Statistical Abstracts of the United States: 1996* (116th edition). Washington, D.C., 1996.

16.9 SW For 18 census tracts randomly selected from the greater metropolitan Shinbone area, data have been collected on rates of child abuse, average level of education, and percentage of intact (two-parent) families. Is there any relationship between the latter two variables and child abuse? Compute r and r^2 and test the correlation for statistical significance. Summarize your findings in terms of the strength, direction, and significance of these relationships.

$$\text{Rate of child abuse} = \frac{\text{Number of reported cases}}{\text{population}} \times 100$$

Rate of Child Abuse	Mean Years of Formal Schooling	Intact Families (%)
1.2	12.1	90
.7	12.2	86
3.5	9.2	80
6.7	11.1	75
5.8	8.5	65
4.2	11.8	76
3.8	10.5	67
1.0	12.3	75
1.0	12.7	74
.5	12.4	88
.3	13.1	85
4.7	10.1	73
4.5	9.8	72
5.3	12.0	61
6.8	11.9	64

Rate of Child Abuse	Mean Years of Formal Schooling	Intact Families (%)
7.1	9.0	60
9.1	11.1	63
9.3	9.2	57

16.10 SOC The table below reports some characteristics of 20 countries in 1995. Compute r and r^2 for each combination of variables. Assuming that these nations are a random sample, test the correlations for their significance. Write a paragraph interpreting the relationship between these variables.

Percent Urban	Population Growth Rate	Birth Rate	Death Rate
15	1.1	45	22
83	1.3	21	9
64	2.1	25	7
29	2.9	41	10
74	.3	13	11
45	2.8	39	25
16	3.1	49	15
76	.5	12	8
60	.3	11	10
4	2.4	41	16
90	.1	14	13
85	.5	10	12
65	1.0	20	11
60	.8	21	15
44	2.2	32	8
66	2.6	35	15
64	3.6	48	10
75	.3	13	11
56	− .1	11	12
22	2.9	46	20

16.11 SOC A researcher is interested in the ways college students become "popular." One possibility is that students who are high academic achievers also tend to be more popular. That is, students who have earned high grade-point averages have, by definition, achieved high status in terms of a central value system of the university. Do they translate this academic success into social success? From the 25 residents of a single dormitory wing, the researcher has collected GPA's and an index of popularity. Are these variables related? (Popularity is measured by a simple count of the number of times each resident was rated as popular by his or her neighbors.)

GPA	Popularity
1.75	10
3.01	3
3.35	5
2.00	5
1.52	1
3.50	8
3.20	11
1.67	10
2.52	5
3.10	14
2.95	10
1.99	11
2.57	22
2.80	2
3.00	19
1.30	12
3.62	21
2.00	19
3.65	18
2.12	15
3.80	12
2.82	10
3.85	15
3.51	7
3.42	14

16.12 Twenty-five individuals were randomly selected from the 1996 General Social Survey data set, and their scores on five variables are reproduced below. Is there a relationship between occupational prestige and age? Between church attendance and number of children? Between number of children and hours of TV watching? Between age and hours of TV watching? Between age and number of children? Between hours of TV watching and occupational prestige? Which of these relationships are significant? Variables and codes are explained in detail in Appendix G.

Occupational Prestige	Number of Children	Age	Church Attendance	Hours of TV per Day
32	3	34	3	1
50	0	41	0	3
17	0	52	7	2
69	3	67	0	5
17	0	40	0	5

Occupa-tional Prestige	Number of Children	Age	Church Attend-ance	Hours of TV per Day	Occupa-tional Prestige	Number of Children	Age	Church Attend-ance	Hours of TV per Day
52	0	22	2	3	20	0	22	4	5
32	3	31	0	4	40	0	40	6	1
50	0	23	8	4	26	3	41	2	5
19	9	64	1	6	50	2	31	3	4
37	4	55	0	2	52	5	70	7	4
14	3	66	5	5	50	0	45	6	12
51	0	22	6	0	46	0	24	4	1
45	0	19	3	7	17	4	72	3	2
44	0	21	4	1	50	3	40	0	7
46	4	58	2	0	22	2	28	1	4

SPSS for Windows

Using *SPSS for Windows* to Produce Pearson's *r*

DEMONSTRATION 16.1 What Are the Correlates of Occupational Prestige?

To what extent is a person's social class, as measured by the prestige of her or his occupation, a result of personal efforts and talents (or lack thereof)? How does the social class of our parents shape our ability to rise in the class structure? We can investigate these issues by analyzing the relationships between three variables: *prestg80,* or the prestige of the respondent's occupation; *educ*, the respondent's years of schooling; and *papres80,* the prestige of the respondent's father's occupation. If a person's social class position reflects her or his own preparation for the job market, there should be a strong positive correlation between *educ* and *prestg80.*

On the other hand, if it's *who* you know and not *what* you know—if class position is greatly affected by the social class of one's family of origin—then there should be a strong positive relationship between *papres80* and *prestg80.* By comparing the strength of these bivariate correlations, we may be able to make some judgment about the relative importance of these factors in determining occupational prestige.

To request Pearson's *r*, click **Statistics, Correlate,** and **Bivariate**. The **Bivariate Correlations** window appears with the variable list on the left. Find *educ, papres80,* and *prestg80* in the list and click the arrow to move them into the **Variables** box. If you wish, you can get descriptive statistics for each variable by clicking the **Options** button and requesting means and standard deviations. Unless you request otherwise, the program will conduct two-tailed tests of significance on the correlations. Note that Spearman's rho (see Chapter 15) is also an option. Click **OK**, and your output should look like this:

Correlations

Correlations

		FATHERS OCCUPATIONAL PRESTIGE SCORE (1980)	RS OCCUPATIONAL PRESTIGE SCORE (1980)	HIGHEST YEAR OF SCHOOL COMPLETED
FATHERS OCCUPATIONAL PRESTIGE SCORE (1980)	Pearson Correlation	1.000	.219**	.319**
	Sig. (2-tailed)	.	.000	.000
	N	1166	1129	1162
RS OCCUPATIONAL PRESTIGE SCORE (1980)	Pearson Correlation	.219**	1.000	.546**
	Sig. (2-tailed)	.000	.	.000
	N	1129	1400	1395
HIGHEST YEAR OF SCHOOL COMPLETED	Pearson Correlation	.319**	.546**	1.000
	Sig. (2-tailed)	.000	.000	.
	N	1162	1395	1446

**Correlation is significant at the 0.01 level (2-tailed).

The output is in the form of a correlation matrix showing the relationships between all variables, including the correlation of a variable with itself. For each possible relationship, Pearson's r is reported in the top row, the results of the test of significance in the second row, and sample size in the third row (N varies because not everyone answered every question).

For our questions, we need to focus on the correlations between prestg80, educ, and papres80. As expected, the relationships are statistically significant ($p < .05$), positive, and moderate-to-strong. The relationship between prestg80 and educ is stronger (.546) than the relationship between prestg80 and papres80 (.219). This suggests that preparation for the job market (educ) is more important than the relative advantage or disadvantage conferred by one's family of origin (papres80). Success may come more easily to people from an advantaged family background, but talent and preparation matter more. To further disentangle these relationships would require more sophisticated statistics and, fortunately, some will become available in Chapter 18.

DEMONSTRATION 16.2 Are the Correlates of Occupational Prestige Affected by Gender?

Are the pathways to success and higher prestige open to all equally? We can begin to address this question by analyzing the relationships reported in Demonstration 16.1 for men and women separately. If gender has no effect on a person's opportunities for success, prestg80, papres80, and educ should be related in the same way for both men and women.

To observe the effect of sex, we will split the GSS sample into two subfiles. Men and women will then be processed separately, and SPSS will produce one correlation matrix for men and another for women.

Click **Data** from the main menu and then click **Split file**. On the **Split File** window, click the **radio** button next to organize output by groups. This will generate separate outputs for men and women. Select *sex* from the variable list and click the arrow to move the variable name into the **Groups Based On** window. Click **OK**, and all procedures requested will be done separately for men and women. To restore the full sample, call up the **Split File** window again and move *sex* back into the variable list. For now, click **Statistics, Correlate,** and **Bivariate** and rerun the correlation matrix for *prestg80, papres80,* and *educ.* The output will look like this:

Correlations
RESPONDENTS SEX = MALE

Correlations[a]

		FATHERS OCCUPATIONAL PRESTIGE SCORE (1980)	RS OCCUPATIONAL PRESTIGE SCORE (1980)	HIGHEST YEAR OF SCHOOL COMPLETED
FATHERS OCCUPATIONAL PRESTIGE SCORE (1980)	Pearson Correlation	1.000	.288**	.342**
	Sig. (2-tailed)	.	.000	.000
	N	524	522	524
RS OCCUPATIONAL PRESTIGE SCORE (1980)	Pearson Correlation	.288**	1.000	.539**
	Sig. (2-tailed)	.000	.	.000
	N	522	633	632
HIGHEST YEAR OF SCHOOL COMPLETED	Pearson Correlation	.342**	.539**	1.000
	Sig. (2-tailed)	.000	.000	.
	N	524	632	635

**Correlation is significant at the 0.01 level (2-tailed).
a. RESPONDENTS SEX = MALE

RESPONDENTS SEX = FEMALE

Correlations[a]

		FATHERS OCCUPATIONAL PRESTIGE SCORE (1980)	RS OCCUPATIONAL PRESTIGE SCORE (1980)	HIGHEST YEAR OF SCHOOL COMPLETED
FATHERS OCCUPATIONAL PRESTIGE SCORE (1980)	Pearson Correlation	1.000	.153**	.297**
	Sig. (2-tailed)	.	.000	.000
	N	642	607	638
RS OCCUPATIONAL PRESTIGE SCORE (1980)	Pearson Correlation	.153**	1.000	.552**
	Sig. (2-tailed)	.000	.	.000
	N	607	767	763
HIGHEST YEAR OF SCHOOL COMPLETED	Pearson Correlation	.297**	.552**	1.000
	Sig. (2-tailed)	.000	.000	.
	N	638	763	811

**Correlation is significant at the 0.01 level (2-tailed).
a. RESPONDENTS SEX = FEMALE

The correlations for both groups are significant, positive, and moderately strong. The correlations between *papres80* and *prestg80* are much stronger for males than those for females, but the correlations between *prestg80* and *educ* are about the same for both genders. This suggests that the social-class position of males is more affected by the social class of their family of origin. On the other hand, education is about equally beneficial. Why might this be so? What additional tests could be conducted to clarify these patterns?

Exercises

16.1 Run the analysis in Demonstration 16.1 again with *income91* rather than *prestg80* as the indicator of social class standing. Use the same independent variables (*papres80, educ*) and see if the patterns are similar to those identified in Demonstration 16.1. (Ignore the fact that *income91* is only ordinal in level of measurement.) Write up your conclusions.

16.2 Run the analysis in Demonstration 16.2 again but substitute *income91* for *prestg80*. Split the sample (use the **Split File** command) by sex. Is there a gender difference in the correlates of *income91* as there was with *prestg80*?

16.3 Switch topics entirely and see if you can explain the variation in television viewing habits. What are the correlates of *tvhours*? Choose three or four possible independent variables (perhaps *age* and *educ*) and run the **Bivariate Correlation** program. Write up your results.

PART III CUMULATIVE EXERCISES

1. For each situation, compute and interpret the appropriate measure of association. Also, compute and interpret percentages for bivariate tables. Describe
relationships in terms of the strength and pattern or direction.

a. For 10 cities, data has been gathered on total crime rate (major felonies
per 100,000 population) and the percentage of people who are new immigrants (arrived in the United States within the past five years). Are the variables related?

City	Total Crime Rate	Percent Immigrants
A	1500	10
B	1200	18
C	2000	9
D	1700	11
E	1600	15
F	1000	20
G	1700	9
H	1300	22
I	900	10
J	700	15

b. There is some evidence that people's involvement in their communities
(membership in voluntary organizations, participation in local politics, and
so forth) has been declining, and television has been cited as the cause. Do
the data below support the idea that TV is responsible for the decline?

Hours of Community Service:	Television Viewing			Totals
	Low	Moderate	High	
Low	5	10	18	33
Moderate	10	12	10	32
High	15	8	7	30
Totals	30	30	35	95

c. A national magazine has rated the states in terms of "quality of life" (a scale
that includes health care, availability of leisure facilities, unemployment rate,
and a number of other variables) and the quality of the state system of higher
education. Both scales range from 1 (low) to 20 (high). Is there a correlation
between these two variables for the 10 states listed below?

State	Quality of Life	Quality of Higher Education
A	10	10
B	12	13
C	15	18
D	18	20
E	10	15

State	Quality of Life	Quality of Higher Education
F	9	11
G	11	12
H	8	6
I	13	9
J	5	8

d. Racial intermarriages are increasing in the United States, and the number of mixed-race people is growing. The table below shows the relationship between racial mixture of parents and the racial category, if any, with which a sample of people of mixed race identifies. Is there a relationship between these variables?

Do You Consider Yourself to Be:	Racial Mixture of Parents			Totals
	Black/ White	White/ Asian	Black/ Asian	
Black	2	0	3	5
White	8	4	0	12
Asian	0	3	3	6
None of the above	10	8	4	22
	20	15	10	45

2. A number of research questions are stated below. Each can be answered by at least one of the techniques presented in Chapters 13 through 16. For each research situation, compute the most appropriate measure of association and write a sentence or two of response to the question. The questions are presented in random order.

In selecting a measure of association, you need to consider the number of possible values and the level of measurement of the variables. The flowcharts at the beginning of each chapter may be helpful.

The research questions refer to the database below, which is taken from the 1996 General Social Survey (GSS). The actual questions asked and the complete response codes for the GSS are presented in Appendix G. Abbreviated codes are listed below. Some variables have been recoded for this exercise.

a. Are scores on "frequency of sex during the last year" associated with income or age? Compute a measure of association, assuming that the variables are interval-ratio in level of measurement.
b. Is fear associated with sex? Is it associated with marital status?
c. Is support for sex education in the public schools associated with church attendance? Is it associated with marital status?

Survey Items:

1. Marital status of respondent (MARITAL):
   1. Married
   2. Not married (includes widowed, divorced, etc.)

2. How often do you attend church? (ATTEND) See Appendix G for original codes:
   0. Never
   1. Rarely
   2. Often
3. Age of respondent (AGE) (Values are actual numbers.):
4. Respondent's total family income (INCOME91) See Appendix G for codes:
5. Sex:
   1. Male
   2. Female
6. Support for sex education in the public schools (SEXEDUC):
   1. Favor
   2. Oppose
7. Fear of walking alone at night (FEAR):
   1. Yes
   2. No
8. Frequency of sex during past year. See Appendix G for codes:

Case	Marital Status	Church Atten- dance	Age	Income	Sex	Sex Educ.	Fear	Frequency of Sex
1	2	0	22	12	2	1	2	5
2	1	0	52	13	1	1	2	2
3	1	2	44	16	1	2	1	3
4	1	0	56	10	1	1	2	1
5	1	2	61	8	2	2	2	0
6	1	0	28	19	2	1	2	5
7	2	1	59	9	1	2	1	0
8	1	1	69	11	1	2	2	5
9	2	0	23	4	1	1	2	6
10	2	2	31	20	2	1	1	6
11	1	2	67	21	2	2	1	1
12	1	1	46	9	1	1	1	2
13	2	0	19	10	2	1	1	5
14	1	2	34	11	1	2	2	2
15	2	0	29	18	2	1	1	4
16	1	1	31	16	1	1	1	5
17	2	1	88	6	2	2	2	1
18	1	0	24	15	2	1	1	4
19	2	2	69	11	1	1	1	2
20	1	2	60	14	1	2	2	3
21	1	1	29	12	2	1	2	4
22	1	2	43	13	2	2	1	2
23	1	0	35	20	1	1	1	4
24	2	2	38	9	2	2	2	4
25	2	0	83	19	1	2	1	0

Part IV Multivariate Techniques

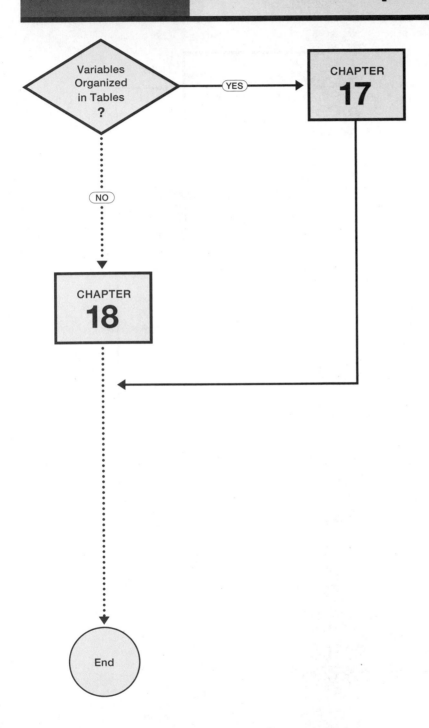

17 Elaborating Bivariate Tables

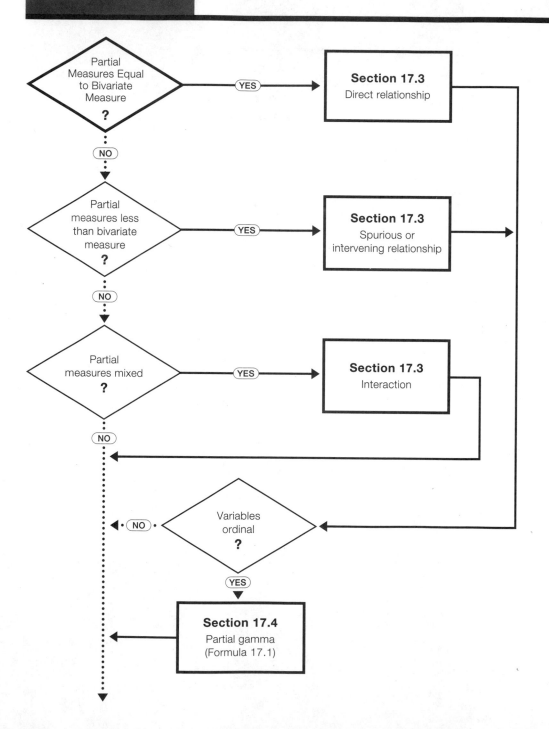

17.1 INTRODUCTION

Few research questions can be answered through a statistical analysis of only two variables. Most, if not all, research projects will require the analysis of many different variables, even when the researcher is principally concerned with a specific bivariate relationship. In Chapters 13–16, we have seen how various statistical techniques can be applied to the analysis of bivariate relationships. In this chapter and Chapter 18, we will see how some of these techniques can be utilized to probe the relationships among three or more variables. This chapter will present some multivariate techniques appropriate for variables that have been measured at the nominal or ordinal level and organized into table format. Chapter 18 will present some techniques that can be used when the variables have been measured at the interval-ratio level.

Before considering the techniques themselves, we should briefly consider why they are important and what they might be able to tell us. There are two general reasons for utilizing multivariate techniques. First, and most fundamental, is the goal of simply gathering additional information about a specific bivariate relationship by observing how that relationship is affected (if at all) by the presence of a third variable (or a fourth or a fifth variable). Multivariate techniques will increase the amount of information we have on the basic bivariate relationship and (we hope) will enhance our understanding of that relationship.

A second, and very much related, rationale for multivariate statistics involves the issue of causation. While multivariate statistical techniques cannot prove the existence of causal connections between variables, they can provide valuable evidence in support of causal arguments. Needless to say, their ability to find evidence for or against causal relationships between variables makes these techniques valuable for purposes of testing and revising theory.

17.2 CONTROLLING FOR A THIRD VARIABLE

For variables arrayed in bivariate tables, multivariate analysis proceeds by systematically observing the effects of other variables on a bivariate relationship. These effects are measured by controlling for the additional variables. The additional variable is fixed so that its values are no longer free to vary (thus, it is "controlled"), and the impact of this procedure on the bivariate relationship is then assessed.*

To illustrate, suppose that a researcher wishes to analyze the relationship between how well an individual is integrated into a group or organization and that individual's level of satisfaction with the group. The researcher has decided to focus on students and their level of satisfaction with the college as a whole. The necessary information on satisfaction (Y) is gathered from a sample of 270 students; all are asked to list the student clubs or

---

*For the sake of brevity, we will focus on the simplest application of multivariate statistical analysis, the case where the relationship between an independent (X) and dependent (Y) variable is analyzed in the presence of a single control variable (Z). Once the basic techniques are grasped, they can be readily expanded to situations involving more than one control variable.

TABLE 17.1   OVERALL SATISFACTION WITH COLLEGE BY NUMBER OF
MEMBERSHIPS IN STUDENT ORGANIZATIONS (frequencies)

Satisfaction (Y)	Memberships (X)		Totals
	None	At Least One	
Low	57	56	113
High	48	109	157
Totals	105	165	270

Gamma = .40

TABLE 17.2   OVERALL SATISFACTION WITH COLLEGE BY NUMBER OF
MEMBERSHIPS IN STUDENT ORGANIZATIONS (percentages)

Satisfaction (Y)	Memberships (X)	
	None	At Least One
Low	54.3	33.9
High	45.7	66.1
	100.0%	100.0%
	(105)	(165)

organizations to which they currently belong. Integration into the organization (X) is measured by dividing the students into two categories. The first category includes students who are not members of any organizations (nonmembers), and the second includes students who are members of at least one organization (members). The researcher suspects that membership and satisfaction are positively related and that students who are members of at least one organization will report higher levels of satisfaction than students who are nonmembers. The basic relationship between these two variables is displayed in Tables 17.1 and 17.2.

Inspection of these tables suggests that these two variables are associated. The conditional distributions of satisfaction (Y) change across the two conditions of membership (X), and membership in at least one organization is associated with high satisfaction while nonmembership is associated with low satisfaction. The existence and direction of the relationship are confirmed by the computation of a gamma of +.40 for this table.* In short, these tables strongly suggest the existence of a relationship between integration and morale, and these results can be taken as evidence for a causal or direct relation-

---

*We will not display the computation of gamma here. See Chapter 15 for a review of this measure of association.

FIGURE 17.1   A DIRECT RELATIONSHIP BETWEEN TWO VARIABLES

$$X \longrightarrow Y$$

ship between the two variables. The causal relationship is summarized symbolically in Figure 17.1, where the arrow represents the effect of X on Y.

The researcher recognizes, of course, that membership is not the sole factor associated with satisfaction (if it were, the gamma noted above would be +1.00). Other variables may well alter this relationship, and the researcher will need to consider the effects of these third variables (**Z's**) in a systematic and logical way. By doing so, the researcher will accumulate further information and detail about the bivariate relationship and can further probe the possible causal connection between X and Y.

For example, perhaps levels of satisfaction are affected by the rewards received by students from the university. Perhaps the most significant reward the university can dispense is recognition of academic achievements (that is, grades). Perhaps students who are more highly rewarded (have higher grade-point averages) are more satisfied with the university and are also more likely to participate actively in the life of the campus by joining student organizations. How can the effects of this third variable on the bivariate relationship be investigated?

Consider Table 17.1 again. This table provides information about the distribution of 270 cases on two variables. By analyzing the table, we can see that 165 students hold at least one membership, that 48 students are both nonmembers and highly satisfied, that the majority of the students are highly satisfied, and so forth. What the bivariate table cannot show us, of course, is the distribution of these students on GPA. For all we know at this point, the 109 highly satisfied members could all have high GPA's, low GPA's, or any combination of scores on this third variable. GPA is "free to vary" in this table, since the distribution of the cases on this variable is not accounted for.

We can control for the effect (if any) of third variables by fixing their distributions so that they are no longer free to vary. We do so by sorting all cases in the sample according to their score on the third variable (Z) and then observing the relationship between X and Y for each value (or score) of Z. In the example at hand, we will construct separate tables displaying the relationship between membership and satisfaction for each category of GPA. The tables so produced are called **partial tables** and are displayed in Tables 17.3 and 17.4. The students have been classified as "high" or "low" on GPA and there are exactly 135 students in each category. Within each category of GPA, students have been reclassified by membership (X) and satisfaction (Y) in order to produce the two partial tables.

Note that these partial tables present the original bivariate relationship in a more detailed form; that is, the partial tables elaborate the relationships displayed in Tables 17.1 and 17.2. The cell frequencies in the partial tables are,

TABLE 17.3 SATISFACTION BY MEMBERSHIP CONTROLLING FOR GPA (frequencies)

A. High GPA:

Satisfaction	Memberships		Totals
	None	At Least One	
Low	29	28	57
High	24	54	78
Totals	53	82	135

Gamma = .40

B. Low GPA:

Satisfaction	Memberships		Totals
	None	At Least One	
Low	28	28	56
High	24	55	79
Totals	52	83	135

Gamma = .39

[handwritten annotation: replication when Gamma the same, assoc is general.]

in this sense, subdivisions of the cell frequencies reported in Table 17.1. For example, if the frequencies of the cells in the partial tables are added together, the original frequencies of Table 17.1 will be reproduced.*

Also note the way this method of controlling for other variables can be extended. First, there will be a partial table for each value of the control variable. In our example above, the control variables had two categories and, thus, generated two partial tables. Had there been three categories, there would have been three partial tables. Second, we can control for more than one variable at a time by sorting the cases on all scores of all control variables and producing partial tables for each combination of scores on the control variables. Thus, if we had controlled for both GPA and gender, we would have had four partial tables to consider. There would have been one partial table for males with low GPA's, a second for males with high GPA's, and two partial tables for females with high and low GPA's. To summarize, for nominal and ordinal variables, multivariate analysis proceeds by first constructing partial tables—tables displaying the relationship between X and Y for each value of Z.

---

*The total of the cell frequencies in the partial tables will always equal the corresponding cell frequencies in the bivariate table except when, as often happens in "real life" research situations, the researcher is missing scores on the third variable for some cases. These cases must be deleted from the analysis and, as a consequence, the partial tables will have fewer cases than the bivariate table.

TABLE 17.4   SATISFACTION BY MEMBERSHIP CONTROLLING FOR GPA
(percentages)

A. High GPA:

| Satisfaction | Memberships | |
	None	At Least One
Low	54.7	34.2
High	45.3	65.9
	100.0%	100.1%
	(53)	(82)

B. Low GPA:

| Satisfaction | Memberships | |
	None	At Least One
Low	53.9	33.7
High	46.2	66.3
	100.0%	100.0%
	(52)	(83)

The next step is to trace the effect of Z by comparing the partial tables with each other and with the original bivariate table. By analyzing the partial tables, we can trace and understand the effects (if any) of the control variable on the original bivariate relationship.

17.3 INTERPRETING PARTIAL TABLES

The cell frequencies in the partial tables may follow a variety of forms, but we will concentrate on three basic patterns. The patterns, determined by comparing the partial tables with each other and with the original bivariate table, are

1. **Direct relationships** (The relationship between X and Y is the same in all partial tables and in the original bivariate table.)
2. **Spurious relationships** or **intervening relationships** (The relationship between X and Y is the same in all partial tables but much weaker than in the bivariate table.)
3. **Interaction** (The relationships between X and Y in the partial tables are different from each other and from the bivariate table.)

Each of these three basic patterns has different implications for the causal interrelationships among the three variables and for the subsequent course of the statistical analysis. I will now describe each pattern in detail and then summarize our discussion in Table 17.9.

Direct Relationships.   This pattern is often called **replication** because the partial tables reproduce (or replicate) the bivariate table; the cell frequencies are the same in the partial tables and the bivariate table. Measures of association calculated on the partial tables have the same value as the measure of association calculated for the bivariate table.

This outcome indicates that the control variable, Z, has no effect on the relationship between X and Y. Tables 17.3 and 17.4 provide an example of this outcome. In these two tables, the relationship between membership in student organizations (X) and overall satisfaction (Y) was investigated with GPA (Z) controlled. Tables 17.3A and 17.4A show the relationship for high-GPA students, and Tables 17.3B and 17.4B show the relationship for low-GPA students. The partial tables show the same conditional distributions of Y: for both high- and low-GPA students, about 45% of the nonmembers are highly satisfied, versus about 66% of the members. This same pattern was observed in the bivariate table (see Tables 17.1 and 17.2). Thus, the conditional distributions of Y are the same in each partial table as they were in the bivariate table.

The pattern of the cell frequencies will be easier to detect if you calculate measures of association. Working from the cell frequencies presented in Table 17.3, the gamma for high-GPA students is .40, and the gamma for low-GPA students is .39. The bivariate gamma (from Table 17.1) is .40, and the essential equivalence of these gammas reinforces our finding that the relationship between X and Y is essentially the same in the partial tables and the bivariate table.

This pattern of outcomes indicates that the control variable has no important impact on the bivariate relationship (if it did, the pattern in the partial tables would be different from the bivariate table) and may be ignored in any further analysis. In terms of the original research problem, the researcher may conclude that students who are members are more likely to express high satisfaction with the university regardless of their GPA. The level of rewards dispensed to students by the institution (as measured by GPA) has no effect on the relationship between membership and satisfaction. Low-GPA students who are members of at least one organization are just as likely to report high satisfaction as high-GPA students who are members of at least one organization. *(For practice in dealing with direct relationships, see Problems 17.1 and 17.2.)*

Spurious or Intervening Relationships.   In this pattern, the relationship between X and Y is much weaker in the partial tables than in the bivariate table but the same across all partials. Measures of association for the partial tables are much lower in value (perhaps even dropping to 0.00) than the measure computed for the bivariate table. This outcome is consistent with two different causal relationships among the three variables. The first is called a spurious relationship or **explanation**; in this situation, Z is conceptualized as being antecedent to both X and Y (that is, Z is thought to occur before the other two variables in time). In this pattern, Z is a common cause of both X and Y, and the original bivariate relationship is said to be spurious. The

---

Application 17.1

Seventy-eight juvenile males in a sample have been classified as high or low on a scale that measures involvement in delinquency. Also, each subject has been classified, using school records, as either a good or poor student. The following table displays a strong relationship between these two variables for this sample ($G = -.69$).

Academic Record

Delinquency	Poor	Good	Totals
Low	13	20	33
High	35	10	45
Totals	48	30	78

Gamma = $-.69$

Juvenile males with poor academic records seem to be especially prone to delinquency.

Is this relationship between delinquency and academic record affected by whether the subject resides in an urban or nonurban area?

Urban areas:

Academic Record

Delinquency	Poor	Good	Totals
Low	10	3	13
High	26	7	33
Totals	36	10	46

Gamma = $-.05$

Nonurban areas:

Academic Record

Delinquency	Poor	Good	Totals
Low	3	17	20
High	9	3	12
Totals	12	20	32

Gamma = $-.89$

For urban juvenile males, the relationship between academic record and delinquency disappears ($G = -.05$). For this group, delinquency is not associated with experience in school.

For nonurban males, there is a very strong relationship between the two variables ($G = -.89$), with poor students being especially prone to delinquency.

Comparing the partial tables with each other and with the bivariate table reveals an interactive relationship among these three variables. Although urban juvenile males are more delinquent than nonurban males (33 out of the 46, or 71.74%, of the urban males were highly delinquent, as compared to 12 out of 32, or 37.50%, of the nonurban males), their delinquency is not associated with academic record. For nonurban males, academic record is very strongly associated with delinquency. Urban males are more delinquent than nonurban males but not because of their experience in school. Nonurban males who are also poor students are especially likely to become involved in delinquency.

---

apparent bivariate relationship between X and Y is due to the effect of Z. Once Z is controlled, the association between X and Y disappears (the value of the measures for the partial tables drops to 0).

To illustrate, suppose that the researcher in our example had also controlled for class standing by dividing the sample into upperclasses (seniors and juniors) and underclasses (sophomores and freshmen). The reasoning of the researcher might be that upperclass students, as a function of simple longevity, display higher levels of satisfaction with the college than do underclass students. Self-selection processes may be operating. Students who are dissatisfied may have transferred to another college or dropped out before they attain upperclass standing. Students who have been on campus longer will be more likely to locate an organization of sufficient appeal or interest to join. This might especially be the case for organizations based on major field (such as the Accounting Club), which underclass students are less likely to join, or honorary organizations for which underclass students are unlikely to qualify.

These thoughts about the possible relationships among these variables are expressed in diagram form in Figure 17.2. The absence of an arrow from membership (X) to satisfaction (Y) indicates the possibility that these variables are not truly associated with each other but rather that both are linked to class standing.

If this causal diagram is a correct description of the relationship of these three variables (if the association between X and Y is spurious), then the association between membership and satisfaction should disappear once class standing has been controlled. That is, even though the bivariate gamma was .40 (Table 17.1), the gammas computed on the partial tables will approach 0. Tables 17.5 and 17.6 display the partial tables generated by controlling for class standing.

The partial tables indicate that, once class standing is controlled, membership is no longer related to satisfaction. Upperclass students are likely to express high satisfaction and underclass students are likely to express low satisfaction regardless of their number of memberships. In the partial tables, the distributions of Y no longer vary by the conditions of X, and the gammas computed on the partial tables are virtually 0. These results indicate that the bivariate association is spurious and that X and Y have no direct relationship. Class standing (Z) is a key factor in accounting for varying levels of satisfaction, and the analysis must be reoriented with class standing as an independent variable.

This outcome (partial measures much weaker than the original measure but equal to each other) is consistent with another conception of the causal links among the variables. In addition to a causal scheme where Z is antecedent to both X and Y, Z may also intervene between the two variables. This pattern is also called **interpretation** and is illustrated in Figure 17.3, where X is causally linked to Z, which is in turn linked to Y. This pattern indicates

FIGURE 17.2   A SPURIOUS RELATIONSHIP

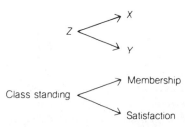

TABLE 17.5 SATISFACTION BY MEMBERSHIP CONTROLLING FOR
CLASS STANDING (frequencies)

A. Upperclass students:

	Memberships		
Satisfaction	None	At Least One	Totals
Low	8	32	40
High	24	97	121
Totals	32	129	161

Gamma = .01

B. Underclass students:

	Memberships		
Satisfaction	None	At Least One	Totals
Low	49	24	73
High	24	12	36
Totals	73	36	109

Gamma = .01

[handwritten margin notes: "Spurious assoc btwn membership & satisfaction"; "explained relationship to upperclass students"; "no satisfaction not memberships"; "so low Gamma"]

TABLE 17.6 SATISFACTION BY MEMBERSHIP CONTROLLING FOR
CLASS STANDING (percentages)

A. Upperclass students:

	Memberships	
Satisfaction	None	At Least One
Low	25.0	24.8
High	75.0	75.2
	100.0%	100.0%
	(32)	(129)

B. Underclass students:

	Memberships	
Satisfaction	None	At Least One
Low	67.1	66.7
High	32.9	33.3
	100.0%	100.0%
	(73)	(36)

that, although X and Y are related, they are associated primarily through the control variable Z. This particular outcome does not allow the researcher to distinguish between spurious relationships (Figure 17.2) and intervening relationships (Figure 17.3). The differentiation between these two types of

FIGURE 17.3 AN INTERVENING RELATIONSHIP

FIGURE 17.4 AN INTERACTIVE RELATIONSHIP

causal patterns may be made on temporal or theoretical grounds, but not on statistical grounds. *(For practice in dealing with spurious or intervening relationships see problem 17.6b.)*

Interaction. In this pattern, also called **specification**, the relationship between X and Y changes markedly, depending on the value of the control variable. The partial tables differ from each other and from the bivariate table. Interaction can be manifested in various ways in the partial tables. One possible pattern, for example, is for one partial table to display a stronger relationship between X and Y than that displayed in the bivariate table, while in a second partial table, the relationship between X and Y drops to 0. Symbolically, this outcome could be represented as in Figure 17.4, which would indicate that X and the first category of Z (Z_1) have strong effects on Y; but, for the second category of Z (Z_2), there is no association between X and Y.

Interaction might be found, for example, in a situation in which all employees of a corporation were required to attend a program (X) designed to reduce racial prejudice (Y). Such a program would be likely to have stronger effects on white employees (Z_1) than African American employees (Z_2). That is, the program would have an effect on prejudice only for certain categories of subjects. The partial tables for white employees (Z_1) might show a strong relationship between program attendance and reduction in prejudice, while the partial tables for African American employees showed no relationship. African American employees, being relatively unprejudiced against themselves in the first place, would be less likely to be affected by the program.

Interaction can take other forms. Let us investigate a case where the relationship between X and Y varies not only in strength but also in direction between the partial tables. This causal relationship is symbolically represented in Figure 17.5, which indicates a situation where X and Y are positively related for the first category of Z (Z_1) and negatively related for the second category of Z (Z_2).

The researcher investigating the relationship between club membership and satisfaction with college life establishes a final control for race and di-

FIGURE 17.5   AN INTERACTIVE RELATIONSHIP

TABLE 17.7   SATISFACTION BY MEMBERSHIP CONTROLLING FOR RACE
(frequencies)

A. White students:

Satisfaction	Memberships		Totals
	None	At Least One	
Low	40	20	60
High	40	100	140
Totals	80	120	200

Gamma = .67

specification

strong relationship for white students, but reverse for black students.

B. Black students:

Satisfaction	Memberships		Totals
	None	At Least One	
Low	17	36	53
High	8	9	17
Totals	25	45	70

Gamma = −.31

does not persist 2° neg. gamma

vides the sample into white and black students. The partial tables are displayed in Tables 17.7 and 17.8.

These partial tables display an interactive relationship. The relationship
between membership and satisfaction is different for white students (Z_1) and
for black students (Z_2), and each partial table is different from the bivariate
table. For white students, the relationship is positive and stronger than in the
bivariate table. White students who are also members are much more likely
to express high overall satisfaction with the university, as indicated by both
the percentage distribution (83% of the white students who are members are
highly satisfied) and the measure of association (gamma = +.67 for white
students).

For black students, the relationship between membership and satisfaction is very different—nearly the reverse of the pattern evinced by white students. The great majority (80%) of the black students who are members report low satisfaction, and the gamma for this partial table is negative (−.31).
Thus, for white students, satisfaction increases with membership, while, for

TABLE 17.8   SATISFACTION BY MEMBERSHIP CONTROLLING FOR RACE
(percentages)

A. White students:

Satisfaction	Memberships	
	None	At Least One
Low	50.0	16.7
High	50.0	83.3
	100.0%	100.0%
	(80)	(120)

B. Black students:

Satisfaction	Memberships	
	None	At Least One
Low	68.0	80.0
High	32.0	20.0
	100.0%	100.0%
	(25)	(45)

black students, satisfaction decreases with membership. Based on these results, one might conclude that the social meanings and implications of joining student clubs and organizations vary by race and that the function of
joining has different effects for black students. It may be, for example, that
black students join different kinds of organizations than do white students. If
the black students belonged primarily to a black student association that had
an antagonistic relationship with the university, then belonging to such an
organization could increase dissatisfaction with the university as it increased
awareness of racial problems. *(For practice in dealing with interaction, see
problem 17.4.)*

In closing this section, let me stress that, for the sake of clarity, our examples have been unusually "clean." In any given research project, the results of controlling for third variables will probably be considerably more ambiguous and open to varying interpretation than are the examples presented
here. In the case of spurious relationships, for example, the measures of association computed for the partial tables will probably not actually drop to zero,
even though they may be dramatically lower than the bivariate measure. (The
pattern where the partial measures are roughly equivalent and much lower
than the bivariate measure, but not zero, is sometimes referred to as attenua-
tion.) It is probably best to consider the examples above as ideal types against
which any given empirical result can be compared.

Table 17.9 summarizes this discussion by outlining guidelines for decision making for each of the three outcomes discussed in this section. Since

TABLE 17.9   A SUMMARY OF THE POSSIBLE RESULTS OF CONTROLLING FOR THIRD VARIABLES

Partial Tables (compared with bivariate table) Show	Pattern	Implications for Further Analysis	Likely Next Step in Statistical Analysis	Theoretical Implications
Same relationship between X and Y	Direct relationship, replication	Disregard Z	Analyze another control variable	Theory that X causes Y is supported
Weaker relationship between X and Y	Spurious relationship	Incorporate Z	Focus on relationship between Z and Y	Theory that X causes Y is not supported
	Intervening relationship	Incorporate Z	Focus on relationships between X, Y, and Z	Theory that X causes Y is partially supported but must be revised to take Z into account
Mixed	Interaction	Incorporate Z	Analyze subgroups (categories of Z) separately	Theory that X causes Y is partially supported but must be revised to take Z into account

your own results will probably be more ambiguous than the ideal types presented here, you should regard this table as a set of suggestions and not as a substitute for your own creativity and sensitivity to the problem under consideration.

17.4 PARTIAL GAMMA (G_p)   When the results of controlling for a third variable indicate a direct, spurious, or intervening relationship, it is often useful to compute an additional measure that indicates the overall strength of the association between X and Y after the effects of the control variable (Z) have been removed. This statistic, called **partial gamma (G_p)**, is somewhat easier to compare with the bivariate gamma than are the gammas computed on the partial tables separately. G_p is computed across all partial tables by Formula 17.1:

FORMULA 17.1

$$G_p = \frac{\Sigma N_s - \Sigma N_d}{\Sigma N_s + \Sigma N_d}$$

where ΣN_s = the number of pairs of cases ranked the same across all partial tables

ΣN_d = the number of pairs of cases ranked differently across all partial tables

In words, ΣN_s is the total of all N_s's from the partial tables, and ΣN_d is the total of all N_d's from all partial tables. (See Chapter 15 to review the computation of N_s and N_d.)

To illustrate the computation of G_p, let us return to Tables 17.3 and 17.5. In Table 17.3, the relationship between satisfaction and membership while controlling for GPA was displayed. The gammas computed on the partial tables were essentially equal to each other and to the bivariate gamma. Our conclusion was that the control variable had no effect on the relationship, and this conclusion can be confirmed by also computing partial gamma.

From Table 17.3A (high GPA)	From Table 17.3B (low GPA)
$N_s = (29)(54) = 1566$	$N_s = (28)(55) = 1540$
$N_d = (28)(24) = 672$	$N_d = (28)(24) = 672$

$$\Sigma N_s = 1566 + 1540 = 3106$$
$$\Sigma N_d = 672 + 672 = 1344$$

$$G_p = \frac{\Sigma N_s - \Sigma N_d}{\Sigma N_s + \Sigma N_d}$$

$$G_p = \frac{3106 - 1344}{3106 + 1344}$$

$$G_p = \frac{1762}{4450}$$

$$G_p = .40$$

The partial gamma measures the strength of the association between X and Y once the effects of Z have been removed. In this instance, the partial gamma is the same value as the bivariate gamma ($G_p = G = .40$) and indicates that GPA has no effect on the relationship between satisfaction and membership.

When class standing was controlled (Table 17.5), clear evidence of a spurious relationship was found, since the gammas computed on the partial tables dropped almost to zero. Let us see what the value of partial gamma (G_p) would be for this second control:

From Table 17.5A (upperclasses)	From Table 17.5B (underclasses)
$N_s = (8)(97) = 776$	$N_s = (49)(12) = 588$
$N_d = (32)(24) = 768$	$N_d = (24)(24) = 576$

$$\Sigma N_s = 776 + 588 = 1363$$
$$\Sigma N_d = 768 + 576 = 1344$$

$$G_p = \frac{\Sigma N_s - \Sigma N_d}{\Sigma N_s + \Sigma N_d}$$

$$G_p = \frac{1364 - 1344}{1364 + 1344}$$

$$G_p = \frac{20}{2708}$$

$$G_p = .01$$

Once the effects of the control variable are removed, there is no relationship between X and Y. The very low value of G_p confirms our previous conclusion that the bivariate relationship between membership and satisfaction is spurious and actually due to the effects of class standing.

In a sense, G_p tells us no more about the relationships than we can see for ourselves from a careful analysis of the percentage distributions of Y in the partial tables or by a comparison of the measures of association computed on the partial tables. The advantage of G_p is that it tells us, in a single number (that is, in a compact and convenient way), the precise effects of Z on the relationship between X and Y. While G_p is no substitute for the analysis of the partial tables per se, it is a convenient way of stating our results and conclusions when working with direct or spurious relationships.

Although G_p can be calculated in cases of interactive relationships (see Table 17.7), it is rather difficult to interpret in these instances. If substantial interaction is found in the partial tables, this indicates that the control variable has a profound effect on the bivariate relationship. Thus, the analyst should not attempt to separate the effects of Z from the bivariate relationship and, since G_p involves exactly this kind of separation, it should not be computed.

17.5 WHERE DO CONTROL VARIABLES COME FROM?

In one sense, this question is quite easy to answer. Although control variables can arise from various sources, they arise especially from theory. Social research proceeds in many different ways and is begun in response to a variety of problems. However, virtually all research projects are guided by a more-or-less-explicit theory or by some question about the relationship between two or more variables. The ultimate goal of social research is to develop defensible generalizations that improve our understanding of the variables under consideration and link back to theory at some level.

Thus, research projects are anchored in theory; and the concepts of interest, which will later be operationalized as variables, are first identified and their interrelationships first probed at the theoretical level. Since the social world is exceedingly complex, very few theories attempt to encompass that world in only two variables. Theories are, almost by definition, multivariate even when they focus on a bivariate relationship. Thus, to the extent that a research project is anchored in theory, the theory itself will suggest the control variables that need to be incorporated into the analysis. In the example used throughout this chapter, I tried to suggest that any researcher attempting to probe the relationship between involvement in an organization and satisfaction would, in the course of thinking over the possibilities, identify a number of additional variables that needed to be explicitly incorporated into the analysis.

Of course, textbook descriptions of the research process are oversimplified and tend to imply that research flows smoothly from conceptualization to operationalization to quantification to generalization. In reality, research is commonly characterized by surprises, unexpected outcomes, and unanticipated results. Research in "real life" is typically more loosely structured and requires more imagination and creativity than textbooks can fully convey. My point is that the control variables that might be appropriate to incorporate in the data-analysis phase will be suggested or implied in the theoretical backdrop of the research project. They will flow from the researcher's imagination and sensitivity to the problem being addressed as much as from any other source.

These considerations have taken us well beyond the narrow realm of statistics and back to the planning stages of the research project. At this early time the researcher must make decisions about which variables to measure during the data-gathering phase and, thus, which variables might be incorporated as potential controls. Careful thinking and an extended consideration of possible outcomes at the planning stage will pay significant dividends during the data-analysis phase. Ideally, all relevant control variables will be incorporated and readily available for statistical analysis. Thus, control variables come from the theory underlying the research project and from creative and imaginative thinking and planning during the early phases of the project. Nonetheless, it is not unheard of for a researcher to realize during data analysis that the control variable now so obviously relevant was never measured during data gathering and is thus unavailable for statistical analysis.

17.6 THE LIMITATIONS OF ELABORATING BIVARIATE TABLES

The basic limitation of this technique involves sample size. Elaboration is a relatively inefficient technique for multivariate analysis because it requires that the researcher divide the sample into a series of partial tables. If the control variable has more than two or three possible values, or if we attempt to control for more than one variable at a time, many partial tables will be produced. The greater the number of partial tables, the more likely we are to run out of cases to fill all the cells of each partial table. Empty or small cells, in turn, can create serious problems in terms of generalizability and confidence in our findings.

To illustrate, the example used throughout this chapter began with two dichotomized variables and a four-cell table (Table 17.1). Each of the control variables was also dichotomized, and we never confronted more than two partial tables with four cells each, or eight cells for each control variable. If we had used a control variable with three values, we would have had 12 cells to fill up, and if we had attempted to control for two dichotomized variables, we would have had 16 cells in four different partial tables. Clearly, as control variables become more elaborate and/or as the process of controlling becomes more complex, the phenomenon of empty or small cells will increasingly become a problem.

Two potential solutions to this dilemma immediately suggest themselves. The easy solution is to reduce the number of cells in the partial tables by collapsing categories within variables. If all variables are dichotomized, for example, the number of cells will be kept to a minimum. The best solution is to work with only very large samples. Unfortunately, the easy solution will often violate common sense (since the more categories are collapsed, the more likely that dissimilar elements will be grouped under a common label); and the best solution is not always feasible (mundane matters of time and money rear their ugly heads).

A third solution to the problem of empty cells requires the (sometimes risky) assumption that the variables of interest are measured at the interval-ratio level. At that level the techniques of partial and multiple correlation and regression, to be introduced in Chapter 18, are available. These multivariate techniques are more efficient than elaboration because they utilize all cases simultaneously and do not require that the sample be divided among the various partial tables.

SUMMARY

1. Most research questions require the analysis of the interrelationship among many variables, even when the researcher is primarily concerned with a specific bivariate relationship. Multivariate statistical techniques provide the researcher with a set of tools by which additional information can be gathered about the variables of interest and by which causal interrelationships can be probed.
2. When variables have been organized in bivariate tables, multivariate analysis proceeds by controlling for a third variable. Partial tables are constructed and compared with each other and with the original bivariate table. Comparisons are made easier if appropriate measures of association are computed for all tables.
3. A direct relationship exists between the independent (X) and dependent (Y) variables if, after controlling for the third variable (Z), the relationship between X and Y is the same across all partial tables and the same as in the bivariate table. This pattern suggests a causal relationship between X and Y.
4. If the relationship between X and Y is the same across all partial tables but much weaker than in the bivariate table, the relationship is either spurious (Z causes both X and Y) or intervening (X causes Z, which causes Y). Either pattern suggests that Z must be explicitly incorporated into the analysis.
5. Interaction exists if the relationship between X and Y varies across the partial tables and between each

partial and the bivariate table. This pattern suggests that no simple or direct causal relationship exists between X and Y and that Z must be explicitly incorporated into the causal scheme.
6. Partial gamma (G_p) is a useful summary statistic that measures the strength of the association between X and Y after the effects of the control variable (Z) have been removed. Partial gamma should not be computed when an analysis of the partial tables shows substantial interaction.
7. Potential control variables must be identified before the data-gathering phase of the research project. The theoretical backdrop of the research project, along with creative thinking and some imagination, will suggest the variables that should be controlled for and measured.
8. Controlling for third variables by constructing partial tables is inefficient in that the cases must be spread out across many cells. If the variables have many categories and/or the researcher attempts to control for more than one variable simultaneously, "empty cells" may become a problem. It may be possible to deal with this problem by either collapsing categories or gathering very large samples. If interval-ratio level of measurement can be assumed, the multivariate techniques presented in the next chapter will be preferred, since they do not require the partitioning of the sample.

SUMMARY OF FORMULAS

Partial gamma   17.1 $G_p = \dfrac{\Sigma N_s - \Sigma N_d}{\Sigma N_s + \Sigma N_d}$

GLOSSARY

Direct relationship. A multivariate relationship in which the control variable has no effect on the bivariate relationship.

Explanation. See **spurious relationship**.

Interaction. A multivariate relationship where a bivariate relationship changes across the categories of the control variable.

Interpretation. See **intervening relationship**.

Intervening relationship. A multivariate relationship where a bivariate relationship becomes substantially weaker after a third variable is controlled for. The independent and dependent variables are linked primarily through the control variable.

Partial gamma (G_p). A statistic that indicates the strength of the association between two variables after the effects of a third variable have been removed.

Partial tables. Tables produced when controlling for a third variable.

Replication. See **direct relationship**.

Specification. See **interaction**.

Spurious relationship. A multivariate relationship in which a bivariate relationship becomes substantially weaker after a third variable is controlled for. The independent and dependent variables are not causally linked. Rather, both are caused by the control variable.

Z. Symbol for any control variable.

PROBLEMS

17.1 $\boxed{\text{SOC}}$ Problem 15.1 concerned the assimilation of a small sample of immigrants. One table showed the relationship between length of residence in the U.S. and facility with the English language:

	Length of Residence		
English Facility	Less than Five Years	More than Five Years	Totals
Low	20	10	30
High	5	15	20
Totals	25	25	50

a. If necessary, find the column percentages and compute gamma for this table. Describe the bivariate relationship in terms of strength and direction.

    Is the relationship between residence and language affected by the gender of the immigrant? The partial tables showing the bivariate relationship for males and females are

A. Males:

	Length of Residence		
English Facility	Less than Five Years	More than Five Years	Totals
Low	10	5	15
High	2	8	10
Totals	12	13	25

B. Females:

	Length of Residence		
English Facility	Less than Five Years	More than Five Years	Totals
Low	10	5	15
High	3	7	10
Totals	13	12	25

b. Find the column percentages and compute gamma for each of the partial tables. Compare these percentages and gammas with each other and with the bivariate table. Does controlling for gender change the bivariate relationship?

Is the relationship between residence and language affected by the origin of the immigrant? The partial tables showing the bivariate relationship for Asian and Hispanic immigrants are

A. Asian immigrants:

English Facility	Length of Residence		Totals
	Less than Five Years	More than Five Years	
Low	8	4	12
High	2	5	7
Totals	10	9	19

B. Hispanic immigrants:

English Facility	Length of Residence		Totals
	Less than Five Years	More than Five Years	
Low	12	6	18
High	3	10	13
Totals	15	16	31

c. Find the column percentages and compute gamma for each of the partial tables. Compare these percentages and gammas with each other and with the bivariate table. Does controlling for the origin of the immigrants change the bivariate relationship?

17.2 [SOC] Data on suicide rates, age structure, and unemployment rates have been gathered for 100 census tracts. Suicide and unemployment rates have been dichotomized at the median so that each tract could be rated as high or low. Age structure is measured in terms of the percentage of the population age 65 and older. This variable has also been dichotomized, and tracts have been rated as high or low. The tables below display the bivariate relationship between suicide rate and age structure and the same relationship controlling for unemployment.

Suicide Rate	Population 65 and Older (%)		Totals
	Low	High	
Low	45	20	65
High	10	25	35
Totals	55	45	100

a. Calculate percentages and gamma for the bivariate table. Describe the bivariate relationship in terms of strength and direction.

Suicide rate by age, controlling for unemployment:

A. High unemployment:

Suicide Rate	Population 65 and Older (%)		Totals
	Low	High	
Low	23	10	33
High	5	12	17
Totals	28	22	50

B. Low unemployment:

Suicide Rate	Population 65 and Older (%)		Totals
	Low	High	
Low	22	10	32
High	5	13	18
Totals	27	23	50

b. Calculate percentages and gamma for each partial table. Compare the partial tables with each other and with the bivariate table. Compute partial gamma (G_p).

c. Summarize the results of the control. Does unemployment rate have any effect on the relationship between age and suicide rate? Describe the effect of the control variable in terms of the pattern of percentages, the value of the gammas, and the possible causal relationships among these variables.

17.3 [SW] Do long-term patients in mental health facilities become more withdrawn and reclusive over time? A sample of 608 institutionalized patients was rated by a standard "reality orientation scale." Is there a relationship with length of institutionalization? Does gender have any effect on the relationship?

Reality orientation by length of institutionalization:

Reality Orientation	Length of Institutionalization		Totals
	Less than 5 Years	More than 5 Years	
Low	200	213	413
High	117	78	195
Totals	317	291	608

Reality orientation by length of institutionalization, controlling for gender:

A. Females:

Reality Orientation	Length of Institutionalization		Totals
	Less than 5 Years	More than 5 Years	
Low	95	120	215
High	60	37	97
Totals	155	157	312

B. Males:

Reality Orientation	Length of Institutionalization		Totals
	Less than 5 Years	More than 5 Years	
Low	105	93	198
High	57	41	98
Totals	162	134	296

17.4 | SOC | Is there a relationship between attitudes on sexuality and age? Are older people more conservative with respect to questions of sexual morality? A national sample of 925 respondents has been questioned about attitudes on premarital sex. Responses have been collapsed into two categories: those who believe that premarital sex is "always wrong" and those who believe it is not wrong under certain conditions ("sometimes wrong"). These responses have been crosstabulated by age, and the results are reported below:

Premarital Sex Is	Age		Totals
	Younger than 35	35 or Older	
Always wrong	90	235	325
Sometimes wrong	420	180	600
Totals	510	415	925

a. Calculate percentages and gamma for the table. Is there a relationship between these two variables? Describe the bivariate relationship in terms of its strength and direction.

Below, the bivariate relationship is reproduced after controlling for the sex of the respondent. Does gender have any effect on the relationship?

Attitude toward premarital sex by gender:

A. Male respondents:

Premarital Sex Is	Age		Totals
	Younger than 35	35 or Older	
Always wrong	70	55	125
Sometimes wrong	190	80	270
Totals	260	135	395

B. Female respondents:

Premarital Sex Is	Age		Totals
	Younger than 35	35 or Older	
Always wrong	20	180	200
Sometimes wrong	230	100	330
Totals	250	280	530

b. Calculate percentages and gamma for each partial table. Compute partial gamma (G_p).

c. Summarize the results of this control. Does gender have any effect on the relationship between age and attitude? If so, describe the effect of the control variable in terms of the pattern of percentages, the value of the gammas, and the possible causal interrelationships among these variables.

17.5 | SOC | A job-training center is trying to justify its existence to its funding agency. To this end, data on four variables have been collected for each of the 403 trainees served over the past three years: (1) whether or not the trainee completed the program, (2) whether or not the trainee got and held a job for at least a year after training, (3) the sex of the trainee, and (4) the race of the trainee. Is employment related to completion of the program? Is the relationship between completion and employment affected by race or sex? Employment by training:

Held Job for at Least One Year?	Training Completed?		Totals
	Yes	No	
Yes	145	60	205
No	72	126	198
Totals	217	186	403

Employment by training, controlling for race:
A. Whites:

Held Job for at Least One Year?	Training Completed?		
	Yes	No	Totals
Yes	85	33	118
No	38	47	85
Totals	123	80	203

B. Blacks:

Held Job for at Least One Year?	Training Completed?		
	Yes	No	Totals
Yes	60	27	87
No	34	79	113
Totals	94	106	200

Employment by training, controlling for sex:
C. Males:

Held Job for at Least One Year?	Training Completed?		
	Yes	No	Totals
Yes	73	30	103
No	36	62	98
Totals	109	92	201

D. Females:

Held Job for at Least One Year?	Training Completed?		
	Yes	No	Totals
Yes	72	30	102
No	36	64	100
Totals	108	94	202

17.6 SOC What are the social sources of support for the environmental movement? A recent survey gathered information on level of concern for such issues as global warming and acid rain. Is concern for the environment related to level of education? What effects do the control variables have? Write a paragraph summarizing your conclusions.

Concern for the environment by level of education:

Concern for the Environment	Level of Education		
	Low	High	Totals
Low	27	35	62
High	22	48	70
Totals	49	83	132

a. Concern for the environment by level of education, controlling for gender:
A. Males:

Concern for the Environment	Level of Education		
	Low	High	Totals
Low	14	17	31
High	11	22	33
Totals	25	39	64

B. Females:

Concern for the Environment	Level of Education		
	Low	High	Totals
Low	13	18	31
High	11	26	37
Totals	24	44	68

b. Concern for the environment by level of education, controlling for "level of trust in the nation's leadership":
A. Low levels of trust:

Concern for the Environment	Level of Education		
	Low	High	Totals
Low	6	22	28
High	10	40	50
Totals	16	62	78

B. High levels of trust:

Concern for the Environment	Level of Education		
	Low	High	Totals
Low	21	13	34
High	12	8	20
Totals	33	21	54

c. Concern for the environment by level of education, controlling for race:

A. White:

Concern for the Environment	Level of Education		Totals
	Low	High	
Low	19	11	30
High	18	44	62
Totals	37	55	92

B. Black:

Concern for the Environment	Level of Education		Totals
	Low	High	
Low	8	24	32
High	4	4	8
Totals	12	28	40

17.7 SOC In a survey, 247 respondents were asked to indicate their overall level of satisfaction with their lot in life. Is satisfaction related to social class? Is the bivariate relationship affected by gender? By marital status?

Satisfaction by social class:

Satisfaction	Social Class		Totals
	Working and Lower	Upper and Middle	
Low	19	12	31
Moderate	80	54	134
High	27	55	82
Totals	126	121	247

a. Calculate percentages for the table so that the effect of social class (X) on satisfaction (Y) is displayed.
b. Compute and interpret gamma for this table.
c. Describe the bivariate relationship in terms of strength and direction.

Satisfaction by social class, controlling for gender:

A. Male respondents:

Satisfaction	Social Class		Totals
	Working and Lower	Upper and Middle	
Low	10	8	18
Moderate	28	20	48
High	12	29	41
Totals	50	57	107

B. Female respondents:

Satisfaction	Social Class		Totals
	Working and Lower	Upper and Middle	
Low	9	4	13
Moderate	52	34	86
High	15	26	41
Totals	76	64	140

d. Calculate percentages for the partial tables so that the effect of social class on satisfaction is displayed.
e. Compute gammas for both partial tables. Compare these with the gamma for the bivariate table.
f. Compute partial gamma (G_p).

Satisfaction by social class, controlling for marital status:

A. Married respondents:

Satisfaction	Social Class		Totals
	Working and Lower	Upper and Middle	
Low	8	8	16
Moderate	42	38	80
High	20	35	55
Totals	70	81	151

B. Nonmarried respondents:

Satisfaction	Social Class		Totals
	Working and Lower	Upper and Middle	
Low	11	4	15
Moderate	38	16	54
High	7	20	27
Totals	56	40	96

g. Calculate percentages for the partial tables so that the effect of social class on satisfaction is displayed.
h. Compute gammas for both partial tables. Compare these with the gamma for the bivariate table.
i. Compute partial gamma (G_p).
j. Summarize the results of the two controls. Are satisfaction and social class related? Does gender or marital status have any effect on the relationship?

17.8 SOC In problem 15.18, we used the 1996 General Social Survey data set to investigate the relationships between income and five dependent variables. Let's return to two of those relationships and see if sex has any effect on the bivariate relationships. The bivariate and partial tables are presented below along with the bivariate gammas that were computed in the previous exercise. Compute percentages and gammas for each of the partial tables and state your conclusions. Does controlling for sex have any effect? Bivariate table (from problem 15.18):

Right to Abortion?	Less than $22,500 (1)	$22,500 to $50,000 (2)	More than $50,000 (3)	Totals
Yes (1)	108	126	134	368
No (2)	159	169	113	441
Totals	267	295	247	809

Gamma = −.18

(HINT: Note that the higher score on ABANY is associated with "No." The negative sign on the gamma means that as score on income increases, score on ABANY—opposition to abortion—decreases. In other words, people with lower incomes were more opposed to the legal right to abortion.)

a. Support for the legal right to abortion by income, controlling for sex:

A. Males:

Right to Abortion?	Less than $22,500 (1)	$22,500 to $50,000 (2)	More than $50,000 (3)	Totals
Yes (1)	29	67	73	169
No (2)	59	76	63	198
Totals	88	143	136	367

B. Females:

Right to Abortion?	Less than $22,500 (1)	$22,500 to $50,000 (2)	More than $50,000 (3)	Totals
Yes (1)	79	59	61	199
No (2)	100	93	50	243
Totals	179	152	111	442

Bivariate table (from problem 15.18):

Right to Suicide?	Less than $22,500 (1)	$22,500 to $50,000 (2)	More than $50,000 (3)	Totals
Yes (1)	149	198	185	532
No (2)	112	108	69	289
Totals	261	306	254	821

Gamma = −.22

b. Approval of suicide for people with an incurable disease (SUICIDE1) by income controlling for sex:

A. Males:

Right to Suicide?	Less than $22,500 (1)	$22,500 to $50,000 (2)	More than $50,000 (3)	Totals
Yes (1)	48	102	102	252
No (2)	34	41	30	105
Totals	82	143	132	357

B. Females:

Right to Suicide?	Less than $22,500 (1)	$22,500 to $50,000 (2)	More than $50,000 (3)	Totals
Yes (1)	101	96	83	280
No (2)	78	67	39	184
Totals	179	163	122	464

17.9 SOC Does support for traditional gender roles vary by social class? Is this relationship affected by sex? By political ideology? By age? The 1996 General Social Survey includes an item (FEFAM) that asked respondents to agree or disagree that it is better if men work and women tend house. Responses to this item have been collapsed for this exercise into (1) agree (support for traditional gender roles) and (2) disagree (opposition to traditional gender roles).

a. Compute percentages and gamma for the bivariate table. Describe the relationship in terms of strength and direction. (HINT: Be careful when interpreting direction. Remember that a higher score on FEFAM means the respondent opposes traditional gender roles.)

b. Compute percentages and gamma for each set of partial tables. What effect did each control variable have on the bivariate relationship? *(NOTE: Since some respondents did not answer all questions, there are fewer cases in the partial tables than in the bivariate table.)*

c. Summarize the results of the analysis in a paragraph.

The bivariate table:

Support for Traditional Gender Roles	Income			
	Less than $22,500 (1)	$22,500 to $50,000 (2)	More than $50,000 (3)	Totals
Agree (1)	153	142	88	383
Disagree (2)	180	244	227	651
Totals	333	386	315	1034

Support for traditional gender roles by income, controlling for sex:

A. Males:

Support for Traditional Gender Roles	Income			
	Less than $22,500 (1)	$22,500 to $50,000 (2)	More than $50,000 (3)	Totals
Agree (1)	58	72	49	179
Disagree (2)	50	118	108	276
Totals	108	190	157	455

B. Females:

Support for Traditional Gender Roles	Income			
	Less than $22,500 (1)	$22,500 to $50,000 (2)	More than $50,000 (3)	Totals
Agree (1)	95	70	39	204
Disagree (2)	130	126	119	375
Totals	225	196	158	579

Support for traditional gender toles by income, controlling for political ideology.*

A. Liberals:

Support for Traditional Gender Roles	Income			
	Less than $22,500 (1)	$22,500 to $50,000 (2)	More than $50,000 (3)	Totals
Agree (1)	28	16	13	57
Disagree (2)	65	75	80	220
Totals	93	91	93	277

B. Moderates:

Support for Traditional Gender Roles	Income			
	Less than $22,500 (1)	$22,500 to $50,000 (2)	More than $50,000 (3)	Totals
Agree (1)	59	46	38	143
Disagree (2)	61	96	70	227
Totals	120	142	108	370

C. Conservatives:

Support for Traditional Gender Roles	Income			
	Less than $22,500 (1)	$22,500 to $50,000 (2)	More than $50,000 (3)	Totals
Agree (1)	52	74	36	162
Disagree (2)	40	68	74	182
Totals	92	142	110	344

Support for traditional gender roles by income, controlling for age (the sample has been divided into three age groups of approximately equal size):

A. Ages 18–36:

Support for Traditional Gender Roles	Income			
	Less than $22,500 (1)	$22,500 to $50,000 (2)	More than $50,000 (3)	Totals
Agree (1)	31	35	21	87
Disagree (2)	103	109	58	270
Totals	134	144	79	357

---

*Partial tables have fewer cases than the bivariate table because of missing scores. Not all respondents answered all questions.

B. Ages 37–50:

Support for Traditional Gender Roles	Income			Totals
	Less than $22,500 (1)	$22,500 to $50,000 (2)	More than $50,000 (3)	
Agree (1)	23	54	37	114
Disagree (2)	45	84	114	243
Totals	68	138	151	357

C. Ages 51–89:

Support for Traditional Gender Roles	Income			Totals
	Less than $22,500 (1)	$22,500 to $50,000 (2)	More than $50,000 (3)	
Agree (1)	98	53	30	181
Disagree (2)	31	51	55	137
Totals	129	104	85	318

SPSS for Windows

Using *SPSS for Windows* to Elaborate Bivariate Tables

DEMONSTRATION 17.1 Analyzing the Effects of Age and Sex on Sexual Attitudes

Since we have been concerned with bivariate tables in this chapter, it will come as no surprise to find that we will once again make use of the **Crosstabs** procedure. To control for a third variable with **Crosstabs**, add the name of the control variable to the box at the bottom of the **Crosstabs** dialog window. In other words, after you click **Statistics**, **Summarize**, and then **Crosstabs**, place the name of your dependent variable(s) in the top (row) box, the name of your independent variable(s) in the middle (column) box, and the name of your control variable(s) in the box at the bottom of the screen.

To illustrate, in Demonstration 13.3 and again in Demonstration 15.1, we looked at the relationships between *premarsx* and recoded *age* or *ager* (see Demonstrations 13.3 and 15.1 for the recoding scheme) and found moderate, negative relationships (gamma = −.26). That is, approval of premarital sex decreased as age increased. Now, let's see if sex has any effect on these bivariate relationships. Are older women more opposed to premarital sex than older men?

If necessary, recode *age* as in Demonstration 13.3. On the **Crosstabs** dialog window, *premarsx* should be in the top (row) box, recoded *age* in the middle (column) box, and *sex* in the box at the bottom. Make sure you request gamma and column percentages. The output will consist of two partial tables, one for males and one for females, and gammas computed for each of these partial tables.

The bivariate relationships and statistics were displayed in Demonstration 15.1. To conserve space, we will concentrate on gamma and not reproduce the partial tables here. For males, the gamma is −.14 while for females the statistic increases in value to −.35. This looks like a case of interaction between the three variables. Compare the percentages in the partial tables and you will see that males are generally less opposed to premarital sex than females. Specifically, compare the top row of the two partial tables. While the percentage of younger males and females who believe premarital sex is "always wrong" is about the same, older females are much more likely than older males to hold this position. Thus, the attitudes of females are much more affected by age than the attitudes of males (hence, the stronger gamma for females). Opposition to premarital sex increases with age, but the increase is more dramatic for females.

DEMONSTRATION 17.2 Are Prestigious Jobs More Satisfying? Another Look

In Demonstration 15.3, we looked at the relationship between recoded *satjob* and *prestg80*. We found a bivariate gamma of $-.23$, indicating that people with higher-prestige jobs had higher levels of satisfaction with their jobs. While not exactly surprising, we might be able to learn a little more about the relationship with some additional controls. In this demonstration, the bivariate relationship is reexamined with *sex* and *degree* as control variables. Does the relationship between satisfaction and prestige vary by sex? Is job satisfaction for the most educated ("at least some college") affected differently by level of prestige than the job satisfaction of people with high school degrees or people with less than a high school education?

The variable *degree* has a total of five values—too many to permit the variable to be used as a control. To simplify the variable, I grouped all respondents with any level of college education into a single category. The recoding scheme I used was

$$0 \rightarrow 0 \text{ (less than high school)}$$
$$1 \rightarrow 1 \text{ (high school diploma)}$$
$$2 \text{ thru } 4 \rightarrow 2 \text{ (at least some college)}$$

I called the recoded version of this variable *rdeg*. If necessary, see Demonstration 15.3 for the recode schemes for *satjob* (*rsat*) and *prestg80* (*rprest*).

Click **Statistics**, **Summarize**, and **Crosstabs** and move *rsat* into the **Row Variable** box, *rprest* into the **Column Variable** box, and *rdeg* and *sex* into the bottom box. Don't forget gamma and column percents. To conserve space, the partial tables are not reproduced here. Instead, we will concentrate on the gammas:

	Gamma
For the bivariate table	$-.23$
For people with less than a high school education	$.11$
For high school graduates	$-.28$
For people with at least some college	$-.17$

There appears to be some interaction between the variables. Satisfaction increases with prestige for both high school and college-educated respondents, although the relationship is weaker for the latter. That is, the job satisfaction for people with at least some college is less dependent on prestige than the job satisfaction of high school graduates.

For people with less than a high school education, surprisingly, the relationship changes direction: members of this group in the more prestigious occupations were less satisfied. How can we explain this? Do people who are less well educated feel "out-of-place" in more prestigious jobs?

Before we make too much of this result, we should note that the number of people in this category (low education/high prestige) is small—a total of 19 cases. At this point we confront a major limitation of the elaboration technique (see Section 17.6): the number of cases is so small that it is hard to know if we have just identified an interesting phenomenon or a statistical anomaly. To what extent can we trust a result based on so few cases? While it's hard to know for sure, it would probably be most prudent to do no more than make note of the pattern.

The control for gender is easier to deal with:

	Gamma
For the bivariate table	−.23
For males	−.26
For females	−.21

These results indicate a direct relationship between prestige and satisfaction. For both males and females, higher-prestige jobs are associated with greater satisfaction.

Exercises

cappun = .03
grass = .238
gunlaw = −.008

spurious

17.1 Use Demonstration 17.1 as a guide and analyze the relationship of recoded *age* on *cappun, grass,* and *gunlaw,* using *sex* as the control variable. Summarize the results in a paragraph or two. Be sure to characterize the relationships as direct, spurious, or interactive.

17.2 Using Demonstration 17.2 as a guide, analyze the relationships between income (*income91*) and job satisfaction, controlling for *sex* and recoded *degree*. Recode *income91* at the median before conducting the analysis. Summarize the results in a paragraph or two. Be sure to characterize the relationships as direct, spurious, or interactive.

18 Partial Correlation and Multiple Regression and Correlation

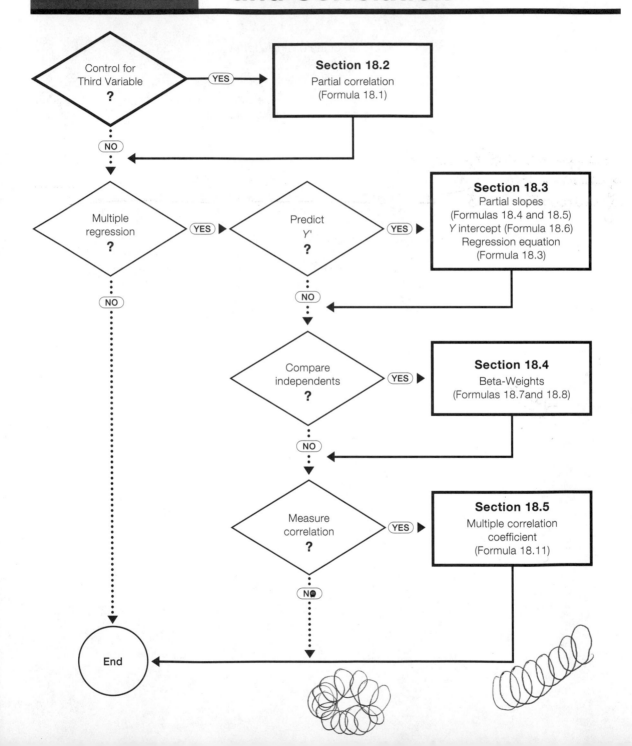

Control for Third Variable ? — YES → **Section 18.2** Partial correlation (Formula 18.1)

NO

Multiple regression ? — YES → **Predict Y' ?** — YES → **Section 18.3** Partial slopes (Formulas 18.4 and 18.5) Y intercept (Formula 18.6) Regression equation (Formula 18.3)

NO

Compare independents ? — YES → **Section 18.4** Beta-Weights (Formulas 18.7 and 18.8)

NO

Measure correlation ? — YES → **Section 18.5** Multiple correlation coefficient (Formula 18.11)

NO

End

18.1 INTRODUCTION

In Chapter 17, I made the point that very few (if any) worthwhile research questions can be answered through a statistical analysis of only two variables. Social science research is, by nature, multivariate and often involves the simultaneous analysis of scores of variables. Some of the most powerful and widely used statistical tools for multivariate analysis are introduced in this chapter. We will cover techniques that are used to analyze causal relationships and to make predictions—crucial endeavors in any science.

These techniques are based on Pearson's r (see Chapter 16). They are generally more flexible than the techniques presented in Chapter 17, produce more information, and provide a wider variety of ways of disentangling the interrelationships among the variables.

The first technique to be introduced will be partial correlation analysis, a technique analogous to controlling for a third variable by constructing partial tables. In fact, the partial correlation coefficient provides much the same kind of information as a partial gamma does. The second technique involves multiple regression and correlation and allows the researcher to assess the effects, separately and in combination, of more than one independent variable on the dependent variable.

Throughout this chapter, we will focus our attention on research situations involving only three variables. This is the least complex application of these techniques, but extensions to situations involving four or more variables are relatively straightforward. To deal efficiently with the computations required by the more complex applications, I refer you to any of the computerized statistical packages probably available at your local computer center.

18.2 PARTIAL CORRELATION

The technique of **partial correlation** can be used when a researcher wishes to observe how a specific bivariate relationship behaves in the presence of a third variable. By observing the partial correlation coefficients, spurious and intervening relationships can be detected (see Chapter 17), and the causal relationships among the variables can be probed. As opposed to the techniques considered in Chapter 17, however, partial correlation analysis does not involve the inspection of the bivariate relationship for each category of the control variable. For this reason, partial correlation techniques will not reveal interactive relationships among the variables (that is, relationships where the nature of the bivariate relationship changes across the categories of the control variable).

Before dealing with partial correlation per se, let me introduce some new terminology. Since we will be dealing with more than one bivariate relationship, we will use subscripts to identify the various correlations. Thus, the symbol r_{yx} will refer to the correlation coefficient between variable Y and variable X. By the same token, r_{yz} will refer to the correlation coefficient between Y and Z, and r_{xz} to the correlation coefficient between X and Z. Correlation coefficients calculated for bivariate relationships are often referred to as **zero-order correlations**.

Partial correlation coefficients, when controlling for a single variable, are called first-order partials and can be symbolized as $r_{yx.z}$. The variable to the right of the dot is the control variable. Thus, $r_{yx.z}$ refers to the partial correlation coefficient that measures the relationship between variables X and Y while controlling for variable Z. The formula for the first-order partial is

FORMULA 18.1

$$r_{yx.z} = \frac{r_{yx} - (r_{yz})(r_{xz})}{\sqrt{1 - r_{yz}^2}\sqrt{1 - r_{xz}^2}}$$

To solve this formula, first calculate the zero-order coefficients between all possible pairs of variables (variables X and Y, X and Z, and Y and Z).

To illustrate the computation of a first-order partial, let us return to the problem introduced in Chapter 16 that examined the relationship between number of children (X) and husband's contribution to housework (Y) for 12 dual-career families. The zero-order r between these two variables ($r_{yx} = .50$) indicated a moderate, positive relationship. Suppose the researcher wished to investigate the possible effects of socioeconomic status (SES) on the bivariate relationship. The original data (from Table 16.1) and the scores of the 12 families on the new variable of SES (measured by the years of education completed by the husband) are presented in Table 18.1.

The three correlations indicate that the husband's contribution to housework is related to number of children ($r_{yx} = .50$), that husbands in higher-SES families tend to do less housework ($r_{yz} = -.30$), and that higher-SES families have fewer children ($r_{xz} = -.47$). Is the relationship between husband's housework and number of children affected by SES? Substituting the above correlations into Formula 18.1, we would have

$$r_{yx.z} = \frac{r_{yx} - (r_{yz})(r_{xz})}{\sqrt{1 - r_{yz}^2}\sqrt{1 - r_{xz}^2}}$$

$$r_{yx.z} = \frac{.50 - (-.30)(-.47)}{\sqrt{1 - (-.30)^2}\sqrt{1 - (-.47)^2}}$$

$$r_{yx.z} = \frac{.50 - (.14)}{\sqrt{1 - .09}\sqrt{1 - .22}}$$

$$r_{yx.z} = \frac{.36}{\sqrt{.91}\sqrt{.78}}$$

$$r_{yx.z} = \frac{.36}{(.95)(.88)}$$

$$r_{yx.z} = \frac{.36}{.84}$$

$$r_{yx.z} = .43$$

The first-order partial ($r_{yx.z} = .43$) is lower in value than the zero-order coefficient ($r_{yx} = .50$), but the difference in the two values is not great. This result

TABLE 18.1   SCORES ON THREE VARIABLES FOR 12 DUAL-WAGE-EARNER FAMILIES
AND ZERO-ORDER CORRELATIONS

Family	Husband's Housework (Y)	Number of Children (X)	Husband's Years of Education (Z)
A	1	1	12
B	2	1	14
C	3	1	16
D	5	1	16
E	3	2	18
F	1	2	16
G	5	3	12
H	0	3	12
I	6	4	10
J	3	4	12
K	7	5	10
L	4	5	16
	$r_{yx} = .50$	$r_{yz} = -.30$	$r_{xz} = -.47$

suggests a direct relationship between variables X and Y (see Section 17.3). That is, when controlling for SES, the statistical relationship between husband's housework and number of children is essentially unchanged. Regardless of SES, the hours devoted to housework by husbands increase with the number of children. As was the case when a direct relationship was found in Chapter 17, the next step in statistical analysis would probably be to discard this control variable (SES) and select another (see Table 17.9). The more the bivariate relationship retains its strength across a series of controls for the third variables (Z's), the stronger the evidence for a direct relationship between X and Y.

In addition to the pattern discussed above, two other possible relationships between the partial and zero-order correlation coefficients exist. If the partial is much lower in value than the zero-order coefficient ($r_{yx.z} < r_{yx}$), the bivariate relationship between X and Y may be spurious, or Z may intervene between X and Y. The interpretation of this outcome follows the lines established in Chapter 17; the probable next step in the statistical analysis would be to incorporate the control variable into the analysis as an independent (see Table 17.9). Let me stress that partial correlation techniques cannot distinguish between the case where the control variable is antecedent (where Z "causes" both X and Y) and the case where the control variable is intervening (when X "causes" Z, which in turn "causes" Y). Judgments about possible causal relationships must be made on the grounds of temporal ordering among the variables and/or theory.

A third possible outcome of the application of partial correlation is for the partial correlation coefficient to be greater in value than the zero-order coefficient ($r_{xy.z} > r_{yx}$). This outcome would be consistent with a causal model

FIGURE 18.1 A POSSIBLE CAUSAL RELATIONSHIP AMONG THREE VARIABLES

where the variable taken as independent and the control variable each had a separate effect on the dependent variable and were uncorrelated with each other. This relationship is depicted in Figure 18.1. The absence of an arrow between X and Z indicates that they have no mutual relationship.

Upon finding this pattern, a researcher would conclude that both X and Z were independent variables, and the next step in the statistical analysis would probably involve multiple correlation and regression. As we shall see in Sections 18.3 and 18.4, these techniques enable the researcher to isolate the separate effects of several independent variables on the dependent variable and thus to make judgments about which independent has the stronger effect on the dependent. *(For practice in computing and interpreting partial correlation coefficients, see problems 18.1 to 18.3.)*

18.3 MULTIPLE REGRESSION: PREDICTING THE DEPENDENT VARIABLE

In Chapter 16, the least-squares regression line was introduced as a way of describing the overall linear relationship between two interval-ratio variables and of predicting scores on Y from scores on X. This line was the best-fitting line to summarize the bivariate relationship and was defined by the formula

FORMULA 18.2

$$Y = a + bX$$

where a = the Y intercept
b = the slope

The least-squares regression line can be modified to include (theoretically) any number of independent variables. This technique is called **multiple regression**. For ease of explication, we will confine our attention to the case involving two independent variables. The least-squares multiple regression equation is

FORMULA 18.3

$$Y = a + b_1 X_1 + b_2 X_2$$

where b_1 = the partial slope of the linear relationship between the first independent variable and Y
b_2 = the partial slope of the linear relationship between the second independent variable and Y

Some new notation and some new concepts are introduced in this formula. First, while the dependent variable is still symbolized as Y, the independent variables are differentiated by subscripts. Thus, X_1 identifies the first independent variable and X_2 the second. The symbol for the slope (b) is also subscripted to identify the independent with which it is associated.

A major difference between the multiple and bivariate regression equations concerns the slopes (b's). In the multiple regression case, the b's are actually **partial slopes**, since they show the amount of change in Y for a unit change in the independent while controlling for the effects of the other independents in the equation. The partial slopes are thus analogous to partial correlation coefficients and represent the direct effect of the associated independent variable on Y.

The partial slopes for the independent variables are determined by Formula 18.4 and Formula 18.5:*

FORMULA 18.4
$$b_1 = \left(\frac{s_y}{s_1}\right)\left(\frac{r_{y1} - r_{y2}r_{12}}{1 - r_{12}^2}\right)$$

FORMULA 18.5
$$b_2 = \left(\frac{s_y}{s_2}\right)\left(\frac{r_{y2} - r_{y1}r_{12}}{1 - r_{12}^2}\right)$$

where b_1 = the partial slope of X_1 on Y
b_2 = the partial slope of X_2 on Y
s_y = the standard deviation of Y
s_1 = the standard deviation of the first independent variable (X_1)
s_2 = the standard deviation of the second independent variable (X_2)
r_{y1} = the bivariate correlation between Y and X_1
r_{y2} = the bivariate correlation between Y and X_2
r_{12} = the bivariate correlation between X_1 and X_2

To illustrate the computation of the partial slopes, we will assess the combined effects of number of children (X_1) and SES (X_2) on husband's contribution to housework. All the relevant information can be calculated from Table 18.1 and is reproduced below:

Husband's Housework	Number of Children	SES
$\bar{Y} = 3.3$	$\bar{X}_1 = 2.7$	$\bar{X}_2 = 13.7$
$s_y = 2.1$	$s_1 = 1.5$	$s_2 = 2.6$

Zero-order correlations
$r_{y1} = .50$
$r_{y2} = -.30$
$r_{12} = -.47$

*Partial slopes can be computed from zero-order slopes, but Formulas 18.4 and 18.5 are somewhat easier to use.

The partial slope for the first independent variable (X_1) is

$$b_1 = \left(\frac{s_y}{s_1}\right)\left(\frac{r_{y1} - r_{y2}r_{12}}{1 - r_{12}^2}\right)$$

$$b_1 = \left(\frac{2.1}{1.5}\right)\left(\frac{.50 - (-.30)(-.47)}{1 - (-.47)^2}\right)$$

$$b_1 = (1.4)\left(\frac{.50 - .14}{1 - .22}\right)$$

$$b_1 = (1.4)\left(\frac{.36}{.78}\right)$$

$$b_1 = (1.4)(.46)$$

$$b_1 = .65$$

For the second independent variable, SES or X_2, the partial slope is

$$b_2 = \left(\frac{s_y}{s_2}\right)\left(\frac{r_{y2} - r_{y1}r_{12}}{1 - r_{12}^2}\right)$$

$$b_2 = \left(\frac{2.1}{2.6}\right)\left(\frac{-.30 - (-.24)}{1 - .22}\right)$$

$$b_2 = (.81)\left(\frac{-.30 + .24}{.78}\right)$$

$$b_2 = (.81)\left(\frac{-.06}{.78}\right)$$

$$b_2 = (.81)(-.08)$$

$$b_2 = -.07$$

Now that partial slopes have been determined for both independent variables, the final unknown in the least-squares multiple regression equation can be found. Note that the Y intercept (a) is calculated from the mean of the dependent variable (symbolized as $\bar{Y}$) and the means of the two independent variables ($\bar{X}_1$ and $\bar{X}_2$).

FORMULA 18.6
$$a = \bar{Y} - b_1\bar{X}_1 - b_2\bar{X}_2$$

Substituting in the proper values for the example problem at hand, we would have

$$a = \bar{Y} - b_1\bar{X}_1 - b_2\bar{X}_2$$
$$a = 3.3 - (.65)(2.7) - (-.07)(13.7)$$
$$a = 3.3 - (1.8) - (-1.0)$$
$$a = 3.3 - 1.8 + 1.0$$
$$a = 2.5$$

For our example problem, the full least-squares multiple regression equation would be

$$Y = a + b_1 X_1 + b_2 X_2$$
$$Y = 2.5 + (.65)X_1 + (-.07)X_2$$

As was the case with the bivariate regression line, this formula can be used to predict scores on the dependent variable from scores on the independent variables. For example, what would be our best prediction of husband's housework (Y') for, say, a family of four children ($X_1 = 4$) where the husband had completed 11 years of schooling ($X_2 = 11$)? Substituting these values into the least-squares formula, we would have

$$Y' = 2.5 + (.65)(4) + (-.07)(11)$$
$$Y' = 2.5 + 2.6 - .8$$
$$Y' = 4.3$$

Our prediction would be that, in a family of four children when the husband had 11 years of education, the husband would contribute 4.3 hours per week to housework. This prediction is, of course, a kind of "educated guess." While our predictions of Y are unlikely to be perfectly accurate, we will make fewer errors of prediction using the least-squares line (and, thus, incorporating information from the independent variables) than we would using any other method of prediction (assuming, of course, that a linear association exists between the independents and the dependent). *(For practice in predicting Y scores and in computing slopes and the Y intercept, see problems 18.1 to 18.6.)*

18.4 MULTIPLE REGRESSION: ASSESSING THE EFFECTS OF THE INDEPENDENT VARIABLES

The least-squares multiple regression equation (Formula 18.3) is used to isolate the separate effects of the independents and to predict scores on the dependent variable. However, in many situations, using this formula to determine the relative importance of the various independents will be awkward—especially when the independent variables differ in terms of units of measurement. In our example problem, the first independent is number of children, while the second is years of education. In situations where the units of measurement differ, we will not necessarily be able to tell from the partial slopes which independent variable has the strongest effect on the dependent variable and is thus the most important of the independent variables. Comparing the partial slopes of variables that differ in units of measurement is a little like comparing apples and oranges.

The independent variables can be made more nearly comparable by converting all of the variables in the equation to a common scale and thereby eliminating variations in the values of the partial slopes that are solely a function of differences in units of measurement. We can, for example, standardize all distributions by converting all scores into Z scores. Each distribution of scores for each variable would then have a mean of 0 and a standard deviation of 1, and comparisons between the standardized independent variables would become much more meaningful. This standardization could be accomplished by converting all scores for each variable into the equivalent Z scores (that is, subtracting the mean of the distribution from each score and dividing by the standard deviation of the distribution) and then recomputing the slopes and the Y intercept by using the Z scores. This procedure requires a good deal of work and, fortunately, a shortcut is available for directly computing the slopes of the standardized scores. These **standardized partial slopes** are called **beta-weights** and are symbolized b^*. The beta-weights will show the amount of change in the standardized scores of Y for a one-unit change in the standardized scores of the independent while controlling for the effects of the other independents. For the case where we have two independents, the beta-weight for each is found by using Formula 18.7 and Formula 18.8:

FORMULA 18.7
$$b_1{}^* = b_1 \left(\frac{s_1}{s_y}\right)$$

FORMULA 18.8
$$b_2{}^* = b_2 \left(\frac{s_2}{s_y}\right)$$

where $b_1{}^*$ = the standardized partial slope of X_1 on Y
$b_2{}^*$ = the standardized partial slope of X_2 on Y

Using standardized scores, the least-squares regression equation can be written as

FORMULA 18.9
$$Z_y = a_z + b_1{}^* Z_1 + b_2{}^* Z_2$$

where the symbol Z indicates that all scores
have been standardized to the normal curve

The standardized regression equation can be further simplified by dropping the term for the Y intercept, since this term will always be zero when scores have been standardized. This value is the point where the regression line crosses the Y axis and is equal to the mean of Y when all independents equal 0. This relationship can be seen by substituting 0 for all independent variables in Formula 18.6:

$$a = \bar{Y} - b_1\bar{X}_1 - b_2\bar{X}_2$$
$$a = \bar{Y} - b_1\,(0) - b_2\,(0)$$
$$a = \bar{Y}$$

Since the mean of any standardized distribution of scores is zero, the mean of the standardized Y scores will be zero and the Y intercept will also be zero ($a = \bar{Y} = 0$). Thus, Formula 18.9 simplifies to

FORMULA 18.10
$$Z_y = b_1{}^* Z_1 + b_2{}^* Z_2$$

We can now compute the beta-weights for our sample problem to see which of the two independents has the stronger effect on the dependent. For the first independent variable, number of children (X_1):

$$b_1^* = b_1 \left(\frac{s_1}{s_y} \right)$$

$$b_1^* = (.65) \left(\frac{1.5}{2.1} \right)$$

$$b_1^* = (.65)(.71)$$

$$b_1^* = .46$$

For the second independent variable, SES (X_2):

$$b_2^* = b_2 \left(\frac{s_2}{s_y} \right)$$

$$b_2^* = (-.07) \left(\frac{2.6}{2.1} \right)$$

$$b_2^* = (-.07)(1.24)$$

$$b_2^* = -.09$$

Thus, the standardized regression equation, with beta-weights noted, would be

$$Z_y = (.46)Z_1 + (-.09)Z_2$$

and it is immediately obvious that the first independent variable has a much stronger direct effect on Y than the second independent variable.

In summary, multiple regression analysis permits the researcher to summarize the linear relationship among two or more independents and a dependent variable. The unstandardized regression equation (Formula 18.2) permits values of Y to be predicted from the independent variables in the original units of the variables. The standardized regression equation (Formula 18.10) allows the researcher to assess the relative importance of the various independents more easily by noting which of the beta-weights is greatest in value. *(For practice in computing and interpreting beta-weights, see any of the problems at the end of this chapter. It is probably a good idea to start with problem 18.1 since it has the smallest data set and the least complex computations.)*

18.5 MULTIPLE CORRELATION

We use the multiple regression equations to disentangle the separate direct effects of each independent variable on the dependent. Using **multiple correlation** techniques, we can also ascertain the combined effects of all independents on the dependent variable. We do so by computing the **multiple**

READING STATISTICS 12: Regression and Correlation

Research projects that analyze the interrelationships among many variables are particularly likely to employ regression and correlation as central statistical techniques. The results of these projects will typically be presented in summary tables that report the multiple correlations, slopes, and, if applicable, the significance of the results. The zero-order correlations are often presented in the form of a matrix that displays the value of Pearson's r for every possible bivariate relationship in the data set. An example of such a matrix can be found in Section 18.6.

Usually, tables that summarize the multivariate analysis will report R^2 and the slope for each independent variable in the regression equation. An example of this kind of summary table would look like this:

Independents	Multiple R^2	Beta-Weights
X_1	.17	.47
X_2	.23	.32
X_3	.27	.16
.	.	.
.	.	.

This table reports that the first independent variable, X_1, has the strongest direct relationship with the dependent variable and explains 17% of the variance in the dependent by itself ($R^2 = 0.17$). The second independent, X_2, adds 6% to the explained variance ($R^2 = 0.23$ after X_2 is entered into the equation). The third independent, X_3, adds 4% to the explained variance of the dependent ($R^2 = 0.27$ after X_3 is entered into the equation).

Statistics in the Professional Literature

Besides the obvious burden of the academic workload, what are the sources of stress for students? Professor Anne Fortune asked this question of a sample of social work graduate students.

She was particularly interested in the possibility that playing multiple social roles (for example, parent, spouse, and worker) might increase stress and decrease psychological well-being for these students. One theory of role conflict, which she calls the "scarcity model," suggests that multiple roles will overload the individual. All of the roles will be played poorly, and stress and feelings of inner conflict will increase.

An alternative theory, however, ". . . suggests that more roles increase the individual's sense of identity, privileges from role-statuses, and interpersonal relationships, and consequently increases self-esteem and decreases stress" (82). She labels this second theory the "expansion model." The table below presents some of her results for second-year social work graduate students. The entries in the table are beta-weights.

Variable:	Well-being ($N = 255$)	General Stress ($N = 260$)	Stress as Student ($N = 283$)
Married	.054	−.074	−.007
Children	.066	−.001	−.013
Working	.017	.009	−.161*
Age	−.096	−.107	−.051
Male	−.047	.068	−.057
Locus of Control	.288*	−.368*	−.329*
.	.	.	.
ADJUSTED R^2	.06*	.15*	.11*

*Probability less than or equal to .05.
NOTE: Four independent variables have been eliminated from the table.

The author uses three dependent variables to measure stress and well-being and a total of ten independent variables (only six of which are reproduced here). Based on this table and other evidence, Professor Fortune finds support for the "expansion" rather than the "scarcity" model. She notes that only one of the role-status variables was significant: working . . . students experienced less stress. . . . Overall, only one factor, Locus of Control, was consistently related to the dependent variables. . . . Students who have additional roles . . . are not at greater risk than "traditional" students (p. 87).

Strong betas for the variables that measure multiple roles would have supported the "scarcity" model. The evidence is that stress and well-being are not related to playing multiple roles. Professor Fortune goes on to suggest that factors such as support groups, quality of role experience, and the like are more important than the number of roles.

Anne E. Fortune: 1987. "Multiple Roles, Stress and Well-Being Among MSW Students." *Journal of Social Work Education*, 23:3, pp. 81–90. Reprinted by permission.

correlation coefficient *(R)* and the **coefficient of multiple determination** *(R²)*. The value of the latter statistic represents the proportion of the variance in Y that is explained by all the independent variables combined. In terms of zero-order correlation, we have seen that "number of children" (X_1) explains a proportion of .25 of the variance in Y $(r_{y1}^2 = (.50)^2 = .25)$ by itself and that SES explains a proportion of .09 of the variance in Y $(r_{y2}^2 = (-.30)^2 = .09)$. The two zero-order correlations cannot be simply added together to ascertain their combined effect on Y, because the two independents are also correlated with each other and, therefore, will "overlap" in their effects on Y and explain some of the same variance. This overlap is eliminated in Formula 18.11:

FORMULA 18.11

$$R^2 = r_{y1}^2 + r_{y2.1}^2 (1 - r_{y1}^2)$$

where R^2 = the multiple correlation coefficient
r_{y1}^2 = the zero-order correlation between Y and X_1, the quantity squared
$r_{y2.1}^2$ = the partial correlation of Y and X_2 while controlling for X_1, the quantity squared
r_{y1}^2 = the zero-order correlation between Y and X_1, the quantity squared

The first term in this formula (r_{y1}^2) is the coefficient of determination for the relationship between Y and X_1. It represents the amount of variation in Y explained by X_1 by itself. To this quantity we add the amount of the variation remaining in Y (given by $1 - r_{y1}^2$) that can be explained by X_2 after the effect of X_1 is controlled $(r_{y2.1}^2)$. Basically, Formula 18.11 allows X_1 to explain as much of Y as it can and then adds in the effect of X_2 after X_1 is controlled (thus eliminating the "overlap" in the variance of Y that X_1 and X_2 have in common).

To observe the combined effects of number of children (X_1) and SES (X_2) on husband's housework (Y), we need two quantities. The correlation between X_1 and Y $(r_{y1} = .50)$ has already been found; but, before we can solve Formula 18.11, we must first calculate the partial correlation of Y and X_2 while controlling for X_1 $(r_{y2.1})$:

$$r_{y2.1} = \frac{r_{y2} - (r_{y1})(r_{12})}{\sqrt{1 - r_{y1}^2}\ \sqrt{1 - r_{12}^2}}$$

$$r_{y2.1} = \frac{(-.30) - (.50)(-.47)}{\sqrt{1 - (.50)^2}\ \sqrt{1 - (-.47)^2}}$$

$$r_{y2.1} = \frac{(-.30) - (-.24)}{\sqrt{.75}\ \sqrt{.78}}$$

$$r_{y2.1} = \frac{-.06}{.77}$$

$$r_{y2.1} = -.08$$

Formula 18.11 can now be solved for our sample problem:

$$R^2 = r_{y1}^2 + r_{y2.1}^2\,(1 - r_{y1}^2)$$
$$R^2 = (.50)^2 + (-.08)^2(1 - .50^2)$$
$$R^2 = (.25) + (.006)(1 - .25)$$
$$R^2 = (.25) + (.006)(.75)$$
$$R^2 = .25 + .005$$
$$R^2 = .255$$

The first independent variable (X_1), number of children, explains 25% of the variance in Y by itself. To this total, the second independent (X_2), SES, adds only a half a percent, for a total explained variance of 25.5%. In combination, the two independents explain a total of 25.5% of the variation in the dependent variable. *(For practice in computing and interpreting R and R², see any of the problems at the end of this chapter. It is probably a good idea to start with problem 18.1 since it has the smallest data set and the least complex computations.)*

18.6 A NON-COMPUTATIONAL EXAMPLE

The last several sections have presented a great deal of technical material. Now, it is time to pause and consider what all this means. Probably the best way to add to your understanding of partial and multiple correlation and regression is to consider how these techniques might be used in a research situation. In this section, we'll consider such an application, but calculations and equations will be minimized in favor of interpretation and meaning. I will also take this opportunity, the final "sample problem" of the text, to indulge myself in some thinly disguised statistical moralizing.

A researcher wonders about the causes of "success" in statistics courses. Why do some students achieve higher grades than others? Is it simply a matter of aptitude or level of preparation for college-level mathematics? What are the effects, if any, of motivation and willingness to work hard? Can level of success in statistics be reasonably predicted? With the appropriate data and the help of the techniques introduced in this chapter, we can answer these questions.

To begin, assume that the researcher has three pieces of information about a large number of statistics students: final grade in the course (success in statistics), score on the college board mathematics test (a measure of level of preparation for college-level statistics courses), and the average number of hours each student spent studying statistics each week (a measure of motivation or willingness to work hard). Statistics grade will be taken as the dependent variable (Y) and the other two as independents (X_1 and X_2).

The first step in the analysis would probably involve the careful inspection of the three possible zero-order relationships. Scattergrams would be produced, regression equations found, and correlation coefficients calculated. Assume that the bivariate relationships are all roughly linear and that the relationship between the "preparation" variable (College Board score) and statistics grade is stronger than the relationship between the "motivation" variable (hours studying) and statistics grade. This result might lead us to conclude that students with higher grades are simply better prepared. The fictitious zero-order relationships are presented in Table 18.2.

These zero-order relationships display the effects of each independent, one at a time, on the dependent variable. To assess the combined effects of both independent variables and the direct effects of each while controlling for the other, we must use the techniques presented in the previous sections.

To begin with prediction, we need to find the least-squares multiple regression equation as defined by Formula 18.3. After calculating the partial slopes and the Y intercept, we can predict a student's final grade in statistics. If the partial slopes were .09 for preparation ($b_1 = .09$) and 1.17 for motivation ($b_2 = 1.17$), with the Y intercept equal to 12 ($a = 12$), the prediction equation would be

$$Y' = 12 + (.09)X_1 + (1.17)X_2$$

and the researcher would be ready to make some educated guesses about statistics grades. For example, how well would a very well-prepared student

TABLE 18.2 ZERO-ORDER CORRELATION COEFFICIENTS (fictitious data)

	College Board Score (preparation)	Hours of Study (motivation)
Statistics grade (success)	.60	.32
College board score (preparation)		−.15

Application 18.1

Five recently divorced men have been asked to rate subjectively the success of their adjustment to single life on a scale ranging from 5 (very successful adjustment) to 1 (very poor adjustment). Is adjustment related to the length of time married? Is adjustment related to socioeconomic status as measured by yearly income?

Case	Adjustment (Y)	Years Married (X_1)	Income (dollars) (X_2)
A	5	5	30,000
B	4	7	45,000
C	4	10	25,000
D	3	2	27,000
E	1	15	17,000

$\bar{Y} = 3.4$ $\bar{X}_1 = 7.8$ $\bar{X}_2 = 28,800.00$
$s = 1.4$ $s = 4.5$ $s = 9,173.88$

The zero-order correlations among these three variables are

	Years Married (X_1)	Income (X_2)
Adjustment (Y)	−.62	.62
Years married (X_1)		−.49

These results suggest strong but opposite relationships between each independent and adjustment. Adjustment decreases as years married increases, and increases as income increases.

To find the multiple regression equation, we must find the partial slopes.

For years married (X_1):

$$b_1 = \frac{s_y}{s_1}\left(\frac{r_{y1} - r_{y2}r_{12}}{1 - r_{12}^2}\right)$$

$$b_1 = \frac{1.4}{4.5}\left(\frac{(-.62) - (.62)(-.49)}{1 - (-.49)^2}\right)$$

$$b_1 = (.31)\left(\frac{(-.62) - (-.30)}{1 - .24}\right)$$

$$b_1 = (.31)\left(\frac{-.32}{.76}\right)$$

$$b_1 = (.31)(-.42)$$

$$b_1 = -.13$$

For income (X_2):

$$b_2 = \frac{s_y}{s_2}\left(\frac{r_{y2} - r_{y1}r_{12}}{1 - r_{12}^2}\right)$$

$$b_2 = \frac{1.4}{9173.88}\left(\frac{(.62) - (-.62)(-.49)}{1 - (-.49)^2}\right)$$

$$b_2 = (.00015)\left(\frac{.62 - .30}{1 - .24}\right)$$

$$b_2 = (.00015)\left(\frac{.32}{.76}\right)$$

$$b_2 = (.00015)(.42)$$

$$b_2 = .000063$$

The Y intercept would be

$a = \bar{Y} - b_1\bar{X}_1 - b_2\bar{X}_2$
$a = 3.4 - (-.13)(7.8) - (.000063)(28,800.00)$
$a = 3.4 - (-1.01) - (1.81)$
$a = 3.4 + 1.01 - 1.81$
$a = 2.60$

The multiple regression equation is

$Y = a + b_1X_1 + b_2X_2$
$Y = 2.60 + (-.13)X_1 + (.000063)X_2$

What adjustment score could we predict for a male who had been married 30 years ($X_1 = 30$) and had an income of $50,000 ($X_2 = 50,000$)?

$Y' = 2.60 + (-.13)(30) + (.000063)(50,000)$
$Y' = 2.60 + (-3.9) + (3.15)$
$Y' = 1.85$

To assess which of the two independents has the stronger effect on adjustment, the standardized partial slopes must be computed. For years married (X_1):

$$b_1^* = b_1\left(\frac{s_1}{s_y}\right)$$

$$b_1^* = (-.13)\left(\frac{4.5}{1.4}\right)$$

$$b_1^* = -0.42$$

For income (X_2):

$$b_2^* = b_2\left(\frac{s_2}{s_y}\right)$$

$$b_2^* = (.000063)\left(\frac{9173.88}{1.4}\right)$$

$$b_2^* = 0.41$$

The standardized regression equation is

$$Z_y = b_1^* Z_1 + b_2^* Z_2 = (-0.42)Z_1 + (0.41)Z_2$$

and the independents have nearly equal but opposite effects on adjustment. To assess the combined effects of the two independents on adjustment, the coefficient of multiple determination must be computed.

$$R^2 = r_{y1}^2 + r_{y2.1}^2(1 - r_{y1}^2)$$
$$R^2 = (-.62)^2 + (.46)^2(1 - (-.62)^2)$$
$$R^2 = .38 + (.21)(1 - .38)$$
$$R^2 = .38 + (.21)(.62)$$
$$R^2 = .38 + .13$$
$$R^2 = .51$$

The first independent, years married, explains 38% of the variation in adjustment by itself. To this quantity, income explains an additional 13% of the variation in adjustment. Taken together, the two independents explain a total of 51% of the variation in adjustment.

($X_1 = 650$) do in the course if he or she did not study ($X_2 = 1$)? The multiple regression equation can help us answer this question by providing a predicted grade for such a student:

$$Y' = 12 + (.09)(650) + (1.17)(1)$$
$$Y' = 12 + 58.5 + 1.17$$
$$Y' = 71.67$$

At the very best, a grade in the low 70s would be average. The prediction would be that a well-prepared student might be able to slide by without working but would be extremely unlikely to do well in the course. How about a not-so-well-prepared student ($X_1 = 400$) who compensates by working hard and studying an average of 40 hours per week ($X_2 = 40$)? Again, the regression equation can supply a prediction:

$$Y' = 12 + (.09)(400) + (1.17)(40)$$
$$Y' = 12 + 36 + 46.8$$
$$Y' = 94.8$$

A score in the middle 90s would indicate a high level of success and would suggest that any lack of preparation could indeed be overcome by hard work. Finally, what would our prediction be for a student of average preparation ($X_1 = 500$) who studied an average number of hours per week ($X_2 = 15$)?

$$Y' = 12 + (.09)(500) + (1.17)(15)$$
$$Y' = 12 + 45 + 17.6$$
$$Y' = 74.6$$

An average student with average motivation would be likely to end the course with an average grade.

Let us stress several points about these predictions. First, we can make no claim to fortune telling. In the absence of perfect relationships, we will be unlikely to predict final grade perfectly. On the other hand, the stronger the relationships between the independents and the dependent, the more accurate our predictions will be.

Second, we know intuitively that other factors besides the two identified in our regression equation have an effect on grades. For example, the efficiency of study time is probably as important as the amount, and there are affective dimensions (for example, whether or not the student likes statistics) that should be explored as well. The accuracy of our predictions would be enhanced if we measured these other variables and incorporated them into the equation as additional independents. As a general rule, increasing the number of independents will increase the amount of variation explained in the dependent. However, to the extent that the independents are correlated with each other, they will overlap in the portion of the variation that they explain in the dependent. Each additional independent variable will likely explain a smaller and smaller portion of the variation in the dependent variable, and the researcher will reach a point of diminishing returns.

Third, although the unstandardized partial slopes permit predictions of the dependent in the original units of measurement (final average grade), they do not permit an assessment of the relative importance of the two independents. Since the independent variables have different scales (College Board scores and hours per week), the unstandardized partial slopes cannot be directly compared to assess which has the stronger effect.

To make a comparison, the beta-weights (standardized partial slopes) must be computed for each independent variable. Since the beta-weights are based on standardizing all distributions to Z scores, they eliminate the difficulties that arise from differences in scale among the independents. Suppose that the standardized regression equation for the present relationship was (based on Formula 18.10)

$$Z_y = b_1^* Z_1 + b_2^* Z_2$$
$$Z_y = (.64)Z_1 + (.42)Z_2$$

These results would indicate that both independents had fairly strong direct effects on statistics grade and that level of preparation (Z_1) was the stronger of the two.

As a final step in the analysis, we would no doubt want to assess the combined effects of these two independents on final grade in statistics. This can be accomplished by computing the coefficient of multiple determination (R^2) as defined in Formula 18.11:

$$R^2 = r_{y1}^2 + r_{y2.1}^2 (1 - r_{y1}^2)$$
$$R^2 = .60^2 + (.44^2)(1 - .60^2)$$
$$R^2 = .36 + (.19)(.64)$$
$$R^2 = .36 + .12$$
$$R^2 = .48$$

The first independent variable, level of preparation, explains 36% of the variation in statistics grade by itself. To this quantity, the second independent, study time, adds another 12%. In combination, the two independents explain almost half of the variation in the dependent variable. About half of the variation remains unexplained, presumably because of some combination of the other, unmeasured variables mentioned above (efficiency of study time, affective factors), measurement error, and random chance.

To summarize in terms of our original questions, we have seen that both independents have important relationships with final grade in statistics. The unstandardized regression equation was used to predict grades for various combinations of preparation and motivation. The standardized regression equation showed that both independent variables had substantial direct effects on the dependent variable and that preparation had a stronger effect on final grade than study time. Finally, we saw that, in combination, the two independents account for about half of the variation in final grade.

The analysis shows that, although well-prepared students may be able to slide by without too much effort, students who are not well prepared can compensate and even achieve high levels of success by increased study time. High levels of preparation do not guarantee high levels of success in the absence of a willingness to work hard. Regardless of background, highly motivated students can acquit themselves with distinction.

18.7 THE LIMITATIONS OF MULTIPLE REGRESSION AND CORRELATION

Multiple regression and correlation are very powerful tools for analyzing the interrelationships among three or more variables. The techniques presented in this chapter permit the researcher to predict scores on one variable from two or more other variables, to distinguish between independent variables in terms of the importance of their direct effects on a dependent, and to ascertain the total effect of a set of independent variables on a dependent variable. In terms of the flexibility of the techniques and the volume of information they can supply, multiple regression and correlation represent some of the most powerful statistical techniques available to social science researchers.

Such powerful tools are not, of course, cheap. They demand high-quality data; and, for many concepts and variables, measurement at the interval-ratio

READING STATISTICS 13: The Causes of Juvenile Delinquency

Professor Siu Kwong Wong recently conducted a study of delinquency among Chinese-Canadian youth living in Winnipeg, Manitoba. Along with other theories of delinquency, Professor Wong examined assimilation as a possible cause of delinquency for this group. Chinese culture is often said to emphasize conformity and strict adherence to authority and convention. Does

delinquency increase as people assimilate from Chinese to Canadian culture? Do more acculturated Chinese-Canadian young people have higher rates of delinquency? Table A seems to suggest that there is a relationship between these variables.

The measure of acculturation (the Behavior Acculturation Scale) had a positive and significant effect on delinquency, suggesting that ". . . the more the respondent was acculturated into Canadian society, the more likely they had committed delinquency." This finding is quite consistent with the idea that Chinese culture inhibits deviance, but note that, when the dependent variable is divided into minor and serious offenses, the effect of acculturation is much stronger for the former. Also, the effect of having Chinese friends is quite weak and suggests that integration into the Chinese-Canadian community (as opposed to adherence to Chinese culture) is not a particular insulation against delinquency. Incidentally, after examining other results, Professor Wong concludes that the most significant cause of delinquency for this group is not acculturation per se but conflict with less acculturated immigrant parents.

TABLE A REGRESSION OF DELINQUENCY ON ACCULTURATION

Independent	Beta-Weight	Serious Offenses	Minor Offenses
Sex[1]	−.25***	−.11	−.28***
Age	.25***	.03	.30***
Behavior Acculturation Scale	.21**	.10	.22**
Association with Chinese friends	−.01	−.08	.02
N	285	285	285
R^2	.17	.04	.22

NOTE: All regression coefficients are standardized.
*p < .05   **P < .01   ***p < .001

[1] It may surprise you to see a nominal-level variable like sex in a regression equation. Actually, it is quite common to include nominal-level variables—called "dummy" variables—as long as they have only two categories. In this case, sex was coded so that males = 1 and females = 2. The negative sign of the beta-weight thus indicates that higher rates of delinquency are associated with male respondents.

Siu Kwong Wong. "Delinquency of Chinese-Canadian Youth: A Test of Opportunity, Control and Intergenerational Conflict Theories." *Youth and Society*. (29) pp. 112–113, copyright © 1997 by Sage Publications. Reprinted by permission of Sage Publications, Inc.

level may be very difficult considering the present stage of development of the social sciences. Furthermore, these techniques assume that the interrelationships among the variables follow a particular form. First, they assume that each independent variable is related to the dependent in a linear fashion. How well a given set of variables meets this assumption can be quickly checked by use of scattergrams.

Second, the techniques presented in this chapter assume that there is no interaction among the variables in the equation. If there is interaction among the variables, it will not be possible to accurately estimate or predict the dependent variable by simply adding the effects of the independents. There are techniques for handling interaction among the variables in the set, but these techniques are beyond the scope of this text.

Third, the techniques of multiple regression and correlation assume that the independent variables are uncorrelated with each other. Strictly speaking, this condition means that the zero-order correlation among all pairs of independents should be zero; but, practically, we act as if this assumption has been met if the intercorrelations among the independents are low.

To the extent that these assumptions are violated, the regression coefficients (especially partial and standardized slopes) and the coefficient of multiple determination (R^2) become less and less trustworthy and the techniques less and less useful. If the assumptions of the model cannot be met, the alternative might be to turn to the multivariate techniques described in the previous chapter. Unfortunately, those techniques, in general, supply a lower volume of less precise information about the interrelationships among the variables.

Finally, we should note that, in this chapter, we have been concerned only with the simplest applications of the techniques of partial correlation and multiple regression and correlation. In terms of logic and interpretation, the extensions to situations involving more than one control variable or more than two independent variables are relatively straightforward. However, the computational routines required to handle such situations can become extremely complex. If you are faced with a situation involving more than three variables, turn to the services offered by your local computer center. Several excellent preprogrammed statistical packages are commonly available, and all can be successfully utilized with minimal computer literacy and can handle these complex calculations in, literally, the blink of an eye. Efficient use of these packages will enable you to avoid drudgery and will also virtually guarantee the accuracy of your results. Furthermore, these statistical packages will free you to do what social scientists everywhere enjoy doing most: pondering the meaning of your results and, by extension, the nature of social life.

SUMMARY

1. Partial correlation involves controlling for third variables in a manner analogous to that introduced in the previous chapter. Partial correlations permit the detection of direct and spurious or intervening relationships between X and Y.

2. Multiple regression includes statistical techniques by which predictions of the dependent variable from more than one independent variable can be made (by partial slopes and the multiple regression equation) and by which we can disentangle the relative

importance of the independents (by standardized partial slopes).

3. The multiple correlation coefficient (R^2) summarizes the combined effects of all independents on the dependent variable in terms of the proportion of the total variation in Y that is explained by all of the independents.

4. Partial correlation and multiple regression and correlation are some of the most powerful tools available to the researcher and demand high-quality

measurement and relationships among the variables that are linear and noninteractive. Further, correlations among the independents must be low (preferably zero). Although the price is high, these techniques pay considerable dividends in the volume of precise and detailed information they generate about the interrelationships among the variables.

SUMMARY OF FORMULAS

Partial correlation coefficient:

18.1 $\quad r_{yx.z} = \dfrac{r_{yx} - (r_{yz})(r_{xz})}{\sqrt{1 - r_{yz}^2}\,\sqrt{1 - r_{xz}^2}}$

Least-squares regression line (bivariate):

18.2 $\quad Y = a + bX$

Least-squares multiple regression line:

18.3 $\quad Y = a + b_1 X_1 + b_2 X_2$

Partial slope for X_1:

18.4 $\quad b_1 = \left(\dfrac{s_y}{s_1}\right)\left(\dfrac{r_{y1} - r_{y2}\,r_{12}}{1 - r_{12}^2}\right)$

Partial slope for X_2:

18.5 $\quad b_2 = \left(\dfrac{s_y}{s_2}\right)\left(\dfrac{r_{y2} - r_{y1}\,r_{12}}{1 - r_{12}^2}\right)$

Y intercept:

18.6 $\quad a = \bar{Y} - b_1 \bar{X}_1 - b_2 \bar{X}_2$

Standardized partial slope (beta-weight) for X_1:

18.7 $\quad b_1^* = b_1\left(\dfrac{s_1}{s_y}\right)$

Standardized partial slope (beta-weight) for X_2:

18.8 $\quad b_2^* = b_2\left(\dfrac{s_2}{s_y}\right)$

Standardized least-squares regression line:

18.9 $\quad Z_y = a_z + b_1^* Z_1 + b_2^* Z_2$

Standardized least-squares regression line (simplified):

18.10 $\quad Z_y = b_1^* Z_1 + b_2^* Z_2$

Coefficient of multiple determination:

18.11 $\quad R^2 = r_{y1}^2 + r_{y2.1}^2 (1 - r_{y1}^2)$

GLOSSARY

Beta-weights (b^*). Standardized partial slopes.

Coefficient of multiple determination (R^2). A statistic that equals the total variation explained in the dependent variable by all independent variables combined.

Multiple correlation. A multivariate technique for examining the combined effects of more than one independent variable on a dependent variable.

Multiple correlation coefficient (R). A statistic that indicates the strength of the correlation between a dependent variable and two or more independent variables.

Multiple regression. A multivariate technique that breaks down the separate effects of the independent variables on the dependent variable; used to make predictions of the dependent variable.

Partial correlation. A multivariate technique for examining a bivariate relationship while controlling for other variables.

Partial correlation coefficient. A statistic that shows the relationship between two variables while controlling for other variables; $r_{yx.z}$ is the symbol for the partial correlation coefficient when controlling for one variable.

Partial slopes. In a multiple regression equation, the slope of the relationship between a particular independent variable and the dependent variable while controlling for all other independents in the equation.

Standardized partial slopes (beta-weights). The slope of the relationship between a particular independent variable and the dependent when all scores have been normalized.

Zero-order correlations. Correlation coefficients for bivariate realtionships.

PROBLEMS

18.1 [PS] In problem 16.1 data regarding voter turnout in five cities was presented. For the sake of convenience, the data for three of the variables is presented again here along with descriptive statistics and zero-order correlations.

City	Turnout	Rate of Un-employment	% of Negative Advertisements
A	55	5	60
B	60	8	63
C	65	9	55
D	68	9	53
E	70	10	48

Univariate statistics:

	Turnout	Unemployment	Negative Ads
Mean =	63.6	8.2	55.8
Standard deviation =	5.5	1.7	5.3

Bivariate correlations:

	Unemployment	Negative Ads
Turnout	.95	−.87
Unemployment		−.70

a. Compute the partial correlation coefficient for the relationship between turnout (Y) and unemployment (X) while controlling for the effect of negative advertising (Z). What effect does this control variable have on the bivariate relationship? Is the relationship between turnout and unemployment direct? *(HINT: Use Formula 18.1 and see Section 18.2.)*

b. Compute the partial correlation coefficient for the relationship between turnout (Y) and negative advertising (X) while controlling for the effect of unemployment (Z). What effect does this have on the bivariate relationship?

Is the relationship between turnout and unemployment direct? *(HINT: Use Formula 18.1 and see Section 18.2. You will need this partial correlation to compute the multiple correlation coefficient.)*

c. Find the unstandardized multiple regression equation with unemployment (X_1) and negative ads (X_2) as the independent variables. What turnout would be expected in a city in which the unemployment rate was 10% and 75% of the campaign ads were negative? *(HINT: Use Formulas 18.4 and 18.5 to compute the partial slopes and then use Formula 18.6 to find a, the Y intercept. The regression line is stated in Formula 18.3. Substitute 10 for X_1 and 75 for X_2 to compute predicted Y.)*

d. Compute beta-weights for each independent variable. Which has the stronger impact on turnout? *(HINT: Use Formulas 18.7 and 18.8 to calculate the beta-weights.)*

e. Compute the multiple correlation coefficient (R) and the coefficient of multiple determination (R^2). How much of the variance in voter turnout is explained by the two independent variables? *(HINT: Use Formula 18.11. You calculated $r^2_{y2.1}$ in part b of this problem.)*

f. Write a paragraph summarizing your conclusions about the relationships among these three variables.

18.2 [SOC] A scale measuring support for increases in the national defense budget has been administered to a sample. The respondents have also been asked to indicate how many years of school they have completed and how many years, if any, they served in the military. Take "support" as the dependent variable.

Case	Support	Years of School	Years of Service
A	20	12	2
B	15	12	4
C	20	16	20
D	10	10	10
E	10	16	20
F	5	8	0
G	8	14	2
H	20	12	20
I	10	10	4
J	20	16	0

a. Compute the partial correlation coefficient for the relationship between support (Y) and years of school (X) while controlling for the effect of years of service (Z). What effect does this have on the bivariate relationship? Is the relationship between support and years of school direct?

b. Compute the partial correlation coefficient for the relationship between support (Y) and years of service (X) while controlling for the effect of years of school (Z). What effect does this have on the bivariate relationship? Is the relationship between support and years of service direct? *(HINT: You will need this partial correlation to compute the multiple correlation coefficient.)*

c. Find the unstandardized multiple regression equation with school (X_1) and service (X_2) as the independent variables. What level of support would be expected in a person with 13 years of school and 15 years of service?

d. Compute beta-weights for each independent variable. Which has the stronger impact on turnout?

e. Compute the multiple correlation coefficient (R) and the coefficient of multiple determination (R^2). How much of the variance in support is explained by the two independent variables? *(HINT: You calculated $r^2_{y2.1}$ in part b of this problem.)*

f. Write a paragraph summarizing your conclusions about the relationships among these three variables.

18.3 $\boxed{\text{SOC}}$ Data on civil strife (number of incidents), unemployment, and urbanization have been gathered for 10 nations. Take civil strife as the dependent variable. Compute the zero-order correlations between all three variables.

Number of Incidents of Civil Strife	Unem-ployment Rate	Percentage of Population Living in Urban Areas
0	5.3	60
1	1.0	65
5	2.7	55
7	2.8	68
10	3.0	69
23	2.5	70
25	6.0	45
26	5.2	40
30	7.8	75
53	9.2	80

a. Compute the partial correlation coefficient for the relationship between strife (Y) and unemployment (X) while controlling for the effect of urbanization (Z). What effect does this have on the bivariate relationship? Is the relationship between strife and unemployment direct?

b. Compute the partial correlation coefficient for the relationship between strife (Y) and urbanization (X) while controlling for the effect of unemployment (Z). What effect does this have on the bivariate relationship? Is the relationship between strife and unemployment direct? *(HINT: You will need this partial correlation to compute the multiple correlation coefficient.)*

c. Find the unstandardized multiple regression equation with unemployment (X_1) and urbanization (X_2) as the independent variables. What level of strife would be expected in a nation in which the unemployment rate was 10% and 90% of the population lived in urban areas?

d. Compute beta-weights for each independent variable. Which has the stronger impact on turnout?

e. Compute the multiple correlation coefficient (R) and the coefficient of multiple determination (R^2). How much of the variance in strife is explained by the two independent variables?

f. Write a paragraph summarizing your conclusions about the relationships among these three variables.

18.4 SOC/CJ In problem 16.5, crime and population data were presented for each of 10 states. For each of the three crime variables:

a. Find the multiple regression equations (unstandardized) with growth and urbanization as the independent variables.

b. Make a prediction for each crime variable for a state with a 5% growth rate and a population that is 90% urbanized.

c. Compute beta-weights for each independent in each equation and compare their relative effect on each dependent.

d. Compute R and R^2 for each crime variable, using the population variables as independents.

e. Write a paragraph summarizing your findings.

18.5 PS Problem 16.4 presented some data on 10 precincts. Take voter turnout as the dependent variable.

a. Find the multiple regression equations (unstandardized).

b. What turnout would you expect for a precinct in which 0% of the voters were Democrats and 5% were minorities?

c. Compute beta-weights for each independent and compare their relative effect on turnout. Which was the more important factor?

d. Compute R and R^2.

e. Write a paragraph summarizing your findings.

18.6 SW Twelve families have been referred to a counselor and she has rated each of them on a cohesiveness scale. Also, she has information on family income and number of children currently living at home. Take family cohesion as the dependent variable.

Family	Cohesion Score	Family Income	Number of Children
A	10	30,000	5
B	10	70,000	4

Family	Cohesion Score	Family Income	Number of Children
C	9	35,000	4
D	5	25,000	0
E	1	55,000	3
F	7	40,000	0
G	2	60,000	2
H	5	30,000	3
I	8	50,000	5
J	3	25,000	4
K	2	45,000	3
L	4	50,000	0

a. Find the multiple regression equations (unstandardized).

b. What level of cohesion would be expected in a family with an income of $20,000 and 6 children?

c. Compute beta-weights for each independent and compare their relative effect on cohesion. Which was the more important factor?

d. Compute R and R^2.

e. Write a paragraph summarizing your findings.

18.7 Problem 16.8 presented per capita expenditures on education for 15 states, along with information on income per capita and school dropout rate. Take educational expenditures as the dependent variable.

a. Compute beta-weights for each independent and compare their relative effect on expenditures. Which was the more important factor?

b. Compute R and R^2.

c. Write a paragraph summarizing your findings.

18.8 SW In problem 16.9, child-abuse rates were presented along with information on education and family structure for each of 18 census tracts. Taking child-abuse rates as the dependent variable:

a. Compute beta-weights for each independent and compare their relative effect on child abuse. Which was the more important factor?

b. Compute R and R^2.

c. Write a paragraph summarizing your findings.

18.9 SOC Problem 16.10 presented some data for 20 nations. Take "percent urban" as the dependent variable and birth rate and death rate as the independent variable.

a. Compute beta-weights for each independent and compare their relative effect on cohesion. Which was the more important factor?

b. Compute R and R^2.

c. Write a paragraph summarizing your findings.

18.10 For the data presented in problem 16.10, take death rate as the dependent variable and percent urban and growth rate as the independent variables.

a. Compute beta-weights for each independent and compare their relative effect on cohesion. Which was the more important factor?

b. Compute R and R^2.

c. Write a paragraph summarizing your findings.

18.11 | SOC | The fatality rate for motor vehicle accidents is reported below for 20 states along with information on some potential independent variables.

Take fatality rate as the dependent variable and choose two of the remaining variables as independents.

a. Compute beta-weights for each of the independent variables you selected and compare their relative effect on the fatality rate from motor vehicle accidents. Which was the more important factor?

b. Compute R and R^2.

c. Write a paragraph summarizing your findings.

18.12 | SOC | Twenty individuals were randomly selected from the 1996 General Social Survey data set, and their scores on four variables are reported below.

Take "Hours of TV Watching per Day" as the dependent variable and select two of the remaining variables as independents.

State	Motor Vehicle Fatality Rate[1]	Urbanization[2]	Average Commute to Work[3]	% 65 Years of Age and Older
Maine	1.5	36	19	13.9
Mass.	.9	96	18	14.2
New York	1.4	92	26	13.4
Penn.	1.6	85	25	15.9
Indiana	1.6	68	21	12.6
Wisconsin	1.4	68	21	13.3
S. Dakota	2.0	33	13	14.4
Kansas	1.8	55	17	13.7
Virginia	1.4	78	24	11.1
S. Car.	2.3	70	21	12.0
Florida	2.2	93	22	18.6
Tenn.	2.3	68	22	12.5
Miss.	2.9	35	21	12.3
Texas	1.8	84	22	10.2
Wyoming	2.1	30	15	11.1
Colorado	1.7	84	21	10.0
Arizona	2.3	87	22	13.3
Wash.	1.1	83	22	11.6
Calif.	1.5	97	25	11.0
Alaska	2.1	42	17	4.9

[1]Number of fatalities from motor vehicle accidents per 100 million vehicle miles, 1994.
[2]Percentage of the population living in metropolitan areas, 1994.
[3]For 1990. In minutes.
Source: United States Bureau of the Census, *Statistical Abstracts of the United States: 1996* (116th edition). Washington, D.C., 1996.

Hours of TV Watching per Day (TVHOURS)	Occupational Prestige (PRESTG80)	Number of Children (CHILDS)	Age (AGE)
4	50	2	43
3	36	3	58
3	36	1	34
4	50	2	42
2	45	2	27
3	50	5	60
4	50	0	28
7	40	3	55
1	57	2	46
3	33	2	65
1	46	3	56
3	31	1	29
1	19	2	41
0	52	0	50
2	48	1	62
4	36	1	24
3	48	0	25
1	62	1	87
5	50	0	45
1	27	3	62

a. Compute beta-weights for each of the independent variables you selected and compare their relative effect on the hours of television watching. Which was the more important factor?

b. Compute R and R^2.

c. Write a paragraph summarizing your findings.

Using *SPSS for Windows* for Regression Analysis

DEMONSTRATION 18.1 What Are the Correlates of Occupational Prestige? Another Look

In Demonstration 16.1, we used the **Correlate** procedure to calculate zero-order correlation coefficients between *prestg80*, *papres80*, and *educ*. In this section, we will use a significantly more complex and flexible procedure (**Regression**) to analyze the effects of these same independent variables on *prestg80*. The **Regression** procedure permits the user to control many aspects of the regression formula, and it can produce a much greater volume of output than the **Correlate** procedure. Among other things, **Regression** displays the slope (*b*) and the *Y* intercept (*a*) so we can use this procedure to find least-squares regression lines. This demonstration represents a very sparing use of the power of this command and an extremely economical use of all the options available. I urge you to explore some of the variations and capabilities of this powerful data-analysis procedure.

With the 1996 GSS loaded, click **Statistics**, **Regression**, and **Linear**, and the **Linear Regression** window will appear. Move *prestg80* into the **Dependent** box and *educ* and *papres80* into the **Independent(s)** box. Click the **Statistics** button and then click **Descriptives** to get zero-order correlations, means, and standard deviations for the variables. Click **Continue** and **OK**, and the following output will appear.

Regression

Descriptive Statistics

	Mean	Std. Deviation	N
RS OCCUPATIONAL PRESTIGE SCORE (1980)	43.87	13.91	1125
HIGHEST YEAR OF SCHOOL COMPLETED	13.60	2.92	1125
FATHERS OCCUPATIONAL PRESTIGE SCORE (1980)	43.20	12.62	1125

Correlations

		RS OCCUPATIONAL PRESTIGE SCORE (1980)	HIGHEST YEAR OF SCHOOL COMPLETED	FATHERS OCCUPATIONAL PRESTIGE SCORE (1980)
Pearson Correlation	RS OCCUPATIONAL PRESTIGE SCORE (1980)	1.000	.529	.219
	HIGHEST YEAR OF SCHOOL COMPLETED	.529	1.000	.313
	FATHERS OCCUPATIONAL PRESTIGE SCORE (1980)	.219	.313	1.000
Sig. (1-tailed)	RS OCCUPATIONAL PRESTIGE SCORE (1980)	.	.000	.000
	HIGHEST YEAR OF SCHOOL COMPLETED	.000	.	.000
	FATHERS OCCUPATIONAL PRESTIGE SCORE (1980)	.000	.000	.
N	RS OCCUPATIONAL PRESTIGE SCORE (1980)	1125	1125	1125
	HIGHEST YEAR OF SCHOOL COMPLETED	1125	1125	1125
	FATHERS OCCUPATIONAL PRESTIGE SCORE (1980)	1125	1125	1125

Variables Entered/Removed[b]

Model	Variables Entered	Variables Removed	Method
1	FATHERS OCCUPATIONAL PRESTIGE SCORE (1980), HIGHEST YEAR OF SCHOOL COMPLETED[a]	.	Enter

a. All requested variables entered.
b. Dependent Variable: RS OCCUPATIONAL PRESTIGE SCORE (1980)

Model Summary

Model	R	R Square	Adjusted R Square	Std. Error of the Estimate
1	.532[a]	.283	.282	11.79

a. Predictors: (Constant), FATHERS OCCUPATIONAL PRESTIGE SCORE (1980), HIGHEST YEAR OF SCHOOL COMPLETED

ANOVA[b]

Model		Sum of Squares	df	Mean Square	F	Sig.
1	Regression	61601.425	2	30800.713	221.712	.000[a]
	Residual	155870.367	1122	138.922		
	Total	217471.792	1124			

a. Predictors: (Constant), FATHERS OCCUPATIONAL PRESTIGE SCORE (1980), HIGHEST YEAR OF SCHOOL COMPLETED
b. Dependent Variable: RS OCCUPATIONAL PRESTIGE SCORE (1980)

Coefficients[a]

Model		Unstandardized Coefficients		Standardized Coefficients	t	Sig.
		B	Std. Error	Beta		
1	(Constant)	8.005	1.824		4.387	.000
	HIGHEST YEAR OF SCHOOL COMPLETED	2.430	.127	.511	19.195	.000
	FATHERS OCCUPATIONAL PRESTIGE SCORE (1980)	6.501E-02	.029	.059	2.217	.027

a. Dependent Variable: RS OCCUPATIONAL PRESTIGE SCORE (1980)

The output begins with the descriptive statistics and a correlation matrix showing the zero-order correlations between all pairs of variables. This section essentially repeats the output of Demonstration 16.1. The next output blocks list the variables entered into the regression equation, the multiple R (.532), R Square (.283), and an ANOVA test of significance (Sig. = .0000). So far, we know that the independent variables explain about 28% of the variance in PRESTG80 and that this result is statistically significant.

In the last output block, we see the slopes (B) of the independent variables on *prestg80*, the standardized partial slopes (Beta), and the *Y* intercept (reported as a constant of 8.005). From this information, we can build a regression equation to predict scores on *prestg80*. The beta for *educ* is greater than the beta for *papres80* so education is the more important independent variable.

What does all this mean? At least for this sample, a person's occupational prestige is more affected by education than by the social class of his or her family of origin.

DEMONSTRATION 18.2 What Are the Correlates of Sexual Activity?

The 1996 GSS includes an item (*sexfreq*) that asks respondents about the frequency of their sexual activity over the past year. What causes variations in sexual activity? Some obvious correlates include gender and age, since it is commonly supposed that men are more sexually active than women and younger people more active than older. What are the effects of social class? Does the level of sexual activity—like so many other aspects of the social world—vary by a person's status?

Note that gender is, of course, a nominal-level variable, and its inclusion in a regression procedure may be surprising. Actually, nominal variables with only two categories—sometimes called "dummy variables"—are commonly used in regression.

Click **Statistics**, **Regression**, and **Linear** and name *sexfreq* as the dependent variable and *age*, *income91*, and *sex* as the independent variables. Don't forget to request descriptive statistics. The output from this procedure will be as shown on pages 473–474.

The R and R square indicate that the independent variables account for about 24% of the variation in frequency of sexual activity. The slopes (B) indicate that the dependent variable increases with income (.071710) and decreases with age (−.053115). The slope for gender is also negative (−.234) and, since a male is coded as a score of 1 and a female as a score of 2, this indicates that the

---

[1]You may notice some differences in the descriptive statistics. These are caused by the fact that the two procedures use different methods of deleting cases with missing values. The bivariate procedure uses "pairwise" deletion—statistics are calculated on all pairs of cases with valid values on each pair of variables. The multivariate procedure uses "listwise" deletion—a case is deleted from all calculations if it is missing a score on ANY of the variables included in the procedure. To change the handling of missing values, click the **Options** button before executing the procedure.

Regression

Descriptive Statistics

	Mean	Std. Deviation	N
FREQUENCY OF SEX DURING LAST YEAR	3.07	1.89	1128
AGE OF RESPONDENT	43.86	15.73	1128
TOTAL FAMILY INCOME	15.13	4.99	1128
RESPONDENTS SEX	1.55	.50	1128

Correlations

		FREQUENCY OF SEX DURING LAST YEAR	AGE OF RESPONDENT	TOTAL FAMILY INCOME	RESPONDENTS SEX
Pearson Correlation	FREQUENCY OF SEX DURING LAST YEAR	1.000	−.448	.210	−.084
	AGE OF RESPONDENT	−.448	1.000	−.029	−.012
	TOTAL FAMILY INCOME	.210	−.029	1.000	−.147
	RESPONDENTS SEX	−.084	−.012	−.147	1.000
Sig. (1-tailed)	FREQUENCY OF SEX DURING LAST YEAR	.	.000	.000	.002
	AGE OF RESPONDENT	.000	.	.166	.349
	TOTAL FAMILY INCOME	.000	.166	.	.000
	RESPONDENTS SEX	.002	.349	.000	.
N	FREQUENCY OF SEX DURING LAST YEAR	1128	1128	1128	1128
	AGE OF RESPONDENT	1128	1128	1128	1128
	TOTAL FAMILY INCOME	1128	1128	1128	1128
	RESPONDENTS SEX	1128	1128	1128	1128

Variables Entered/Removed[b]

Model	Variables Entered	Variables Removed	Method
1	RESPONDENTS SEX, AGE OF RESPONDENT, TOTAL FAMILY INCOME[a]	.	Enter

a. All requested variables entered.
b. Dependent Variable: FREQUENCY OF SEX DURING LAST YEAR

Model Summary

Model	R	R Square	Adjusted R Square	Std. Error of the Estimate
1	.493[a]	.243	.241	1.65

a. Predictors: (Constant), RESPONDENTS SEX, AGE OF RESPONDENT, TOTAL FAMILY INCOME

ANOVA[b]

Model		Sum of Squares	df	Mean Square	F	Sig.
1	Regression	981.819	3	327.273	120.408	.000[a]
	Residual	3055.073	1124	2.718		
	Total	4036.893	1127			

a. Predictors: (Constant), RESPONDENTS SEX, AGE OF RESPONDENT, TOTAL FAMILY INCOME
b. Dependent Variable: FREQUENCY OF SEX DURING LAST YEAR

Coefficients[a]

Model		Unstandardized Coefficients		Standardized Coefficients	t	Sig.
		B	Std. Error	Beta		
1	(Constant)	4.699	.276		17.002	.000
	AGE OF RESPONDENT	−5.334E-02	.003	−.443	−17.079	.000
	TOTAL FAMILY INCOME	7.118E-02	.010	.188	7.150	.000
	RESPONDENTS SEX	−.234	.100	−.062	−2.352	.019

a. Dependent Variable: FREQUENCY OF SEX DURING LAST YEAR

frequency of sex is greater for men. In other words, the frequency of sex increases as gender "decreases"—its rate is higher for people with the "lower" score on gender. The beta-weights (Beta) suggest that age has the greatest influence, followed by income and gender. The rate of sexual activity increases as affluence increases, decreases as age increases, and is more associated with males than females.

Exercises

18.1 Conduct the analysis in Demonstration 18.1 again with *income91* as the dependent variable. Compare your conclusions with those that you made in Demonstration 16.1. Write a paragraph summarizing the relationships.

18.2 Conduct the analysis on Demonstration 18.2 again with *partnrs5* (number of different sexual partners over the past five years) as the dependent variable. Compare and contrast with the analysis of *sexfreq*.

18.3 Conduct a regression analysis of *tvhours*, *childs*, or *attend*. Choose two or three potential independent variables and follow the instructions in Demonstrations 18.1 and 18.2. Ideally, your independent variables should be interval-ratio in level of measurement, but ordinal variables with a broad range of scores and nominal variables with only two scores will work as well.

1. Two research questions that can be answered by one of the techniques presented in Chapters 17 and 18 are stated below. For each research situation, choose either the elaboration technique or regression analysis. The level of measurement of the variables should have a great deal of influence on your decision.

   a. For a sample of college graduates, did having a job during the school year interfere with academic success? For 20 students, data have been gathered on college GPA, the average number of hours the student worked each week, and College Board scores (a measure of preparedness for college-level work). Take GPA as the dependent variable.

   b. Only about half of this sample graduated within four years (1 = yes, 2 = no). Was their progress affected by their level of social activity (1 = low, 2 = high)? Is the relationship between these variables the same for both males (1) and females (2)?

GPA	Hours of Work Each Week	Average College Board Scores	Graduated in 4 years	Social Activity	Sex
3.14	13	550	1	1	1
2.00	20	375	1	2	2
2.11	22	450	1	1	2
3.00	10	575	1	2	1
3.75	0	600	1	1	1
3.11	21	650	1	1	2
3.22	7	605	1	2	1
2.75	25	630	1	1	2
2.50	30	680	1	1	1
2.10	32	610	1	1	2
2.45	20	580	2	2	2
2.01	40	590	2	1	1
3.90	0	675	2	2	2
3.45	0	650	2	2	1
2.30	25	550	2	2	1
2.20	18	470	2	1	2
2.60	25	600	2	2	1
3.10	15	525	2	2	1
2.60	27	480	2	1	2
2.20	20	500	2	2	1

2. A research project has been conducted on the audiences of Christian religious television programs. Several research questions have been developed with regard to the following variables:

   How many hours per week do you watch religious programs on television? _____ (actual hours)

   What is your age? _____ (years)

   How many years of formal schooling have you completed? _____ (years)

Have you ever donated money to a religious televison program?
___ 1. Yes
___ 2. No
What is your denomination?
___ 1. Protestant
___ 2. Catholic
What is your sex?
___ 1. Female
___ 2. Male

a. The amount of time devoted to religious television will increase with age and decrease with education, but age will have the strongest effect.
b. Protestants will be more likely to donate money. Protestant females will be especially likely to donate money.
Does the data support these hypotheses?

Case	Hours	Age	Education	Ever Donate?	Denomi- nation	Sex
1	14	52	10	1	1	1
2	10	45	12	1	1	2
3	8	18	12	1	1	1
4	5	45	12	1	1	1
5	10	57	10	1	1	2
6	14	65	8	2	1	1
7	3	23	12	1	1	1
8	12	47	11	2	1	1
9	2	30	14	2	1	2
10	1	20	16	2	1	2
11	21	60	16	1	2	1
12	15	55	12	1	2	2
13	12	47	9	2	2	1
14	8	32	14	1	2	2
15	15	50	12	1	2	1
16	10	45	10	2	2	2
17	20	72	16	1	2	1
18	12	40	16	2	2	2
19	14	42	14	2	2	2
20	10	38	12	1	2	1

Appendix A   Area Under the Normal Curve

Column (a) lists *Z* scores from 0.00 to 4.00. Only positive scores are displayed, but, since the normal curve is symmetrical, the areas for negative scores will be exactly the same as areas for positive scores. Column (b) lists the proportion of the total area between the *Z* score and the mean. Figure A.1 displays areas of this type. Column (c) lists the proportion of the area beyond the *Z* score, and Figure A.2 displays this type of area.

FIGURE A.1 AREA BETWEEN MEAN AND Z	**FIGURE A.2** AREA BEYOND Z
	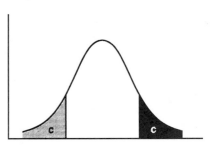

(a)	(b)	(c)	(a)	(b)	(c)
Z	Area Between Mean and *Z*	Area Beyond *Z*	*Z*	Area Between Mean and *Z*	Area Beyond *Z*
0.00	0.0000	0.5000	0.21	0.0832	0.4168
0.01	0.0040	0.4960	0.22	0.0871	0.4129
0.02	0.0080	0.4920	0.23	0.0910	0.4090
0.03	0.0120	0.4880	0.24	0.0948	0.4052
0.04	0.0160	0.4840	0.25	0.0987	0.4013
0.05	0.0199	0.4801	0.26	0.1026	0.3974
0.06	0.0239	0.4761	0.27	0.1064	0.3936
0.07	0.0279	0.4721	0.28	0.1103	0.3897
0.08	0.0319	0.4681	0.29	0.1141	0.3859
0.09	0.0359	0.4641	0.30	0.1179	0.3821
0.10	0.0398	0.4602			
0.11	0.0438	0.4562	0.31	0.1217	0.3783
0.12	0.0478	0.4522	0.32	0.1255	0.3745
0.13	0.0517	0.4483	0.33	0.1293	0.3707
0.14	0.0557	0.4443	0.34	0.1331	0.3669
0.15	0.0596	0.4404	0.35	0.1368	0.3632
0.16	0.0636	0.4364	0.36	0.1406	0.3594
0.17	0.0675	0.4325	0.37	0.1443	0.3557
0.18	0.0714	0.4286	0.38	0.1480	0.3520
0.19	0.0753	0.4247	0.39	0.1517	0.3483
0.20	0.0793	0.4207	0.40	0.1554	0.3446

(a) Z	(b) Area Between Mean and Z	(c) Area Beyond Z	(a) Z	(b) Area Between Mean and Z	(c) Area Beyond Z
0.41	0.1591	0.3409	0.91	0.3186	0.1814
0.42	0.1628	0.3372	0.92	0.3212	0.1788
0.43	0.1664	0.3336	0.93	0.3238	0.1762
0.44	0.1700	0.3300	0.94	0.3264	0.1736
0.45	0.1736	0.3264	0.95	0.3289	0.1711
0.46	0.1772	0.3228	0.96	0.3315	0.1685
0.47	0.1808	0.3192	0.97	0.3340	0.1660
0.48	0.1844	0.3156	0.98	0.3365	0.1635
0.49	0.1879	0.3121	0.99	0.3389	0.1611
0.50	0.1915	0.3085	1.00	0.3413	0.1587
0.51	0.1950	0.3050	1.01	0.3438	0.1562
0.52	0.1985	0.3015	1.02	0.3461	0.1539
0.53	0.2019	0.2981	1.03	0.3485	0.1515
0.54	0.2054	0.2946	1.04	0.3508	0.1492
0.55	0.2088	0.2912	1.05	0.3531	0.1469
0.56	0.2123	0.2877	1.06	0.3554	0.1446
0.57	0.2157	0.2843	1.07	0.3577	0.1423
0.58	0.2190	0.2810	1.08	0.3599	0.1401
0.59	0.2224	0.2776	1.09	0.3621	0.1379
0.60	0.2257	0.2743	1.10	0.3643	0.1357
0.61	0.2291	0.2709	1.11	0.3665	0.1335
0.62	0.2324	0.2676	1.12	0.3686	0.1314
0.63	0.2357	0.2643	1.13	0.3708	0.1292
0.64	0.2389	0.2611	1.14	0.3729	0.1271
0.65	0.2422	0.2578	1.15	0.3749	0.1251
0.66	0.2454	0.2546	1.16	0.3770	0.1230
0.67	0.2486	0.2514	1.17	0.3790	0.1210
0.68	0.2517	0.2483	1.18	0.3810	0.1190
0.69	0.2549	0.2451	1.19	0.3830	0.1170
0.70	0.2580	0.2420	1.20	0.3849	0.1151
0.71	0.2611	0.2389	1.21	0.3869	0.1131
0.72	0.2642	0.2358	1.22	0.3888	0.1112
0.73	0.2673	0.2327	1.23	0.3907	0.1093
0.74	0.2703	0.2297	1.24	0.3925	0.1075
0.75	0.2734	0.2266	1.25	0.3944	0.1056
0.76	0.2764	0.2236	1.26	0.3962	0.1038
0.77	0.2794	0.2206	1.27	0.3980	0.1020
0.78	0.2823	0.2177	1.28	0.3997	0.1003
0.79	0.2852	0.2148	1.29	0.4015	0.0985
0.80	0.2881	0.2119	1.30	0.4032	0.0968
0.81	0.2910	0.2090	1.31	0.4049	0.0951
0.82	0.2939	0.2061	1.32	0.4066	0.0934
0.83	0.2967	0.2033	1.33	0.4082	0.0918
0.84	0.2995	0.2005	1.34	0.4099	0.0901
0.85	0.3023	0.1977	1.35	0.4115	0.0885
0.86	0.3051	0.1949	1.36	0.4131	0.0869
0.87	0.3078	0.1922	1.37	0.4147	0.0853
0.88	0.3106	0.1894	1.38	0.4162	0.0838
0.89	0.3133	0.1867	1.39	0.4177	0.0823
0.90	0.3159	0.1841	1.40	0.4192	0.0808

(a)	(b) Area Between Mean and Z	(c) Area Beyond Z	(a)	(b) Area Between Mean and Z	(c) Area Beyond Z
Z			Z		
1.41	0.4207	0.0793	1.91	0.4719	0.0281
1.42	0.4222	0.0778	1.92	0.4726	0.0274
1.43	0.4236	0.0764	1.93	0.4732	0.0268
1.44	0.4251	0.0749	1.94	0.4738	0.0262
1.45	0.4265	0.0735	1.95	0.4744	0.0256
1.46	0.4279	0.0721	1.96	0.4750	0.0250 ←
1.47	0.4292	0.0708	1.97	0.4756	0.0244
1.48	0.4306	0.0694	1.98	0.4761	0.0239
1.49	0.4319	0.0681	1.99	0.4767	0.0233
1.50	0.4332	0.0668	2.00	0.4772	0.0228
1.51	0.4345	0.0655	2.01	0.4778	0.0222
1.52	0.4357	0.0643	2.02	0.4783	0.0217
1.53	0.4370	0.0630	2.03	0.4788	0.0212
1.54	0.4382	0.0618	2.04	0.4793	0.0207
1.55	0.4394	0.0606	2.05	0.4798	0.0202
1.56	0.4406	0.0594	2.06	0.4803	0.0197
1.57	0.4418	0.0582	2.07	0.4808	0.0192
1.58	0.4429	0.0571	2.08	0.4812	0.0188
1.59	0.4441	0.0559	2.09	0.4817	0.0183
1.60	0.4452	0.0548	2.10	0.4821	0.0179
1.61	0.4463	0.0537	2.11	0.4826	0.0174
1.62	0.4474	0.0526	2.12	0.4830	0.0170
1.63	0.4484	0.0516	2.13	0.4834	0.0166
1.64	0.4495	0.0505	2.14	0.4838	0.0162
1.65	0.4505	0.0495	2.15	0.4842	0.0158
1.66	0.4515	0.0485	2.16	0.4846	0.0154
1.67	0.4525	0.0475	2.17	0.4850	0.0150
1.68	0.4535	0.0465	2.18	0.4854	0.0146
1.69	0.4545	0.0455	2.19	0.4857	0.0143
1.70	0.4554	0.0446	2.20	0.4861	0.0139
1.71	0.4564	0.0436	2.21	0.4864	0.0136
1.72	0.4573	0.0427	2.22	0.4868	0.0132
1.73	0.4582	0.0418	2.23	0.4871	0.0129
1.74	0.4591	0.0409	2.24	0.4875	0.0125
1.75	0.4599	0.0401	2.25	0.4878	0.0122
1.76	0.4608	0.0392	2.26	0.4881	0.0119
1.77	0.4616	0.0384	2.27	0.4884	0.0116
1.78	0.4625	0.0375	2.28	0.4887	0.0113
1.79	0.4633	0.0367	2.29	0.4890	0.0110
1.80	0.4641	0.0359	2.30	0.4893	0.0107
1.81	0.4649	0.0351	2.31	0.4896	0.0104
1.82	0.4656	0.0344	2.32	0.4898	0.0102
1.83	0.4664	0.0336	2.33	0.4901	0.0099
1.84	0.4671	0.0329	2.34	0.4904	0.0096
1.85	0.4678	0.0322	2.35	0.4906	0.0094
1.86	0.4686	0.0314	2.36	0.4909	0.0091
1.87	0.4693	0.0307	2.37	0.4911	0.0089
1.88	0.4699	0.0301	2.38	0.4913	0.0087
1.89	0.4706	0.0294	2.39	0.4916	0.0084
1.90	0.4713	0.0287	2.40	0.4918	0.0082

(a)	(b)	(c)	(a)	(b)	(c)
Z	Area Between Mean and Z	Area Beyond Z	Z	Area Between Mean and Z	Area Beyond Z
2.41	0.4920	0.0080	2.91	0.4982	0.0018
2.42	0.4922	0.0078	2.92	0.4982	0.0018
2.43	0.4925	0.0075	2.93	0.4983	0.0017
2.44	0.4927	0.0073	2.94	0.4984	0.0016
2.45	0.4929	0.0071	2.95	0.4984	0.0016
2.46	0.4931	0.0069	2.96	0.4985	0.0015
2.47	0.4932	0.0068	2.97	0.4985	0.0015
2.48	0.4934	0.0066	2.98	0.4986	0.0014
2.49	0.4936	0.0064	2.99	0.4986	0.0014
2.50	0.4938	0.0062	3.00	0.4986	0.0014
2.51	0.4940	0.0060	3.01	0.4987	0.0013
2.52	0.4941	0.0059	3.02	0.4987	0.0013
2.53	0.4943	0.0057	3.03	0.4988	0.0012
2.54	0.4945	0.0055	3.04	0.4988	0.0012
2.55	0.4946	0.0054	3.05	0.4989	0.0011
2.56	0.4948	0.0052	3.06	0.4989	0.0011
2.57	0.4949	0.0051	3.07	0.4989	0.0011
2.58	0.4951	0.0049	3.08	0.4990	0.0010
2.59	0.4952	0.0048	3.09	0.4990	0.0010
2.60	0.4953	0.0047	3.10	0.4990	0.0010
2.61	0.4955	0.0045	3.11	0.4991	0.0009
2.62	0.4956	0.0044	3.12	0.4991	0.0009
2.63	0.4957	0.0043	3.13	0.4991	0.0009
2.64	0.4959	0.0041	3.14	0.4992	0.0008
2.65	0.4960	0.0040	3.15	0.4992	0.0008
2.66	0.4961	0.0039	3.16	0.4992	0.0008
2.67	0.4962	0.0038	3.17	0.4992	0.0008
2.68	0.4963	0.0037	3.18	0.4993	0.0007
2.69	0.4964	0.0036	3.19	0.4993	0.0007
2.70	0.4965	0.0035	3.20	0.4993	0.0007
2.71	0.4966	0.0034	3.21	0.4993	0.0007
2.72	0.4967	0.0033	3.22	0.4994	0.0006
2.73	0.4968	0.0032	3.23	0.4994	0.0006
2.74	0.4969	0.0031	3.24	0.4994	0.0006
2.75	0.4970	0.0030	3.25	0.4994	0.0006
2.76	0.4971	0.0029	3.26	0.4994	0.0006
2.77	0.4972	0.0028	3.27	0.4995	0.0005
2.78	0.4973	0.0027	3.28	0.4995	0.0005
2.79	0.4974	0.0026	3.29	0.4995	0.0005
2.80	0.4974	0.0026	3.30	0.4995	0.0005
2.81	0.4975	0.0025	3.31	0.4995	0.0005
2.82	0.4976	0.0024	3.32	0.4995	0.0005
2.83	0.4977	0.0023	3.33	0.4996	0.0004
2.84	0.4977	0.0023	3.34	0.4996	0.0004
2.85	0.4978	0.0022	3.35	0.4996	0.0004
2.86	0.4979	0.0021	3.36	0.4996	0.0004
2.87	0.4979	0.0021	3.37	0.4996	0.0004
2.88	0.4980	0.0020	3.38	0.4996	0.0004
2.89	0.4981	0.0019	3.39	0.4997	0.0003
2.90	0.4981	0.0019	3.40	0.4997	0.0003

(a) Z	(b) Area Between Mean and Z	(c) Area Beyond Z	(a) Z	(b) Area Between Mean and Z	(c) Area Beyond Z
3.41	0.4997	0.0003	3.60	0.4998	0.0002
3.42	0.4997	0.0003			
3.43	0.4997	0.0003	3.70	0.4999	0.0001
3.44	0.4997	0.0003			
3.45	0.4997	0.0003	3.80	0.4999	0.0001
3.46	0.4997	0.0003			
3.47	0.4997	0.0003	3.90	0.4999	<0.0001
3.48	0.4997	0.0003			
3.49	0.4998	0.0002	4.00	0.4999	<0.0001
3.50	0.4998	0.0002			

Appendix B   Distribution of *t*

Degrees of freedom (df)	Level of significance for one-tailed test					
	.10	.05	.025	.01	.005	.0005
	Level of significance for two-tailed test					
	.20	.10	.05	.02	.01	.001
1	3.078	6.314	12.706	31.821	63.657	636.619
2	1.886	2.920	4.303	6.965	9.925	31.598
3	1.638	2.353	3.182	4.541	5.841	12.941
4	1.533	2.132	2.776	3.747	4.604	8.610
5	1.476	2.015	2.571	3.365	4.032	6.859
6	1.440	1.943	2.447	3.143	3.707	5.959
7	1.415	1.895	2.365	2.998	3.499	5.405
8	1.397	1.860	2.306	2.896	3.355	5.041
9	1.383	1.833	2.262	2.821	3.250	4.781
10	1.372	1.812	2.228	2.764	3.169	4.587
11	1.363	1.796	2.201	2.718	3.106	4.437
12	1.356	1.782	2.179	2.681	3.055	4.318
13	1.350	1.771	2.160	2.650	3.012	4.221
14	1.345	1.761	2.145	2.624	2.977	4.140
15	1.341	1.753	2.131	2.602	2.947	4.073
16	1.337	1.746	2.120	2.583	2.921	4.015
17	1.333	1.740	2.110	2.567	2.898	3.965
18	1.330	1.734	2.101	2.552	2.878	3.922
19	1.328	1.729	2.093	2.539	2.861	3.883
20	1.325	1.725	2.086	2.528	2.845	3.850
21	1.323	1.721	2.080	2.518	2.831	3.819
22	1.321	1.717	2.074	2.508	2.819	3.792
23	1.319	1.714	2.069	2.500	2.807	3.767
24	1.318	1.711	2.064	2.492	2.797	3.745
25	1.316	1.708	2.060	2.485	2.787	3.725
26	1.315	1.706	2.056	2.479	2.779	3.707
27	1.314	1.703	2.052	2.473	2.771	3.690
28	1.313	1.701	2.048	2.467	2.763	3.674
29	1.311	1.699	2.045	2.462	2.756	3.659
30	1.310	1.697	2.042	2.457	2.750	3.646
40	1.303	1.684	2.021	2.423	2.704	3.551
60	1.296	1.671	2.000	2.390	2.660	3.460
120	1.289	1.658	1.980	2.358	2.617	3.373
∞	1.282	1.645	1.960	2.326	2.576	3.291

Source: Table III of Fisher & Yates: *Statistical Tables for Biological, Agricultural and Medical Research*, published by Longman Group Ltd., London (1974), 6th edition (previously published by Oliver & Boyd Ltd., Edinburgh).

$p = .01$

n_1 n_2	1	2	3	4	5	6	8	12	24	∞
1	4052	4999	5403	5625	5764	5859	5981	6106	6234	6366
2	98.49	99.01	99.17	99.25	99.30	99.33	99.36	99.42	99.46	99.50
3	34.12	30.81	29.46	28.71	28.24	27.91	27.49	27.05	26.60	26.12
4	21.20	18.00	16.69	15.98	15.52	15.21	14.80	14.37	13.93	13.46
5	16.26	13.27	12.06	11.39	10.97	10.67	10.27	9.89	9.47	9.02
6	13.74	10.92	9.78	9.15	8.75	8.47	8.10	7.72	7.31	6.88
7	12.25	9.55	8.45	7.85	7.46	7.19	6.84	6.47	6.07	5.65
8	11.26	8.65	7.59	7.01	6.63	6.37	6.03	5.67	5.28	4.86
9	10.56	8.02	6.99	6.42	6.06	5.80	5.47	5.11	4.73	4.31
10	10.04	7.56	6.55	5.99	5.64	5.39	5.06	4.71	4.33	3.91
11	9.65	7.20	6.22	5.67	5.32	5.07	4.74	4.40	4.02	3.60
12	9.33	6.93	5.95	5.41	5.06	4.82	4.50	4.16	3.78	3.36
13	9.07	6.70	5.74	5.20	4.86	4.62	4.30	3.96	3.59	3.16
14	8.86	6.51	5.56	5.03	4.69	4.46	4.14	3.80	3.43	3.00
15	8.68	6.36	5.42	4.89	4.56	4.32	4.00	3.67	3.29	2.87
16	8.53	6.23	5.29	4.77	4.44	4.20	3.89	3.55	3.18	2.75
17	8.40	6.11	5.18	4.67	4.34	4.10	3.79	3.45	3.08	2.65
18	8.28	6.01	5.09	4.58	4.25	4.01	3.71	3.37	3.00	2.57
19	8.18	5.93	5.01	4.50	4.17	3.94	3.63	3.30	2.92	2.49
20	8.10	5.85	4.94	4.43	4.10	3.87	3.56	3.23	2.86	2.42
21	8.02	5.78	4.87	4.37	4.04	3.81	3.51	3.17	2.80	2.36
22	7.94	5.72	4.82	4.31	3.99	3.76	3.45	3.12	2.75	2.31
23	7.88	5.66	4.76	4.26	3.94	3.71	3.41	3.07	2.70	2.26
24	7.82	5.61	4.72	4.22	3.90	3.67	3.36	3.03	2.66	2.21
25	7.77	5.57	4.68	4.18	3.86	3.63	3.32	2.99	2.62	2.17
26	7.72	5.53	4.64	4.14	3.82	3.59	3.29	2.96	2.58	2.13
27	7.68	5.49	4.60	4.11	3.78	3.56	3.26	2.93	2.55	2.10
28	7.64	5.45	4.57	4.07	3.75	3.53	3.23	2.90	2.52	2.06
29	7.60	5.42	4.54	4.04	3.73	3.50	3.20	2.87	2.49	2.03
30	7.56	5.39	4.51	4.02	3.70	3.47	3.17	2.84	2.47	2.01
40	7.31	5.18	4.31	3.83	3.51	3.29	2.99	2.66	2.29	1.80
60	7.08	4.98	4.13	3.65	3.34	3.12	2.82	2.50	2.12	1.60
120	6.85	4.79	3.95	3.48	3.17	2.96	2.66	2.34	1.95	1.38
∞	6.64	4.60	3.78	3.32	3.02	2.80	2.51	2.18	1.79	1.00

Values of n_1 and n_2 represent the degrees of freedom associated with the between and within estimates of variance respectively.

Appendix E A Table of Random Numbers

The numbers in this table were generated by a random process and have no pattern or order to them. In the social sciences, tables such as this are used mainly in the selection of random samples (see Chapter 6). To enhance readability, the numbers have been organized into 14 columns of five digits each and then subdivided into groups of five rows. This format is used solely to clarify the presentation and make the table easier to read. Columns and rows have no other significance with respect to the randomness of the digits.

To select a random sample from a numbered population list, begin anywhere in the table and select groups of digits the same length as the largest number on your population list. That is, if the list numbers in the 1000s, select digits in groups of four. If the list numbers in the 100s, select in groups of three, and so on. Every time a number selected from the table corresponds to the identification number of a case on the population list, select that case for the sample. Ignore repeated numbers and numbers that are not on the population list. Continue the process until you have achieved your desired sample size.

	1	2	3	4	5	6	7	8	9	10	11	12	13	14
1	10480	15011	01536	02011	81647	91646	69179	14194	62590	36207	20969	99570	91291	90700
	22368	46573	25595	85393	30995	89198	27982	53402	93965	34095	52666	19174	39615	99505
	24130	48360	22527	97265	76393	64809	15179	24830	49340	32081	30680	19655	63348	58629
	42167	93093	06243	61680	07856	16376	39440	53537	71341	57004	00849	74917	97758	16379
5	37570	39975	81837	16656	06121	91782	60468	81305	49684	60672	14110	06927	01263	54613
	77921	06907	11008	42751	27756	53498	18602	70659	90655	15053	21916	81825	44394	42880
	99562	72905	56420	69994	98872	31016	71194	18738	44013	48840	63213	21069	10634	12952
	96301	91977	05463	07972	18876	20922	94595	56869	69014	60045	18425	84903	42508	32307
	89579	14342	63661	10281	17453	18103	57740	84378	25331	12566	58678	44947	05585	56941
10	85475	36857	53342	53988	53060	59533	38867	62300	08158	17983	16439	11458	18593	64952
	28918	69578	88231	33276	70997	79936	56865	05859	90106	31595	01547	85590	91610	78188
	63553	40961	48235	03427	49626	69445	18663	72695	52180	20847	12234	90511	33703	90322
	09429	93969	52636	92737	88974	33488	36320	17617	30015	08272	84115	27156	30613	74952
	10365	61129	87529	85689	48237	52267	67689	93394	01511	26358	85104	20285	29975	89868
15	07119	97336	71048	08178	77233	13916	47564	81056	97735	85977	29372	74461	28551	90707
	51085	12765	51821	51259	77452	16308	60756	92144	49442	53900	70960	63990	75601	40719
	02368	21382	52404	60268	89368	19885	55322	44819	01188	65255	64835	44919	05944	55157
	01011	54092	33362	94904	31273	04146	18594	29852	71585	85030	51132	01915	92747	64951
	52162	53916	46369	58586	23216	14513	83149	98736	23495	64350	94738	17752	35156	35749
20	07056	97628	33787	09998	42698	06691	76988	13602	51851	46104	88916	19509	25625	58104
	48663	91245	85828	14346	09172	30168	90229	04734	59193	22178	30421	61666	99904	32812
	54164	58492	22421	74103	47070	25306	76468	26384	58151	06646	21524	15227	96909	44592
	32639	32363	05597	24200	13363	38005	94342	28728	35806	06912	17012	64161	18296	22851
	29334	27001	87637	87308	58731	00256	45834	15398	46557	41135	10367	07684	36188	18510
25	02488	33062	28834	07351	19731	92420	60952	61280	50001	67658	32586	86679	50720	94953
	81525	72295	04839	96423	24878	82651	66566	14778	76797	14780	13300	87074	79666	95725
	29676	20591	68086	26432	46901	20849	89768	81536	86645	12659	92259	57102	80428	25280
	00742	57392	39064	66432	84673	40027	32832	61362	98947	96067	64760	64584	96096	98253
	05366	04213	25669	26422	44407	44048	37937	63904	45766	66134	75470	66520	34693	90449
30	91921	26418	64117	94305	26766	25940	39972	22209	71500	64568	91402	42416	07844	69618
	00582	04711	87917	77341	42206	35126	74087	99547	81817	42607	43808	76655	62028	76630
	00725	69884	62797	56170	86324	88072	76222	36086	84637	93161	76038	65855	77919	88006
	69011	65795	95876	55293	18988	27354	26575	08625	40801	59920	29841	80150	12777	48501
	25976	57948	29888	88604	67917	48708	18912	82271	65424	69774	33611	54262	85963	03547
35	09763	83473	73577	12908	30883	18317	28290	35797	05998	41688	34952	37888	38917	88050
	91567	42595	27958	30134	04024	86385	29880	99730	55536	84855	29080	09250	79656	73211
	17955	56349	90999	49127	20044	59931	06115	20542	18059	02008	73708	83517	36103	42791
	46503	18584	18845	49618	02304	51038	20655	58727	28168	15475	56942	53389	20562	87338
	92157	89634	94824	78171	84610	82834	09922	25417	44137	48413	25555	21246	35509	20468
40	14577	62765	35605	81263	39667	47358	56873	56307	61607	49518	89656	20103	77490	18062
	98427	07523	33362	64270	01638	92477	66969	98420	04880	45585	46565	04102	46880	45709
	34914	63976	88720	82765	34476	17032	87589	40836	32427	70002	70663	88863	77775	69348
	70060	28277	39475	46473	23219	53416	94970	25832	69975	94884	19661	72828	00102	66794
	53976	54914	06990	67245	68350	82948	11398	42878	80287	88267	47363	46634	06541	97809
45	76072	29515	40980	07391	58745	25774	22987	80059	39911	96189	41151	14222	60697	59583
	90725	52210	83974	29992	65831	38857	50490	83765	55657	14361	31720	57375	56228	41546
	64364	67412	33339	31926	14883	24413	59744	92351	97473	89286	35931	04110	23726	51900
	08962	00358	31662	25388	61642	34072	81249	35648	56891	69352	48373	45578	78547	81788
	95012	68379	93526	70765	10592	04542	76463	54328	02349	17247	28865	14777	62730	92277
50	15664	10493	20492	38391	91132	21999	59516	81652	27195	48223	46751	22923	32261	85653

	15	16	17	18	19	20	21	22	23	24	25	26	27	28
51	16408	81899	04153	53381	79401	21438	83035	92350	36693	31238	59649	91754	72772	02338
	18629	81953	05520	91962	04739	13092	97662	24822	94730	06496	35090	04822	86774	98289
	73115	35101	47498	86737	99016	71060	88824	71013	18735	20286	23153	72924	35165	43040
	57491	16703	23167	49323	45021	33132	12544	41035	80780	45393	44812	12515	98931	91202
55	30405	83946	23792	14422	15059	45799	22716	19792	09983	74353	68668	30429	70735	25499
	16631	35006	85900	98275	32388	52390	16815	69298	82732	38480	73817	32523	41961	44437
	96773	20206	42559	78985	05300	22164	24369	54224	35083	19687	11052	91491	60383	19746
	38935	64202	14349	82674	66523	44133	00697	35552	35970	19124	63318	29686	03387	59846
	31624	76384	17403	53363	44167	64486	64758	75366	76554	31601	12614	33072	60332	92325
60	78919	19474	23632	27889	47914	02584	37680	20801	72152	39339	34806	08930	85001	87820
	03931	33309	57047	74211	63445	17361	62825	39908	05607	91284	68833	25570	38818	46920
	74426	33278	43972	10119	89917	15665	52872	73823	73144	88662	88970	74492	51805	93378
	09066	00903	20795	95452	92648	45454	09552	88815	16553	51125	79375	97596	16296	66092
	42238	12426	87025	14267	20979	04508	64535	31355	86064	29472	47689	05974	52468	16834
65	16153	08002	26504	41744	81959	65642	74240	56302	00033	67107	77510	70625	28725	34191
	21457	40742	29820	96783	29400	21840	15035	34537	33310	06116	95240	15957	16572	06004
	21581	57802	02050	89728	17937	37621	47075	42080	97403	48626	68995	43805	33386	21597
	55612	78095	83197	33732	05810	24813	86902	60397	16489	03264	88525	42786	05269	92532
	44657	66999	99324	51281	84463	60563	79312	93454	68876	25471	93911	25650	12682	73572
70	91340	84979	46949	81973	37949	61023	43997	15263	80644	43942	89203	71795	99533	50501
	91227	21199	31935	27022	84067	05462	35216	14486	29891	68607	41867	14951	91696	85065
	50001	38140	66321	19924	72163	09538	12151	06878	91903	18749	34405	56087	82790	70925
	65390	05224	72958	28609	81406	39147	25549	48542	42627	45233	57202	94617	23772	07896
	27504	96131	83944	41575	10573	08619	64482	73923	36152	05184	94142	25299	84387	34925
75	37169	94851	39117	89632	00959	16487	65536	49071	39782	17095	02330	74301	00275	48280
	11508	70225	51111	38351	19444	66499	71945	05422	13442	78675	84081	66938	93654	59894
	37449	30362	06694	54690	04052	53115	62757	95348	78662	11163	81651	50245	34971	52924
	46515	70331	85922	38329	57015	15765	97161	17869	45349	61796	66345	81073	49106	79860
	30986	81223	42416	58353	21532	30502	32305	86482	05174	07901	54339	58861	74818	46942
80	63798	64995	46583	09785	44160	78128	83991	42865	92520	83531	80377	35909	81250	54238
	82486	84846	99254	67632	43218	50076	21361	64816	51202	88124	41870	52689	51275	83556
	21885	32906	92431	09060	64297	51674	64126	62570	26123	05155	59194	52799	28225	85762
	60336	98782	07408	53458	13564	59089	26445	29789	85205	41001	12535	12133	14645	23541
	43937	46891	24010	25560	86355	33941	25786	54990	71899	15475	95434	98227	21824	19585
85	97656	63175	89303	16275	07100	92063	21942	18611	47348	20203	18534	03862	78095	50136
	03299	01221	05418	38982	55758	92237	26759	86367	21216	98442	08303	56613	91511	75928
	79626	06486	03574	17668	07785	76020	79924	25651	88325	88428	85076	72811	22717	50585
	85636	68335	47539	03129	65651	11977	02510	26113	99447	68645	34327	15152	55230	93448
	18039	14367	61337	06177	12143	46609	32989	74014	64708	00533	35398	58408	13261	47908
90	08362	15656	60627	36478	65648	16764	53412	09013	07832	41574	17639	82163	60859	75567
	79556	29068	04142	16268	15387	12856	66227	38358	22478	73373	88732	09443	82558	05250
	92608	82674	27072	32534	17075	27698	98204	63863	11951	34648	88022	56148	34925	57031
	23982	25835	40055	67006	12293	02753	14827	23235	35071	99704	37543	11601	35503	85171
	09915	96306	05908	97901	28395	14186	00821	80703	70426	75647	76310	88717	37890	40129
95	59037	33300	26695	62247	69927	76123	50842	43834	86654	70959	79725	93872	28117	19233
	42488	78077	69882	61657	34136	79180	97526	43092	04098	73571	80799	76536	71255	64239
	46764	86273	63003	93017	31204	36692	40202	35275	57306	55543	53203	18098	47625	88684
	03237	45430	55417	63282	90816	17349	88298	90183	36600	78406	06216	95787	42579	90730
	86591	81482	52667	61582	14972	90053	89534	76036	49199	43716	97548	04379	46370	28672
100	38534	01715	94964	87288	65680	43772	39560	12918	86537	62738	19636	51132	25739	56947

Abridged from *Handbook of Tables for Probability and Statistics,* Second Edition, edited by William H. Beyer (Cleveland: The Chemical Rubber Company, 1968).

Appendix F   Introduction to SPSS for Windows

Computers have affected virtually every aspect of human society and, as you would expect, their impact on the conduct of social research has been profound. Researchers routinely use computers to record data and compute statistics—activities that humans often find dull, tedious, and difficult but which computers accomplish with accuracy and ease. This division of labor allows social scientists to spend more time on analysis and interpretation—activities that humans typically enjoy but which are beyond the power of computers (so far, at least).

These days, the skills needed to use computers successfully are quite accessible, even for people with little or no experience. This appendix will prepare you to use a statistics program called **SPSS for Windows**. If you have used a mouse to "point and click" and run a computer program, you are ready to learn to use *SPSS for Windows*. Even if you are completely unfamiliar with computers, you will find the program accessible. After you finish this appendix, you will be ready to do the SPSS exercises found at the end of most chapters of this text.

A word of caution before we begin: This appendix is intended only as an *introduction* to SPSS. It will give you an overview of the program and enough information so that you can complete the SPSS assignments in the text. It is unlikely, however, that this appendix will answer all your questions or provide solutions to all the problems you might encounter. So, this is a good place to tell you that *SPSS for Windows* has an extensive and easy-to-use "help" facility that will provide assistance as you request it. I will periodically remind you of this facility and suggest some ways in which it can be used.

SPSS stands for "Statistical Package for the Social Sciences," and the program is designed to work with either the Windows 95 operating system or the earlier Windows 3.1 system. SPSS also comes in a version for Apple, and the instructions specified here will work for that version as well. A **statistical package** (or **statpak**) is a set of computer programs that work with data and compute statistics as requested by the user (you). Once you have entered the data for a particular group of observations, you can easily and quickly produce an abundance of statistical information without doing any computations yourself.

Why bother to learn this technology? The truth is that the labor-saving capacity of computers is sometimes exaggerated, and there are research situations in which they are unnecessary. If you are working with a small number of observations or need only a few, uncomplicated statistics, then computerized statpaks are probably not going to be helpful. However, as the number of cases increases and as your requirements for statistics become more sophisticated, computers and statpaks will become more and more useful.

An example should make this point clearer. Suppose you have gathered a sample of 150 respondents and the *only* thing you want to know about these people is their average age. To compute an average, as you know, you add the scores and divide by the total number of cases. How long do you think it would take you to add 150 two-digit numbers (ages) with a hand calculator? (Don't even think about doing it by hand.) If you entered the scores at the (pretty fast) rate of one per second, or sixty scores a minute, it would take about 3 or 4 minutes to enter the ages and get the average. Even if you did the addition a second and third time to check your math, you could probably complete the calculations in less than 20 minutes. If this were all the information you needed, computers and statpaks would not save you any time.

Such a simple research project is not very common or realistic, however. Typically, researchers deal with not one but scores or even hundreds of variables, and samples have hundreds or thousands of cases. While you could add 150 numbers in perhaps 3 or 4 minutes, how long would it take to add the scores for 1500 cases? What are the chances of processing 1500 numbers without making significant errors of arithmetic? As you can see, the more complex the research situation, the more valuable and useful statpaks become. *SPSS for Windows* can produce statistical information in a few keystrokes or clicks of the mouse that might take you minutes, hours, or even days to produce with a hand calculator.

Clearly, this is technology worth mastering by any social researcher. With statpaks, you can avoid the drudgery of mere computation, spend more time on analysis and interpretation, and conduct research projects with very large data sets. Mastery of this technology might be very handy indeed in your senior-level courses, in a wide variety of jobs, or in graduate school.

F.1 GETTING STARTED— Databases and Computer Files

Before statistics can be calculated, SPSS must first have some data to process. A **database** is an organized collection of related information such as the responses to a survey. For purposes of computer analysis, a database is organized into a **file**: a collection of information that is stored under the same name in the memory of the computer or on a diskette. Words as well as numbers can be saved in files. If you've ever used a word processing program to type a letter or term paper, you probably saved your work in a file so that you could update or make corrections at a later time. Data can be stored in files indefinitely. Since it often takes months to conduct a thorough data analysis, the ability to save a database is another advantage of using computers.

For the SPSS exercises in this text, we will use a database that contains some of the results of the General Social Survey (GSS) for 1996. This database contains the responses of a sample of adult Americans to questions about a

FIGURE F.1 THE DATA ANALYSIS PROCESS

DataBase → **Statpak** → **Output** → **Analysis**
(raw   (computer   (statistics   (interpretation)
information)   programs)   and graphs)

wide variety of social issues. The GSS has been conducted almost every year since 1972 and has been the basis for hundreds of research projects by professional social researchers. It is a rich source of information about public opinion in the United States and includes data on everything from attitudes about abortion to opinions on assisted suicide.

The GSS is especially valuable because the respondents are chosen so that the sample as a whole is representative of the entire U.S. population. A representative sample reproduces, in miniature form, the characteristics of the population from which it was taken (see Chapters 6 and 7). So, when you analyze the 1996 General Social Survey database, you are in effect analyzing U.S. society as of 1996. The data is real, and the relationships you will analyze reflect some of the most important and sensitive issues in American life.

It's important that you understand the difference between a statpak and a database and what we are ultimately after here. A database consists of information. A statpak such as *SPSS for Windows* organizes the information in the database and produces statistics. Our goal is to apply the statpak to the database to produce output (for example, statistics and graphs) that we can analyze and use to answer questions. The process might be diagrammed as in Figure F.1.

A statpak such as *SPSS for Windows* is a general research tool that can be used to analyze databases of all sorts: it is not limited to the 1996 GSS. In the same way, the 1996 GSS could be analyzed with statpaks other than *SPSS for Windows*. Other widely used statpaks include SAS, Microcase, and Stata—all of which may be available on your campus.

F.2 STARTING *SPSS FOR WINDOWS* AND LOADING THE 1996 GSS

If you are using the complete, professional version of *SPSS for Windows*, you will probably be working in a computer lab, and you can begin running the program immediately. If you are using the student version of the program on your personal computer, the first thing you need to do is install the software. Follow the instructions that came with the program and return to this appendix when installation is complete.

To start *SPSS for Windows*, find the icon (or picture) on the screen of your monitor that has an "SPSS" label attached to it. Use the computer mouse to move the arrow on the monitor screen over this icon and then double-click the left button on the mouse. This will start up the SPSS program.

After a few seconds the *SPSS for Windows* screen will appear and ask, at the top of the screen, "What would you like to do?" Of the choices listed, the

TABLE F.1 SUMMARY OF COMMANDS

To start SPSS for Windows:	Click the SPSS icon on the screen of the computer monitor.
To open a data file:	Double click the data file name listed in the window at the bottom of the *SPSS for Windows* screen.

radio button next to 'Open an existing file' will be checked (or preselected), and there will probably be a number of data sets listed in the window at the bottom of the screen. Find the 1996 General Social Survey data set, probably labeled *GSS96.sav* or something similar. If you have the data set on a disk, you will probably need to specify the correct drive. Check with your instructor to make sure that you know where to find the 1996 GSS.

Once you've located the data set, click on the name with the left-hand button on the mouse and SPSS will load the data. The next screen you will see is the SPSS Data Editor screen.

Note that there is a list of commands across the very top of the screen. These commands begin with **File** at the far left and end with **Help** at the far right. This is the main menu bar for SPSS. When you click any of these words, a **menu** of commands and choices will drop down. Basically, you tell SPSS what to do by clicking on your desired choices from these menus. Sometimes, submenus will appear, and you will need to specify your choices further.

In this section, you learned how to start up *SPSS for Windows* and load a data file. These procedures are summarized in Table F.1.

F.3 WORKING WITH DATABASES AND CODE LIBRARIES

The complete General Social Survey for 1996 includes hundreds of items of information (age, sex, opinion about social issues like capital punishment, and so forth) for almost 3000 respondents. Some of you will be using the student version of *SPSS for Windows*, which can deal with only a limited number of cases and variables, so I have reduced the database to about 50 items of information and fewer than 1500 respondents.

Note that in the SPSS Data Editor Window the data are organized into a two-dimensional grid with columns running up and down (vertically) and rows running across (horizontally). Each column is a variable or item of information from the survey. The names of the variables are listed at the tops of the columns, and you can see that the first column contains the scores for a variable named *abany*, the second column is the variable *abhlth*, the third column is the variable *age*, and so forth.

In SPSS, the names of variables must be no more than eight characters long. In many cases, the resultant need for brevity is not a problem, and variable names (e.g., *age*) are easy to figure out. In other cases, the eight-character limit necessitates extreme abbreviation, and some variable names (like *abany* and *abhlth*) are not so obvious. You have two ways to decipher these names. First, Appendix G in this text lists all variables in the database in alphabetical order and shows the wording of the item that generated the variable. For example, the *abany* variable consists of responses to a question about legal abortion: Should it be possible for a woman to have an abortion for "any reason"? Note that the variable name is formed from the question: Should an *ab*ortion be possible for *any* reason? The *abhlth* variable consists of responses to a different scenario: Should legal abortion be possible if the woman's health is in danger? Appendix G is an example of a code book for the database since it lists all the codes (or scores) for the survey items along with their meanings.

Notice in Appendix G that some of the possible responses to *abany* and *abhlth* are labeled 'Not applicable,' NA, or DK. The first of these responses means that the item in question was not given to a respondent. The full GSS is very long and, to keep the time frame for completing the survey reasonable, not all respondents are asked every question. NA stands for 'No Answer' and means that the respondent was asked the question but refused to answer. DK stands for 'Don't Know,' which means that the respondent did not have the requested information. All three of these scores are *Missing Values* and, as "noninformation," they are eliminated from statistical analysis. Missing values are common on surveys and, as long as they are not too numerous, they are not a particular problem.

A second method for deciphering variable names is provided by SPSS, which has a "built-in" or on-line code book—a duplicate of Appendix G in electronic form. To use this facility, click **Utilities** on the menu bar and then click **Variables**. The Variables window opens. This window has two parts. On the left is a list of all variables in the database arranged in alphabetical order with the first variable highlighted. On the right is the Variable Information window with information about the highlighted variable. The first variable in our database is *abany*. The Variable Information window displays a fragment of the question that was actually asked during the survey ("ABORTION IF WOMAN WANTS FOR ANY REASON") and shows the possible scores on this variable (a score of 1 = yes, a score of 2 = no, etc.) along with some other information. The same information can be displayed for any variable in the data set. Close this window by clicking the **Close** button at the bottom of the window.

Examine the window displaying the 1996 GSS a little more. Each row of the window (reading across or from left to right) contains the scores of a particular respondent on all the variables in the database. Note that the upper-left-hand cell is highlighted (outlined in a darker border) than the other cells. This cell contains the score of respondent #1 on the variable *abany*. Recall that we saw in the variable information window and in Appendix G that this score means that respondent #1 believes that a legal abortion for any reason should be available.

You can move the highlight around in this window with the **arrow keys** on your keyboard. The highlight moves in the direction of the arrow, one cell

TABLE F.2   SUMMARY OF COMMANDS

To move around in the Data window:	1. Click on the cell you want to highlight or 2. Move the slider buttons on the right-hand and bottom margins or 3. Click the arrows on the right-hand and bottom margins
To get information about a variable:	1. From the menu bar, click **Utilities** and then click **Variables**. 2. Scroll through the list of variables in the window on the left until you highlight the name of the variable in which you are interested. Variable information will appear in the window on the right. 3. Close the window by clicking on the **Close** button.
Or	4. See Appendix G.

at a time. Press the down arrow, and the score of respondent #2 on *abany* will be highlighted. Press the right arrow, and you will highlight the score of respondent #2 on *abhlth*.

There are two other ways you can move around in this window. First, note the slider buttons on the right-hand and bottom margins of the window. Use the mouse to move the arrow over one of these buttons. The slider button moves when you move the mouse while holding down its left-hand button. If you move the slider button on the botton margin to the right, the window scrolls to the right, and you can quickly review all of the variables arrayed in the columns. If you move the slider button on the right-hand margin up and down, the window scrolls in the same direction, and you can quickly review the scores of the respondents.

Second, the right-hand margin has arrows at the top and bottom, and the bottom margin has arrows at the left and right. Click on any of these arrows, and the screen will scroll one row or one column in that direction. These buttons allow you to move through the database at a more controlled space than the slider buttons.

In this section, you learned to read information in the data display window and to decipher the meaning of variable names and scores. These commands are summarized in Table F.2, and we are now prepared to actually perform some statistical operations with the 1996 GSS database.

F.4 PUTTING SPSS TO WORK: PRODUCING STATISTICS

At this point, the database on the screen is just a mass of numbers with little meaning for you. That's okay because you will not have to actually read any information from this screen. Virtually all of the statistical operations you will conduct will begin by clicking the **Statistics** command from the menu bar, selecting a procedure or statistic, and then naming the variable or variables you would like to process.

To illustrate, let's have *SPSS for Windows* produce a frequency distribution for the variable *sex*. Frequency distributions are tables that display the number of times each score of a variable occurred in the sample (see Chapter 2). So, when we complete this procedure, we will know the number of males and females in the 1996 GSS sample.

With the 1996 GSS loaded, begin by clicking the **Statistics** command on the menu bar. From the menu that drops down, click **Summarize** and then **Frequencies**. The Frequencies window appears with the variables listed in alphabetical order in the box on the left. The first variable (*abany*) will be highlighted, but all you will be able to see is the beginning of the descriptive variable label "ABORTION IF WOMAN WANTS FOR ANY REASON". If you move the arrow over the label with the mouse, the entire label and the variable name will become visible. Use the slider button or the arrow keys on the right-hand margin of this box to scroll through the variable list until you highlight the variable *sex*. You can also move to the approximate location of the variable you want to access by typing the first letter of the variable label. If you type 's,' for example, the first variable label to begin with that letter will be highlighted. From there, use the arrow or slider buttons to move to your desired variable.

Once the variable you want to process has been highlighted, click the arrow button in the middle of the screen to move the variable name to the box on the right-hand side of the screen. SPSS will produce frequency distributions for all variables listed in this box but, for now, we will confine our attention to *sex*. Click the **OK** button in the upper-right-hand corner of the Frequencies window and, in seconds, a frequency distribution will be produced.

SPSS sends all tables and statistics to the Output window or SPSS viewer. This window is now "closest" to you, and the Data Editor window is "behind" the Output window. If you wanted to return to the Data Editor, click on any part of it, and it will move to the "front" and the Output window will be "behind" it. To switch positions again, click on any part of the Output window.

The Output window will display only a portion of the frequency distribution. In Windows 95, you can open, close, and change the size of windows by clicking the symbols on their upper-right-hand corners. The middle symbol is shaped like a square, and the window will increase in size when you click it. Do this now and you will see more of the frequency distribution for *sex*. Note that the symbol has changed to two squares, one on top of the other. To return the window to its original size, click this button again. The entire table will look like this:

Frequencies

Statistics

RESPONDENTS SEX

N	Valid	1452
	Missing	0

RESPONDENTS SEX

		Frequency	Percent	Valid Percent	Cumulative Percent
Valid	MALE	637	43.9	43.9	43.9
	FEMALE	815	56.1	56.1	100.0
	Total	1452	100.0	100.0	

What can we tell from this table? The score labels (male and female) are printed at the left with the number of cases (frequency) in each category of the variable one column to the right. As you can see, there are 637 males and 815 females in the sample. The next two columns give information about percentages, and the last column to the right displays cumulative percentages. We will defer a discussion of this last column until a later exercise.

One of the percentage columns is labeled Percent and the other is labeled Valid Percent. The difference between these two columns lies in the handling of missing values. The Percent column is based on all cases, including people who did not respond to the item (NA) and people who said they did not have the requested information (DK). The Valid Percent column excludes all missing scores. Since we will almost always want to ignore missing scores, we will pay attention only to the Valid Percent column. Note that for *sex*, there are no missing scores (gender was determined by the interviewer), and the two columns are identical.

F.5 PRINTING AND SAVING OUTPUT

Once you've gone to the trouble of producing statistics, a table, or a graph, you will probably want to keep a permanent record. There are two ways to do this. First, you can print a copy of the contents of the Output Window to take with you. To do this, click on **File** and then click **Print** from the **File** menu. Alternatively, find the icon of a printer (third from the left) in the row of icons just below the menu bar and click on it.

The other way to create a permanent record of SPSS output is to save the Output window to the computer's memory or to a diskette. To do this, click **Save** from the **File** menu. The Save dialog box opens. Give the output a name (some abbreviations such as 'freqsex' might do) and, if necessary, specify the name of the drive in which your diskette is located. Click **OK**, and the table will be permanently saved.

F.6 ENDING YOUR *SPSS FOR WINDOWS* SESSION

Once you have saved or printed your work, you are ready to end your SPSS session. Click on **File** from the menu bar and then click **Exit**. The program will close, and you will be returned to the screen from which you began.

Appendix G   Code Book for the General Social Survey, 1996

The General Social Survey is a public opinion poll conducted yearly by the National Opinion Research Council. The data set supplied with this text includes about 50 variables for a randomly selected subsample of about half of the 2900 original respondents. This code book lists each item in the data set. The variable names are those used in the SPSS data file. The questions have been reproduced exactly as they were asked (with a few exceptions in order to conserve space), and the numbers beside each response are the scores recorded in the data file.

The data set includes variables that measure demographic or background characteristics of the respondents, including sex, age, race, religion, and several indicators of socioeconomic status. Also included are items that measure opinion on such current and controversial topics as abortion, capital punishment, and homosexuality.

Please tell me if you think it should be possible for a woman to get a legal abortion if . . .

abany
She wants it for any reason.
    1. Yes
    2. No
    0. Not applicable
    8. DK
    9. NA

abhlth
The woman's health is seriously endangered.
    (Same scoring as abany)

age
Age of respondent
    18–89. Actual age in years
    98. DK
    99. NA

attend
How often do you attend religious services?
    0. Never
    1. Less than once per year
    2. Once or twice a year
    3. Several times per year
    4. About once a month

     5. 2–3 times a month
     6. Nearly every week
     7. Every week
     8. Several times a week
     9. DK or no answer

cappun   Do you favor or oppose the death penalty for persons convicted of murder?
     1. Favor
     2. Oppose
     0. Not applicable
     8. DK
     9. NA

childs   How many children have you ever had? Please count all that were born alive at any time (including any from a previous marriage).
     0–7. Actual number
     8. Eight or more
     9. NA

class   Subjective class identification
     1. Lower class
     2. Working class
     3. Middle class
     4. Upper class
     0. Not applicable
     8. DK
     9. NA

degree   Respondent's highest degree
     0. Less than HS
     1. High school
     2. Assoc./Junior college
     3. Bachelor's
     4. Graduate
     7. Not applicable
     8. DK
     9. NA

educ   Highest year of school completed
     0–20. Actual number of years
     97. Not applicable
     98. DK
     99. NA

fear   Is there any area right around here—that is, within a mile—where you would be afraid to walk alone at night?
     1. Yes
     2. No

0. Not applicable
8. DK
9. NA

fefam   It is much better for everyone involved if the man is the achiever outside the home and the woman takes care of the home and family.
1. Strongly agree
2. Agree
3. Disagree
4. Strongly disagree
0. Not applicable
8. DK
9. NA

fehelp   It is more important for a wife to help her husband's career than to have one herself.
(Same scoring as fefam)

getahead   Some people think that people get ahead by their own hard work; others say that lucky breaks or help from other people are more important. Which do you think is most important?
1. Hard work most important
2. Hard work, luck equally important
3. Luck most important
0. Not applicable
8. DK
9. NA

grass   Do you think the use of marijuana should be made legal or not?
1. Should
2. Should not
0. Not applicable
8. DK
9. NA

gunlaw   Would you favor or oppose a law that requires a person to obtain a police permit before he or she could buy a gun?
1. Favor
2. Oppose
0. Not applicable
8. DK
9. NA

hapmar   Taking things all together, how would you describe your marriage? Would you say that your marriage is very happy, pretty happy, or not too happy?
1. Very happy
2. Pretty happy
3. Not too happy

0. Not applicable
8. DK
9. NA

happy   Taken all together, how would you say things are these days—would you say that you are very happy, pretty happy, or not too happy?
1. Very happy
2. Pretty happy
3. Not too happy
0. Not applicable
8. DK
9. NA

homosex   What about relations between two adults of the same sex: Do you think it is always wrong, almost always wrong, wrong only sometimes, or not wrong at all?
1. Always wrong
2. Almost always wrong
3. Wrong only sometimes
4. Not wrong at all
5. Other
0. Not applicable
8. DK
9. NA

income91   Respondent's total family income from all sources for 1996

1. Less than 1000	13. 20,000 to 22,499
2. 1000 to 2999	14. 22,500 to 24,999
3. 3000 to 3999	15. 25,000 to 29,999
4. 4000 to 4999	16. 30,000 to 34,999
5. 5000 to 5999	17. 35,000 to 39,999
6. 6000 to 6999	18. 40,000 to 49,999
7. 7000 to 7999	19. 50,000 to 59,999
8. 8000 to 9999	20. 60,000 to 74,999
9. 10,000 to 12,499	21. 75,000 or over
10. 12,500 to 14,999	0. Not applicable
11. 15,000 to 17,499	98. DK
12. 17,500 to 19,999	99. NA

marital   Are you currently married, widowed, divorced, separated, or have you never been married?
1. Married
2. Widowed
3. Divorced
4. Separated
5. Never married
9. NA

news   How often do you read the newspaper?
1. Every day
2. A few times a week
3. Once a week
4. Less than once a week
5. Never
0. Not applicable
8. DK
9. NA

paeduc   Father's highest year of school completed
0–20. Actual number of years
97. Not applicable
98. DK
99. NA

papres80   Prestige of father's occupation
17–86. Actual score
0. Not applicable, DK, NA

partnrs5   How many sex partners have you had over the past five years?
0. No partners
1. 1 partner
2. 2 partners
3. 3 partners
4. 4 partners
5. 5–10 partners
6. 11–20 partners
7. 21–100 partners
8. More than 100 partners
9. 1 or more, don't know the number
95. Several
98. DK
99. NA
−1. Not applicable

polviews   I'm going to show you a seven-point scale on which the political views that people might hold are arranged from extremely liberal to extremely conservative. Where would you place yourself on this scale?
1. Extremely liberal
2. Liberal
3. Slightly liberal
4. Moderate
5. Slightly conservative
6. Conservative
7. Extremely conservative
0. Not applicable

8. DK
9. NA

premarsx

There's been a lot of discussion about the way morals and attitudes about sex are changing in this country. If a man and a woman have sex relations before marriage, do you think it is always wrong, almost always wrong, wrong only sometimes, or not wrong at all?
1. Always wrong
2. Almost always wrong
3. Wrong only sometimes
4. Not wrong at all
0. Not applicable
8. DK
9. NA

pres92

In 1992, you remember that Clinton ran for president on the Democratic ticket against Bush for the Republicans and Perot as an Independent. Did you vote for Clinton, Bush, or Perot? (Includes only those who said they voted in this election)
1. Clinton
2. Bush
3. Perot
4. Other
6. No presidential vote
0. Not applicable
8. DK
9. NA

prestg80

Prestige of respondent's occupation
17–86. Actual score
0. Not applicable, DK, NA

race

Race of respondent
1. White
2. Black
3. Other

region

Region of interview
1. New England
2. Mid-Atlantic
3. East N. Cent.
4. West N. Cent.
5. So. Atlantic
6. East So. Cent.
7. West So. Cent.
8. Mountain
9. Pacific

relig	What is your religious preference? Is it Protestant, Catholic, Jewish, some other religion, or no religion?
	1. Protestant
	2. Catholic
	3. Jewish
	4. None
	5. Other
	8. DK
	9. NA

reliten   Would you call yourself a strong (preference named on *relig*) or a not very strong (preference named on *relig*)?
        1. Strong
        2. Not very strong
        3. Somewhat strong (volunteered)
        4. No religion
        0. Not applicable
        8. DK
        9. NA

satfin   So far as you and your family are concerned, would you say that you are pretty well satisfied with your present financial situation, more or less satisfied, or not satisfied at all?
        1. Pretty well satisfied
        2. More or less satisfied
        3. Not satisfied at all
        8. DK
        9. NA

satjob   On the whole, how satisfied are you with the work you do?
        1. Very satisfied
        2. Moderately satisfied
        3. A little dissatisfied
        4. Very dissatisfied
        0. Not applicable
        8. DK
        9. NA

sex   Respondent's gender
        1. Male
        2. Female

sexeduc   Would you be for or against sex education in your public schools?
        1. Favor
        2. Oppose
        3. Depends
        0. Not applicable

8. DK
9. NA

sexfreq   About how many times did you have sex during the last 12 months?
- 0. Not at all
- 1. Once or twice
- 2. About once a month
- 3. 2 or 3 times a month
- 4. About once a week
- 5. 2 or 3 times a week
- 6. More than 3 times a week
- −1. Not applicable
- 8. DK
- 9. NA

size   Size of place in thousands. Population figures from U.S. Census. Add three zeros to code for actual values.
- −1. Not assigned

spanking   Do you strongly agree, agree, disagree, or strongly disagree that it is sometimes necessary to discipline a child with a good, hard spanking?
- 1. Strongly agree
- 2. Agree
- 3. Disagree
- 4. Strongly disagree
- 0. Not applicable
- 8. DK
- 9. NA

suicide1   Do you think a person has a right to end his or her own life if this person has an incurable disease?
- 1. Yes
- 2. No
- 0. Not applicable
- 8. DK
- 9. NA

trust   Generally speaking, would you say that most people can be trusted or that you can't be too careful in dealing with people?
- 1. Most people can be trusted
- 2. Can't be too careful
- 3. Other, depends (volunteered)
- 0. Not applicable
- 8. DK
- 9. NA

tvhours

On the average day, about how many hours do you personally watch television?

    00–24. Actual hours
   −1. Not applicable, DK, NA
    98. DK
    99. NA

wrkstat

Last week, were you working full time, part time, going to school, keeping house, or what?

1. Working full time
2. Working part time
3. With a job but not at work (ill, vacation, etc.)
4. Unemployed
5. Retired
6. In school
7. Keeping house
8. Other

xmovie

Have you seen an X-rated movie in the last year?

1. Yes
2. No
0. Not applicable
8. DK
9. NA

Appendix H   Basic Mathematics Review

H.1 OF ARITHMETIC, CALCULATORS, AND COMPUTERS

You will probably be relieved to hear that first courses in statistics are not particularly mathematical and do not stress computation per se. While you will encounter many numbers to work with and numerous formulas to use, the major emphasis will be on understanding the role of statistics in research and the logic by which we attempt to answer research questions empirically. You will also find that, at least in this text, the example problems and many of the homework problems have been intentionally simplified so that the computations will not unduly distract you from the task of understanding the statistics themselves.

On the other hand, you will probably regret to hear that there is, inevitably, some arithmetic that you simply cannot avoid if you want to master this material. It is likely that some of you haven't had any math in a long time, others have convinced themselves that they just cannot do math under any circumstances, and still others are just rusty and out of practice. All of you will find that even the most complex and intimidating operations and formulas can be broken down into simpler steps. If you have forgotten how to cope with some of these simpler steps or are unfamiliar with these operations, this appendix is designed to ease you into the skills you will need to do all of the computation in this textbook.

Before we begin the review, let me point out that a calculator is a virtual necessity for this text. While you could do all the arithmetic by hand, the calculator will save you time and effort and is definitely worth the small investment. Incidentally, you do not need to invest in any of the super-sophisticated, more expensive models you might see for sale. A square root function is the only extra feature you really need. Of course, if you want the additional features, such as memories and preprogrammed functions, by all means spend the extra money and learn how to use them. However, a simple, inexpensive calculator will work fine for all of the problems in this text.

Along the same lines, many of you probably have access to computers and statistical packages. If so, take the time now to learn how to use them, because they will eventually save you time and effort (not to mention that they will guarantee the accuracy of your answers). This text includes a guide to a statistical package called SPSS, but many other programs are available that will accomplish the goals of saving time and avoiding drudgery while generating precise and accurate results.

In summary, you should find a way at the beginning of this course—with a calculator, a computer, or both—to minimize the tedium and hassle of mere computing. This will permit you to devote maximum effort to the truly important goal of increasing your understanding of the meaning of statistics in particular and social research in general.

H.2 VARIABLES AND SYMBOLS

Statistics are a set of techniques by which we can describe, analyze, and manipulate **variables**. A variable is any attribute or trait that can change value from case to case or from time to time. Examples of variables would include height, weight, level of prejudice, and political party preference. The possible values associated with a given variable might be numerous (for example, income) or relatively few (for example, gender). I will often use symbols, usually the letter X, to refer to variables in general or to a specific variable.

Sometimes we will need to refer to a specific value or set of values of a variable. This is usually done with the aid of subscripts. So, the symbol X_1 (read "X-sub-one") would refer to the first score in a set of scores, X_2 ("X-sub-two") to the second score, and so forth. On occasion, we will use the subscript i to refer to all the scores in a set. Thus, the symbol X_i ("X-sub-eye") would be taken to mean all of the scores associated with a given variable (for example, the test grades of a particular class).

H.3 OPERATIONS

You are all familiar with the four basic mathematical operations of addition, subtraction, multiplication, and division and the standard symbols ($+$, $-$, $\times$, $\div$) used to denote them. Some of you may not be aware, however, that the latter two operations can be symbolized in a variety of ways. For example, the operation of multiplying some number a by some number b may be symbolized in (at least) six different ways:

$$a \times b$$
$$a \cdot b$$
$$a * b$$
$$ab$$
$$a(b)$$
$$(a)(b)$$

In this text, we will commonly use the "adjacent symbols" format (that is, ab), the conventional times sign ($\times$), or adjacent parentheses to indicate multiplication. On most calculators and computers, the asterisk (*) is the symbol for multiplication.

The operation of division is also indicated by multiple symbols:

$$a \div b$$
$$a/b$$
$$\frac{a}{b}$$

In this text, the latter two symbols are used almost exclusively.

Several of the formulas with which we will be working require us to find the square of a number. To do this, simply multiply the number by itself. This operation is symbolized as X^2 (read "X squared"), which is the same thing as $(X)(X)$. If X has a value of 4, then

$$X^2 = (X)(X) = (4)(4) = 16$$

or we could say that "4 squared is 16."

The square root of a number is the value that, when multiplied by itself, results in the original number. So the square root of 16 is 4 because $(4)(4)$ is 16. The operation of finding the square root of a number is symbolized as

$$\sqrt{X}$$

Be sure that you have access to a calculator with a built-in square root function.

A final operation with which you should be familiar is summation, or the addition of the scores associated with a particular variable. When a formula requires the addition of a series of scores, this operation is usually symbolized as ΣX_i, where Σ is the uppercase Greek letter sigma and stands for "the summation of." So the combination of symbols ΣX_i means "the summation of all the scores" and directs us to add the value of all the scores for that variable. If four people had family sizes of 2, 4, 5, and 7, then the summation of these four scores for this variable could be symbolized as

$$\Sigma X_i = 2 + 4 + 5 + 7 = 18$$

The symbol Σ is an operator, just like the $+$ or $\times$ signs. It directs us to add all of the scores on the variable indicated by the X symbol.

There are two other common uses of the summation sign and, unfortunately, the symbols denoting these uses are not, at first glance, sharply different from each other or from the symbol used above. A little practice and some careful attention to these various meanings should minimize the confusion.

Beginning with Chapter 4, you will frequently encounter the set of symbols ΣX_i^2, which means "the sum of the squared scores." This quantity is found by *first squaring* each of the scores and *then adding* the squared scores together. A second common set of symbols will be $(\Sigma X_i)^2$, which means "the sum of the scores, squared." This quantity is found by *first summing* the scores and *then squaring* the total.

These distinctions might be confusing at first, so let's see if an example helps to clarify the situation. Suppose we had a set of three scores: 10, 12, and 13. So,

$$X_i = 10, 12, 13$$

The sum of these scores would be indicated as

$$\Sigma X_i = 10 + 12 + 13 = 35$$

The sum of the squared scores would be

$$\Sigma X_i^2 = (10)^2 + (12)^2 + (13)^2 = 100 + 144 + 169 = 413$$

Take careful note of the order of operations here. First, the scores are squared one at a time and then the squared scores are added. This is a completely different operation from squaring the sum of the scores:

$$(\Sigma X_i)^2 = (10 + 12 + 13)^2 = (35)^2 = 1225$$

To find this quantity, first the scores are summed and then the total of all the scores is squared. The value of the sum of the scores, squared (1225) is not at all the same as the value of the sum of the squared scores (413). In summary, the table below indicates the operations associated with each set of symbols.

Symbols	Operations
ΣX_i	Add the scores.
ΣX_i^2	First square the scores and then add the squared scores.
$(\Sigma X_i)^2$	First add the scores and then square the total.

H.4 OPERATIONS WITH NEGATIVE NUMBERS

A number can be either positive (if it is preceded by a $+$ sign or by no sign at all) or negative (if it is preceded by a $-$ sign). Positive numbers are greater than zero, and negative numbers are less than zero. It is very important to keep track of signs because they will affect the outcome of virtually every mathematical operation. This section will briefly summarize the relevant rules for dealing with negative numbers. First, adding a negative number is the same as subtraction. For example,

$$3 + (-1) + 4 = 3 - 1 + 4 = 2 + 4 = 6$$

Second, subtraction changes the sign of a negative number:

$$3 - (-1) - 4 = 3 + 1 - 4 = 4 - 4 = 0$$

Note the importance of keeping track of signs here. It would be relatively easy for an inexperienced person to forget to change the sign of the negative number in the expression above and arrive at the wrong answer.

For multiplication and division, there are various combinations of negative and positive numbers you should be aware of. For purposes of this text, you will rarely have to multiply or divide more than two numbers at a time, and we will confine our attention to this situation. Ignoring the case of all positive numbers, this leaves several possible combinations. A negative number times a positive number results in a negative value:

$$(-3)(4) = -12$$

or

$$(3)(-4) = -12$$

A negative number times a negative number is always positive:

$$(-3)(-4) = 12$$

Division follows the same patterns. If there is a single negative number in the calculations, the answer will be negative. If both numbers are negative, the answer will be positive. So,

$$(-4)/(2) = -2$$
$$\text{and}$$
$$(4)/(-2) = -2$$
$$\text{but}$$
$$(-4)/(-2) = 2$$

Negative numbers do not have square roots, since multiplying a number by itself cannot result in a negative value. Squaring a negative number always results in a positive value (see the multiplication rules above).

H.5 ACCURACY AND ROUNDING OFF

A possible source of confusion in computation involves the issues of accuracy and rounding off. People work at different levels of accuracy and precision and, for this reason alone, may arrive at different answers to problems. This is important because, if you work at one level of precision and I (or your instructor or your study partner) work at another, we will often arrive at solutions that are at least slightly different. You may sometimes think you've gotten the wrong answer when all you've really done is round off at a different place in the calculations or in a different way.

There are two issues here: when to round off and how to round off. In this text, I have followed the convention of working in as much accuracy as my calculator or statistics package will allow and then rounding off to two places of accuracy (two places beyond the decimal point) at the very end. If a set of calculations is lengthy and requires the reporting of intermediate sums or subtotals, I will round the subtotals off to two places also.

In terms of how to round off, begin by looking at the digit immediately to the right of the last digit you want to retain. If you want to round off to 100ths (two places beyond the decimal point), look at the digit in the 1000ths place (three places beyond the decimal point). If that digit is greater than 5, round up. For example, 23.346 would round off to 23.35. If the digit to the right is less than 5, round down. So, 23.343 would become 23.34. If the digit to the right is 5, round up if the digit immediately to the left is even and round down if the digit is odd. So, 23.345 would become 23.35 and 23.355 would round to 23.35.

Let's look at some more examples of how to follow the rounding rules stated above. If you are calculating the mean value of a set of test scores and your calculator shows a final value of 83.459067, and you want to round off to two places beyond the decimal point, look at the digit three places beyond the decimal point. In this case the value is 9 (greater than 5), so we would round the second digit beyond the decimal point up and report the mean as 83.46. If the value had been 83.453067, we would have reported our final answer as 83.45. A value of 83.455067 would round to 83.45, and a value of 83.445067 would be 83.45.

H.6 FORMULAS, COMPLEX OPERATIONS, AND THE ORDER OF OPERATIONS

A mathematical formula is a set of directions, stated in general symbols, for calculating a particular statistic. To "solve a formula" means that you must replace the symbols with the proper values and then manipulate the values through a series of calculations. Even the most complex formula can be rendered manageable if it is broken down into smaller steps. Working through these steps requires some knowledge of general procedure and the rules of precedence of mathematical operations. This is because the order in which you perform calculations may affect your final answer. Consider the following expression:

$$2 + 3(4)$$

Note that if you do the addition first, you will evaluate the expression as

$$5(4) = 20$$

but if you do the multiplication first, the expression becomes

$$2 + 12 = 14$$

Obviously, it is crucial to complete the steps of a calculation in the correct order.

The basic rules of precedence are to find all squares and square roots first, then do all multiplication and division, and finally complete all addition and subtraction. So the following expression:

$$8 + 2 \times 2^2/2$$

would be evaluated as

$$8 + 2 \times 4/2$$
$$= 8 + 8/2$$
$$= 8 + 4$$
$$= 12$$

The rules of precedence may be overridden when an expression contains parentheses. Solve all expressions within parentheses before applying the rules stated above. For most of the complex formulas in this text, the order of calculations will be controlled by the parentheses. Consider the following expression:

$$(8 + 2) - 4(3)^2/(8 - 6)$$

Resolving the parenthetical expressions first, we would have

$$(10) - 4 \times 9/(2)$$
$$= 10 - 36/2$$
$$= 10 - 18$$
$$= -8$$

Without the parentheses, the same expression would be evaluated as

$$8 + 2 - 4 \times 3^2/8 - 6$$
$$= 8 + 2 - 4 \times 9/8 - 6$$
$$= 8 + 2 - 36/8 - 6$$
$$= 8 + 2 - 4.5 - 6$$
$$= 10 - 10.5$$
$$= -.5$$

A final operation you will encounter in some formulas in this text involves denominators of fractions that themselves contain fractions. In this situation, solve the fraction in the denominator first and then complete the division. For example,

$$\frac{15 - 9}{6/2}$$

would become

$$\frac{15 - 9}{3} = \frac{6}{3} = 2$$

When you are confronted with complex expressions such as these, don't be intimidated. If you're patient with yourself and work through them step by step, beginning with the parenthetical expression, even the most imposing formulas can be managed.

H.7 EXERCISES

You can use the problems below as a "self-test" on the material presented in this review. If you can handle these problems, you're ready to do all of the arithmetic in this text. If you have difficulty with any of these problems, please review the appropriate section of this appendix. You might also want to use this section as an opportunity to become more familiar with your calculator. Answers are given on the next page, along with some commentary and some reminders.

1. Complete each of the following:

    a. $17 \times 3 =$ _____   **c.** $(17)(3) =$ _____

    b. $17(3) =$ _____   **d.** $17/3 =$ _____

e. $(42)^2 =$ _____

f. $\sqrt{113} =$ _____

2. For the set of scores (X_i) of 50, 55, 60, 65, and 70, evaluate each of the expressions below:

$\Sigma X_i \quad =$ _____

$\Sigma X_i^2 \quad =$ _____

$(\Sigma X_i)^2 \quad =$ _____

3. Complete each of the following:

a. $17 + (-3) + (4) + (-2) =$ _____

b. $15 - 3 - (-5) + 2 =$ _____

c. $(-27)(54) =$ _____

d. $(113)(-2) =$ _____

e. $(-14)(-100) =$ _____

f. $-34/-2 =$ _____

g. $322/-11 =$ _____

h. $\sqrt{-2} =$ _____

i. $(-17)^2 =$ _____

4. Round off each of the following to two places beyond the decimal point:

a. 17.17532 _____

b. 43.119 _____

c. 1076.77337 _____

d. 32.4651152301 _____

e. 32.4751152301 _____

5. Evaluate each of the following:

a. $(3 + 7)/10 =$ _____

b. $3 + 7/10 =$ _____

c. $((4 - 3) + (7 + 2))/((4 + 5)(10)) =$ _____

d. $\sqrt{(7(5 - 3)^2)/((17/3)(4))} =$ _____

e. $\dfrac{22 + 44}{15/3} =$ _____

H.8 ANSWERS

1. **a.** 51

b. 51

c. 51
(The obvious purpose of these first three problems is to remind you that there are several different ways of expressing multiplication.)

d. 5.67 (Note the rounding off.)

e. 1764

f. 10.63

2. The first expression translates to "the sum of the scores," so this operation would be

$\Sigma X_i = 50 + 55 + 60 + 65 + 70 = 300$

The second expression is the "sum of the squared scores." So

$\Sigma X_i^2 = (50)^2 + (55)^2 + (60)^2 + (65)^2 + (70)^2$

$\Sigma X_i^2 = 2500 + 3025 + 3600 + 4225 + 4900$

$\Sigma X_i^2 = 18,250$

The third expression is "the sum of the scores, squared":

$(\Sigma X_i)^2 = (50 + 55 + 60 + 65 + 70)^2$

$(\Sigma X_i)^2 = (300)^2$

$(\Sigma X_i)^2 = 90,000$

Remember that ΣX_i^2 and $(\Sigma X_i)^2$ are two completely different expressions with very different values.

3. **a.** 16

b. 19 (Remember to change the sign of -5.)

c. -1458

d. -226

e. 1400 (A negative number times a negative number results in a positive number.)

f. 17

g. -29.27

h. Your calculator probably gave you some sort of error message for this problem, since negative numbers do not have square roots.

i. 289

4. **a.** 17.17

b. 43.12

c. 1076.77

d. 32.47

e. 32.47

5. **a.** 1

b. 3.7 (Note again the importance of parentheses.)

c. 0.11

d. Whoa!—this looks complicated and should be worked out carefully, step by step. Note that complex expressions such as this are solved from the "inside out." That is, we begin within the innermost parentheses, follow the rules of precedence until we have resolved the numbers under the square root sign into a single number, and the *final* thing we do is take the square root:

$$\sqrt{(7(5-3)^2)/((17/3)(4))}$$
$$= \sqrt{7(2)^2/(5.6666667)(4)}$$
$$= \sqrt{7(4)/22.666667}$$
$$= \sqrt{28/22.666667}$$
$$= \sqrt{1.2352941}$$
$$= 1.11143790 \text{ or, rounded off}$$
$$= 1.11$$

e. 13.2

Answers to Odd-Numbered Computational Problems

In addition to answers, this section suggests some problem-solving strategies and provides examples of how to interpret the numerical answers. You should try to solve and interpret the problems on your own before consulting this section.

In solving these problems, I let my calculator or computer do most of the work. I worked with whatever level of precision these devices permitted. I didn't round off until the end or until I had to record an intermediate sum. I always rounded off to two places of accuracy (or, two places beyond the decimal point, or to 100ths). If you follow these same conventions, your answers will almost always match mine. However, there is no guarantee that our answers will always be exact matches, and you should be aware that small discrepancies might occur and that they are almost always trivial. If the difference between your answer and mine doesn't seem trivial, you should double-check to make sure you haven't made an error or solve the problem again using a greater degree of precision.

Finally, please allow me a brief disclaimer about mathematical errors in this section. Let me assure you, first of all, that I know how important this section is for most students and that I worked hard to be certain that these answers are correct. Human fallibility being what it is, however, I know that I cannot make absolute guarantees. Should you find any errors, please let me know so I can make corrections in the future.

Chapter 2

2.1 **a.** Complex A: $(5/20) \times 100 = 25.00\%$
Complex B: $(10/20) \times 100 = 50.00\%$
  b. Complex A: $4:5 = 0.80$
Complex B: $6:10 = 0.60$
  c. Complex A: $(0/20) = 0.00$
Complex B: $(1/20) = 0.05$
  d. $(6/(4 + 6)) = (6/10) = 60.00\%$
  e. Complex A: $8:5 = 1.60$
Complex B: $2:10 = 0.20$

2.3 Bank robbery rate $= (47/211732) \times 100000 = 22.20$
Homicide rate $= (13/211732) \times 100000 = 6.14$
Auto theft rate $= (23/211732) \times 100000 = 10.86$

Chapter 3

3.1 "Region of birth" is a nominal-level variable, "support for legalization" is ordinal, and "expenses" is interval-ratio. The mode, the most common score, is the only measure of central tendency available for nominal-level variables. Don't forget to array the scores from high to low before locating the median. There are ten freshmen (N is even), so the median will be the score halfway between the scores of the two middle cases. For this variable, both middle cases have scores of 3, so the median is three. There are 11 seniors (N is odd), so the median will be the score of the middle case. To find average expenses, add the scores and divide by the number of cases.

Freshmen:
  Region of birth: Mode = North
  Legalization: Median = 3
  Expenses: Mean = 38.50
Seniors:
  Region of birth: Mode = North
  Legalization: Median = 5
  Expenses: Mean = 53.00

3.3   For gender, the modal category is male. For social class, the median is "medium" (i.e., the middle case is in this category). The mean number of years in the party is 26.15.

3.5   For marital status, the modal category is "married." For race, the mode is "white." For age, the mean is 27.53, and the median is 7 for attitude on abortion

3.7   In Table 2.4, there are 10 males and 10 females, so sex is bimodal. The mode of "marital status" is "single," with 10 cases. The median of "satisfaction" is 3, and the mean age is 20.05.

3.9   Attitude and opinion scales almost always generate ordinal-level data, so the appropriate measure of central tendency would be the median. For the students, the median is 9 and, for the neighbors, the median is 2. Incidentally, the means are 7.80 and 4.00, respectively.

3.11   The mean is 520.55, and the median is 512.50. The greater value for the mean indicates a positive skew (a few high scores).

3.13   For freshmen, the median is 35, and the mean is 31.72. For seniors, the median is 30, and the mean is 28.4.

3.15   For the full sample, the mean is 22.75, and the median is 20.25. This distribution has a positive skew. By inspection, we can see that New York's score is much higher than any other score in the sample, and this accounts for the skew. Sure enough, if you remove New York and recalculate, the median falls to 19.20, and the mean falls to 20.89. There is still a skew in the distribution, but it is less pronounced (as indicated by the fact that the difference between the median and mean is less with New York removed).

3.17   For problem 2.8, the mode is 7, the median is the score associated with the 13th case, or 5, and the mean is 150/25, or 6. The mean is higher in value than the median, indicating a slight positive skew. The third quartile is the score associated with the

19th case, or 7. The fourth decile is the score associated with the 10th case, or 3. The 23rd percentile is the score associated with the sixth case, or 2. For problem 2.9, the mode is 17. The median is the score halfway between the scores of the 25th and 26th cases. Both cases have scores of 47, so the median is also 47. The mean is 2280/50, or 45.60. Since the mean is lower than the median, this distribution has a negative skew. The third quartile is 66, the fourth decile is 29, and the 23rd percentile is 19.

3.19   A total of 50 cases were tested, so the median is the score halfway between the scores of the 25th and 26th cases. For the "release date" test, both of these cases have a score of 22, so the median is 22. The mean for this first test is 1198/50, or 23.96. The median and mean for the second test are 27.5 and 28.58, respectively. The average scores are higher for the second test, indicating that the sample as a whole experienced an improvement in level of functioning and competence.

Chapter 4

4.1   The IQV is .89 for complex A, .99 for complex B, .71 for complex C, and .74 for complex D. Complex B is the most heterogeneous and complex C is the least.

4.3   The range is $50 - 10$ or 40. The standard deviation is 12.28.

4.5   For problem 2.8, the range is $21 - 0$, or 21. Q_1 is the score of the $(.25)(25) = 6.25$th or, rounding off, the 6th case. This value is 2. Q_3 is the score of the $(.75)(25) = 18.75$, or 19th case, or 7. $Q = Q_3 - Q_1 = 7 - 2 = 5$. The standard deviation is 5.49, and the variance is $(5.49)^2$, or 30.14.

4.7   Enrollment: $s = 290.90$
% college bound: $s = 16.79$
Condition: $s = 2.21$

4.9   Pretest: $s = 3.50$
Posttest: $s = 4.40$

4.11

1990	1995
Mean = 101	Mean = 158.60
Standard deviation = 29.81	Standard deviation = 45.17

4.13

1973	1975
Mean = 12.21	Mean = 19.06
Standard deviation = 12.20	Standard deviation = 9.63

Over the 2-year period, the rate of legal abortion rose, and the states became less diverse. If you look at the scores state-by-state, you'll see that they increased in every state except New York. The Supreme Court decision was *Roe v. Wade*.

4.15 Number of children: Mean = 1.24, $s = 1.21$
Educational level: Mean = 11.56, $s = 3.23$
Liberalism: Mean = 4.28, $s = 1.48$
Attendance: Mean = 3.92, $s = 2.67$

4.17 For 1987, $s = 7.27$
For 1997, $s = 9.09$

4.19 $s = 14.50$

4.21 For the respondents: $R = 78 - 12 = 66$
$s = 16.22$
For the fathers: $R = 70 - 16 = 54$
$s = 12.77$

4.23 For 1975, the standard deviation is 4.02, and for 1994, the standard deviation is 4.10.

Chapter 5

5.1

X_i	Z score	% Area Above	% Area Below
5	−1.67	95.25	4.75
6	−1.33	90.82	9.18
7	−1.00	84.13	15.87
8	−.67	74.86	25.14
9	−.33	62.93	37.07
11	.33	37.07	62.93
12	.67	25.14	74.86
14	1.33	9.18	90.82
15	1.67	4.75	95.25
16	2.00	2.28	97.72
18	2.67	.38	99.62

5.3

Math and Science			Language Skills and Writing Ability			History and Social Sciences		
X_i	Z	% Area Below	X_i	Z	% Area Below	X_i	Z	% Area Below
55	−1.00	15.87	44	1.79	96.33	64	−1.07	14.23
52	−1.77	3.84	45	2.15	98.42	60	−1.46	7.21
49	−2.54	.55	41	0.70	75.80	58	−1.66	4.85
48	−2.80	.26	43	1.43	92.36	55	−1.96	2.50
45	−3.56	.02	42	1.07	85.77	50	−2.45	.71
44	−3.82	.01	38	−0.38	35.20	46	−2.85	.22
43	−4.08	.00	35	−1.47	7.08	40	−3.44	.03
47	−3.05	.11	50	3.96	99.99	75	0.02	50.80
58	−0.23	40.90	40	0.34	63.31	77	0.22	58.71
60	0.28	61.03	37	−0.75	22.66	80	0.52	69.85
61	0.54	70.54	39	−0.02	49.20	81	0.62	73.24
65	1.56	94.06	34	−1.83	3.36	89	1.41	92.07
67	2.08	98.12	33	−2.20	1.39	90	1.51	93.45
70	2.85	99.78	30	−3.28	.05	94	1.90	97.13
71	3.10	99.90	29	−3.65	.01	88	1.31	90.49

5.5 **a.** 22.34% **b.** 61.79% **c.** 61.79% **d.** 86.64%
e. 1.39% **f.** 5.74% **g.** 19.34% **h.** 34.62%

5.7 **a.** 5.05 **b.** 23.27 **c.** 57.14 **d.** 86.21
e. 6.85 **f.** 41.69 **g.** 70.67 **h.** 29.07
i. 2.81 **j.** 15.87 **k.** 60.64 **l.** 88.10

5.9 **a.**

X_i	Z score	% Below	% Above
19	.67	74.86	25.14
10	−2.33	.99	99.01
14	−1.00	15.87	84.13
15	−.67	25.14	74.86
18	.33	62.93	37.07
20	1.00	84.13	15.87
22	1.67	95.25	4.75
23	2.00	97.72	2.28

b.

Scores	Z scores	% Area Between
8 and 12	−3.00 and −1.67	4.61
9 and 13	−2.67 and −1.33	8.80
11 and 17	−2.00 and 0.00	47.72
15 and 19	−.67 and .67	49.72
16 and 20	−.33 and 1.00	47.06
17 and 23	0.00 and 2.00	47.72
18 and 19	.33 and .67	11.93
19 and 22	.67 and 1.67	20.39

c.

Scores	Z scores	Probability
less than 17	0.00	.5000
less than 24	2.33	.9901
less than 10	−2.33	.0099
less than 8	−3.00	.0014
between 8 and 12	−3.00 and −1.67	.0461
between 11 and 17	−2.00 and 0.00	.4772
between 16 and 18	−.33 and .33	.2586
between 20 and 24	1.00 and 2.33	.1488
more than 24	2.33	.0099
more than 20	1.00	.1587
more than 15	−.67	.7486
more than 9	−2.67	.9962

5.11 a. .5328 **b.** .0638 **c.** .8944
d. .3085 **e.** .0062 **f.** .9332

5.13

	Score	Z score	Area
a.	55	−1.00	
	65	1.56	.7819
b.	60	.28	
	65	1.56	.3303
c.	65	1.56	.0594
d.	60	.28	
	50	−2.28	.5990
e.	55	−1.00	
	50	−2.28	.1474
f.	55	−1.00	.1587

5.15 Yes. The raw score of 110 translates into a Z score of +2.88. 99.80% of the area lies below this score, so this individual was in the top 1% on this test.

5.17 For the first event, the probability is .0919 and, for the second, the probability is .0655. The first event is more likely.

Part I Cumulative Exercises

1. First, determine the level of measurement of each variable. This will help you to set up the frequency distributions and determine the appropriate mea-

sures of central tendency and dispersion. There is one nominal variable (religion), two ordinal variables (strength and comfort), and two interval-ratio-level variables (pray and age). For nominal-level variables, the mode is the category with the most cases. For this sample, there are more Protestants than any other religion. The index of qualitative variation (IQV), the only measure of dispersion presented in this text for nominal-level variables, is .86. This is a high value (the maximum would be 1.00) and indicates a great deal of dispersion.

To find the median for "Strength" and "Comfort," the scores must first be arrayed from high to low. There are fifteen cases in the sample, so the median is the score associated with the 8th case. The median for "Strength" is 7 and 1 for "Comfort." The ranges for these variables are 9 and 4, respectively.

As an alternative, it would be reasonable to use the standard deviation as a measure of dispersion for "Strength." It is not uncommon in the social science literature for the standard deviation to be reported for ordinal-level variables with a broad range of scores. The standard deviation for this variable is 2.77. As opposed to "Strength," "Comfort" has only 5 categories, so an alternative to the range for this variable would be the IQV, which is .80.

For "Pray," the mean is 1.40, and the standard deviation is 1.62. For age, the mean is 41.53, and the standard deviation is 12.54.

3. N of children: Mean = 2.33, $s = 1.91$
Yrs of school: Median = 1, Range = 4
Race: Mode = white, IQV = .49
Birth control: Median = 2, Range = 3
TV: Mean = 3.13, $s = 2.39$
Religion: Mode = Protestant, IQV = .61

Chapter 7

7.1 a. 5.2 ± 0.11 **b.** 100 ± 0.71 **c.** 20 ± 0.40
d. 1020 ± 5.41 **e.** 7.3 ± 0.23 **f.** 33 ± 0.80

7.5 a. 2.30 ± 0.04 **b.** $2.10 \pm 0.01, 0.78 \pm .07$
c. 6.00 ± 0.37

7.7 a. 178.23 ± 1.97 The estimate is that students spent between $176.26 and $180.20 on books.
b. $1.5 \pm .04$ The estimate is that students visited

the clinic between 1.46 and 1.54 times on the average.

c. $2.8 \pm .13$ **d.** $3.5 \pm .19$

7.9 $.14 \pm .07$ The estimate is that between 7% and 21% of the population consists of unmarried couples living together.

7.11 a. $P_s = (823/1496) = .55$
Confidence interval: $.55 \pm .02$
Between 53% and 57% of the population agrees with the statement.
b. $P_s = (650/1496) = .44$
Confidence interval: $.44 \pm .02$
c. $P_s = (375/1496) = .25$
Confidence interval: $.25 \pm .02$
d. $P_s = (1023/1496) = .68$
Confidence interval: $.68 \pm .02$
e. $P_s = (800/1496) = .54$
Confidence interval: $.54 \pm .02$

7.13 For alpha = .10: $100 \pm .74$
For alpha = .05: $100 \pm .88$
For alpha = .01: 100 ± 1.16
For alpha = .001: 100 ± 1.47

7.15 The confidence interval is $.51 \pm .05$. The estimate would be that between 46% and 56% of the population prefer candidate A. The population parameter (P_u) is equally likely to be anywhere in the interval (that is, it's just as likely to be 46% as it is to be 56%), so a winner cannot be predicted.

7.17 The confidence interval is 0.23 ± 0.08. At the 95% confidence level, the estimate would be that between 240 (15%) and 496 (31%) of the 1600 freshmen would be extremely interested. The estimated numbers are found by multiplying N (1600) by the upper (.31) and lower (.15) limits of the interval.

*7.19 With alpha = .05 and $P_u = .5$, N should be about 1000.

7.21 a. $43.35 \pm .73$ **b.** $3.07 \pm .24$ **c.** $1.86 \pm .09$
d. $.25 \pm .03$ (Catholics make up about 24% of the population.)

e. $.22 \pm .03$ (About 23% of the population has never married.)
f. $.51 \pm .03$ (About 43% of the electorate voted for Clinton in 1992.)
g. $.71 \pm .03$

Chapter 8

8.3 a. Z (obtained) = -41.00
b. Z (obtained) = 29.09

8.5 Z (obtained) = 6.04

8.7 a. Z (obtained) = -13.66
b. Z (obtained) = 25.50

8.9 t (obtained) = 4.50

8.11 Z (obtained) = 3.06

8.13 Z (obtained) = -1.48

8.15 a. Z (obtained) = -0.73
b. Z (obtained) = 2.19
c. Z (obtained) = -8.55
d. Z (obtained) = -18.07
e. Z (obtained) = 2.09
f. Z (obtained) = -53.33

8.17 Z (obtained) = 1.73

Chapter 9

9.1 a. $\sigma_{\bar{x}-\bar{x}} = 1.39$ Z (obtained) = -2.53
b. $\sigma_{\bar{x}-\bar{x}} = 1.61$ Z (obtained) = 2.49

9.3 a. Z (obtained) = 1.70
b. Z (obtained) = -2.48

9.5 a. $\sigma_{\bar{x}-\bar{x}} = 0.08$ Z (obtained) = 11.25
b. $\sigma_{\bar{x}-\bar{x}} = 0.12$ Z (obtained) = -3.33
    $\sigma_{\bar{x}-\bar{x}} = 0.15$ Z (obtained) = 20.00

9.7 These are small samples (combined N's of less than 100), so be sure to use Formulas 9.5 and 9.6 in step 4.
a. $\sigma_{\bar{x}-\bar{x}} = 0.12$ t (obtained) = -1.33
b. $\sigma_{\bar{x}-\bar{x}} = 0.13$ t (obtained) = 14.85

9.9 t (obtained) = 3.52

9.11 t (obtained) = 2.33

9.13 a. $P_u = 0.19$ $\sigma_{p-p} = 0.05$ Z(obtained) = -0.60
   b. $P_u = 0.61$ $\sigma_{p-p} = 0.03$ Z(obtained) =   0.67

9.15 a. Z(obtained) = $-.76$
   b. Z(obtained) = $-.83$

9.17 $P_u = 0.30$ $\sigma_{p-p} = 0.05$ Z(obtained) = 2.00

Chapter 10

10.1 a. The overall mean is 12.17, SST = 230.69, SSB = 173.19, and SSW = 57.5. The F ratio is 13.55.
   b. The overall mean is 6.87, SST = 455.05, SSB = 78.54, and SSW = 376.51. The F ratio is 1.25.
   c. The overall mean is 31.65, SST = 8362.565, SSB = 5053.35, and SSW = 3309.20. The F ratio is 8.14.

10.3   The overall mean is 4.12, SST = 132.64, SSB = 71.28, and SSW = 61.36. The obtained F ratio is 12.77. With alpha set at .05 and 2 and 22 degrees of freedom, the critical F ratio would be 3.44, so the null may be rejected. Decision making does vary significantly by type of relationship. By inspection of the group means, it would appear that the "cohabitational" category accounts for most of the differences.

10.5   SST = 342.99
   SSB =   19.50
   SSW = 323.50
   F =   .81

10.7   SST = 10609.31
   SSB = 3342.99
   SSW = 7265.32
   F =   7.59

10.9 $F = 2.92$

10.11 $F = 3.25$

Chapter 11

11.1 a. Mann-Whitney U: $U = 48.5$, Z (obtained) = $-.11$
   Runs test: $R = 13$ or 15, Z(obtained) = -1.83 or $-.92$
   b. Mann-Whitney U: $U = 22$, Z (obtained) = -2.12
   Runs test: $R = 10$, Z(obtained) = $-.46$
   c. Mann-Whitney U: $U = 20.5$, Z (obtained) = -2.23
   Runs test: $R = 8$, Z(obtained) = -1.38

11.3   Mann-Whitney U test: $U = 59.5$, Z (obtained) = -3.80
   Runs test: $R = 9$ or 10, Z (obtained) = $-.61$ or $-.56$

11.5 $U = 187.5$ Z (obtained) = $-.93$
   $R = 23$ Z (obtained) = .34

11.7 $U = 66.5$ Z (obtained) = -1.91

11.9 $U = 43$ Z (obtained) = -1.16

11.11 $U = 20$ Z (obtained) = -2.67

Chapter 12

12.1 a. 1.11   **b.** 0.00   **c.** 1.52   **d.** 1.46

12.3   There is 1 degree of freedom in a 2 × 2 table, so—with alpha set at .05—the critical value for the chi square would be 3.841. The obtained chi square is 0.65, so we fail to reject the null hypothesis of independence between the variables. There is no statistically significant relationship between race and services received.

12.5   With 1 degree of freedom and alpha set at .05, the critical region will begin at 3.841. The obtained chi square of 2.15 does not fall within this area, so the null cannot be rejected. No statistically significant relationship exists between unionization and salary.

12.7   The obtained chi square is 3.37, which is not significant (df = 1, alpha = .05).

12.9   The obtained chi square is 0.04.

12.11 The obtained chi square is 0.30.

12.13 The obtained chi square is 12.58.

12.15 The obtained chi square is 19.34.

12.17 Expected frequencies would be found by dividing the total N equally across all three categories. So, $f_e = 182/3 = 60.67$. The obtained chi square is 4.43.

12.19 a. Chi square is 1.56   **b.** Chi square is 1.17
      c. Chi square is 3.50   **d.** Chi square is 13.57
      e. Chi square is 2.02

Part II Cumulative Exercises

1. Fitting an appropriate test to a research situation is one of the continuing challenges of using statistics reasonably. As you are now aware, many tests are available, and it can be difficult to decide which is more suitable. In case it's any comfort, skilled researchers with years of experience get confused about this too. If you approach the decision systematically and consider the selection criteria carefully, the confusion and ambiguity can be minimized.

Let's use this first problem of the exercise to consider some ways in which reasonable decisions can be made. Begin by considering this research situation carefully. The situation calls for a test of hypotheses (is the difference significant?), so our choice of procedures will be limited to Chapters 8–12. Next, determine the types of variables you are working with. Number of minutes seems like an interval-ratio-level variable, and gender is definitely nominal. Note also that gender has only two categories. Which test should we use?

Since we are comparing one group (women) with another (men), the techniques in Chapter 8 (one-sample tests) and Chapter 10 (tests involving more than two samples or categories) are irrelevant. Chi square (Chapter 12) won't work unless we collapse the scores on internet use. This leaves Chapter 9. A test of sample means fits the situation, and we have a large sample (combined N's are greater than 100), so it looks like we're going to wind up in Section 9.2.

Another way to approach test selection would be to use the flowcharts at the beginning of Part II and each chapter. The Part II flowchart would quickly lead to Chapter 9, since we are conducting a test of hypotheses with two samples and are using means. Turn to the flowchart at the beginning of Chapter 9, respond to the questions (YES, we want to test for the significance of the difference between population means; NO, the samples are not matched; YES, sample size is large), and you should wind up again in Section 9.2.

a. Using the formulas presented in Section 9.2 to test for the difference in minutes by gender:

$$\sigma_{\bar{x}-\bar{x}} = .141 \qquad Z \text{ (obtained)} = -35.51$$

The difference in Internet minutes is significant. Given the sign of the obtained Z score, these results would suggest that men, on the average, use this technology more frequently.

b. For this problem, the table format is a sure tip-off that the chi square test is appropriate. The obtained chi square is 0.87, which is not significant at the .05 level. There is no statistically significant relationship between involvement and social class.

Can you think of other ways to deal with this research situation? For example, could you test the proportion of middle-class people who are "high" in involvement against the proportion of lower-class people in the same category?

c. "Number of partners" sounds like an interval-ratio-level variable, and education has three categories. If you follow the flowcharts, this will lead you to the analysis of variance as an appropriate test for this situation. The F ratio is .13—not at all significant—so we must conclude that this dimension of sexuality does not vary by level of education.

d. The problem asks for a characteristic of a population ("how many times do adult Americans move?") but gives information only for a "random sample." The estimation procedures presented in Chapter 7 fit the situation and, since the information is presented in the form of a mean, you should use Formula 7.2 to form the estimate. At an alpha level of .05, the confidence interval would be 3.5 ± 0.02.

e. The research question focuses on the difference between a single sample and a population, so Chapter 8 is relevant. The population standard deviation is unknown, and we have a large sample, so the flowcharts will direct us to Sections 8.1–8.5 and Formula 8.1. Z (obtained) is -2.50, which would be significant at alpha $= .05$. The sample is significantly different from the population—rural school districts are different from the universe of school districts in this state.

Chapter 13

Because of space constraints, I've provided answers to only two problems here.

13.1

Efficiency	Authoritarianism Low	Authoritarianism High
Low	37.04% (10)	70.59% (12)
High	62.96% (17)	29.41% (5)
Totals	100.00% (27)	100.00% (17)

The conditional distributions change, so there is a relationship between the variables. The change from column to column is quite large, and the relationship is strong. Efficiency decreases as authoritarianism increases—workers produce less for dictatorial bosses (or, maybe, bosses become more dictatorial when workers are nonproductive) so this relationship is negative in direction.

13.9

GPA	Attractiveness Low	Attractiveness Moderate	Attractiveness High
Low	28.00%	26.67%	33.33%
Moderate	40.00%	33.33%	35.56%
High	32.00%	40.00%	31.11%
Totals	100.00%	100.00%	100.00%

The conditional distributions change, so there is a relationship between these two variables. The change from column to column is quite minimal, and the relationship is weak. There is a slight tendency for GPA to increase with attractiveness (I think—it's hard to tell when the relationship is so weak), but these patterns do not support the hypothesis.

Chapter 14

14.1 **a.** $\phi = 0.00$ $\lambda = 0.00$ **b.** $\phi = 0.09$ $\lambda = 0.00$
  c. $\phi = 0.25$ $\lambda = 0.14$

14.3 $\phi = .17$ $\lambda = .00$

14.5 $\phi = 0.31$ $\lambda = 0.03$

14.7 $\phi = .00$ $\lambda = .00$

14.9 **a.** $\phi = .05$ $\lambda = .00$ **b.** $\phi = .54$ $\lambda = .33$
  c. $\phi = .39$ $\lambda = .05$

14.11 Cramer's $V = .05$ $\lambda = .00$

14.13 $V = 0.11$ $\lambda = 0.00$

14.15 Measures of association for the tables presented in problem 13.15 are as follows:

a. Cramer's $V = .23$ $\lambda = .14$
b. Cramer's $V = .20$ $\lambda = .00$
c. Cramer's $V = .16$ $\lambda = .00$
d. Cramer's $V = .20$ $\lambda = .00$
e. Cramer's $V = .20$ $\lambda = .00$

Lambda is zero for all tables except for the first. This is a result of unequal row totals, and we should probably select Cramer's V as the most appropriate measure of association. The values for V suggest, at best, moderately strong relationships between political ideology and these items.

For the tables presented in problem 14.15:

a. $\phi = .02$ $\lambda = .00$ **b.** $\phi = .13$ $\lambda = .00$
c. $\phi = .11$ $\lambda = .00$ **d.** $\phi = .04$ $\lambda = .00$
e. $\phi = .12$ $\lambda = .00$

The measures suggest weak relationships between the variables. The only relationships that seem at all interesting are those between sex and support for capital punishment and sex and support for the right to die. In both cases males are more supportive.

Comparing these tables with those in problem 13.15, political ideology has stronger relationships with these variables than sex.

Chapter 15

15.1 **a.** $G = 0.71$ **b.** $G = 0.71$ **c.** $G = -0.88$

These relationships are strong and indicate that facility in English and income increase with length of residence ($+0.71$). Use the percentages to help interpret the direction of a relationship. In the first table, 80% of the "newcomers" (length of residence = less than five years) had "Low" facility in English, and 60% of the "old-timers" had "High" facility. In this relationship, low scores on one variable are associated with low scores on the other, and scores increase together (as one increases, the other increases), so this is a positive relationship.

Contact with the old country decreases with length of residence (-0.88). Most newcomers have high levels of contact (frequent), and most oldtimers have low levels (rare). As length of residence increases, contact decreases.

15.3 $G = -0.61$

15.5 $G = 0.27$

15.7 $G = 0.22$ Z (obtained) = 0.93

15.9 $G = -0.14$

15.11 $G = -.33$ Z (obtained) = -2.35

15.13 $r_s = -0.46$ t (obtained) = -1.55

15.15 $r_s = 0.70$

There is a strong positive relationship between ethnic heterogeneity and strife for this sample. The greater the diversity, the greater the strife.

15.17 $r_s = .90$ Z (obtained) = 8.76

Chapter 16

16.1 (*HINT*: When finding the slope, remember that "Turnout" is the dependent or Y variable.)

Turnout and Unemployment:
$b = 3.00$ $a = 39.00$
$Y = (39) + (3)X$
$r = 0.95$ $r^2 = 0.90$
t (obtained) = 4.96

Turnout and Education:
$b = 12.67$ $a = -94.73$
$Y = (-94.73) + (12.67)X$
$r = 0.98$ $r^2 = 0.96$
t (obtained) = 9.80

Turnout and Negative Campaigning:
$b = -0.90$ $a = 114.01$
$Y = (114.01) + (-0.90)X$
$r = -0.87$ $r^2 = 0.76$
t (obtained) = -3.08

16.3 (*HINT*: When finding the slope, remember that "Number of visitors" is the dependent or Y variable.)

$b = -0.26$ $a = 12.13$
$r = -0.31$ $r^2 = 0.10$

16.5 **b.**

		Growth Rate	Population Density	% Urban
Homicide	$b =$	.34	.01	.14
	$a =$	5.58	5.97	-2.78
Robbery	$b =$	-2.08	.92	4.39
	$a =$	185.38	42.62	-148.24
Auto theft	$b =$	37.18	.64	12.22
	$a =$	280.91	422.17	-380.98

c. For a growth rate of -1, the predicted homicide rate would be 5.24.

For a population density of 250, the predicted robbery rate would be 272.62.

For a state with 50% urbanization, the predicted rate of auto theft would be 230.02.

d. Pearson r's and (r^2s) are

	Growth Rate	Population Density	% Urban
Homicide	.46	.40	.87
	(.21)	(.16)	(.76)
Robbery	-.08	.83	.75
	(.01)	(.69)	(.56)
Auto theft	.55	.22	.81
	(.30)	(.05)	(.66)

16.7 $b = 0.05$ $a = 53.18$
$r = 0.40$ $r^2 = 0.16$

16.9 For years of schooling and abuse:
$r = -0.62$ $r^2 = 0.38$ t (obtained) = -3.15

For % intact and abuse:
$r = -0.84$ $r^2 = 0.71$ t (obtained) $= -6.24$

16.11 $r = 0.20$ $r^2 = 0.04$

Part III Cumulative Exercises

1. a. The Part III flowchart shows that the choice of a measure of association is largely dependent on the level of measurement of the variables involved. Assuming that crime rates and "percent immigrant" are both measured at the interval-ratio level, the flowchart leads to Chapter 16 and Pearson's r. For this relationship, $r = -.47$. This is a moderate to strong relationship. The percentage of immigrants explains 22% ($-.47^2 \times 100$) of the variation in crime rate for these ten cities. The relationship is negative, which means that as the percentage of immigrants increases, the crime rate decreases.

b. These variables are ordinal in level of measurement, and gamma or Somer's d or Kendall's tau-b would be appropriate measures of association. Gamma is $-.43$, indicating a moderate to strong negative relationship. As TV viewing increases, involvement decreases.

c. The appropriate measure for ordinal-level variables with many scores is Spearman's rho, which, for these variables, is .74. There is a strong positive relationship between these variables. States with higher quality of life also have superior systems of higher education.

d. Race is a nominal-level variable, and the table is larger than 2 × 2. Cramer's V is .40, indicating a moderate relationship between the variables. The column percentages show that this sample tends to reject all of the singular racial categories in favor of "None of the above." A second, weaker trend is for people with one white parent to identify with that group. Because of the uneven row totals, lambda is zero even though there is an association between these variables.

Chapter 17

17.1 a. $G = 0.71$
b. $G = 0.77$ (males)
    $G = 0.65$ (females)
c. $G = 0.67$ (Asians)
    $G = 0.74$ (Hispanics)

The bivariate relationship is strong and positive. The longer the residence, the greater the facility in English. The relationship between these two variables is not affected by the sex or origin of the immigrants. These results would be taken as strong evidence of a direct (causal) relationship between length of residence and English facility.

17.3 $G = -.23$
    $G = -.34$ (females)
    $G = -.10$ (males)

This is an interactive relationship. Length of institutionalization has a greater effect on the reality orientation of females than of males.

17.5 $G = 0.62$ (bivariate table)
    $G = 0.52$ (whites)
    $G = 0.68$ (blacks)
    $G_P = 0.60$ (controlling for race)

    $G = 0.62$ (males)
    $G = 0.62$ (females)
    $G_P = 0.62$ (controlling for sex)

The bivariate relationship is strong and positive. Completion of the training program is closely associated with holding a job for at least one year. There is some interaction with race. The training has less impact for whites than for blacks. Blacks who did not complete the training were less likely to have held a job for at least one year. Gender has no impact at all on the relationship. Overall, there is a direct relationship between completion of the training and holding a job for at least a year, though there is some interaction with race.

17.7 $G = 0.41$
    $G = 0.40$ (males)
    $G = 0.42$ (females)
    $G_P = 0.41$

    $G = 0.24$ (married)
    $G = 0.60$ (nonmarried)
    $G_P = 0.35$

17.9 $G = 0.25$
    $G = 0.28$ (males)
    $G = 0.25$ (females)

$G = 0.33$ (Liberals)
$G = 0.20$ (Moderates)
$G = 0.31$ (Conservatives)

$G = -0.06$ (18–36)
$G = 0.20$ (37–50)
$G = 0.53$ (51–89)

Chapter 18

18.1 a. For turnout (Y) and unemployment (X) while controlling for negative advertising (Z), $r_{yx.z} = 0.95$. The relationship between X and Y is not affected by the control variable Z.

   b. For turnout (Y) and negative advertising (X) while controlling for unemployment (Z), $r_{yx.z} = -0.89$. The relationship between X and Y is not affected by the control variable Z.

   c. Turnout $(Y) = 70.25 + (2.09)$ unemployment $(X_1) + (-.43)$ negative advertising (X_2). For unemployment $(X_1) = 10$ and negative advertising $(X_2) = 75$, turnout $(Y) = 58.90$.

   d. For unemployment (X_1): $b_1^* = .66$. For negative advertising (X_2): $b_2^* = -.41$.

Unemployment has a stronger effect on turnout than negative advertising. Note that the independent variables' effect on turnout is in opposite directions.

   e. $R^2 = .98$

18.3 a. For strife (Y) and unemployment (X), controlling for urbanization (Z), $r_{yx.z} = .79$.

   b. For strife (Y) and urbanization (X), controlling for unemployment (Z), $r_{yx.z} = .20$.

   c. Strife $(Y) = (-14.60) + (4.94)$ unemployment $(X_1) + (.16)$ urbanization (X_2). With unemployment $= 10$ and urbanization $= 90$, strife (Y') would be 49.2.

   d. For unemployment (X_1): $b_1^* = .78$. For urbanization (X_2): $b_2^* = .13$.

   e. $R^2 = .65$

18.5 a. Turnout $(Y) = 83.80$
$+ (-1.16)$ Democrat (X_1)
$+ (2.89)$ minority (X_2)

   b. For $X_1 = 0$ and $X_2 = 5$, $Y' = 98.2$

   c. $Z_y = (-1.27)Z_1 + (.84)Z_2$

   d. $R^2 = 0.51$

18.7 a. $Z_y = (-.43)$ dropout $(Z_1) + (.54)$ income (Z_2)

   b. $R^2 = .61$

18.9 a. $Z_y = (-.18)$ death rate (Z_1)
$+ (-.68)$ birth rate (Z_2)

   b. $R^2 = 0.63$

18.11a. $Z_y = (.05)$ age $(Z_1) + (-.47)$ urban (Z_2)
$Z_y = (-.03)$ age $(Z_1) + (-.19)$ commute (Z_2)
$Z_y = (-.68)$ urban $(Z_1) + (.30)$ commute (Z_2)

   b. With age and urbanization: $R^2 = 0.21$
   With age and average commute: $R^2 = 0.04$
   With urbanization and average commute: $R^2 = 0.25$

Part IV Cumulative Exercises

1. a. The choice of multivariate procedures will depend on level of measurement and the number of possible scores for each variable. The powerful and efficient regression techniques presented in Chapter 18 are appropriate for interval-ratio, continuous variables. In this situation, we have three variables: GPA, hours of work, and College Board scores. College Board scores are arguably only ordinal in level of measurement, but the scores have a wide range and seem continuous.

GPA is the dependent variable, and the zero-order correlation with "hours worked" is −.83, a strong and negative relationship which indicates that having a job did interfere with academic success for this sample. The zero-order correlation with College Board scores is .55, a relationship that is consistent with the idea that College Board scores do predict success in college.

The results of the regression analysis with "hours worked" (X_1) and College Board scores (X_2) as independents:

$$Y = 1.71 + (-.04)X_1 + (.003)X_2$$
$$Z_y = (-.75)Z_1 + (.43)Z_2$$
$$R^2 = .86$$

The beta-weights indicate that "hours worked" has a stronger direct effect on GPA than College Board scores. Even for students who were well prepared for college (had high College Board scores), having a job had a negative impact on GPA. The high value for R^2

means that only a small percentage of the variance in GPA (14%) is left unexplained by these two independent variables.

b. In this situation, we have three dichotomous variables, two ordinal-level and one nominal. With limited variation possible, the elaboration technique is appropriate to analyze these relationships. The bivariate table shows that there is a relationship between graduating and level of social activity for this sample: 70% of the students with low levels graduated in four years vs. only 30% of the students with high levels. Bivariate gamma is .69, indicating a strong relationship.

Controlling for sex reveals some interaction. The gamma for males (.77) is stronger than the bivariate gamma, and the gamma for females (.60) is weaker. In other words, while level of social activity has an effect on graduation rates, the effect is stronger for males than for females. This suggests that sex should be incorporated into the analysis with a view to developing an understanding of why it would have different effects for males and females.

Index

Glossary

Each entry includes a brief definition and notes the chapter in which the term was introduced.

Alpha error. *See Type I error.* Chapter 8

Alpha level (α). In inferential statistics, the probability of error. (1) In estimation, the probability that a confidence interval does not contain the population value. Chapter 7 (2) In hypothesis testing, the proportion of the area under the sampling distribution that contains unlikely sample outcomes if the null is true. The probability of Type I error. Chapter 8

Analysis of variance. A test of significance appropriate when testing for the differences among more than two sample means. Chapter 10

ANOVA. *See Analysis of variance.* Chapter 10

Arrow keys. When using SPSS, the keys used to move the cursor around the screen. Appendix F

Association. The relationship between two (or more) variables. Two variables are said to be associated if the distribution of one variable changes for the various categories or scores of the other variable. Chapter 13

Average deviation. The average of the absolute deviations of the scores around the mean. Chapter 4

Bar chart. A graphic display device for nominal- and ordinal-level variables. Chapter 2

Beta error. *See Type II error.* Chapter 8

Beta-weights (b^*). Standardized partial slopes. Chapter 18

Bias. A criterion used to select sample statistics for estimation procedures. A statistic is unbiased if the mean of its sampling distribution is equal to the population value of interest. Chapter 7

Bivariate normal distributions. The model assumption in the test of significance for Pearson's r that both variables are normally distributed. Chapter 16

Bivariate table. A table that displays the joint frequency distribution of two variables. Chapter 12

Cells. The cross-classification categories of the variables in a bivariate table. Chapter 12

Central Limit Theorem. A theorem that specifies the mean, standard deviation, and shape of the sampling distribution, given that the sample is large. Chapter 6

χ^2 (critical). The chi square score that marks the beginnings of the critical region of the chi square sampling distribution. Chapter 12

χ^2 (obtained). The test statistic computed in step 4 of the five-step model. Chapter 12

Chi square test. A nonparametric test of hypothesis for variables organized in a bivariate table. Chapter 12

Class intervals. The categories used in frequency distributions for interval-ratio level variables. Chapter 2

Cluster sampling. A method of EPSEM sampling that is based on selecting groups such as geographical areas rather than cases for a list of the population. Chapter 6

Coefficient of determination (r^2). The proportion of all variation in Y that is explained by X. Chapter 16

Coefficient of multiple determination (R^2). A statistic that equals the total variation explained in the dependent variable by all independent variables combined. Chapter 18

Column. The vertical dimension of a bivariate table. Chapter 12

Conditional distribution of Y. The distribution of scores on the dependent variable for a specific score or category of the independent variable. Chapter 13

Conditional means of Y. The mean of all scores on Y for each value of X. Chapter 16

Confidence interval. An estimate of a population value in which a range of values is specified. Chapter 7

Confidence level. An alternative way to express alpha, the probability that a confidence interval will not contain the population value. Chapter 7

Continuous variable. A variable with a unit of measurement that can be subdivided infinitely. Chapter 1

Cramer's V. A chi-square–based measure of association. Chapter 14

Critical region (region of rejection). The area under the sampling distribution that includes all unlikely sample outcomes. Chapter 8

Cumulative frequency. A column in a frequency distribution that displays the number of cases in an interval and all preceding intervals. Chapter 2

Cumulative percentage. A column in a frequency distribution that displays the percentage of cases in an interval and all preceding intervals. Chapter 2

Data. In social science research, information that is represented by numbers. Chapter 1

Database. An organized collection of related information. Appendix F

Data reduction. Summarizing many scores with a few statistics. A major goal of descriptive statistics. Chapter 1

Deciles. The points that divide a distribution of scores into tenths. Chapter 3

Dependent variable. A variable that is identified as an effect, result, or outcome variable. The dependent variable is thought to be caused by the independent variable. Chapters 1 and 13

Descriptive statistics. The branch of statistics concerned with (1) summarizing the distribution of a single variable or (2) measuring the relationship between two or more variables. Chapter 1

Deviations. The distances between the scores and the mean. Chapter 4

Direct relationship. A multivariate relationship in which a control variable has no effect on the bivariate relationship. Chapter 17

Discrete variable. A variable with a basic unit of measurement that cannot be subdivided. Chapter 1

Dispersion. The amount of variety or heterogeneity in a distribution of scores. Chapter 4

E_1. For lambda, the number of errors of prediction made when predicting which category of the dependent variable cases will fall into while ignoring the independent variable. Chapter 14

E_2. For lambda, the number of errors of prediction made when predicting which category of the dependent variable cases will fall into while taking the independent variable into account. Chapter 14

Efficiency. The extent to which sample outcomes are clustered around the mean of the sampling distribution. Chapter 7

EPSEM. Equal probability of selection method. A technique for selecting samples in which every element or case in the population has an equal probability of being selected for the sample. Chapter 6

Expected frequency (f_e). The cell frequencies that would be expected in a bivariate table if the variables were independent. Chapter 12

Explained variation. The proportion of all variation in Y that is attributed to the effect of X. Chapter 16

Explanation. *See **Spurious relationship**.* Chapter 17

F ratio. For the analysis of variance, the test statistic computed in step 4 of the five-step model. Chapter 10

File. A database (or any other information) that is stored under the same name in the memory of the computer or on a floppy disk. Appendix F

Five-step model. A step-by-step guideline for conducting tests of hypotheses. Chapter 8

Frequency distribution. A table that displays the number of cases in each category of a variable. Chapter 2

Frequency polygon (Line chart). A graphic display device for interval-ratio variables. Chapter 2

Gamma (G). A measure of association for ordinal variables organized in table format. Chapter 15

Goodness-of-fit test. A chi square test to see if a variable is randomly distributed across a series of categories. Chapter 12

Histogram. A graphic display device for interval-ratio variables. Chapter 2

Homoscedasticity. The model assumption in the test of significance for Pearson's r that the variance of the Y scores is uniform across all values of X. Chapter 16

Hypothesis. A statement about the relationship between variables that is derived from a theory. Hypotheses are more specific than theories, and all terms and concepts are fully defined. Chapter 1

Independence. The null hypothesis in the chi square test. Two variables are independent if the classification of a case on one variable has no effect on the probability that the case will be classified in any particular category of the second variable. Chapter 12

Independent random samples. Random samples gathered so that the selection of a case for one sample has no effect on the probability that any particular case will be selected for the other samples. Chapter 9

Independent variable. A variable that is identified as a causal variable. The independent variable is thought to cause the dependent variable. Chapters 1 and 13

Index of qualitative variation (IQV). A measure of dispersion for variables that have been organized into frequency distributions. Chapter 4

Inferential statistics. The branch of statistics concerned with making generalizations from samples to populations. Chapter 1

Interaction. A multivariate relationship in which a bivariate relationship changes across the categories of the control variable. Chapter 17

Interpretation. *See Intervening relationship.* Chapter 17

Interquartile range (Q). The distance from the third quartile to the first. Chapter 4

Interval estimate. *See Confidence interval.* Chapter 7

Intervening relationship. A multivariate relationship in which the independent and dependent variables are linked primarily through the control variable. Chapter 17

Kendall's tau-b. A measure of association for ordinal variables organized in table format. Especially appropriate for square tables. Chapter 15

Lambda (λ). A measure of association for nominal-level variables that have been organized into a bivariate table. Lambda is based on the logic of PRE. Chapter 14

Level of measurement. The mathematical characteristics of a variable as determined by the measurement process. A major criterion for selecting statistical techniques. See Table 1.2. Chapter 1

Linear relationship. A relationship between two variables in which the observation points (dots) in the scattergram can be approximated with a straight line. Chapter 16

Line chart. *See Frequency polygon.*

Mann-Whitney U. A nonparametric test of significance for the two-sample case with ordinal-level variables. Chapter 11

Marginals. The row and column totals of a bivariate table. Chapter 12

Mean ($\bar{X}$). The arithmetic average of a set of scores. Chapter 3

Mean square. In the analysis of variance, an estimate of the variance calculated by dividing the sum of squares within (SSW) or the sum of squares between (SSB) by the appropriate degrees of freedom. Chapter 10

Measures of association. Statistics that summarize the strength and direction of the relationship between variables. Chapters 1 and 13

Measures of central tendency. Statistics that summarize a distribution of scores by reporting the most typical or representative value of the distribution. Chapter 3

Measures of dispersion. Statistics that indicate the amount of variety or heterogeneity in a distribution of scores. Chapter 4

Median (Md). The point in a distribution of scores above and below which exactly half of the cases fall. Chapter 3

Menu. A list of options in a statistical package. Appendix F

Midpoint. The point halfway between the upper and lower limits of a class interval. Chapter 2

Mode. The most common value in a distribution or the largest category of a variable. Chapter 3

Multiple correlation. A multivariate technique for examining the combined effects of more than one independent variable on a dependent variable. Chapter 18

Multiple correlation coefficient (R). A statistic that indicates the strength of the correlation between a dependent variable and two or more independent variables. Chapter 18

Multiple regression. A multivariate technique that breaks down the separate effects of the independent variables on the dependent variable. Chapter 18

Negative association. A bivariate relationship in which the variables vary in opposite directions. As one variable increases, the other decreases, and high scores on one variable are associated with low scores on the other. Chapter 13

Nonparametric test. A type of significance test in which no assumptions about the shape of the sampling distribution are made. Chapter 11

Normal curve. A theoretical distribution of scores that is symmetrical, unimodal, and bell shaped. The standard normal curve always has a mean of 0 and a standard deviation of 1. Chapter 5

Normal curve table. Appendix A; a detailed description of the area between a Z score and the mean of a standardized normal distribution. Chapter 5

Null hypothesis (H_0). A statement of "no difference." The specific form varies from test to test. Chapter 8

Observed frequency (f_o). The cell frequencies actually observed in a bivariate table. Chapter 12

One-tailed test. A type of hypothesis test that can be used when (1) the direction of the difference can be predicted or (2) concern is focused on only one tail of the sampling distribution. Chapter 8

One-way analysis of variance. An application of ANOVA in which the effect of a single variable on another is observed. Chapter 10

Partial correlation. A multivariate technique for examining a bivariate relationship while controlling for other variables. Chapter 18

Partial correlation coefficient. A statistic that shows the relationship between two variables while controlling for other variables; $r_{yx.z}$ is the symbol for the partial correlation coefficient when controlling for one variable. Chapter 18

Partial gamma (G_p). A statistic that indicates the strength of the association between two variables after the effects of a third variable have been removed. Chapter 17

Partial slopes. In a multiple regression equation, the slope of the relationship between a particular independent variable and the dependent variable while controlling for all other independents in the equation. Chapter 18

Partial tables. Tables produced when controlling for a third variable. Chapter 17

Pearson's r (r). A measure of association for variables that have been measured at the interval-ratio level. Chapter 16

Percentage (%). The number of cases in a category divided by the number of cases in all categories, the entire quantity multiplied by 100. Chapter 2

Percentile. The point in a distribution of scores below which a specific percentage of the cases fall. Chapter 3

Phi (ϕ). A chi square-based measure of association. Chapter 14

Pie chart. A graphic display device for nominal- and ordinal-level variables. Chapter 2

Point estimate. An estimate of a population value in which a single value is specified. Chapter 7

Population. The total collection of all cases in which the researcher is interested. Chapter 1

Positive association. A bivariate relationship in which the variables vary in the same direction. As one variable increases, the other also increases, and high scores on one variable are associated with high scores on the other. Chapter 13

Proportion (p). The number of cases in a category divided by the number of cases in all categories. Chapter 2

Proportional reduction in error (PRE). The logic that underlies the definition and computation of several different measures of association. Statistics are derived by comparing the number of errors made in predicting the dependent variable while ignoring the independent variable with the number of errors made while taking the independent variable into account. Chapter 14

Quartiles. The points in a distribution of scores that divide the distribution into quarters. Chapter 3

Random samples. *See EPSEM.* Chapter 6

Range (R). The highest score minus the lowest score. Chapter 4

Rate. The number of actual occurrences divided by the number of possible occurrences per some unit of time. Chapter 2

Ratio. The number of cases in one category divided by the number of cases in another category. Chapter 2

Real class limits. The upper and lower limits of a class interval used when the distribution is conceptualized as a continuous series of categories. Chapter 2

Region of rejection. *See Critical region.* Chapter 8

Regression line. The best-fitting straight line that summarizes the relationship between two variables. The regression line is fitted to the data points by the least-squares criterion whereby the line touches all conditional means of Y or comes as close to doing so as possible. Chapter 16

Replication. *See Direct relationship.* Chapter 17

Representative. A characteristic a random sample or a sample drawn according to the rule of EPSEM. A sample that accurately reproduces the major characteristics of the population from which it was drawn is said to be representative of the population. Chapter 6

Research hypothesis (H_1). A statement that contradicts the null hypothesis. The specific form varies from test to test. Chapter 8

Row. The horizontal dimension of a table. Chapter 12

Runs test. A nonparametric test of significance for the two-sample case with ordinal-level variables. Chapter 11

Sample. A subset of a population. In inferential statistics, information is gathered from random or EPSEM samples and then generalized to populations. Chapter 1

Sampling distribution. The distribution of all possible sample outcomes of a given statistic. Chapter 6

Scattergram. A graphic display device that depicts the relationship between two variables. Chapter 16

Simple random sample. A sample drawn from a population so that every case has an equal chance of being included. Chapter 6

Skew. The extent to which a distribution of scores has a few cases that are extremely high (positive skew) or extremely low (negative skew). Chapter 3

Slope (b). The amount of change in a variable per unit change in the other variable. Chapter 16

Somer's d. A measure of association for ordinal variables organized in table format. Chapter 15

SPSS for Windows. A statistical package designed for the analysis of social science data. Appendix F

Spearman's rho (r_s). A measure of association for ordinal variables that are in "continuous" format. Chapter 15

Specification. *See Interaction.* Chapter 17

Spurious relationship. A multivariate relationship in which both the independent and dependent variables are actually caused by the control variable. The independent and dependent are not causally related. Chapter 17

Standard deviation (s or σ). The square root of the squared deviations of the scores around the mean, divided by N. The most commonly used measure of dispersion; s represents the standard deviation of a sample, and σ represents the standard deviation of a population. Chapter 4

Standard error of the mean. The standard deviation of a sampling distribution of sample means. Chapter 6

Standardized partial slopes (beta-weights). The slope of the relationship between a particular independent variable and the dependent when all scores are expressed as Z scores. Chapter 18

Stated class limits. The upper and lower limits of a class interval used in a frequency distribution. Chapter 2

Statistical package (statpak). A set of computer programs designed to manipulate and statistically analyze data. Appendix F

Statistics. A set of mathematical techniques for organizing and analyzing data. Chapter 1

Stratified random sample. A random sample drawn from a population by selecting cases from sublists of groups in proportion to the representation of the groups in the population. Chapter 6

Sum of squares between (SSB). The sum of the squared deviations of the sample means from the overall mean, weighted by sample size. Chapter 10

Sum of squares total (SST). The sum of the squared deviations of the scores from the overall mean. Chapter 10

Sum of squares within (SSW). The sum of the squared deviations from the category means. Chapter 10

Systematic random sample. A sample selected by choosing the first case from a list of the population randomly and then choosing every kth case. Chapter 6

t (critical). The t score that marks the beginnings of the critical region of a t distribution. Chapter 8

t distribution. A distribution used to find the critical region for tests of sample means when N is small and σ is unknown. Chapter 8

t (obtained). The test statistic computed in step 4 of the five-step model for tests of sample means when N is small and σ is unknown. Chapter 8

Test statistic. The value computed in step 4 of the five-step model that places the sample outcome on the sampling distribution. Chapter 8

Theory. A generalized explanation of the relationship between two or more variables. Chapter 1

Total variation. The spread of the Y scores around the mean of Y. Chapter 16

Two-tailed test. A type of hypothesis test that can be used when (1) the direction of the difference cannot be predicted or (2) concern is focused on both tails of the sampling distribution. Chapter 8

Type I error (alpha error). The probability of rejecting a null hypothesis that is true. Chapter 8

Type II error (beta error). The probability of failing to reject a null hypothesis that is false. Chapter 8

Unexplained variation. The proportion of the total variation in Y that is not accounted for by X. Chapter 16

Variable. Any trait that can change values from case to case. Chapter 1 and Appendix H

Variance (s^2 or σ^2). The squared deviations of the scores around the mean, divided by N. A measure of dispersion used in inferential statistics and in regression techniques; s^2 represents the variance of a sample, and σ^2 represents the variance of a population. Chapter 4

Y intercept (a). The point where the regression line crosses the Y axis. Chapter 16

Z (critical). The Z score that marks the beginnings of the critical region of a Z distribution. Chapter 8

Z (obtained). The test statistic computed in step 4 of the five-step model. Chapter 8

Z scores. Standard scores; the way scores are expressed after they have been standardized to the theoretical normal curve. Chapter 5

Appendix C   Distribution of Chi Square

p value (handwritten, with .05 column circled)

df	.99	.98	.95	.90	.80	.70	.50	.30	.20	.10	.05	.02	.01	.001
1	$.0^3157$	$.0^3628$	.00393	.0158	.0642	.148	.455	1.074	1.642	2.706	3.841	5.412	6.635	10.827
2	.0201	.0404	.103	.211	.446	.713	1.386	2.408	3.219	4.605	5.991	7.824	9.210	13.815
3	.115	.185	.352	.584	1.005	1.424	2.366	3.665	4.642	6.251	7.815	9.837	11.341	16.268
4	.297	.429	.711	1.064	1.649	2.195	3.357	4.878	5.989	7.779	9.488	11.668	13.277	18.465
5	.554	.752	1.145	1.610	2.343	3.000	4.351	6.064	7.289	9.236	11.070	13.388	15.086	20.517
6	.872	1.134	1.635	2.204	3.070	3.828	5.348	7.231	8.558	10.645	12.592	15.033	16.812	22.457
7	1.239	1.564	2.167	2.833	3.822	4.671	6.346	8.383	9.803	12.017	14.067	16.622	18.475	24.322
8	1.646	2.032	2.733	3.490	4.594	5.527	7.344	9.524	11.030	13.362	15.507	18.168	20.090	26.125
9	2.088	2.532	3.325	4.168	5.380	6.393	8.343	10.656	12.242	14.684	16.919	19.679	21.666	27.877
10	2.558	3.059	3.940	4.865	6.179	7.267	9.342	11.781	13.442	15.987	18.307	21.161	23.209	29.588
11	3.053	3.609	4.575	5.578	6.989	8.148	10.341	12.899	14.631	17.275	19.675	22.618	24.725	31.264
12	3.571	4.178	5.226	6.304	7.807	9.034	11.340	14.011	15.812	18.549	21.026	24.054	26.217	32.909
13	4.107	4.765	5.892	7.042	8.634	9.926	12.340	15.119	16.985	19.812	22.362	25.472	27.688	34.528
14	4.660	5.368	6.571	7.790	9.467	10.821	13.339	16.222	18.151	21.064	23.685	26.873	29.141	36.123
15	5.229	5.985	7.261	8.547	10.307	11.721	14.339	17.322	19.311	22.307	24.996	28.259	30.578	37.697
16	5.812	6.614	7.962	9.312	11.152	12.624	15.338	18.418	20.465	23.542	26.296	29.633	32.000	39.252
17	6.408	7.255	8.672	10.085	12.002	13.531	16.338	19.511	21.615	24.769	27.587	30.995	33.409	40.790
18	7.015	7.906	9.390	10.865	12.857	14.440	17.338	20.601	22.760	25.989	28.869	32.346	34.805	42.312
19	7.633	8.567	10.117	11.651	13.716	15.352	18.338	21.689	23.900	27.204	30.144	33.687	36.191	43.820
20	8.260	9.237	10.851	12.443	14.578	16.266	19.337	22.775	25.038	28.412	31.410	35.020	37.566	45.315
21	8.897	9.915	11.591	13.240	15.445	17.182	20.337	23.858	26.171	29.615	32.671	36.343	38.932	46.797
22	9.542	10.600	12.338	14.041	16.314	18.101	21.337	24.939	27.301	30.813	33.924	37.659	40.289	48.268
23	10.196	11.293	13.091	14.848	17.187	19.021	22.337	26.018	28.429	32.007	35.172	38.968	41.638	49.728
24	10.856	11.992	13.848	15.659	18.062	19.943	23.337	27.096	29.553	33.196	36.415	40.270	42.980	51.179
25	11.524	12.697	14.611	16.473	18.940	20.867	24.337	28.172	30.675	34.382	37.652	41.566	44.314	52.620
26	12.198	13.409	15.379	17.292	19.820	21.792	25.336	29.246	31.795	35.563	38.885	42.856	45.642	54.052
27	12.879	14.125	16.151	18.114	20.703	22.719	26.336	30.319	32.912	36.741	40.113	44.140	46.963	55.476
28	13.565	14.847	16.928	18.939	21.588	23.647	27.336	31.391	34.027	37.916	41.337	45.419	48.278	56.893
29	14.256	15.574	17.708	19.768	22.475	24.577	28.336	32.461	35.139	39.087	42.557	46.693	49.588	58.302
30	14.953	16.306	18.493	20.599	23.364	25.508	29.336	33.530	36.250	40.256	43.773	47.962	50.892	59.703

Source: Table IV of Fisher & Yates: *Statistical Tables for Biological, Agricultural and Medical Research*, published by Longman Group Ltd., London (1974), 6th edition (previously published by Oliver & Boyd Ltd., Edinburgh). Reprinted by permission of Addison Wesley Longman Ltd.

Appendix D   Distribution of *F*

$p = .05$

n_1 n_2	1	2	3	4	5	6	8	12	24	∞
1	161.4	199.5	215.7	224.6	230.2	234.0	238.9	243.9	249.0	254.3
2	18.51	19.00	19.16	19.25	19.30	19.33	19.37	19.41	19.45	19.50
3	10.13	9.55	9.28	9.12	9.01	8.94	8.84	8.74	8.64	8.53
4	7.71	6.94	6.59	6.39	6.26	6.16	6.04	5.91	5.77	5.63
5	6.61	5.79	5.41	5.19	5.05	4.95	4.82	4.68	4.53	4.36
6	5.99	5.14	4.76	4.53	4.39	4.28	4.15	4.00	3.84	3.67
7	5.59	4.74	4.35	4.12	3.97	3.87	3.73	3.57	3.41	3.23
8	5.32	4.46	4.07	3.84	3.69	3.58	3.44	3.28	3.12	2.93
9	5.12	4.26	3.86	3.63	3.48	3.37	3.23	3.07	2.90	2.71
10	4.96	4.10	3.71	3.48	3.33	3.22	3.07	2.91	2.74	2.54
11	4.84	3.98	3.59	3.36	3.20	3.09	2.95	2.79	2.61	2.40
12	4.75	3.88	3.49	3.26	3.11	3.00	2.85	2.69	2.50	2.30
13	4.67	3.80	3.41	3.18	3.02	2.92	2.77	2.60	2.42	2.21
14	4.60	3.74	3.34	3.11	2.96	2.85	2.70	2.53	2.35	2.13
15	4.54	3.68	3.29	3.06	2.90	2.79	2.64	2.48	2.29	2.07
16	4.49	3.63	3.24	3.01	2.85	2.74	2.59	2.42	2.24	2.01
17	4.45	3.59	3.20	2.96	2.81	2.70	2.55	2.38	2.19	1.96
18	4.41	3.55	3.16	2.93	2.77	2.66	2.51	2.34	2.15	1.92
19	4.38	3.52	3.13	2.90	2.74	2.63	2.48	2.31	2.11	1.88
20	4.35	3.49	3.10	2.87	2.71	2.60	2.45	2.28	2.08	1.84
21	4.32	3.47	3.07	2.84	2.68	2.57	2.42	2.25	2.05	1.81
22	4.30	3.44	3.05	2.82	2.66	2.55	2.40	2.23	2.03	1.78
23	4.28	3.42	3.03	2.80	2.64	2.53	2.38	2.20	2.00	1.76
24	4.26	3.40	3.01	2.78	2.62	2.51	2.36	2.18	1.98	1.73
25	4.24	3.38	2.99	2.76	2.60	2.49	2.34	2.16	1.96	1.71
26	4.22	3.37	2.98	2.74	2.59	2.47	2.32	2.15	1.95	1.69
27	4.21	3.35	2.96	2.73	2.57	2.46	2.30	2.13	1.93	1.67
28	4.20	3.34	2.95	2.71	2.56	2.44	2.29	2.12	1.91	1.65
29	4.18	3.33	2.93	2.70	2.54	2.43	2.28	2.10	1.90	1.64
30	4.17	3.32	2.92	2.69	2.53	2.42	2.27	2.09	1.89	1.62
40	4.08	3.23	2.84	2.61	2.45	2.34	2.18	2.00	1.79	1.51
60	4.00	3.15	2.76	2.52	2.37	2.25	2.10	1.92	1.70	1.39
120	3.92	3.07	2.68	2.45	2.29	2.17	2.02	1.83	1.61	1.25
∞	3.84	2.99	2.60	2.37	2.21	2.09	1.94	1.75	1.52	1.00

Values of n_1 and n_2 represent the degrees of freedom associated with the between and within estimates of variance respectively.

Source: Table V of Fisher and Yates: *Statistical Tables for Biological, Agricultural and Medical Research*, published by Longman Group Ltd., London (1974), 6th edition (previously published by Oliver and Boyd Ltd., Edinburgh). Reprinted by permission of Addison Wesley Longman Ltd.

Glossary of Symbols

(The number in parentheses indicates the chapter in which the symbol is introduced.)

a	Point at which the regression line crosses the Y axis (16)
AD	Average deviation (4)
ANOVA	The analysis of variance (10)
b	Slope of the regression line (16)
b_i	Partial slope of the linear relationship between the ith independent variable and the dependent variable (18)
b_i^*	Standardized partial slope of the linear relationship between the ith independent variable and the dependent variable (18)
d	Somer's d (15)
df	Degrees of freedom (8)
f	Frequency (2)
F	The F ratio (10)
f_e	Expected frequency (12)
f_o	Observed frequency (12)
G	Gamma for a sample (15)
G_p	Partial gamma (17)
H_0	Null hypothesis (8)
H_1	Research or alternate hypothesis (8)
IQV	Index of qualitative variation (4)
Md	Median (3)
Mo	Mode (3)
N	Number of cases (2)

N_d	Number of pairs of cases ranked in different order on two variables (15)
N_s	Number of pairs of cases ranked in the same order on two variables (15)
%	Percentage (2)
P	Proportion (2)
P_s	A sample proportion (7)
P_u	A population proportion (7)
PRE	Proportional reduction in error (14)
Q	Interquartile range (4)
r	Pearson's correlation coefficient for a sample (16)
r^2	Coefficient of determination (16)
R	Range (4)
r_s	Spearman's rho for a sample (15)
$r_{xy \cdot z}$	Partial correlation coefficient (18)
R^2	Multiple correlation coefficient (18)
s	Sample standard deviation (4)
SSB	The sum of squares between (10)
SST	The total sum of squares (10)
SSW	The sum of squares within (10)
s^2	Sample variance (4)
t	Student's t score (8)
T_x	Number of pairs of cases tied on the independent variable (15)
T_y	Number of pairs of cases tied on the dependent variable (15)
tau-b	Kendall's tau-b (15)